Ehlers • Bluhm (Eds.)

Porous Media

Springer

Berlin
Heidelberg
New York
Barcelona
Hong Kong
London
Milan
Paris
Tokyo

W. Ehlers · J. Bluhm (Eds.)

Porous Media

Theory, Experiments and Numerical Applications

With 152 Figures

Springer

Prof. Dr.-Ing. Wolfgang Ehlers
Universität Stuttgart
Institut für Mechanik (Bauwesen)
Lehrstuhl II
Pfaffenwaldring 7
D - 70569 Stuttgart
e-mail: ehlers@mechbau.uni-stuttgart.de

Priv. Doz. Dr.-Ing. Joachim Bluhm
Universität Essen
Institut für Mechanik
Universitätsstr. 15
D - 45141 Essen
e-mail: joachim.bluhm@uni-essen.de

ISBN 978-3-642-07843-9

Library of Congress Cataloging-in-Publication-Data applied for

Die Deutsche Bibliothek – Cip-Einheitsaufnahme
Porous media : theory, experiments and numerical applications /
W. Ehlers ; J. Bluhm (ed.) - Berlin ; Heidelberg ; New York ;
Barcelona ; Hong Kong ; London ; Milan ; Paris ; Tokyo : Springer, 2002
(Engineering online library)

Springer-Verlag Berlin Heidelberg New York
a member of BertelsmannSpringer Science+Business Media GmbH
http://www.springer.de

Printed in Germany

Cover design: de'blik, Berlin
Printed on acid free paper

to *Reint de Boer*, our teacher

Contents

III. Experiments and Numerical Applications

Preface

The present volume offers a *state-of-the-art* report on the various recent scientific developments in the Theory of Porous Media (TPM) comprehending the basic theoretical concepts in continuum mechanics on porous and multiphasic materials as well as the wide range of experimental and numerical applications. Following this, the volume does not only address the sophisticated reader but also the interested beginner in the area of Porous Media by presenting a collection of articles. These articles written by experts in the field concern the fundamental approaches to multiphasic and porous materials as well as various applications to engineering problems.

In many branches of engineering just as in applied natural sciences like bio- and chemomechanics, one often has to deal with continuum mechanical problems which cannot be uniquely classified within the well-known disciplines of either *"solid mechanics"* or *"fluid mechanics"*. These problems, characterized by the fact that they require a unified treatment of volumetrically coupled solid-fluid aggregates, basically fall into the categories of either mixtures or porous media. Following this, there is a broad variety of problems ranging in this category as for example the investigation of reacting fluid mixtures or solid-fluid suspensions as well as the investigation of the coupled solid deformation and pore-fluid flow behaviour of liquid- and gas-saturated porous solid skeleton materials like geomaterials (soil, rock, concrete, etc.), polymeric and metallic foams or biomaterials (hard and soft tissues, etc). In particular, the reader may be interested in the solution of geotechnical problems like the well-known consolidation problem of soil mechanics, or in applications concerning the exploitation of natural gas and oil reservoirs. Furthermore, the interest may be directed towards the investigation of the deformation of foamed shock absorbers as well as towards biomechanical problems like the investigation of bones, cartilage, intervertebral disks or charged hydrated tissues, etc.

Porous Media generally concern an immiscible mixture of a solid skeleton and a fluid pore content which itself can either be a single fluid, gas or liquid, an immiscible fluid mixture of gas and liquids or a miscible fluid mixture of different reacting aggregates. The basic tool for a successful description of this kind of media in the framework of a macroscopic approach is the well-known Theory of Mixtures. On the other hand, since there is no measure to incorporate any kind of microscopic information in the Theory of Mixtures, it was found convenient to combine the *Theory of Mixtures* with the *Concept of Volume Fractions*. By use of this procedure, basically defining the *Theory of Porous Media*, one obtains an excellent tool for the macroscopic description of general immiscible or miscible multiphasic aggregates, where the volume fractions and the saturations, respectively, are the measures of the local portions of the individual constituents of the overall medium and where, furthermore, all incorporated fields can be understood as the local averages of the corresponding quantities of an underlying microstructure.

It is the goal of the present contribution to exhibit the fundamental concepts of the Theory of Porous Media and to apply these concepts to a variety of interesting problems in engineering and natural sciences. Thus, the first chapter of this volume concerns the foundations of modelling multiphasic materials represented by two articles by the editors, whereas the following chapters include five articles on different aspects of the TPM and eight articles on experiments and numerical applications.

Chapter *I. Foundations* starts with an article on the general description of the basic concepts of the Theory of Porous Media, where, in addition to the standard kinematics of the classical continuum mechanical approach to multiphasic and porous materials, also extended kinematics in the sense of the *Cosserat* brothers are taken into consideration. Apart from these fundamental concepts leading to both the standard and the extended framework of balance relations of multiphasic systems, the article furthermore presents two different constitutive models describing, on the one hand, a simple biphasic material of an elastic solid skeleton which is saturated by a single pore-liquid and, on the other hand, a triphasic material consisting of a micropolar elasto-plastic or elasto-viscoplastic solid skeleton saturated by a binary immiscible pore-content of a pore-liquid and a pore-gas. Furthermore, the basic numerical tools in treating volumetrically coupled solid-fluid aggregates are explained and extended by the application of time- and space-adaptive methods. Finally, the article contains a variety of numerical examples exhibiting the solution of civil engineering problems of geotechnical relevance.

The second article of Chapter I describes a fluid-saturated thermo-elastic porous solid, where the solid and the fluid materials are assumed to undergo different phase temperatures. Apart from the governing balance relations, the article includes a general setting of developing a constitutive theory on the basis of thermodynamical restrictions. In addition, the stress states of the thermo-elastic solid and the non-viscous pore-fluid are discussed in detail.

Chapter *II. Theory of Porous Media* presents several extensions in the general description of porous materials. In particular, this chapter starts with an article by *Stefan Diebels* on micropolar mixture models followed by a paper by *Nina Kirchner* and *Kolumban Hutter* on the elasto-plastic behaviour of granular materials with an additional scalar degree of freedom. Furthermore, the contribution by *Stefan Kowalski* deals with the mechanical aspects on drying of wet porous media, whereas *Renato Lancelotta* studies the coupling between the evolution of a deformable porous medium and the motion of fluids in the connected porosity. In contrast to these articles which are more or less related to engineering problems, the last article of this chapter by *Van C. Mow* et al. concerns a very interesting problem of biomechanical engineering, namely a triphasic paradigm by investigating fixed negative charges modulating mechanical behaviours and electrical signals in articular cartilage under unconfined compression.

Chapter *III. Experiments and Numerical Applications* includes eight articles on various applications of the Theory of Porous Media. In particular, *Harald Cramer* et al. describe a time-adaptive analysis of saturated soil by a discontinuous Galerkin method. The following paper by *Wolfgang Ehlers* et al. concerns a biphasic description of viscoelastic foams by use of an extended Ogden-type formulation. The article by *Jacques Huyghe* et al. exhibits an experimental measurement of electrical conductivity and electro-osmotic permeability of ionised porous media and thus presents the second paper of this volume dealing with biomechanical problems. The following three contributions concern civil engineering problems. The first one written by *Ragnar Larsson* et al. describes the theory and numerics of localization in a fluid-saturated elasto-plastic porous medium followed by an article by *Lorenzo Sanavia* et al. on a geometrical and material non-linear analysis of fully and partially saturated porous media. Finally, the paper by *Martin Schanz* and *Heinz Antes* on waves in a poroelastic half-space proceeds from the *Biot*'s theory, which can be understood as a variation of the Theory of Porous Media. In contrast to the preceding articles which are commonly based on the application of the finite element method, this paper furthermore proceeds from the boundary element analysis. Directing the focus of interest to another topic, the article by *Reem Freij-Ayoub* et al. describes a multicomponent reactive transport applied to ore body genesis and environmental hazards. The final paper of this volume by *D. J. R. Owen* et al. concerns a numerical model and its finite element solution for multiphase flow situations with an interesting application to pulp and paper processing.

The foregoing description clearly exhibits the wide range of applications of the Theory of Porous Media to various problems in the fields of engineering and natural sciences. Making use of the quotations given in the individual contributions, the interested reader is in the position to nearly arbitrarily extend the list of possible topics for an application of the Theory of Porous Media.

Finally, it is the wish of the editors to thank Professor Dr.-Ing. *Reint de Boer*, our scientific teacher, for his valuable contributions to the Theory of Porous Media and, particularly, its historical background. His engagement in this topic gave the research in the field of Porous Media a strong input and effectively promoted the modern approach to porous and multiphasic materials.

Stuttgart and Essen, April 2002

Wolfgang Ehlers
Joachim Bluhm

I. Foundations

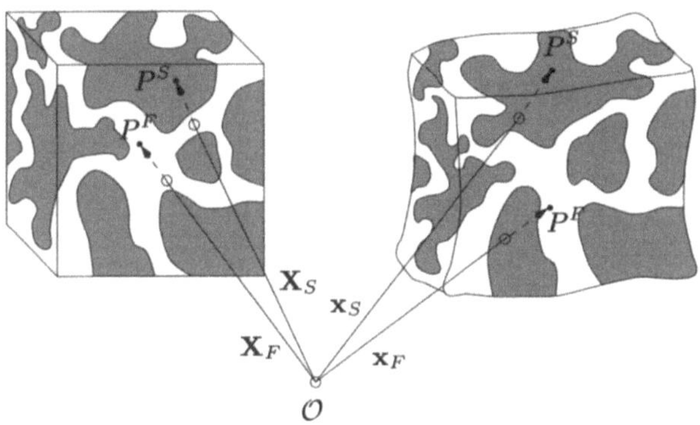

Foundations of multiphasic and porous materials

Wolfgang Ehlers

University of Stuttgart, Institute of Applied Mechanics (CE),
70550 Stuttgart, Germany

Abstract. Miscible multiphasic materials like classical mixtures as well as immiscible materials like saturated and partially saturated porous media can be successfully described on the common basis of the well-founded Theory of Mixtures (TM) or the Theory of Porous Media (TPM). In particular, both the TM and the TPM provide an excellent frame for a macroscopic description of a broad variety of engineering applications and further problems in applied natural sciences. The present article portrays both the standard and the micropolar approaches to multiphasic materials reflecting their mechanical and their thermodynamical frameworks. Including some constitutive models and various illustrative numerical examples, the article can be understood as a reference paper to all the following articles of this volume on theoretical, experimental and numerical investigations in the Theory of Porous Media.

1 Motivation

In many branches of engineering as well as in applied natural sciences, one often has to deal with continuum mechanical problems which cannot be uniquely classified within the well-known disciplines of either *"solid mechanics"* or *"fluid mechanics"*. These problems, characterized by the fact that they require a unified treatment of volumetrically coupled solid-fluid aggregates, basically fall into the categories of either mixtures or porous media. Following this, there is a broad variety of problems ranging in this category as for example the investigation of reacting fluid mixtures or solid-fluid suspensions as well as the investigation of the coupled solid deformation and pore-fluid flow behaviour of liquid- and gas-saturated porous solid skeletons like geomaterials (soil, rock, concrete, etc.), polymeric as well as metallic foams or biomaterials (hard and soft tissues, etc), cf. Figure 1. In particular, one may be interested in the solution of geotechnical applications like the well-known consolidation problem of soil mechanics, or in applications concerning the exploitation of natural gas and oil reservoirs. Furthermore, the interest may be directed towards the investigation of the deformation of foamed shock absorbers or automotive seat cushions as well as towards biomechanical problems like the investigation of bones, cartilage or intervertebral disks, etc.

The treatment of this kind of problems consisting of an immiscible mixture of a solid skeleton and a fluid pore content on the basis of a continuum

Fig. 1. Three typical porous solid materials.

mechanical method, i. e. within the framework of a macroscopic approach, first of all concerns the question of the necessary or the convenient degree of homogenization. Thus, one has to decide *a priori*, whether (1) each material of the multiphasic aggregate has to be treated by use of a micromechanical approach on its own domain as a single body where, additionally, the interaction mechanisms at the internal interfaces between the individual materials have to be taken into account carefully, or if one (2) wants to proceed from homogenization methods where the real microstructure is statistically smeared out through the considered domain on the basis of a real or a virtual averaging process. Both procedures are based on a considerable scientific tradition and are characterized by certain advantages and disadvantages.

The advantage of the micromechanical procedure lies, on the one hand, in the fact that each individual aggregate can be described on the basis of its own motion (cf. Figure 2) considering the usually known information on the constitutive assumptions of single continua mechanics. On the other hand, however, one has to define convenient interaction and contact relations to catch the coupling mechanisms at the internal interfaces. And just this is the basic problem of the first procedure, since the internal geometry of porous solid materials is, apart from some very few industrial products, of an arbitrary and irregular shape which, in general, is and remains completely unknown. Based on this enormous disadvantage of the first procedure, the necessity of using macroscopic strategies is obvious. Proceeding from macroscopic approaches, one has to be aware that the local information on the material behaviour does only reflect the physical properties of the individual aggregates in the sense of a local average. However, in order to obtain local averages as representative statements, one has to require that (a) the subdomains under study, in the sense of a representative elementary volume (REV), must be large enough to allow for a statistical statement and that (b)

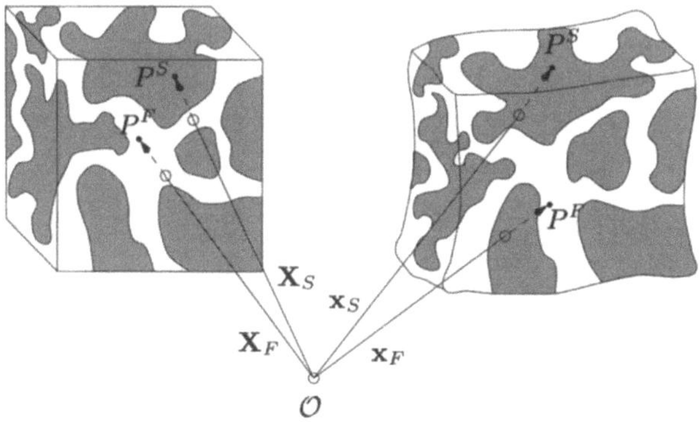

Fig. 2. Biphasic solid-fluid aggregate with individual motion functions; reference configuration (left) and actual configuration (right).

the local microscopic subdomains must be small enough in comparison to the macroscopic domain so that the arguments of scale separation can be applied. Furthermore, it is basically insubstantial whether or not the homogenization process is really carried out by applying a mathematical averaging theory to a typical REV of a multiphasic aggregate or if one assumes the existence of a previous virtual averaging process and furthermore deals with macroscopic (homogenized) quantities. No matter if the aggregates of the multiphasic continuum are miscible or immiscible, one obtains in both cases a continuum with statistically distributed constituents, where it is assumed that the individual aggregates of the multiphasic body are completely smeared out through the considered domain in the sense of superimposed and interacting continua, cf. Figure 3. Proceeding from this image directly leads to the axioms of the classical Theory of Mixtures assuming heterogeneously composed continua consisting of an arbitrary number of miscible and interacting constituents. Concerning the evolution of the Theory of Mixtures, the reader is referred to the work by Truesdell [100], Truesdell and Toupin [103], Bowen [14] or Truesdell [101] and the quotations therein.

Since in the Theory of Mixtures, there is no measure to incorporate any kind of microscopic information, it was found convenient to combine the *Theory of Mixtures* with the *Concept of Volume Fractions*. By use of this procedure, basically defining the *Theory of Porous Media*, cf. e. g. the work by Bowen [15, 16], de Boer [7, 8], de Boer and Ehlers [9–11] and Ehlers [27, 31, 32], one obtains an excellent tool for the macrocopic description of general immiscible multiphasic aggregates, where the volume fractions are the measures of the local portions of the individual constituents of the overall medium and where, furthermore, all incorporated fields are understood as local averages of the corresponding quantities of an underlying microstructure.

Following the classical literature, the *Concept of Volume Fractions* goes back to the year 1794, when Woltman [107] who was a harbour construction

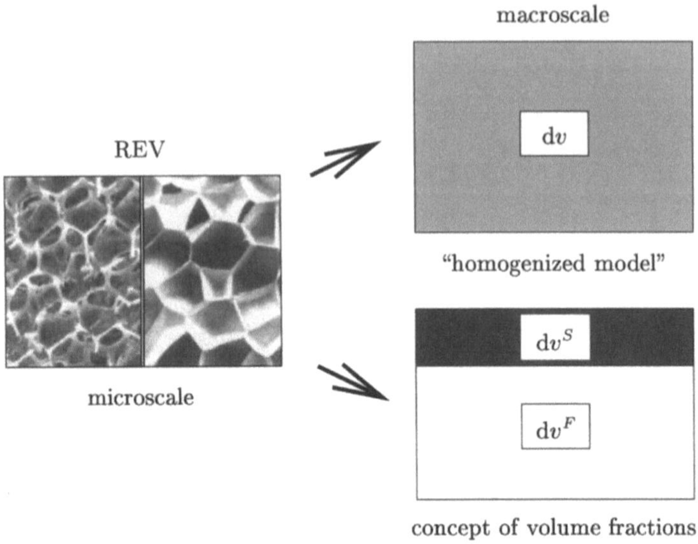

Fig. 3. Micro- versus macroscopic view on fluid-saturated porous material (here: microstructure of open- and closed-celled polymeric foams).

director from Hamburg was interested in the description of a water-saturated mud. Delesse [22] who was a mining engineer recovered the volume fraction concept in 1848 when he was investigating the relation between surface and volume fractions of multi-aggregate rocks. This concept has then been applied by a lot of scientists investigating the consolidation problem, like Biot [1–3] or Heinrich and Desoyer [65–67]. An excellent survey on the historical development of the Theory of Porous Media can be found in the work by de Boer [7, 8] and quotations therein. In particular, the consolidation problem (cf. Figure 4) is characterized by the fact that the application of an additional load (e. g. through a building) onto a fluid-saturated soil half-space leads to time-depending deformations accompanied by a draining process of the pore-water. The consolidation process only comes to an end when the effective soil stress, defined as the total stress minus the excess pore-pressure and obtained from the soil deformation by a convenient constitutive equation, yields an equilibrium state with the applied external load. Concerning the concept of effective stresses, the reader is referred to the work by Terzaghi [97], Bishop [4], Skempton [93], de Boer and Ehlers [10], Ehlers [31], Lade and de Boer [72] and de Boer [7, 8].

It is the goal of the present contribution to exhibit the fundamental concepts of the Theory of Porous Media, where all incorporated field quantities are assumed as the local averages of their microscopic representatives obtained by convenient averaging or homogenization processes. Concerning averaging and homogenization techniques, the reader is referred to the work by Hassanizadeh and Gray [61, 62], Suquet [95], de Boer et al. [12], Plischka

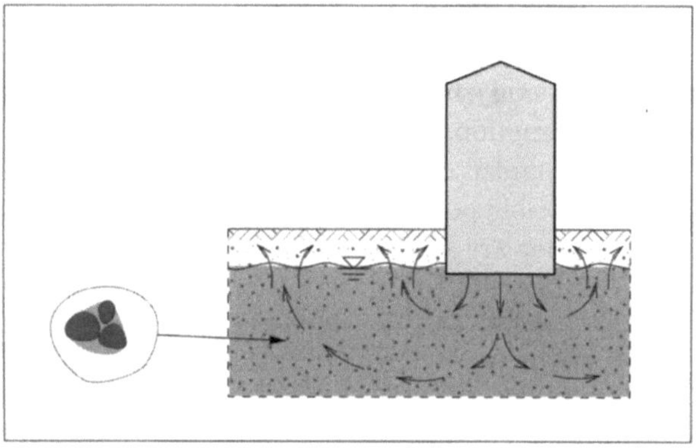

Fig. 4. Consolidation problem.

[86], Miehe et al. [78] and Schröder [91]. Furthermore, since the basic behaviour of a lot of materials is strongly affected by microstructural effects in the sense of micropolar kinematics, the following treatment is based on general micropolar constituents which are, of course, downward compatible to standard continua. For example, the advantage of including micropolar degrees of freedom into the general description of porous materials results in the fact that, on the one side, micropolarity is the natural tool to describe both the translational and the rotational deformation behaviour of granular materials like, e. g., non-cohesive or slightly cohesive soils, and, on the other side, it can be used to excellently regularize shear band localization phenomena occurring, e. g., in soil mechanical situations like the well-known slope and base failure problems. Concerning a broad review of micropolarity of single and multiphasic materials, the reader is referred to the work by the Cosserat brothers [18], Günther [58], Eringen and Kafadar [56], Steinmann [94], Diebels [23], Diebels and Ehlers [25], Ehlers and Volk [51–53] and Volk [106] and the quotations therein.

Apart from a discussion of the general frame of the Theory of Porous Media, two constitutive models are presented describing, on the one hand, a simple binary aggregate of a materially incompressible non-polar elastic skeleton saturated by a materially incompressible viscous pore-liquid and, on the other hand, a triphasic model of a materially incompressible elasto-plastic or elasto-viscoplastic micropolar skeleton saturated by two viscous pore-fluids, a materially incompressible pore-liquid and a pore-gas. Concerning the wide range of numerical applications to any kind of initial boundaty-value problem, lots of numerical examples have been treated by the author and coworkers [33–35, 39–42, 46, 47, 45, 51–53]. Some of these examples are presented at the end of this article to show the wide range of applications of the Theory of Porous Media to various engineering problems.

Concerning the direct tensor notation applied in the following sections, the reader is referred to the work by de Boer [6, 8].

2 The concept of volume fractions

Based on the fundamental concepts of the Theory of Porous Media, one proceeds from the assumption that all k individual materials composing the multiphasic aggregate under consideration (one porous solid skeleton and $k-1$ miscible or immiscible pore-fluids) are assumed to be in a state of ideal disarrangement, i. e. they are statistically distributed over the control space. This assumption together with the prescription of a real or a virtual averaging process leads to a model φ of superimposed and interacting continua φ^α ($\alpha = 1, \ldots, k$), cf. Figure 3:

$$\varphi = \sum_{\alpha=1}^{k} \varphi^\alpha . \tag{1}$$

Thus, each spatial point $\mathbf{x}$ of the control space Ω is simultaneously occupied by particles P^α of all k constituents. Consequently, the mathematical functions for the description of the geometrical and physical properties of the individual materials are field functions defined all over the control space. The volume V of the overall multiphasic aggregate $\mathcal{B}$ results from the sum of the partial volumes of the constituents φ^α in $\mathcal{B}$:

$$V = \int_{\mathcal{B}} \mathrm{d}v = \sum_{\alpha=1}^{k} V^\alpha , \qquad \text{where} \qquad V^\alpha = \int_{\mathcal{B}} \mathrm{d}v^\alpha =: \int_{\mathcal{B}} n^\alpha \, \mathrm{d}v . \tag{2}$$

Following this, the quantity n^α is defined as the local ratio of the volume element $\mathrm{d}v^\alpha$ of a given constituent φ^α with respect to the volume element $\mathrm{d}v$ of the overall medium φ:

$$n^\alpha = \frac{\mathrm{d}v^\alpha}{\mathrm{d}v} . \tag{3}$$

The relations (2) and (3) represent the concept of volume fractions. Since, in general, there is no vacant space in the overall medium, equation (2) directly leads to the saturation condition

$$\sum_{\alpha=1}^{k} n^\alpha = 1 . \tag{4}$$

In contrast to (2) and (4), it may appear convenient for some particular applications to neglect the existence of gaseous pore-phases in comparison with the solid material. In this case, one talks about empty or partially empty porous materials, where the total volume V is greater than the sum of the partial volumes V^α. Thus,

$$V > \sum_{\alpha=1}^{k} V^\alpha \qquad \text{and} \qquad \sum_{\alpha=1}^{k} n^\alpha < 1 . \tag{5}$$

Although the above assumption is a possibility to simplify a porous media model, the following material proceeds from the saturated case.

By use of the definition of the volume fractions n^α, two different density functions of each constituent φ^α can be introduced:

$$\begin{aligned} &\bullet \text{ material density: } \rho^{\alpha R} = \frac{\mathrm{d}m^\alpha}{\mathrm{d}v^\alpha}\,, \\ &\bullet \text{ partial density: } \rho^{\alpha} = \frac{\mathrm{d}m^\alpha}{\mathrm{d}v}\,. \end{aligned} \tag{6}$$

The material (realistic or effective) density $\rho^{\alpha R}$ relates the local mass $\mathrm{d}m^\alpha$ to the volume element $\mathrm{d}v^\alpha$, whereas the partial (global or bulk) density ρ^α relates the same mass to the volume element $\mathrm{d}v$. Following this, the density functions are related to each other by

$$\mathrm{d}m^\alpha = \left\{ \begin{matrix} \rho^{\alpha R}\,\mathrm{d}v^\alpha \\ \rho^\alpha\,\mathrm{d}v \end{matrix} \right\} \quad \longrightarrow \quad \rho^\alpha = n^\alpha \rho^{\alpha R}\,. \tag{7}$$

Based on the above relation, it is immediately evident that the property of material incompressibility ($\rho^{\alpha R} = \text{const.}$) of any constituent φ^α is not equivalent to the property of bulk incompressibility of this constituent, since the partial density functions ρ^α can still change through changes in the volume fractions n^α.

In addition to the volume fractions, it appears convenient for various applications to introduce the so-called saturation functions s^β. For example, if the porous medium under study contains k constituents φ^α consisting of a single porous solid skeleton constituents φ^S and $k-1$ pore-fluids φ^β, the saturation functions are given by

$$s^\beta = \frac{n^\beta}{n^F}, \qquad \text{where} \qquad n^F = \sum_{\beta=1}^{k-1} n^\beta. \tag{8}$$

Thus, n^F is the porosity, equivalent to the sum of the volume fractions n^β of all $k-1$ pore-fluid constituents. Furthermore, one easily concludes to

$$\sum_{\alpha=1}^{k-1} s^\beta = 1\,. \tag{9}$$

Following this, it is seen that (8) relates the saturation condition to the pore content, whereas (4) is related to the overall material.

3 Kinematical relations

3.1 Basic setting

Proceeding from mixture theories as the fundamental basis of the Theory of Porous Media, one directly makes use of the concept of superimposed continua with internal interactions and individual states of motion. In the framework of this concept, each spatial point $\mathbf{x}$ of the current configuration is, at any time t, simultaneously occupied by material particles (material points) P^α of all constituents φ^α ($\alpha = S$: solid skeleton, $\alpha = \beta$: $k-1$ porefluids). These particles proceed from different reference positions at time t_0, cf. Figure 5. Thus, each constituent is assigned its own motion function

$$\mathbf{x} = \boldsymbol{\chi}_\alpha(\mathbf{X}_\alpha, t)\,. \tag{10}$$

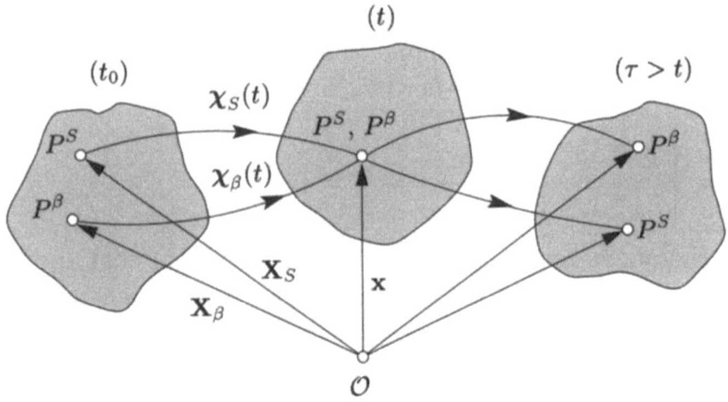

Fig. 5. Motion of a multiphasic mixture.

As a result, each spatial point $\mathbf{x}$ can only be occupied by one single material point P^α of each constituent φ^α. The assumption of unique motion functions, where each material point P^α of the current configuration has a unique reference position $\mathbf{X}_\alpha$ at time t_0, requires the existence of unique inverse motion functions $\boldsymbol{\chi}_\alpha^{-1}$ based on non-singular *Jacobi*an determinants J_α:

$$\mathbf{X}_\alpha = \boldsymbol{\chi}_\alpha^{-1}(\mathbf{x}, t)\,, \qquad J_\alpha = \det \frac{\partial \boldsymbol{\chi}_\alpha}{\partial \mathbf{X}_\alpha} \neq 0\,. \tag{11}$$

It follows from (10) that each constituent has an own velocity and acceleration field. In the basic *Lagrange*an setting, these fields are given by

$$\overset{\prime}{\mathbf{x}}_\alpha = \frac{\partial \boldsymbol{\chi}_\alpha(\mathbf{X}_\alpha, t)}{\partial t}\,, \qquad \overset{\prime\prime}{\mathbf{x}}_\alpha = \frac{\partial^2 \boldsymbol{\chi}_\alpha(\mathbf{X}_\alpha, t)}{\partial t^2}\,. \tag{12}$$

With the aid of the inverse motion function $(11)_1$, an alternative formulation of (12) leads to the *Euler*ian description

$$\overset{\prime}{\mathbf{x}}_\alpha = \overset{\prime}{\mathbf{x}}_\alpha(\mathbf{x}, t)\,, \qquad \overset{\prime\prime}{\mathbf{x}}_\alpha = \overset{\prime\prime}{\mathbf{x}}_\alpha(\mathbf{x}, t)\,. \tag{13}$$

The so-called mixture velocity

$$\dot{\mathbf{x}} = \frac{1}{\rho} \sum_{\alpha=1}^{k} \rho^\alpha \, \overset{\prime}{\mathbf{x}}_\alpha \,, \qquad \text{where} \qquad \rho = \sum_{\alpha=1}^{k} \rho^\alpha \tag{14}$$

is the so-called mixture density, indicates the local barycentric velocity of the overall medium.

Suppose that Γ is an arbitrary, steady and sufficiently often steadily differentiable scalar function of $(\mathbf{x}, t)$, then, the material time derivative of Γ following the motion of φ^α reads

$$(\Gamma)'_\alpha = \frac{\mathrm{d}_\alpha}{\mathrm{d}t} \Gamma = \frac{\partial \Gamma}{\partial t} + \operatorname{grad} \Gamma \cdot \overset{\prime}{\mathbf{x}}_\alpha \,. \tag{15}$$

Therein, the operator "grad $(\,\cdot\,)$" denotes the partial derivative of $(\,\cdot\,)$ with respect to the local position $\mathbf{x}$. In analogy to (15), one obtains the so-called mixture derivative of Γ by

$$\dot{\Gamma} = \frac{\mathrm{d}}{\mathrm{d}t} \Gamma = \frac{\partial \Gamma}{\partial t} + \operatorname{grad} \Gamma \cdot \dot{\mathbf{x}} \,. \tag{16}$$

Describing coupled solid-fluid problems, it is generally convenient to proceed from a *Lagrange*an description of the solid matrix φ^S using the solid displacement vector $\mathbf{u}_S$ as the primary kinematic variable, whereas the $k-1$ pore-fluids φ^β are better described in a modified *Euler*ian setting by use of the seepage velocities $\mathbf{w}_\beta$ describing the fluid motions with respect to the deforming skeleton material:

$$\mathbf{u}_S = \mathbf{x} - \mathbf{X}_S \,, \qquad \mathbf{w}_\beta = \overset{\prime}{\mathbf{x}}_\beta - \overset{\prime}{\mathbf{x}}_S \,. \tag{17}$$

Given this framework together with the corresponding velocity-displacement relation $\overset{\prime}{\mathbf{x}}_S = (\mathbf{u}_S)'_S$, the fluid acceleration can be substituted by

$$\overset{\prime\prime}{\mathbf{x}}_\beta = [\,(\mathbf{u}_S)'_S + \mathbf{w}_\beta\,]'_S + \operatorname{grad}[\,(\mathbf{u}_S)'_S + \mathbf{w}_\beta\,]\,\mathbf{w}_\beta \,. \tag{18}$$

From (10) and $(11)_1$, one obtains the material deformation gradient $\mathbf{F}_\alpha$ and its inverse $\mathbf{F}_\alpha^{-1}$ by

$$\mathbf{F}_\alpha = \operatorname{Grad}_\alpha \mathbf{x} \,, \qquad \mathbf{F}_\alpha^{-1} = \operatorname{grad} \mathbf{X}_\alpha \,. \tag{19}$$

Therein, the operator "$\operatorname{Grad}_\alpha(\,\cdot\,)$" denotes the partial derivative of $(\,\cdot\,)$ with respect to the reference position $\mathbf{X}_\alpha$ of φ^α. In addition to the deformation gradient $\mathbf{F}_\alpha$, the description of deformation processes sometimes proceeds from the displacement gradient

$$\mathbf{H}_\alpha = \operatorname{Grad}_\alpha \mathbf{u}_\alpha \,, \qquad \text{where} \qquad \mathbf{F}_\alpha = \mathbf{I} + \mathbf{H}_\alpha \,. \tag{20}$$

Since the motion of any constituent φ^α was assumed to be unique and uniquely invertible with a non-zero *Jacobian* J_α, the domain of $\det \mathbf{F}_\alpha$ is restricted to positive values

$$\det \mathbf{F}_\alpha = J_\alpha > 0\,, \tag{21}$$

where, in the undeformed state, $\det \mathbf{F}_\alpha(t_0) = 1$ has been considered.

By use of the multiplicative decomposition of the deformation gradient $\mathbf{F}_\alpha$ into a purely volumetric part characterized by $\det \mathbf{F}_\alpha$ and an isochoric part $\tilde{\mathbf{F}}_\alpha$,

$$\mathbf{F}_\alpha = (\det \mathbf{F}_\alpha)^{1/3}\, \tilde{\mathbf{F}}_\alpha, \qquad \text{where} \qquad \det \tilde{\mathbf{F}}_\alpha \equiv 1\,, \tag{22}$$

it is easily shown by use of the symmetry group of fluids (isotropic fluid crystals) that the deformation gradient $\mathbf{F}_\beta$ of fluids φ^β must have the following form (Cross [21], Ehlers [28]):

$$\mathbf{F}_\beta = (\det \mathbf{F}_\beta)^{1/3}\, \mathbf{I}\,. \tag{23}$$

Therein, $\mathbf{I}$ is the fundamental tensor of second order (second order identity).

In case of arbitrary constituents (solids or fluid crystals) governed by standard kinematical relations, the polar decomposition theorem can be applied to the deformation gradient. Thus, one obtains the unique decomposition of $\mathbf{F}_\alpha$ into a proper orthogonal rotation tensor $\mathbf{R}_\alpha$ and either a symmetric and positive definite right or left stretch tensor $\mathbf{U}_\alpha$ or $\mathbf{V}_\alpha$ via

$$\mathbf{F}_\alpha = \begin{cases} \mathbf{R}_\alpha\, \mathbf{U}_\alpha, \\ \mathbf{V}_\alpha\, \mathbf{R}_\alpha. \end{cases} \tag{24}$$

Therein, the so-called continuum rotation tensor $\mathbf{R}_\alpha$ can be expressed by the corresponding continuum rotation vector $\boldsymbol{\varphi}_\alpha$ representing the rotation of material line elements during deformation, e. g. through the *Euler-Rodrigues* formula

$$\mathbf{R}_\alpha = \mathbf{e}_\alpha \otimes \mathbf{e}_\alpha + (\,\mathbf{I} - \mathbf{e}_\alpha \otimes \mathbf{e}_\alpha)\cos\varphi_\alpha + (\mathbf{e}_\alpha \times \mathbf{I}\,)\sin\varphi_\alpha, \tag{25}$$

where φ_α is the value of $\boldsymbol{\varphi}_\alpha$ and $\mathbf{e}_\alpha$ is the axis of rotation. The multiplicative decomposition of $\mathbf{F}_\alpha$ in a pure rotation and a pure stretch allows for the interpretation that a line element $\mathrm{d}\mathbf{x}$ of the actual configuration can be obtained either by a stretch of the referential line element $\mathrm{d}\mathbf{X}_\alpha$ by $\mathbf{U}_\alpha$ followed by a rotation through $\mathbf{R}_\alpha$ or by a rotation of $\mathrm{d}\mathbf{X}_\alpha$ through $\mathbf{R}_\alpha$ followed by a stretch through $\mathbf{V}_\alpha$:

$$\mathrm{d}\mathbf{x} = \begin{cases} \mathbf{R}_\alpha\,(\mathbf{U}_\alpha \mathrm{d}\mathbf{X}_\alpha), \\ \mathbf{V}_\alpha\,(\mathbf{R}_\alpha \mathrm{d}\mathbf{X}_\alpha). \end{cases} \tag{26}$$

Although $\mathbf{U}_\alpha$ and $\mathbf{V}_\alpha$ contain the same amount of stretch, $\mathbf{U}_\alpha$ is defined with respect to the referential frame, whereas $\mathbf{V}_\alpha$ corresponds to the actual configuration. In contrast, $\mathbf{R}_\alpha$ represents a rotation from the reference configuration onto the actual configuration, no matter if the right or the left decomposition is concerned.

Concerning the definition of non-linear strain measures, a standard representation is given, e. g., by the introduction of the *Green-Lagrange*an strain $\mathbf{E}_\alpha$ or the *Karni-Reiner* strain $\mathbf{K}_\alpha$, respectively:

$$\mathbf{E}_\alpha = \begin{cases} \frac{1}{2}(\mathbf{F}_\alpha^T\mathbf{F}_\alpha - \mathbf{I}), \\ \frac{1}{2}(\mathbf{U}_\alpha\mathbf{U}_\alpha - \mathbf{I}), \end{cases} \qquad \mathbf{K}_\alpha = \begin{cases} \frac{1}{2}(\mathbf{F}_\alpha\mathbf{F}_\alpha^T - \mathbf{I}), \\ \frac{1}{2}(\mathbf{V}_\alpha\mathbf{V}_\alpha - \mathbf{I}). \end{cases} \tag{27}$$

Therein, $(\,\cdot\,)^T$ characterizes the transposition of $(\,\cdot\,)$. Furthermore, note that $\mathbf{E}_\alpha$ and $\mathbf{K}_\alpha$ are related to each other by the push-forward rotation

$$\mathbf{K}_\alpha = \mathbf{R}_\alpha\mathbf{E}_\alpha\mathbf{R}_\alpha^T. \tag{28}$$

Considering geometrically linear situations, the rotation tensor and the *Lagrange*an strain tensor yield

$$\begin{aligned} \mathbf{R}_{\alpha\,\text{lin.}} &= \mathbf{I} + \mathbf{H}_{\alpha\,\text{skw}}, \\ \mathbf{E}_{\alpha\,\text{lin.}} &= \mathbf{H}_{\alpha\,\text{sym}}. \end{aligned} \tag{29}$$

Therein, $\mathbf{H}_{\alpha\,\text{skw}}$ and $\mathbf{H}_{\alpha\,\text{sym}}$ are the skew-symmetric and the symmetric parts of the displacement gradient $\mathbf{H}_\alpha$, which, from $(20)_1$, is given by

$$\mathbf{H}_\alpha = \text{Grad}_\alpha\,\mathbf{u}_\alpha \quad \longrightarrow \quad \begin{cases} \mathbf{H}_{\alpha\,\text{sym}} = \frac{1}{2}\,(\mathbf{H}_\alpha + \mathbf{H}_\alpha^T)\,, \\ \mathbf{H}_{\alpha\,\text{skw}} = \frac{1}{2}\,(\mathbf{H}_\alpha - \mathbf{H}_\alpha^T)\,. \end{cases} \tag{30}$$

In the framework of the standard formulation embedded into the geometrically linear approach, $\mathbf{H}_{\alpha\,\text{sym}}$ and $\mathbf{H}_{\alpha\,\text{skw}}$ govern the linear *Lagrange*an strain tensor $\boldsymbol{\varepsilon}_\alpha := \mathbf{E}_{\alpha\,\text{lin.}}$ and, in addition, the continuum rotation vector $\boldsymbol{\varphi}_\alpha$ via

$$\mathbf{H}_{\alpha\,\text{sym}} = \boldsymbol{\varepsilon}_\alpha, \qquad \mathbf{H}_{\alpha\,\text{skw}} = -\overset{3}{\mathbf{E}}\,\boldsymbol{\varphi}_\alpha\,. \tag{31}$$

Therein, $\overset{3}{\mathbf{E}}$ is the *Ricci* permutation tensor or the fundamental tensor of third order. Given the geometrically linear approach, it is seen from $(31)_2$ that the continuum rotation $\boldsymbol{\varphi}_\alpha$ represents the axial vector of the displacement gradient or the rotational displacement, respectively, assigned to the skew-symmetric part of $\mathbf{H}_\alpha$. Thus,

$$\boldsymbol{\varphi}_\alpha = \begin{cases} \frac{1}{2}\,\overset{3}{\mathbf{E}}\,(\mathbf{H}_{\alpha\,\text{skw}})^T, \\ \frac{1}{2}\,\text{rot}_\alpha\,\mathbf{u}_\alpha\,, \end{cases} \tag{32}$$

where $(\overset{3}{\mathbf{E}}\overset{3}{\mathbf{E}})\underset{=}{2} = 2\,\mathbf{I}$ has been used. Note in passing that $(\,\cdot\,)\underset{=}{i}$ defines a contraction of $(\,\cdot\,)$ towards a tensor of i-th order.

In addition to the above strain measures, which are based on the deformation gradient $\mathbf{F}_\alpha$, the different strain rate measures proceed from the spatial velocity gradient

$$\mathbf{L}_\alpha = (\mathbf{F}_\alpha)'_\alpha \mathbf{F}_\alpha^{-1} = \operatorname{grad} \overset{\prime}{\mathbf{x}}_\alpha \quad \longrightarrow \quad \begin{cases} \mathbf{D}_\alpha = \frac{1}{2}\,(\mathbf{L}_\alpha + \mathbf{L}_\alpha^T)\,, \\ \mathbf{W}_\alpha = \frac{1}{2}\,(\mathbf{L}_\alpha - \mathbf{L}_\alpha^T)\,. \end{cases} \tag{33}$$

Therein, the quantities $\mathbf{D}_\alpha$ and $\mathbf{W}_\alpha$ represent the symmetric rate of deformation tensor and the skew-symmetric rate of rotation (spin or vorticity) tensor of φ^α, respectively. Note in passing that the vorticity $\boldsymbol{v}_\alpha$ is related to $\mathbf{W}_\alpha$ like $\boldsymbol{\varphi}_\alpha$, in the framework of the geometrically linear approach, is related to $\mathbf{H}_{\alpha\,\mathrm{skw}}$:

$$\boldsymbol{v}_\alpha \;=\; \begin{cases} \frac{1}{2}\,\overset{3}{\mathbf{E}}\,\mathbf{W}_\alpha^T, \\ \frac{1}{2}\,\mathrm{rot}_\alpha\,\overset{\prime}{\mathbf{x}}_\alpha \end{cases} \tag{34}$$

and

$$\mathbf{W}_\alpha \;=\; -\overset{3}{\mathbf{E}}\,\boldsymbol{v}_\alpha\,. \tag{35}$$

Furthermore, since the vorticity $\boldsymbol{v}_\alpha$ corresponds to the rotational part of the velocity field, it is easily concluded that the time derivative $(\,\cdot\,)'_\alpha$ of the directions $\mathbf{e}_l$ of material line elements $\mathrm{d}\mathbf{x} = \mathrm{d}x\,\mathbf{e}_l$ is governed by

$$(\mathbf{e}_l)'_\alpha \;=\; \mathbf{L}_\alpha \mathbf{e}_l - [\,\mathbf{D}_\alpha \,\cdot\, (\mathbf{e}_l \otimes \mathbf{e}_l)\,]\,\mathbf{e}_l\,. \tag{36}$$

Considering the case where $\mathbf{e}_l$ is an eigenvector of $\mathbf{D}_\alpha$ (cf. e. g. Haupt [64]), a combination of (36) and $(33)_2$ yields

$$\mathbf{D}_\alpha \mathbf{e}_l - [\,\mathbf{D}_\alpha \,\cdot\, (\mathbf{e}_l \otimes \mathbf{e}_l)\,]\,\mathbf{e}_l \;=\; \mathbf{0} \quad \longrightarrow \quad (\mathbf{e}_l)'_\alpha \;=\; \mathbf{W}_\alpha \mathbf{e}_l\,. \tag{37}$$

Thus, the vorticity $\boldsymbol{v}_\alpha$ can be interpreted as the rotational velocity of the eigenvectors of $\mathbf{D}_\alpha$.

Apart from splitting the velocity gradient $\mathbf{L}_\alpha$ into the rate of deformation tensor $\mathbf{D}_\alpha$ and into the rate of rotation tensor $\mathbf{W}_\alpha$, it is convenient for a lot of applications to split $\mathbf{L}_\alpha$ into a non-symmetric deformation rate $\boldsymbol{\Delta}_\alpha$ and into the skew-symmetric gyration tensor $\boldsymbol{\Omega}_\alpha$:

$$\mathbf{L}_\alpha = \boldsymbol{\Delta}_\alpha + \boldsymbol{\Omega}_\alpha, \qquad \text{where} \qquad \begin{cases} \boldsymbol{\Delta}_\alpha = \mathbf{R}_\alpha (\mathbf{U}_\alpha)'_\alpha \mathbf{U}_\alpha^{-1} \mathbf{R}_\alpha^T, \\ \boldsymbol{\Omega}_\alpha = (\mathbf{R}_\alpha)'_\alpha\,\mathbf{R}_\alpha^T. \end{cases} \tag{38}$$

Given the decompositions (33) and (38), the following relations hold:

$$\begin{aligned} \mathbf{D}_\alpha &= \tfrac{1}{2}(\boldsymbol{\Delta}_\alpha + \boldsymbol{\Delta}_\alpha^T), \\ \mathbf{W}_\alpha &= \boldsymbol{\Omega}_\alpha + \tfrac{1}{2}(\boldsymbol{\Delta}_\alpha - \boldsymbol{\Delta}_\alpha^T). \end{aligned} \tag{39}$$

Finally, it is seen from (38) that the rotational velocity of line elements is given by $\boldsymbol{\omega}_\alpha = (\boldsymbol{\varphi}_\alpha)'_\alpha$ where $\boldsymbol{\omega}_\alpha$ is related to the gyration tensor $\boldsymbol{\Omega}_\alpha$ through

$$\boldsymbol{\omega}_\alpha = \tfrac{1}{2}\overset{3}{\mathbf{E}}\,\boldsymbol{\Omega}_\alpha^T \quad \text{and} \quad \boldsymbol{\Omega}_\alpha = -\overset{3}{\mathbf{E}}\,\boldsymbol{\omega}_\alpha. \tag{40}$$

3.2 Extended kinematics of micropolar constituents

If a constituent under study turns out to carry microscopic information in the sense of microrotations, the standard kinematical setting has to be extended by the so-called micromotion

$$\boldsymbol{\xi}_\alpha = \bar{\mathbf{R}}_\alpha\,\boldsymbol{\Xi}_\alpha. \tag{41}$$

Therein, $\boldsymbol{\Xi}_\alpha$ and $\boldsymbol{\xi}_\alpha$ represent two triads of rigid directors representing the particle orientations both in the reference and the actual configuration of φ^α, cf. Figure 6. In addition to the standard kinematical setting, where the

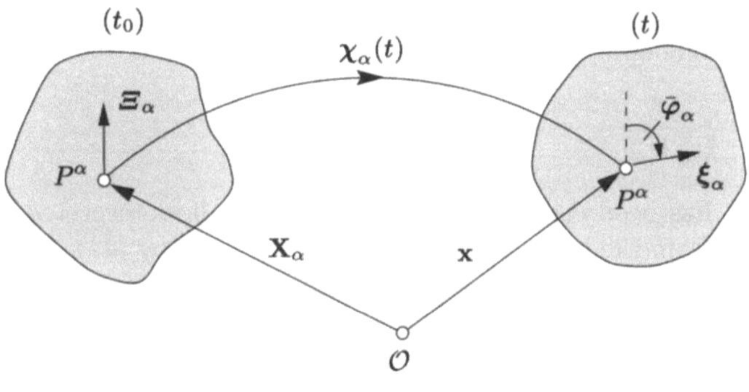

Fig. 6. Micromotion of a micropolar constituent φ^α.

motion function (10) governs the location of material particles P^α and the deformation gradient $\mathbf{F}_\alpha$ contains the information of both the stretch $\mathbf{U}_\alpha$ or $\mathbf{V}_\alpha$ and the continuum rotation via $\mathbf{R}_\alpha$, the quantity $\bar{\mathbf{R}}_\alpha$ contains the continuum *and* the independent micropolar rotation. Consequently, the deformation gradient is now decomposed via

$$\mathbf{F}_\alpha = \begin{cases} \bar{\mathbf{R}}_\alpha\,\bar{\mathbf{U}}_\alpha, \\ \bar{\mathbf{V}}_\alpha\,\bar{\mathbf{R}}_\alpha, \end{cases} \tag{42}$$

where this decomposition must be understood as an alternative to (24). However, although (42) looks similar to (24), the above decomposition is not uniquely defined by the decomposition theorem, namely, as a result of the fact that $\bar{\mathbf{U}}_\alpha$ is not necessarily symmetric although $\bar{\mathbf{R}}_\alpha$ is proper orthogonal. Thus, in comparison to (25), $\bar{\mathbf{R}}_\alpha$ reads

$$\bar{\mathbf{R}}_\alpha = \bar{\mathbf{e}}_\alpha \otimes \bar{\mathbf{e}}_\alpha + (\,\mathbf{I}\, - \bar{\mathbf{e}}_\alpha \otimes \bar{\mathbf{e}}_\alpha) \cos \bar{\varphi}_\alpha + (\bar{\mathbf{e}}_\alpha \times \mathbf{I}\,) \sin \bar{\varphi}_\alpha \,, \tag{43}$$

where $\bar{\varphi}_\alpha$ is the value of $\bar{\boldsymbol{\varphi}}_\alpha$ with the rotation axis $\bar{\mathbf{e}}_\alpha$. Note that in the micropolar case, $\bar{\boldsymbol{\varphi}}_\alpha$ represents the rotation of the directors or the microparticles, respectively, in the sense of a joint rotation of line elements and independent particle rotations. Since the line elements rotate through $\mathbf{R}_\alpha$ given by (25), it is assumed that the additional micropolar rotation is given by

$$\overset{*}{\mathbf{R}}_\alpha = \overset{*}{\mathbf{e}}_\alpha \otimes \overset{*}{\mathbf{e}}_\alpha + (\,\mathbf{I} - \overset{*}{\mathbf{e}}_\alpha \otimes \overset{*}{\mathbf{e}}_\alpha) \cos \overset{*}{\varphi}_\alpha + (\overset{*}{\mathbf{e}}_\alpha \times \mathbf{I}\,) \sin \overset{*}{\varphi}_\alpha \,. \tag{44}$$

Following this, there are two reasonable possibilities to combine $\mathbf{R}_\alpha$ and $\overset{*}{\mathbf{R}}_\alpha$ in order to construct $\bar{\mathbf{R}}_\alpha$, namely

$$\bar{\mathbf{R}}_\alpha = \begin{cases} \mathbf{R}_\alpha \overset{*}{\mathbf{R}}_\alpha & \text{together with} \quad \mathbf{F}_\alpha = \mathbf{R}_\alpha \mathbf{U}_\alpha, \\ \overset{*}{\mathbf{R}}_\alpha \mathbf{R}_\alpha & \text{together with} \quad \mathbf{F}_\alpha = \mathbf{V}_\alpha \mathbf{R}_\alpha. \end{cases} \tag{45}$$

Given (45), one easily concludes to

$$\bar{\mathbf{U}}_\alpha = \overset{*}{\mathbf{R}}{}^T_\alpha \mathbf{U}_\alpha, \qquad \bar{\mathbf{V}}_\alpha = \mathbf{V}_\alpha \overset{*}{\mathbf{R}}{}^T_\alpha, \tag{46}$$

where (42) has been used. The above results can be interpreted as follows: Proceeding from the right decomposition of $\mathbf{F}_\alpha$ given by $(42)_1$ together with $(45)_1$ and $(46)_1$ means that the application of $\bar{\mathbf{U}}_\alpha$ to a line element $\mathrm{d}\mathbf{X}_\alpha$ of the reference configuration of a constituent φ^α yields a pure stretching by $\mathbf{U}_\alpha$ in the given referential frame followed by a back rotation through $\overset{*}{\mathbf{R}}{}^T_\alpha$. Then, the result is rotated forward by $\bar{\mathbf{R}}_\alpha$ to yield the line element $\mathrm{d}\mathbf{x}$ of the actual configuration. Consequently, the direction of $\mathrm{d}\mathbf{x}$ can be interpreted as the direction of the director frame $\boldsymbol{\xi}_\alpha$, whereas the direction of $\bar{\mathbf{U}}_\alpha \mathrm{d}\mathbf{X}_\alpha$ yields the direction of the referential director frame $\boldsymbol{\Xi}_\alpha$.

In contrast, proceeding from the left decomposition of $\mathbf{F}_\alpha$ given by $(42)_2$ together with $(45)_2$ and $(46)_2$ means that the application of $\bar{\mathbf{R}}_\alpha$ to a line element $\mathrm{d}\mathbf{X}_\alpha$ yields a forward rotation by $\mathbf{R}_\alpha$ onto the direction of $\mathrm{d}\mathbf{x}$ followed by an additional rotation through $\overset{*}{\mathbf{R}}_\alpha$. Then, applying $\bar{\mathbf{V}}_\alpha$ to this result finally yields $\mathrm{d}\mathbf{x}$ itself by a combination of a backward rotation through $\overset{*}{\mathbf{R}}{}^T_\alpha$ followed by a stretching with $\mathbf{V}_\alpha$. Consequently, the direction of $\mathrm{d}\mathbf{X}_\alpha$ is now interpreted as the direction of the referential director frame $\boldsymbol{\Xi}_\alpha$, whereas the direction of $\bar{\mathbf{R}}_\alpha \mathrm{d}\mathbf{X}_\alpha$ yields the direction of the director frame $\boldsymbol{\xi}_\alpha$ of the actual configuration.

Strain measures corresponding to the definitions of *Green-Lagrange*an or *Karni-Reiner* strains of the standard continuum formulation, respectively, are given through

$$\bar{\mathbf{E}}_\alpha = \bar{\mathbf{U}}_\alpha - \mathbf{I}, \qquad \bar{\mathbf{K}}_\alpha = \bar{\mathbf{V}}_\alpha - \mathbf{I}. \tag{47}$$

Therein, $\bar{\mathbf{E}}_\alpha$ is usually known as the first *Cosserat* strain[1]. Furthermore, $\bar{\mathbf{K}}_\alpha$ is obtained from $\bar{\mathbf{E}}_\alpha$ by a push-forward rotation with $\bar{\mathbf{R}}_\alpha$ like $\mathbf{K}_\alpha$ was obtained from $\mathbf{E}_\alpha$ by push-forward with $\mathbf{R}_\alpha$:

$$\bar{\mathbf{K}}_\alpha = \bar{\mathbf{R}}_\alpha \bar{\mathbf{E}}_\alpha \bar{\mathbf{R}}_\alpha^T. \tag{48}$$

Apart from the above strain measures, the micropolar approach requires the additional introduction of curvature tensors

$$\begin{aligned} {}^R\overset{3}{\bar{\mathbf{C}}}_\alpha &= (\bar{\mathbf{R}}_\alpha^T \,\mathrm{Grad}_\alpha\, \bar{\mathbf{R}}_\alpha)^{\underline{3}}, \\ \overset{3}{\bar{\mathbf{C}}}_\alpha &= -(\bar{\mathbf{R}}_\alpha \,\mathrm{grad}\, \bar{\mathbf{R}}_\alpha^T)^{\underline{3}}, \end{aligned} \tag{49}$$

which can be given in a referential frame, ${}^R\overset{3}{\bar{\mathbf{C}}}_\alpha$, and in a spatial frame, $\overset{3}{\bar{\mathbf{C}}}_\alpha$, respectively. For a discussion of further possibilities to introduce curvature tensors, the reader is referred to the work by Steinmann [94] and Volk [106]. Both curvature tensors are skew-symmetric with respect to first two basis systems, i. e.:

$$\begin{aligned} {}^R\overset{3}{\bar{\mathbf{C}}}_\alpha &= -({}^R\overset{3}{\bar{\mathbf{C}}}_\alpha)^{\overset{12}{T}}, \\ \overset{3}{\bar{\mathbf{C}}}_\alpha &= -(\overset{3}{\bar{\mathbf{C}}}_\alpha)^{\overset{12}{T}}. \end{aligned} \tag{50}$$

Given the above skew-symmetry, the corresponding axial tensors are defined via

$$\begin{aligned} {}^R\overset{A}{\bar{\mathbf{C}}}_\alpha &= \tfrac{1}{2}(\overset{3}{\mathbf{E}}\; {}^R\overset{3}{\bar{\mathbf{C}}}{}_\alpha^{\overset{12}{T}})^{\underline{2}}, \\ \overset{A}{\bar{\mathbf{C}}}_\alpha &= \tfrac{1}{2}(\overset{3}{\mathbf{E}}\; \overset{3}{\bar{\mathbf{C}}}{}_\alpha^{\overset{12}{T}})^{\underline{2}}. \end{aligned} \tag{51}$$

Note in passing that the above axial tensors are defined as the curvature tensors of second order[2]. Furthermore, in analogy to the computation of an axial vector corresponding to a skew-symmetric tensor of second order

[1] Note that the notion *Cosserat* strain is also used for $\bar{\mathbf{U}}_\alpha$ or for $\bar{\mathbf{U}}_\alpha^T$, respectively, cf. Eringen and Kafadar [56, 69].

and, vice versa, the computation of the skew-symmetric tensor through its axial vector, cf. (34) and (35), the third-order curvature tensors are uniquely defined through

$$\begin{aligned} {}^R\overset{3}{\bar{\mathbf{C}}}_\alpha &= -(\overset{3}{\mathbf{E}}\ {}^R\overset{A}{\bar{\mathbf{C}}}_\alpha)^{\underline{3}}, \\ \overset{3}{\bar{\mathbf{C}}}_\alpha &= -(\overset{3}{\mathbf{E}}\ \overset{A}{\bar{\mathbf{C}}}_\alpha)^{\underline{3}}. \end{aligned} \tag{52}$$

In the frame of micropolar materials, the curvature tensors ${}^R\overset{3}{\bar{\mathbf{C}}}_\alpha$ and $\overset{3}{\bar{\mathbf{C}}}_\alpha$ are not independent but depend on the micropolar strain tensors $\bar{\mathbf{U}}_\alpha$ and $\bar{\mathbf{V}}_\alpha$ and their spatial derivatives through a micropolar compatibility condition. This fact can be seen in analogy to the concept of non-polar materials, where the strain tensor depends on the displacement vector through the usual compatibility conditions. To obtain the additional micropolar compatibility condition, use is made of the so-called second-grade linear group $\mathcal{L}$ defined by

$$\mathcal{L} = \{(\mathbf{J},\ \overset{3}{\bar{\mathbf{G}}})\}\,. \tag{53}$$

Therein, $\mathbf{J}$ is a non-singular linear transformation and $\overset{3}{\bar{\mathbf{G}}}$ a third-order tensor such that (cf. Cross [21], Ehlers [28, 29])

$$\overset{3}{\bar{\mathbf{G}}} = \overset{3}{\bar{\mathbf{G}}}{}^{\overset{23}{T}}. \tag{54}$$

Apart from the pairs $(\mathbf{F}_\alpha\,,\ \mathrm{Grad}_\alpha \mathbf{F}_\alpha)$ and $(\mathbf{F}_\alpha^{-1},\ \mathrm{grad}\,\mathbf{F}_\alpha^{-1})$ defined through the first and second deformation gradients, the following pairs are also elements of $\mathcal{L}$ (Volk [106]):

$$(\mathbf{F}_\alpha\,,{}^R\overset{3}{\bar{\mathbf{G}}}_\alpha), \qquad (\mathbf{F}_\alpha^{-1},\ \overset{3}{\bar{\mathbf{G}}}_\alpha). \tag{55}$$

Therein,

$$\begin{aligned} {}^R\overset{3}{\bar{\mathbf{G}}}_\alpha &= (\mathbf{F}_\alpha^{-1}\,\mathrm{Grad}_\alpha\,\mathbf{F}_\alpha)^{\underline{3}} &\longrightarrow\quad {}^R\overset{3}{\bar{\mathbf{G}}}_\alpha &= {}^R\overset{3}{\bar{\mathbf{G}}}{}_\alpha^{\overset{23}{T}}, \\ \overset{3}{\bar{\mathbf{G}}}_\alpha &= -(\mathbf{F}_\alpha\,\mathrm{grad}\,\mathbf{F}_\alpha^{-1})^{\underline{3}} &\longrightarrow\quad \overset{3}{\bar{\mathbf{G}}}_\alpha &= \overset{3}{\bar{\mathbf{G}}}{}_\alpha^{\overset{23}{T}}. \end{aligned} \tag{56}$$

[2] In the classical literature, cf. Truesdell and Toupin [103], § 61, or Eringen and Kafadar [56, 69], the referential second-order curvature is also called *wryness* of the director frame.

Proceeding from (55) and (56) together with the decomposition of $\mathbf{F}_\alpha$ given by (42), it is concluded that

$$
\begin{aligned}
{}^R\overset{3}{\mathbf{G}}_\alpha &= \{\mathbf{F}_\alpha^{-1}\,[\,(\mathrm{Grad}_\alpha\,\bar{\mathbf{V}}_\alpha)^{\overset{23}{T}}\bar{\mathbf{R}}_\alpha\,]^{\underline{3}}\,\}^{\underline{3}} + {}^R\overset{3}{\bar{\mathbf{C}}}{}_\alpha^{\overset{23}{T}}\,,\\
\overset{3}{\mathbf{G}}_\alpha &= -\{\mathbf{F}_\alpha\,[\,(\mathrm{grad}\,\bar{\mathbf{U}}_\alpha^{-1})^{\overset{23}{T}}\bar{\mathbf{R}}_\alpha^{T}\,]^{\underline{3}}\,\}^{\underline{3}} + \overset{3}{\bar{\mathbf{C}}}{}_\alpha^{\overset{23}{T}}\,.
\end{aligned}
\tag{57}
$$

Thus, based on the symmetry conditions given by (56), one obtains

$$
\begin{aligned}
{}^R\overset{3}{\bar{\mathbf{C}}}_\alpha - {}^R\overset{3}{\bar{\mathbf{C}}}{}_\alpha^{\overset{23}{T}} &= \{\mathbf{F}_\alpha^{-1}\,[\,(\mathrm{Grad}_\alpha\,\bar{\mathbf{V}}_\alpha)^{\overset{23}{T}}\bar{\mathbf{R}}_\alpha\,]^{\underline{3}}\,\}^{\underline{3}} -\\
&\quad -\{\mathbf{F}_\alpha^{-1}\,[\,(\mathrm{Grad}_\alpha\,\bar{\mathbf{V}}_\alpha)^{\overset{23}{T}}\bar{\mathbf{R}}_\alpha\,]^{\underline{3}}\,\}^{\underline{3}\overset{23}{T}}\,,\\
\overset{3}{\bar{\mathbf{C}}}_\alpha - \overset{3}{\bar{\mathbf{C}}}{}_\alpha^{\overset{23}{T}} &= -\{\mathbf{F}_\alpha\,[\,(\mathrm{grad}\,\bar{\mathbf{U}}_\alpha^{-1})^{\overset{23}{T}}\bar{\mathbf{R}}_\alpha^{T}\,]^{\underline{3}}\,\}^{\underline{3}} +\\
&\quad +\{\mathbf{F}_\alpha\,[\,(\mathrm{grad}\,\bar{\mathbf{U}}_\alpha^{-1})^{\overset{23}{T}}\bar{\mathbf{R}}_\alpha^{T}\,]^{\underline{3}}\,\}^{\underline{3}\overset{23}{T}}\,.
\end{aligned}
\tag{58}
$$

These relations are the additional compatibility conditions of micropolar continuum mechanics given with respect to the referential and to the actual frames.

Considering geometrically linear situations, a linearization of the *Cosserat* rotation $\bar{\mathbf{R}}_\alpha$ given by (43) around the natural state expressed by $\bar{\varphi}_\alpha = 0$ yields

$$
\begin{aligned}
\bar{\mathbf{R}}_{\alpha\,\mathrm{lin.}} &= \bar{\mathbf{R}}_\alpha\Big|_{\bar{\varphi}_\alpha = 0} + \frac{\partial\bar{\mathbf{R}}_\alpha}{\partial\bar{\varphi}_\alpha}\Big|_{\bar{\varphi}_\alpha = 0}\;\bar{\varphi}_\alpha\\
&= \mathbf{I} + (\bar{\mathbf{e}}_\alpha \times \mathbf{I})\,\bar{\varphi}_\alpha\,.
\end{aligned}
\tag{59}
$$

Thus, given $(\bar{\mathbf{e}}_\alpha \times \mathbf{I})\,\bar{\varphi}_\alpha = \bar{\boldsymbol{\varphi}}_\alpha \times \mathbf{I} = -\overset{3}{\mathbf{E}}\,\bar{\boldsymbol{\varphi}}_\alpha$, one easily finds the linear *Cosserat* rotation and its transposition as

$$
\begin{aligned}
\bar{\mathbf{R}}_{\alpha\,\mathrm{lin.}} &= \mathbf{I} - \overset{3}{\mathbf{E}}\,\bar{\boldsymbol{\varphi}}_\alpha\,,\\
\bar{\mathbf{R}}^T_{\alpha\,\mathrm{lin.}} &= \mathbf{I} + \overset{3}{\mathbf{E}}\,\bar{\boldsymbol{\varphi}}_\alpha\,.
\end{aligned}
\tag{60}
$$

Note in passing that $\overset{3}{\mathbf{E}}{}^{\overset{12}{T}} = -\overset{3}{\mathbf{E}}$ has been used to obtain $\bar{\mathbf{R}}^T_{\alpha\,\mathrm{lin.}}$ from $\bar{\mathbf{R}}_{\alpha\,\mathrm{lin.}}$.

In addition to the linearization of $\bar{\mathbf{R}}_\alpha$, a linearization of the *Cosserat* strain

$$
\bar{\mathbf{E}}_\alpha = \bar{\mathbf{U}}_\alpha - \mathbf{I} = \bar{\mathbf{R}}_\alpha^T\mathbf{F}_\alpha - \mathbf{I}
\tag{61}
$$

around the natural state given by $\bar{\varphi}_\alpha = 0$ and $\mathbf{F}_\alpha = \mathbf{I}$ yields

$$
\begin{aligned}
\bar{\mathbf{E}}_{\alpha\,\mathrm{lin.}} &= \bar{\mathbf{E}}_\alpha\Big|_{\substack{\mathbf{F}_\alpha = \mathbf{I}\\ \bar{\varphi}_\alpha = 0}} + \frac{\partial \bar{\mathbf{E}}_\alpha}{\partial \mathbf{F}_\alpha}\Big|_{\substack{\mathbf{F}_\alpha = \mathbf{I}\\ \bar{\varphi}_\alpha = 0}} (\mathbf{F}_\alpha - \mathbf{I}) + \frac{\partial \bar{\mathbf{E}}_\alpha}{\partial \bar{\varphi}_\alpha}\Big|_{\substack{\mathbf{F}_\alpha = \mathbf{I}\\ \bar{\varphi}_\alpha = 0}} \bar{\varphi}_\alpha \\
&= \quad \mathbf{0} \quad + \; (\mathbf{I} \otimes \mathbf{I})^{\overset{23}{T}} (\mathbf{F}_\alpha - \mathbf{I}) \; - \; (\bar{\mathbf{e}}_\alpha \times \mathbf{I})\, \bar{\varphi}_\alpha \,.
\end{aligned}
\tag{62}
$$

Thus, it is easily concluded that

$$
\bar{\varepsilon}_\alpha := \bar{\mathbf{E}}_{\alpha\,\mathrm{lin.}} = \mathbf{H}_\alpha + \overset{3}{\mathbf{E}}\, \bar{\varphi}_\alpha \,, \tag{63}
$$

where $\mathbf{F}_\alpha = \mathbf{I} + \mathbf{H}_\alpha$ with $\mathbf{H}_\alpha$ from (30) has been used. In addition to (63), the symmetric and skew symmetric parts of the *Cosserat* strain are given by

$$
\begin{aligned}
\bar{\varepsilon}_{\alpha\,\mathrm{sym}} &= \tfrac{1}{2}\,(\mathbf{H}_\alpha + \mathbf{H}_\alpha^T)\,, \\
\bar{\varepsilon}_{\alpha\,\mathrm{skw}} &= \tfrac{1}{2}\,(\mathbf{H}_\alpha - \mathbf{H}_\alpha^T) + \overset{3}{\mathbf{E}}\, \bar{\varphi}_\alpha \,.
\end{aligned}
\tag{64}
$$

In this representation, $\bar{\varepsilon}_{\alpha\,\mathrm{sym}}$ equals the linearized *Lagrange*an strain ε_α of non-polar materials, whereas recalling $(31)_2$

$$
\bar{\varepsilon}_{\alpha\,\mathrm{skw}} = \overset{3}{\mathbf{E}}\, (\bar{\varphi}_\alpha - \varphi_\alpha) = \overset{3}{\mathbf{E}}\, \overset{*}{\varphi}_\alpha \tag{65}
$$

exhibits the tensorial measure for the additional micropolar rotation $\overset{*}{\varphi}_\alpha$. Note in passing that, in contrast to the geometrically non-linear approach, where the rotation vectors are non-additive pseudo-vectors, an addition of the geometrically linear rotation measures yields

$$
\bar{\varphi}_\alpha = \varphi_\alpha + \overset{*}{\varphi}_\alpha \,. \tag{66}
$$

Additionally, this result is easily concluded from the above linearization of $\bar{\mathbf{R}}_\alpha$ together with (45). Proceeding, e. g., from $(45)_1$ yields

$$
\begin{aligned}
\bar{\mathbf{R}}_{\alpha\,\mathrm{lin.}} &= \mathbf{R}_{\alpha\,\mathrm{lin.}} \overset{*}{\mathbf{R}}_{\alpha\,\mathrm{lin.}} \\
&= (\mathbf{I} - \overset{3}{\mathbf{E}}\, \varphi_\alpha)(\mathbf{I} - \overset{3}{\mathbf{E}}\, \overset{*}{\varphi}_\alpha) \\
&= \mathbf{I} - \overset{3}{\mathbf{E}}\, (\varphi_\alpha + \overset{*}{\varphi}_\alpha)\,,
\end{aligned}
\tag{67}
$$

where $(\overset{3}{\mathbf{E}}\, \varphi_\alpha)(\overset{3}{\mathbf{E}}\, \overset{*}{\varphi}_\alpha)$ has been neglected in the sense of small terms of higher order.

Apart from the linearization of the *Cosserat* rotation and strain measures, a linearization of the third-order curvature tensor given by $(49)_1$ leads to

$$
\begin{aligned}
{}^R\overset{3}{\bar{\mathbf{C}}}_{\alpha\,\mathrm{lin.}} &= {}^R\overset{3}{\bar{\mathbf{C}}}_{\alpha}\Big|_{\bar{\varphi}_\alpha = 0} + \left(\frac{\partial \bar{\mathbf{R}}_\alpha^T}{\partial \bar{\varphi}_\alpha}\,\mathrm{Grad}_\alpha\,\bar{\mathbf{R}}_\alpha\right)^{\underline{3}}\Big|_{\bar{\varphi}_\alpha = 0}\bar{\varphi}_\alpha + \\
&\quad + \left(\bar{\mathbf{R}}_\alpha^T\,\frac{\partial\,\mathrm{Grad}_\alpha\bar{\mathbf{R}}_\alpha}{\partial\bar{\varphi}_\alpha}\right)^{\underline{3}}\Big|_{\bar{\varphi}_\alpha = 0}\bar{\varphi}_\alpha \\
&= \overset{3}{\mathbf{0}} + \overset{3}{\mathbf{0}} - \left((\overset{3}{\mathbf{E}}\,\mathrm{Grad}_\alpha\bar{\mathbf{e}}_\alpha)^{\underline{3}} + (\overset{3}{\mathbf{E}}\,\bar{\mathbf{e}}_\alpha)\otimes\frac{\partial\,\mathrm{Grad}_\alpha\,\bar{\varphi}_\alpha}{\partial\bar{\varphi}_\alpha}\Big|_{\bar{\varphi}_\alpha = 0}\right)\bar{\varphi}_\alpha .
\end{aligned}
\tag{68}
$$

Thus, it is finally concluded that

$$
{}^R\overset{3}{\bar{\mathbf{C}}}_{\alpha\,\mathrm{lin.}} = -\,(\overset{3}{\mathbf{E}}\,\mathrm{Grad}_\alpha\bar{\boldsymbol{\varphi}}_\alpha)^{\underline{3}}. \tag{69}
$$

Furthermore, one easily obtains the linear curvature tensor of second order by combination of $(51)_1$ and (69) to yield

$$
\bar{\boldsymbol{\kappa}}_\alpha := {}^R\overset{A}{\bar{\mathbf{C}}}_{\alpha\,\mathrm{lin.}} = \tfrac{1}{2}(\overset{3}{\mathbf{E}}\;{}^R\overset{3\;12}{\bar{\mathbf{C}}}{}^{T}_{\alpha\,\mathrm{lin.}})^{\underline{2}} = \tfrac{1}{2}(\overset{3}{\mathbf{E}}\,\overset{3}{\mathbf{E}}\,)^{\underline{2}}\,\mathrm{Grad}_\alpha\bar{\boldsymbol{\varphi}}_\alpha\,, \tag{70}
$$

thus recovering the classical result

$$
\bar{\boldsymbol{\kappa}}_\alpha = \mathrm{Grad}_\alpha\bar{\boldsymbol{\varphi}}_\alpha\,. \tag{71}
$$

Apart from the linearization of strains and curvatures, one additionally has to linearize the compatibility condition (58). This can be done by a formal linearization of either the right or the left hand side of (58). Using the latter possibility, one obtains with the aid of (69)

$$
\left({}^R\overset{3}{\bar{\mathbf{C}}}_\alpha - {}^R\overset{3\;23}{\bar{\mathbf{C}}}{}^{T}_{\alpha}\right)_{\mathrm{lin.}} = -\,(\overset{3}{\mathbf{E}}\,\mathrm{Grad}_\alpha\bar{\boldsymbol{\varphi}}_\alpha)^{\underline{3}} + (\overset{3}{\mathbf{E}}\,\mathrm{Grad}_\alpha\bar{\boldsymbol{\varphi}}_\alpha)^{\underline{3}\overset{23}{T}}. \tag{72}
$$

Furthermore, a combination of this result with (63) and (71) yields

$$
(\overset{3}{\mathbf{E}}\,\bar{\boldsymbol{\kappa}}_\alpha)^{\underline{3}} - (\overset{3}{\mathbf{E}}\,\bar{\boldsymbol{\kappa}}_\alpha)^{\underline{3}\overset{23}{T}} = \mathrm{Grad}_\alpha\,\bar{\boldsymbol{\varepsilon}}_\alpha - \mathrm{Grad}_\alpha^{\overset{23}{T}}\,\bar{\boldsymbol{\varepsilon}}_\alpha\,, \tag{73}
$$

where $(\overset{3}{\mathbf{E}}\,\mathrm{Grad}_\alpha\bar{\boldsymbol{\varphi}}_\alpha)^{\underline{3}} = \mathrm{Grad}_\alpha(\overset{3}{\mathbf{E}}\,\bar{\boldsymbol{\varphi}}_\alpha)$ has been used. Note in passing that the above compatibility condition, in a simple index notation, has firstly been proposed by Nowacki in 1969 (cf. Nowacki [84]). In contrast to the geometrically finite approach, the geometrically linear representation of the micropolar compatibility condition can be solved with respect to the curvature $\bar{\kappa}_\alpha$, cf. Ehlers and Volk [52]:

$$\bar{\kappa}_\alpha = \tfrac{1}{2}\overset{3}{\mathbf{E}}\left(\mathrm{Grad}_\alpha\,\bar{\varepsilon}_\alpha + \mathrm{Grad}_\alpha^{\overset{13}{T}}\,\bar{\varepsilon}_\alpha - \mathrm{Grad}_\alpha^{\overset{23}{T}}\,\bar{\varepsilon}_\alpha\right)\underset{-}{2}\,. \tag{74}$$

As a result of (74), it follows immediately that, once $\bar{\varepsilon}_\alpha$ is given, it is straight forward to compute $\bar{\kappa}_\alpha$.

Concerning the definition of translational and rotational velocity measures of *Cosserat* continua, the standard kinematical relations (33)–(40) have to be extended with respect to the additional rotational degrees of freedom. In particular, the spatial velocity gradient $\mathbf{L}_\alpha$ from (33) reads

$$\mathbf{L}_\alpha = \begin{cases} \mathbf{D}_\alpha + \mathbf{W}_\alpha\,, \\ \boldsymbol{\Delta}_\alpha + \boldsymbol{\Omega}_\alpha\,, \\ \bar{\boldsymbol{\Delta}}_\alpha + \bar{\boldsymbol{\Omega}}_\alpha\,. \end{cases} \tag{75}$$

Therein, in addition to $\mathbf{D}_\alpha$, $\mathbf{W}_\alpha$ and $\boldsymbol{\Delta}_\alpha$, $\boldsymbol{\Omega}_\alpha$, the quantities $\bar{\boldsymbol{\Delta}}_\alpha$ and $\bar{\boldsymbol{\Omega}}_\alpha$ are defined by

$$\begin{aligned} \bar{\boldsymbol{\Delta}}_\alpha &= \bar{\mathbf{R}}_\alpha(\bar{\mathbf{U}}_\alpha)'_\alpha\bar{\mathbf{U}}_\alpha^{-1}\bar{\mathbf{R}}_\alpha^T, \\ \bar{\boldsymbol{\Omega}}_\alpha &= (\bar{\mathbf{R}}_\alpha)'_\alpha\,\bar{\mathbf{R}}_\alpha^T. \end{aligned} \tag{76}$$

Thus, in analogy to the decomposition (39), the following relation holds:

$$\begin{aligned} \mathbf{D}_\alpha &= \tfrac{1}{2}(\bar{\boldsymbol{\Delta}}_\alpha + \bar{\boldsymbol{\Delta}}_\alpha^T), \\ \mathbf{W}_\alpha &= \bar{\boldsymbol{\Omega}}_\alpha + \tfrac{1}{2}(\bar{\boldsymbol{\Delta}}_\alpha - \bar{\boldsymbol{\Delta}}_\alpha^T). \end{aligned} \tag{77}$$

Finally, it is seen in analogy to (40) that the rotational velocity of the director frame is governed by

$$\bar{\omega}_\alpha = \tfrac{1}{2}\overset{3}{\mathbf{E}}\,\bar{\boldsymbol{\Omega}}_\alpha^T \quad \text{and} \quad \bar{\boldsymbol{\Omega}}_\alpha = -\overset{3}{\mathbf{E}}\,\bar{\omega}_\alpha\,, \tag{78}$$

where $\bar{\omega}_\alpha = (\bar{\varphi}_\alpha)'_\alpha$ and $\bar{\boldsymbol{\Omega}}_\alpha$ is the micropolar gyration tensor relating the director frame rate $(\xi_\alpha)'_\alpha$ to the director frame ξ_α:

$$(\xi_\alpha)'_\alpha = \bar{\boldsymbol{\Omega}}_\alpha\,\xi_\alpha\,. \tag{79}$$

4 Balance relations

4.1 General frame

The discussion of balance relations for multiphasic materials is based on *Truesdell*'s "metaphysical principles" of mixture theories, cf. Truesdell [102, p. 221]):

1. *All properties of the mixture must be mathematical consequences of properties of the constituents.*
2. *So as to describe the motion of a constituent, we may in imagination isolate it from the rest of the mixture, provided we allow properly for the actions of the other constituents upon it.*
3. *The motion of the mixture is governed by the same equations as is a single body.*

The foundation of *Truesdell*'s principles proceeds from the idea that both the balance relations of the constituents φ^α and the balance relations of the overall medium (the mixture) $\varphi = \sum_{\alpha=1}^{k} \varphi^\alpha$ can be given in analogy to the balance relations of the classical continuum mechanics of single-phase materials, provided one allows for the interaction mechanisms between the constituents by the introduction of so-called production terms.

4.2 General balance relations of the mixture

Following this, the general balance relations for the overall multiphasic material φ can be directly given from the results of the classical continuum mechanics of single-phase media:

$$\boxed{\begin{aligned} \frac{\mathrm{d}}{\mathrm{d}t}\int_{\mathcal{B}} \Psi \,\mathrm{d}v &= \int_{\mathcal{S}} (\boldsymbol{\phi}\cdot\mathbf{n})\,\mathrm{d}a + \int_{\mathcal{B}} \sigma\,\mathrm{d}v + \int_{\mathcal{B}} \hat{\Psi}\,\mathrm{d}v,\\ \frac{\mathrm{d}}{\mathrm{d}t}\int_{\mathcal{B}} \boldsymbol{\Psi} \,\mathrm{d}v &= \int_{\mathcal{S}} (\boldsymbol{\Phi}\,\mathbf{n})\,\mathrm{d}a + \int_{\mathcal{B}} \boldsymbol{\sigma}\,\mathrm{d}v + \int_{\mathcal{B}} \hat{\boldsymbol{\Psi}}\,\mathrm{d}v. \end{aligned}} \tag{80}$$

In the above equations, Ψ or $\boldsymbol{\Psi}$, respectively, are the volume-specific scalar- or vector-valued densities of the mechanical quantities in $\mathcal{B}$ to be balanced. In the framework of a general thermodynamical description, these quantities are given by the mass density, the linear momentum, the moment of momentum (angular momentum), the total energy (internal and kinetic) and the entropy. The quantities $\boldsymbol{\phi}\cdot\mathbf{n}$ and $\boldsymbol{\Phi}\,\mathbf{n}$, respectively, are the densities on the surface $\mathcal{S}$ of $\mathcal{B}$ (effluxes) of the mechanical quantities resulting from the external vicinity. Furthermore, $\mathbf{n}$ is the outward oriented surface normal. σ or $\boldsymbol{\sigma}$, respectively, are the supply terms of the mechanical quantities resulting from the external distance, whereas $\hat{\Psi}$ or $\hat{\boldsymbol{\Psi}}$, respectively, are the production terms of the mechanical quantities as a result of possible couplings of φ with the surrounding of φ. Assuming steady and steadily differentiable integrands, differentiation of the left-hand side of (80) and transformation of the surface integrals incorporated in the right-hand side of (80) into volume integrals yields the local forms of the balance relations, viz.:

$$\boxed{\begin{aligned} \dot{\Psi} + \Psi \operatorname{div} \dot{\mathbf{x}} &= \operatorname{div} \boldsymbol{\phi} + \sigma + \hat{\Psi}, \\ \dot{\boldsymbol{\Psi}} + \boldsymbol{\Psi} \operatorname{div} \dot{\mathbf{x}} &= \operatorname{div} \boldsymbol{\Phi} + \boldsymbol{\sigma} + \hat{\boldsymbol{\Psi}}. \end{aligned}} \tag{81}$$

4.3 Specific balance equations of the mixture

The specific balance equations of mass, linear momentum, moment of momentum (m. o. m.), energy and entropy in both formulations the standard (*sf*) and the micropolar (*mf*) one are introduced, as is usual in continuum thermodynamics, via axioms, thus leading to the representation given in Table 1. Therein, $\mathbf{T}$ is the *Cauchy* stress tensor and $\mathbf{b}$ the external volume force per unit mass or the gravitation, respectively. Furthermore, $\rho\,\dot{\mathbf{x}}$ is the momentum of the overall medium, whereas $\mathbf{x} \times (\rho\,\dot{\mathbf{x}})$ yields the moment of momentum of the standard formulation. This term has to be extended, in the framework of the micropolar formulation, by $\rho\,\bar{\boldsymbol{\Theta}}\,\bar{\omega}$. Therein, $\bar{\boldsymbol{\Theta}}$ is the tensor of microinertia per unit mass, and $\bar{\omega} = \dot{\bar{\varphi}}$ is the total rotational velocity. Note in passing that $\rho\,\bar{\boldsymbol{\Theta}}\,\bar{\omega}$ represents the vector of angular momentum of the

	Ψ, $\boldsymbol{\Psi}$	$\boldsymbol{\phi}$, $\boldsymbol{\Phi}$	σ, $\boldsymbol{\sigma}$	$\hat{\Psi}$, $\hat{\boldsymbol{\Psi}}$
mass	ρ	$\mathbf{0}$	0	0
momentum	$\rho\,\dot{\mathbf{x}}$	$\mathbf{T}$	$\rho\,\mathbf{b}$	$\mathbf{0}$
m. o. m. (*sf*)	$\mathbf{x} \times (\rho\,\dot{\mathbf{x}})$	$\mathbf{x} \times \mathbf{T}$	$\mathbf{x} \times (\rho\,\mathbf{b})$	$\mathbf{0}$
m. o. m. (*mf*)	$\mathbf{x} \times (\rho\,\dot{\mathbf{x}}) + \rho\,\bar{\boldsymbol{\Theta}}\,\bar{\omega}$	$\mathbf{x} \times \mathbf{T} + \mathbf{M}$	$\mathbf{x} \times (\rho\,\mathbf{b}) + \rho\,\mathbf{c}$	$\mathbf{0}$
energy (*sf*)	$\rho\,\varepsilon + \frac{1}{2}\,\dot{\mathbf{x}} \cdot (\rho\,\dot{\mathbf{x}})$	$\mathbf{T}^T\,\dot{\mathbf{x}} - \mathbf{q}$	$\dot{\mathbf{x}} \cdot (\rho\,\mathbf{b}) + \rho\,r$	0
energy (*mf*)	$\rho\,\varepsilon + \frac{1}{2}\,\dot{\mathbf{x}} \cdot (\rho\,\dot{\mathbf{x}})$ $+\frac{1}{2}\,\bar{\omega} \cdot (\rho\,\bar{\boldsymbol{\Theta}}\,\bar{\omega})$	$\mathbf{T}^T\,\dot{\mathbf{x}} - \mathbf{q}$ $+\mathbf{M}^T\bar{\omega}$	$\dot{\mathbf{x}} \cdot (\rho\,\mathbf{b}) + \rho\,r$ $+\bar{\omega} \cdot \rho\,\mathbf{c}$	0
entropy	$\rho\,\eta$	$\boldsymbol{\phi}_\eta$	σ_η	$\hat{\eta}$

Table 1. Balance relations for the overall medium φ.

rigid microparticles of micropolar continua substituting the material points of the standard formulation. Finally, $\mathbf{M}$ and $\mathbf{c}$ are the couple stress tensor and the body couple stress vector per unit mass. Concerning the energy and entropy balance relations, ε is the internal energy, $\mathbf{q}$ is the heat influx vector and r is the external heat supply. In addtion, η is the entropy, $\boldsymbol{\phi}_\eta$ and σ_η are the efflux of entropy and the external entropy supply, whereas $\hat{\eta}$ is the non-negative entropy production [32]. Furthermore, it might be interesting to know that some classical approaches to micropolar materials additionally proceed from a so-called "balance of microinertia" which is not introduced in this article because of its artificial character. This specific balance relation simply uses the fact that, in a rotating frame fixed to the microparticles, the values of the tensor of microinertia are constants so that the *Green-Naghdi* derivative of $\bar{\boldsymbol{\Theta}}$ vanishes. For further details, the interested reader is referred to the work by Eringen and Kafadar [56] or by Volk [106].

Inserting the quantities given in Table 1 into the local balances (81), one obtains with the aid of the respective "lower" balances the following relations known from continuum mechanics of single-phase materials (cf. e. g. [23, 25, 32, 63]).

Non-polar materials (standard formulation):

$$\begin{array}{lrl}
\bullet\ \text{mass:} & \dot{\rho} + \rho \operatorname{div} \dot{\mathbf{x}} = & 0\,, \\
\bullet\ \text{momentum:} & \rho\, \ddot{\mathbf{x}} = & \operatorname{div} \mathbf{T} + \rho\, \mathbf{b}\,, \\
\bullet\ \text{m. o. m.:} & \mathbf{0} = & \mathbf{I} \times \mathbf{T} \quad \longrightarrow \quad \mathbf{T} = \mathbf{T}^T, \\
\bullet\ \text{energy:} & \rho\, \dot{\varepsilon} = & \mathbf{T} \cdot \mathbf{L} - \operatorname{div} \mathbf{q} + \rho\, r\,, \\
\bullet\ \text{entropy:} & \rho \dot{\eta} \geq & \operatorname{div} \boldsymbol{\phi}_\eta + \sigma_\eta\,.
\end{array} \tag{82}$$

Micropolar materials (extended formulation):

$$\begin{array}{lrl}
\bullet\ \text{mass:} & \dot{\rho} + \rho \operatorname{div} \dot{\mathbf{x}} = & 0\,, \\
\bullet\ \text{momentum:} & \rho\, \ddot{\mathbf{x}} = & \operatorname{div} \mathbf{T} + \rho\, \mathbf{b}\,, \\
\bullet\ \text{m. o. m.:} & \rho\, (\bar{\boldsymbol{\Theta}}\, \bar{\boldsymbol{\omega}})^{\boldsymbol{\cdot}} = & \mathbf{I} \times \mathbf{T} + \operatorname{div} \mathbf{M} + \rho\, \mathbf{c} \quad \longrightarrow \quad \mathbf{T} \neq \mathbf{T}^T, \\
\bullet\ \text{energy:} & \rho\, \dot{\varepsilon} = & \mathbf{T} \cdot \dot{\bar{\boldsymbol{\Delta}}} + \mathbf{M} \cdot \operatorname{grad} \bar{\boldsymbol{\omega}} - \operatorname{div} \mathbf{q} + \rho\, r\,, \\
\bullet\ \text{entropy:} & \rho \dot{\eta} \geq & \operatorname{div} \boldsymbol{\phi}_\eta + \sigma_\eta\,.
\end{array} \tag{83}$$

As a result of $(82)_3$ and $(83)_3$, the stress tensor is symmetric in the standard formulation, whereas it is non-symmetric in the micropolar case. To obtain this result, the relations

$$\begin{aligned}
\operatorname{div} (\mathbf{x} \times \mathbf{T}) &= \mathbf{x} \times \operatorname{div} \mathbf{T} + \mathbf{I} \times \mathbf{T}\,, \\
\mathbf{I} \times \mathbf{T} &= \overset{3}{\mathbf{E}}\, \mathbf{T}^T = 2\, \overset{A}{\mathbf{t}}
\end{aligned} \tag{84}$$

have been used. Note that $\overset{A}{\mathbf{t}}$ is the axial vector associated to the skew-symmetric part of $\mathbf{T}$.

In the framework of single-phase materials, it is usual (and correct) to proceed from the following *a priori* constitutive assumptions for the entropy efflux and the entropy supply,

$$\boldsymbol{\phi}_\eta = -\frac{1}{\theta}\mathbf{q}, \quad \sigma_\eta = \frac{1}{\theta}\rho r, \tag{85}$$

where θ is the *Kelvin*'s temperature. Given (85), the entropy relations $(82)_5$ or $(83)_5$, respectively, read

$$\rho\dot{\eta} \geq \operatorname{div}(-\frac{1}{\theta}\mathbf{q}) + \frac{1}{\theta}\rho r. \tag{86}$$

However, transferring this result to multiphasic materials leads to a wrong result and can hence not be used. This problem, by the way, gave rise to considerable irritations in the literature on mixture theories in the sixties of the last century, cf. Ehlers [28].

4.4 General balance relations of the constituents

Following *Truesdell*'s principles, the general balance relations of a constituent φ^α of the overall medium φ yield in analogy to (80):

$$\boxed{\begin{aligned} \frac{\mathrm{d}_\alpha}{\mathrm{d}t}\int_{\mathcal{B}} \Psi^\alpha \,\mathrm{d}v &= \int_{\mathcal{S}} (\boldsymbol{\phi}^\alpha \cdot \mathbf{n})\,\mathrm{d}a + \int_{\mathcal{B}} \sigma^\alpha \,\mathrm{d}v + \int_{\mathcal{B}} \hat{\Psi}^\alpha \,\mathrm{d}v, \\ \frac{\mathrm{d}_\alpha}{\mathrm{d}t}\int_{\mathcal{B}} \boldsymbol{\Psi}^\alpha \,\mathrm{d}v &= \int_{\mathcal{S}} (\boldsymbol{\Phi}^\alpha\, \mathbf{n})\,\mathrm{d}a + \int_{\mathcal{B}} \boldsymbol{\sigma}^\alpha \,\mathrm{d}v + \int_{\mathcal{B}} \hat{\boldsymbol{\Psi}}^\alpha \,\mathrm{d}v. \end{aligned}} \tag{87}$$

Therein, the mechanical quantities $(\,\cdot\,)^\alpha$ have the same physical meaning as the quantities $(\,\cdot\,)$ included into (80). Differentiation of the left-hand side of (87) and transformation of the surface integrals into volume integrals yields the local forms

$$\boxed{\begin{aligned} (\Psi^\alpha)'_\alpha + \Psi^\alpha \operatorname{div} \overset{\prime}{\mathbf{x}}_\alpha &= \operatorname{div} \boldsymbol{\phi}^\alpha + \sigma^\alpha + \hat{\Psi}^\alpha, \\ (\boldsymbol{\Psi}^\alpha)'_\alpha + \boldsymbol{\Psi}^\alpha \operatorname{div} \overset{\prime}{\mathbf{x}}_\alpha &= \operatorname{div} \boldsymbol{\Phi}^\alpha + \boldsymbol{\sigma}^\alpha + \hat{\boldsymbol{\Psi}}^\alpha. \end{aligned}} \tag{88}$$

General constraints. From *Truesdell*'s metaphysical principles, the local balances of the overall medium φ are given, on the one hand, by the balance relations (81) of single-phase media. On the other hand, these balance equations can be obtained by the sum of the balance relations (88) over all k constituents φ^α. This statement leads to constraints expressed by sum relations which, e. g. for scalar-valued mechanical quantities, read:

$$\begin{aligned}
&\bullet \text{ mechanical quantity}: \Psi &&= \textstyle\sum_{\alpha=1}^{k} \Psi^\alpha\,,\\
&\bullet \text{ efflux} &&: \boldsymbol{\phi} \cdot \mathbf{n} = \textstyle\sum_{\alpha=1}^{k} [\, \boldsymbol{\phi}^\alpha - \Psi^\alpha (\overset{\prime}{\mathbf{x}}_\alpha - \dot{\mathbf{x}}\,)] \cdot \mathbf{n}\,,\\
&\bullet \text{ supply} &&: \sigma = \textstyle\sum_{\alpha=1}^{k} \sigma^\alpha\,,\\
&\bullet \text{ production} &&: \hat{\Psi} = \textstyle\sum_{\alpha=1}^{k} \hat{\Psi}^\alpha\,.
\end{aligned} \tag{89}$$

Concerning vector-valued mechanical quantities, one analogously obtains

$$\begin{aligned}
&\bullet \text{ mechanical quantity}: \mathbf{\Psi} &&= \textstyle\sum_{\alpha=1}^{k} \mathbf{\Psi}^\alpha\,,\\
&\bullet \text{ efflux} &&: \mathbf{\Phi}\, \mathbf{n} = \textstyle\sum_{\alpha=1}^{k} [\, \mathbf{\Phi}^\alpha - \mathbf{\Psi}^\alpha \otimes (\overset{\prime}{\mathbf{x}}_\alpha - \dot{\mathbf{x}}\,)]\, \mathbf{n}\,,\\
&\bullet \text{ supply} &&: \boldsymbol{\sigma} = \textstyle\sum_{\alpha=1}^{k} \boldsymbol{\sigma}^\alpha\,,\\
&\bullet \text{ production} &&: \hat{\mathbf{\Psi}} = \textstyle\sum_{\alpha=1}^{k} \hat{\mathbf{\Psi}}^\alpha\,.
\end{aligned} \tag{90}$$

4.5 Specific balance equations of the constituents

The individual balance equations of mass, momentum, moment of momentum, energy and entropy are obtained in direct analogy to those of single-phase materials provided one allows for the interaction mechanisms between the constituents by introducing additional production terms. The quantities $(\,\cdot\,)^\alpha$ included into the balance relations of φ^α, cf. Table 2, have the same physical meaning as the corresponding quantities $(\,\cdot\,)$ of φ incorporated into Table 1. Furthermore, the mass production $\hat{\rho}^\alpha$ allows for mass exchange or phase transition processes between the constituents, $\hat{\mathbf{s}}^\alpha$ is the total momentum production of φ^α, and $\hat{\mathbf{h}}^\alpha$ and $\hat{\mathbf{j}}^\alpha$ represent the total productions of moment of momentum, whereas $\hat{e}^\alpha$ and $\hat{f}^\alpha$ are the total energy production terms both in the standard and the micropolar formulations. Finally, $\hat{\eta}^\alpha$ is the total entropy production of φ^α. In the same way, as it was assumed in the theory of single-phase materials, cf. (85), it is possible (and correct) to specify the efflux of entropy and the external entropy supply of any constituent φ^α as

$$\boldsymbol{\phi}_\eta^\alpha = -\frac{1}{\theta^\alpha}\,\mathbf{q}^\alpha, \qquad \sigma_\eta^\alpha = \frac{1}{\theta^\alpha}\,\rho^\alpha\, r^\alpha. \tag{91}$$

Admitting different *Kelvin*'s temperatures θ^α in the above constitutive assumptions, one allows for the possibility that each constituent has an individual temperature function.

	$\Psi^\alpha,\ \boldsymbol{\Psi}^\alpha$	$\phi^\alpha,\ \boldsymbol{\phi}^\alpha$	$\sigma^\alpha,\ \boldsymbol{\sigma}^\alpha$	$\hat{\Psi}^\alpha,\ \hat{\boldsymbol{\Psi}}^\alpha$
mass	ρ^α	$\mathbf{0}$	0	$\hat{\rho}^\alpha$
momentum	$\rho^\alpha\,\overset{\prime}{\mathbf{x}}_\alpha$	$\mathbf{T}^\alpha$	$\rho^\alpha\mathbf{b}^\alpha$	$\hat{\mathbf{s}}^\alpha$
m.o.m. (sf)	$\mathbf{x}\times(\rho^\alpha\overset{\prime}{\mathbf{x}}_\alpha)$	$\mathbf{x}\times\mathbf{T}^\alpha$	$\mathbf{x}\times(\rho^\alpha\mathbf{b}^\alpha)$	$\hat{\mathbf{h}}^\alpha$
m.o.m. (mf)	$\mathbf{x}\times(\rho^\alpha\overset{\prime}{\mathbf{x}}_\alpha)+\rho^\alpha\bar{\boldsymbol{\Theta}}^\alpha\bar{\boldsymbol{\omega}}_\alpha$	$\mathbf{x}\times\mathbf{T}^\alpha+\mathbf{M}^\alpha$	$\mathbf{x}\times(\rho^\alpha\mathbf{b}^\alpha)+\rho^\alpha\mathbf{c}^\alpha$	$\hat{\mathbf{j}}^\alpha$
energy (sf)	$\rho^\alpha\varepsilon^\alpha+\frac{1}{2}\overset{\prime}{\mathbf{x}}_\alpha\cdot(\rho^\alpha\overset{\prime}{\mathbf{x}}_\alpha)$	$(\mathbf{T}^\alpha)^T\overset{\prime}{\mathbf{x}}_\alpha-\mathbf{q}^\alpha$	$\overset{\prime}{\mathbf{x}}_\alpha\cdot(\rho^\alpha\,\mathbf{b}^\alpha)+\rho^\alpha r^\alpha$	$\hat{e}^\alpha$
energy (mf)	$\rho^\alpha\varepsilon^\alpha+\frac{1}{2}\overset{\prime}{\mathbf{x}}_\alpha\cdot(\rho^\alpha\overset{\prime}{\mathbf{x}}_\alpha)$ $+\frac{1}{2}\bar{\boldsymbol{\omega}}_\alpha\cdot(\rho^\alpha\bar{\boldsymbol{\Theta}}^\alpha\bar{\boldsymbol{\omega}}_\alpha)$	$(\mathbf{T}^\alpha)^T\overset{\prime}{\mathbf{x}}_\alpha-\mathbf{q}^\alpha$ $+(\mathbf{M}^\alpha)^T\bar{\boldsymbol{\omega}}_\alpha$	$\overset{\prime}{\mathbf{x}}_\alpha\cdot(\rho^\alpha\mathbf{b}^\alpha)+\rho^\alpha r^\alpha$ $+\bar{\boldsymbol{\omega}}_\alpha\cdot\rho^\alpha\,\mathbf{c}^\alpha$	$\hat{f}^\alpha$
entropy	$\rho^\alpha\,\eta^\alpha$	$\boldsymbol{\phi}^\alpha_\eta$	σ^α_η	$\hat{\eta}^\alpha$

Table 2. Balance relations for a constituent φ^α of the overall medium φ.

In the framework of mixture theories, the total productions can be split into a direct term and additional terms governed by the "lower" productions. Thus,

$$
\begin{aligned}
\hat{\mathbf{s}}^\alpha &= \hat{\mathbf{p}}^\alpha + \hat{\rho}^\alpha\,\overset{\prime}{\mathbf{x}}_\alpha\,,\\
\hat{\mathbf{h}}^\alpha &= \hat{\mathbf{m}}^\alpha + \mathbf{x}\times(\hat{\mathbf{p}}^\alpha + \hat{\rho}^\alpha\,\overset{\prime}{\mathbf{x}}_\alpha)\,,\\
\hat{\mathbf{j}}^\alpha &= \hat{\mathbf{m}}^\alpha + \mathbf{x}\times(\hat{\mathbf{p}}^\alpha + \hat{\rho}^\alpha\,\overset{\prime}{\mathbf{x}}_\alpha) + \hat{\rho}^\alpha\,\bar{\boldsymbol{\Theta}}^\alpha\,\bar{\boldsymbol{\omega}}_\alpha\,,\\
\hat{e}^\alpha &= \hat{\varepsilon}^\alpha + \hat{\mathbf{p}}^\alpha\cdot\overset{\prime}{\mathbf{x}}_\alpha + \hat{\rho}^\alpha\,(\varepsilon^\alpha + \tfrac{1}{2}\,\overset{\prime}{\mathbf{x}}_\alpha\cdot\overset{\prime}{\mathbf{x}}_\alpha)\,,\\
\hat{f}^\alpha &= \hat{\varepsilon}^\alpha + \hat{\mathbf{p}}^\alpha\cdot\overset{\prime}{\mathbf{x}}_\alpha + \hat{\mathbf{m}}^\alpha\cdot\bar{\boldsymbol{\omega}}_\alpha +\\
&\quad + \hat{\rho}^\alpha\,[\varepsilon^\alpha + \tfrac{1}{2}\,\overset{\prime}{\mathbf{x}}_\alpha\cdot\overset{\prime}{\mathbf{x}}_\alpha + \tfrac{1}{2}(\bar{\boldsymbol{\Theta}}^\alpha\,\bar{\boldsymbol{\omega}}_\alpha)\cdot\bar{\boldsymbol{\omega}}_\alpha]\,,\\
\hat{\eta}^\alpha &= \hat{\zeta}^\alpha + \hat{\rho}^\alpha\eta^\alpha.
\end{aligned}
\qquad (92)
$$

In $(92)_1$, the direct momentum production $\hat{\mathbf{p}}^\alpha$ can be interpreted as the local interaction force per unit volume between φ^α and the other constituents of the overall medium, whereas the second term represents the additional momentum production as a result of the density production term. Analogously, $\hat{\mathbf{m}}^\alpha$ is the direct term of the total moment of momentum productions $\hat{\mathbf{h}}^\alpha$ or $\hat{\mathbf{j}}^\alpha$, respectively, whereas the further terms represent the additional productions of angular momentum resulting from the density production and the direct momentum production. Furthermore, $\hat{\varepsilon}^\alpha$ is the direct energy production term included into the total energy productions $\hat{e}^\alpha$ or $\hat{f}^\alpha$, respectively, whereas the remainder of terms represents the additional energy production stemming from the density production and the momentum as well as the angular momentum productions.

Using the same procedure as to obtain (82) and (83) from Table 1, one obtains the following equations for non-polar and micropolar constituents from Table 2 using the above additive split of the production terms.

Non-polar materials (standard formulation):

$$
\begin{aligned}
&\bullet \text{ mass:} && (\rho^\alpha)'_\alpha + \rho^\alpha \operatorname{div} \overset{\prime}{\mathbf{x}}_\alpha = \hat{\rho}^\alpha, \\
&\bullet \text{ momentum:} && \rho^\alpha \overset{\prime\prime}{\mathbf{x}}_\alpha = \operatorname{div} \mathbf{T}^\alpha + \rho^\alpha \mathbf{b}^\alpha + \hat{\mathbf{p}}^\alpha, \\
&\bullet \text{ m. o. m.:} && \mathbf{0} = \mathbf{I} \times \mathbf{T}^\alpha + \hat{\mathbf{m}}^\alpha, \\
&\bullet \text{ energy:} && \rho^\alpha (\varepsilon^\alpha)'_\alpha = \mathbf{T}^\alpha \cdot \mathbf{L}_\alpha - \operatorname{div} \mathbf{q}^\alpha + \rho^\alpha r^\alpha + \hat{\varepsilon}^\alpha, \\
&\bullet \text{ entropy:} && \rho^\alpha (\eta^\alpha)'_\alpha = \operatorname{div} (-\frac{1}{\theta^\alpha} \mathbf{q}^\alpha) + \frac{1}{\theta^\alpha} \rho^\alpha r^\alpha + \hat{\zeta}^\alpha .
\end{aligned}
\tag{93}
$$

Micropolar materials (extended formulation):

$$
\begin{aligned}
&\bullet \text{ mass:} && (\rho^\alpha)'_\alpha + \rho^\alpha \operatorname{div} \overset{\prime}{\mathbf{x}}_\alpha = \hat{\rho}^\alpha, \\
&\bullet \text{ momentum:} && \rho^\alpha \overset{\prime\prime}{\mathbf{x}}_\alpha = \operatorname{div} \mathbf{T}^\alpha + \rho^\alpha \mathbf{b}^\alpha + \hat{\mathbf{p}}^\alpha, \\
&\bullet \text{ m. o. m.:} && \rho^\alpha (\bar{\mathbf{\Theta}}^\alpha \bar{\boldsymbol{\omega}}_\alpha)'_\alpha = \mathbf{I} \times \mathbf{T}^\alpha + \operatorname{div} \mathbf{M}^\alpha + \rho^\alpha \mathbf{c}^\alpha + \hat{\mathbf{m}}^\alpha, \\
&\bullet \text{ energy:} && \rho^\alpha (\varepsilon^\alpha)'_\alpha = \mathbf{T}^\alpha \cdot \bar{\mathbf{\Delta}}_\alpha + \mathbf{M}^\alpha \cdot \operatorname{grad} \bar{\boldsymbol{\omega}}_\alpha - \\
& && \qquad - \operatorname{div} \mathbf{q}^\alpha + \rho^\alpha r^\alpha + \hat{\varepsilon}^\alpha, \\
&\bullet \text{ entropy:} && \rho^\alpha (\eta^\alpha)'_\alpha = \operatorname{div} (-\frac{1}{\theta^\alpha} \mathbf{q}^\alpha) + \frac{1}{\theta^\alpha} \rho^\alpha r^\alpha + \hat{\zeta}^\alpha .
\end{aligned}
\tag{94}
$$

Specific constraints. By summing up the relations (93) and (94) over all k constituents φ^α, one obtains in comparison with the relations (82) and (83) of the overall medium φ the following constraints of the production terms:

$$
(sf) \left\{
\begin{aligned}
&\textstyle\sum_{\alpha=1}^k \hat{\rho}^\alpha = 0, \\
&\textstyle\sum_{\alpha=1}^k \hat{\mathbf{s}}^\alpha = \mathbf{0}, \\
&\textstyle\sum_{\alpha=1}^k \hat{\mathbf{h}}^\alpha = \mathbf{0}, \\
&\textstyle\sum_{\alpha=1}^k \hat{e}^\alpha = 0, \\
&\textstyle\sum_{\alpha=1}^k \hat{\eta}^\alpha \geq 0,
\end{aligned}
\right.
\qquad
(mf) \left\{
\begin{aligned}
&\textstyle\sum_{\alpha=1}^k \hat{\rho}^\alpha = 0, \\
&\textstyle\sum_{\alpha=1}^k \hat{\mathbf{s}}^\alpha = \mathbf{0}, \\
&\textstyle\sum_{\alpha=1}^k \hat{\mathbf{j}}^\alpha = \mathbf{0}, \\
&\textstyle\sum_{\alpha=1}^k \hat{f}^\alpha = 0, \\
&\textstyle\sum_{\alpha=1}^k \hat{\eta}^\alpha \geq 0 .
\end{aligned}
\right.
\tag{95}
$$

Therein, the left column belongs to the standard formulation *(sf)* and the right column to the micropolar one *(mf)*.

Proceeding from the general constraints given by (92), the explicit relations between the total quantities of Table 1 and the partial quantities of Table 2 read

$$
\begin{aligned}
\rho\,\mathbf{b} &= \textstyle\sum_{\alpha=1}^{k} \rho^{\alpha}\mathbf{b}^{\alpha}\,,\\
\rho\,\mathbf{c} &= \textstyle\sum_{\alpha=1}^{k} \rho^{\alpha}\mathbf{c}^{\alpha}\,,\\
\mathbf{T} &= \textstyle\sum_{\alpha=1}^{k} (\mathbf{T}^{\alpha} - \rho^{\alpha}\mathbf{d}_{\alpha}\otimes\mathbf{d}_{\alpha})\,,\\
\mathbf{M} &= \textstyle\sum_{\alpha=1}^{k} (\mathbf{M}^{\alpha} - \rho^{\alpha}\,\bar{\boldsymbol{\Theta}}^{\alpha}\,\bar{\boldsymbol{\omega}}_{\alpha}\otimes\mathbf{d}_{\alpha})\,,\\
\rho\,\varepsilon &= \textstyle\sum_{\alpha=1}^{k} \rho^{\alpha}(\varepsilon^{\alpha} + \frac{1}{2}\,\mathbf{d}_{\alpha}\cdot\mathbf{d}_{\alpha} + \frac{1}{2}\,\bar{\boldsymbol{\Theta}}^{\alpha}\boldsymbol{\delta}_{\alpha}\cdot\boldsymbol{\delta}_{\alpha})\,,\\
\mathbf{q} &= \textstyle\sum_{\alpha=1}^{k} \{\mathbf{q}^{\alpha} - (\mathbf{T}^{\alpha})^{T}\mathbf{d}_{\alpha} - (\mathbf{M}^{\alpha})^{T}\boldsymbol{\delta}_{\alpha} + \rho^{\alpha}\varepsilon^{\alpha}\,\mathbf{d}_{\alpha} +\\
&\qquad\quad + \tfrac{1}{2}\,\rho^{\alpha}(\mathbf{d}_{\alpha}\cdot\mathbf{d}_{\alpha})\,\mathbf{d}_{\alpha} + \tfrac{1}{2}\,[\rho^{\alpha}\bar{\boldsymbol{\Theta}}^{\alpha}\cdot(\boldsymbol{\delta}_{\alpha}\otimes\boldsymbol{\delta}_{\alpha} - \bar{\boldsymbol{\omega}}_{\alpha}\otimes\bar{\boldsymbol{\omega}}_{\alpha})]\,\mathbf{d}_{\alpha}\}\,,\\
\rho\,r &= \textstyle\sum_{\alpha=1}^{k} \rho^{\alpha}(r^{\alpha} + \mathbf{b}^{\alpha}\cdot\mathbf{d}_{\alpha} + \mathbf{c}^{\alpha}\cdot\boldsymbol{\delta}_{\alpha})\,,\\
\rho\,\eta &= \textstyle\sum_{\alpha=1}^{k} \rho^{\alpha}\,\eta^{\alpha},
\end{aligned}
\tag{96}
$$

where to obtain $(96)_{3-7}$, the translational and the rotational diffusion velocities

$$
\mathbf{d}_{\alpha} = \overset{\prime}{\mathbf{x}}_{\alpha} - \dot{\mathbf{x}} \quad \text{and} \quad \boldsymbol{\delta}_{\alpha} = \bar{\boldsymbol{\omega}}_{\alpha} - \bar{\boldsymbol{\omega}}
\tag{97}
$$

have been used. Note that the rotational velocity $\bar{\boldsymbol{\omega}}$ of an overall micropolar medium, in analogy to the mixture velocity $\dot{\mathbf{x}}$ given through $(14)_1$, is defined via

$$
\bar{\boldsymbol{\omega}} = (\rho\,\bar{\boldsymbol{\Theta}})^{-1}\sum_{\alpha=1}^{k} \rho^{\alpha}\bar{\boldsymbol{\Theta}}^{\alpha}\bar{\boldsymbol{\omega}}_{\alpha}\,.
\tag{98}
$$

Given $(14)_1$ and (98), it is easily proved that the translational and the rotational diffusion velocities satisfy the relations

$$
\sum_{\alpha=1}^{k} \rho^{\alpha}\mathbf{d}_{\alpha} = \mathbf{0} \quad \text{and} \quad \sum_{\alpha=1}^{k} \rho^{\alpha}\bar{\boldsymbol{\Theta}}^{\alpha}\boldsymbol{\delta}_{\alpha} = \mathbf{0}\,.
\tag{99}
$$

Considering the sum relations (96), it is concluded that in case of the standard non-polar description of multiphasic materials, where $\mathbf{c}^{\alpha} \equiv \mathbf{0}$ and $\mathbf{M}^{\alpha} \equiv \mathbf{0}$ and, furthermore, $\bar{\boldsymbol{\Theta}}^{\alpha} \equiv \mathbf{0}$ and $\boldsymbol{\delta}_{\alpha} \equiv \mathbf{0}$, the relations (96) are identical to those of the classical theory of mixtures and the standard Theory of Porous Media, cf. e. g. [32]. In addition, given either non-polar or micropolar constituents, an evaluation of the above sum relations should be based on the following interpretations:

- As far as there is no diffusion process ($\mathbf{d}_\alpha = \mathbf{0}$ and $\boldsymbol{\delta}_\alpha = \mathbf{0}$), all terms of the overall medium are given by summing up the respective terms of the constituents.
- In case that there is a diffusion process ($\mathbf{d}_\alpha \neq \mathbf{0}$ and $\boldsymbol{\delta}_\alpha \neq \mathbf{0}$), summing up $\mathbf{T}^\alpha$, $\mathbf{M}^\alpha$, $\rho^\alpha\varepsilon^\alpha$, $\mathbf{q}^\alpha$ and $\rho^\alpha r^\alpha$ yields the so-called inner parts (kernels) of $\mathbf{T}, \mathbf{M}, \rho\,\varepsilon, \mathbf{q}$ and $\rho\, r$. The remainder of terms included in the sum relations are governed by the diffusion process through $\mathbf{d}_\alpha$ and $\boldsymbol{\delta}_\alpha$. These terms can be interpreted as follows: That part of the stress tensor $\mathbf{T}$ that is initiated, for a given constituent φ^α, by the diffusion process is comparable to the *Reynolds* stress occurring in turbulent flow situations of single-constituent fluids. Concerning the couple stress $\mathbf{M}$, the additional terms governed by $\mathbf{d}_\alpha$ exhibit the diffusive moment of momentum effects of the constituents. Furthermore, the volume specific internal energy $\rho\,\varepsilon$ contains the diffusive kinetic energy of the constituents. The non-mechanical influx vector $\mathbf{q}$ is influenced by the sum of the influx terms generated by the diffusive work of the partial contact forces and contact couples as well as by the influx vectors generated by the diffusive internal and kinetic energies. Finally, the non-mechanical supply term $\rho\, r$ contains additional terms which stem from the diffusive work of the external volume forces and volume couples.

Discussion of the entropy principle. Given the relations (93)–(96), the entropy principle for multiphasic and saturated porous materials yields

$$\sum_{\alpha=1}^{k} [\,\rho^\alpha(\eta^\alpha)'_\alpha + \hat{\rho}^\alpha\eta^\alpha + \operatorname{div}(\frac{1}{\theta^\alpha}\mathbf{q}^\alpha) - \frac{1}{\theta^\alpha}\rho^\alpha r^\alpha\,] \geq 0\,. \tag{100}$$

This inequality is independent from the fact whether the constituents are non-polar or micropolar. Applying the mixture derivative defined by (16) to the entropy functions η^α, one obtains with the aid of $(96)_8$ instead of (100)

$$\rho\dot{\eta} \geq \sum_{\alpha=1}^{k} \operatorname{div}(\frac{1}{\theta^\alpha}\mathbf{q}^\alpha - \rho^\alpha\eta^\alpha\mathbf{d}_\alpha) + \sum_{\alpha=1}^{k} \frac{1}{\theta^\alpha}\rho^\alpha r^\alpha\,. \tag{101}$$

This form of the entropy principle can easily be compared with the entropy inequality (86) for single-phase materials. Following this, it is concluded that

$$\boldsymbol{\phi}_\eta = -\sum_{\alpha=1}^{k} (\frac{1}{\theta^\alpha}\mathbf{q}^\alpha + \rho^\alpha\eta^\alpha\mathbf{d}_\alpha) \quad \text{and} \quad \sigma_\eta = \sum_{\alpha=1}^{k} \frac{1}{\theta^\alpha}\rho^\alpha r^\alpha\,. \tag{102}$$

Given (102), it is immediately seen that the sum relations defined by (89) are also valid for the entropy flux and the entropy supply term. In comparison of the relations (91) and (102), it is furthermore seen that the entropy principles for single-phase and multiphasic materials only leads to the same inequalities if the diffusion process vanishes ($\mathbf{d}_\alpha = \mathbf{0}$ and $\boldsymbol{\delta}_\alpha = \mathbf{0}$) and if all constituents φ^α additionally have the same *Kelvin*'s temperature ($\theta^\alpha \equiv \theta$).

Introducing mass specific constituent free energy functions ψ^α via

$$\psi^\alpha := \varepsilon^\alpha - \theta^\alpha \eta^\alpha \,, \tag{103}$$

one obtains with the aid of the energy balance relations $(93)_4$ for non-polar materials and $(94)_4$ for micropolar materials the mostly used forms of the entropy principle for multiphasic materials, viz.:

Non-polar materials (standard formulation):

$$\begin{aligned} \sum_{\alpha=1}^{k} \frac{1}{\theta^\alpha} [\, & \mathbf{T}^\alpha \cdot \mathbf{L}_\alpha - \rho^\alpha [\, (\psi^\alpha)'_\alpha + (\theta^\alpha)'_\alpha \eta^\alpha \,] - \hat{\mathbf{p}}^\alpha \cdot \overset{\prime}{\mathbf{x}}_\alpha - \\ & - \hat{\rho}^\alpha \, (\psi^\alpha + \tfrac{1}{2} \, \overset{\prime}{\mathbf{x}}_\alpha \cdot \overset{\prime}{\mathbf{x}}_\alpha \,) - \frac{1}{\theta^\alpha} \, \mathbf{q}^\alpha \cdot \operatorname{grad} \theta^\alpha + \hat{e}^\alpha \,] \geq 0 \,. \end{aligned} \tag{104}$$

Micropolar materials (extended formulation):

$$\begin{aligned} \sum_{\alpha=1}^{k} \frac{1}{\theta^\alpha} \{\, & \mathbf{T}^\alpha \cdot \bar{\boldsymbol{\Delta}}_\alpha + \mathbf{M}^\alpha \cdot \operatorname{grad} \bar{\boldsymbol{\omega}}_\alpha - \rho^\alpha [\, (\psi^\alpha)'_\alpha + (\theta^\alpha)'_\alpha \eta^\alpha \,] - \\ & - \hat{\mathbf{p}}^\alpha \cdot \overset{\prime}{\mathbf{x}}_\alpha - \hat{\mathbf{m}}^\alpha \cdot \bar{\boldsymbol{\omega}}_\alpha - \hat{\rho}^\alpha \, [\, \psi^\alpha + \tfrac{1}{2} \, \overset{\prime}{\mathbf{x}}_\alpha \cdot \overset{\prime}{\mathbf{x}}_\alpha + \tfrac{1}{2} (\bar{\boldsymbol{\Theta}}^\alpha \bar{\boldsymbol{\omega}}_\alpha) \cdot \bar{\boldsymbol{\omega}}_\alpha \,] \\ & - \frac{1}{\theta^\alpha} \, \mathbf{q}^\alpha \cdot \operatorname{grad} \theta^\alpha + \hat{\zeta}^\alpha \,\} \geq 0 \,. \end{aligned} \tag{105}$$

In case of single temperature problems, the above inequalities can be multiplied by $\theta^\alpha \equiv \theta$. Thus, these relations reduce to

$$\begin{aligned} \sum_{\alpha=1}^{k} [\, & \mathbf{T}^\alpha \cdot \mathbf{L}_\alpha - \rho^\alpha \, (\psi^\alpha)'_\alpha - \hat{\mathbf{p}}^\alpha \cdot \overset{\prime}{\mathbf{x}}_\alpha - \hat{\rho}^\alpha \, (\, \psi^\alpha + \tfrac{1}{2} \, \overset{\prime}{\mathbf{x}}_\alpha \cdot \overset{\prime}{\mathbf{x}}_\alpha \,) \,] - \\ & - \rho \eta \, \dot{\theta} - \frac{1}{\theta} \mathbf{h} \cdot \operatorname{grad} \theta \geq 0 \end{aligned} \tag{106}$$

for non-polar materials and to

$$\begin{aligned} \sum_{\alpha=1}^{k} \{\, & \mathbf{T}^\alpha \cdot \bar{\boldsymbol{\Delta}}_\alpha + \mathbf{M}^\alpha \cdot \operatorname{grad} \bar{\boldsymbol{\omega}}_\alpha - \rho^\alpha \, (\psi^\alpha)'_\alpha - \hat{\mathbf{p}}^\alpha \cdot \overset{\prime}{\mathbf{x}}_\alpha - \\ & - \hat{\mathbf{m}}^\alpha \cdot \bar{\boldsymbol{\omega}}_\alpha - \hat{\rho}^\alpha \, [\, \psi^\alpha + \tfrac{1}{2} \, \overset{\prime}{\mathbf{x}}_\alpha \cdot \overset{\prime}{\mathbf{x}}_\alpha + \tfrac{1}{2} (\bar{\boldsymbol{\Theta}}^\alpha \bar{\boldsymbol{\omega}}_\alpha) \cdot \bar{\boldsymbol{\omega}}_\alpha \,] \,\} - \\ & - \rho \eta \, \dot{\theta} - \frac{1}{\theta} \mathbf{h} \cdot \operatorname{grad} \theta \geq 0 \end{aligned} \tag{107}$$

for micropolar materials, where $(95)_{4,9}$ and $(96)_8$ together with the definitions

$$\mathbf{h} := \sum_{\alpha=1}^{k} \mathbf{h}^\alpha \quad \text{and} \quad \mathbf{h}^\alpha := \mathbf{q}^\alpha + \theta\,\rho^\alpha \eta^\alpha \mathbf{d}_\alpha \tag{108}$$

have been used. Furthermore, note in passing that the different forms of the entropy principle described above for multiphasic materials are usually known as the *Clausius-Duhem* representation of the entropy inequality. In case of isothermal problems, these inequalities naturally reduce to the so-called *Clausius-Planck* representation.

Finally, the investigation of constitutive equations for any kind of multiphasic problem must be based on a careful evaluation of the respective entropy inequality. In this respect, it may appear that not only non-polar or micropolar aggregates have to be considered, but also combinations of non-polar and micropolar materials. To give an example, imagine a ternary problem governed by a micropolar solid skeleton φ^S and two non-polar pore-fluids, a pore-liquid φ^L and a pore-gas φ^G. Assuming furthermore that the constituents exhibit the same temperature function ($\theta^\alpha \equiv \theta$) and that there is no mass production between the constituents ($\hat{\rho}^S = \hat{\rho}^F = 0$), the entropy principle reads

$$\begin{aligned} &\mathbf{T}^S \cdot \bar{\boldsymbol{\Delta}}_S + \mathbf{M}^S \cdot \operatorname{grad} \bar{\boldsymbol{\omega}}_S + \mathbf{T}^L \cdot \mathbf{L}_L + \mathbf{T}^G \cdot \mathbf{L}_G - \\ &-\rho^S (\psi^S)'_S - \rho^L (\psi^L)'_L - \rho^G (\psi^G)'_G - \rho\eta\,\dot{\theta} - \frac{1}{\theta}\,\mathbf{h} \cdot \operatorname{grad}\theta - \\ &-\hat{\mathbf{p}}^S \cdot \overset{\prime}{\mathbf{x}}_S - \hat{\mathbf{p}}^L \cdot \overset{\prime}{\mathbf{x}}_L - \hat{\mathbf{p}}^G \cdot \overset{\prime}{\mathbf{x}}_G - \hat{\mathbf{m}}^S \cdot \bar{\boldsymbol{\omega}}_S \geq 0\,. \end{aligned} \tag{109}$$

Given the above entropy inequality, it is not difficult to find further reductions to, e. g., isothermal problems or non-polar binary aggregates.

5 Constitutive theory

5.1 Fundamentals

The preceding sections show that there is a variety of possibilities for the definition of different solid-fluid aggregates. In particular, proceeding from a single solid material, the porous skeleton, the pore content can be either a single- or a multiphasic fluid. Furthermore, the solid skeleton may undergo small or finite deformations caused by elastic, viscoelastic, elasto-plastic or elasto-viscoplastic material properties, whereas the pore content is generally governed by viscous material behaviour. However, as is usual in porous media theories, the frictional stresses are neglected in comparison with the interacting momentum production terms. Furthermore, it is generally assumed that the solid skeleton is materially incompressible, thus neglecting the intrinsic compressibility of the solid material itself in comparison to the bulk compressibility of the porous skeleton. This assumption generally holds, except of the range of very small porosity values. Concerning the pore-fluids, the assumption is made that, if the pore-fluid is a real pore-liquid, it behaves materially incompressible, whereas, if the pore-fluid is a real pore-gas or a pore-liquid with dispersed gas particles, it naturally behaves materially compressible.

Once a porous media model is defined, one has to find the set of governing constitutive equations satisfying the entropy principle. Since porous media models are, like mixtures, multiphasic materials, one has to proceed from the general concept of second-grade materials in order to catch the constitutive coupling effects, especially for the various production terms. This procedure, however some authors mean that it is quite arbitrary, is straight forward and proceeds from a basic set of constitutive variables which has to be reduced to describe the particular model under consideration. For a broad review of the basic techniques to construct thermodynamically consistent constitutive models, the interested reader is referred to the work by Bowen [14–16], Ehlers [27–31], Diebels [23], de Boer [7, 8] and Bluhm [5].

To give an example how multiphasic material models can be treated constitutively, the following section concerns a simple binary material consisting of a non-polar elastic skeleton and a single pore-liquid under isothermal conditions. This model can easily be extended towards more sophisticated models and material behaviours.

5.2 A simple biphasic model

Considering a quasi-static problem ($\overset{\prime\prime}{\mathbf{x}}_\alpha = \mathbf{0}$) of a biphasic model consisting of a non-polar elastic skeleton material φ^S and a single viscous pore-liquid φ^F under isothermal conditions ($\theta^\alpha \equiv \theta \equiv 0$) yields the following set of basic relations:

- overall volume balance:

$$\operatorname{div}\,[(\mathbf{u}_S)'_S + n^F \mathbf{w}_F] = 0\,,$$

- momentum balances:

$$\operatorname{div}\mathbf{T}^S + \rho^S\,\mathbf{b} - \hat{\mathbf{p}}^F = \mathbf{0}\,,$$

$$\operatorname{div}\mathbf{T}^F + \rho^F\,\mathbf{b} + \hat{\mathbf{p}}^F = \mathbf{0}\,,$$

- moment of momentum balances: (110)

$$\mathbf{I} \times \mathbf{T}^S - \hat{\mathbf{m}}^F = \mathbf{0}\,,$$

$$\mathbf{I} \times \mathbf{T}^F + \hat{\mathbf{m}}^F = \mathbf{0}\,,$$

- entropy inequality:

$$\mathbf{T}^S_E \cdot \mathbf{L}_S + \mathbf{T}^F_E \cdot \mathbf{L}_F - \rho^S\,(\psi^S)'_S - \rho^F\,(\psi^F)'_F - \hat{\mathbf{p}}^F_E \cdot \mathbf{w}_F \geq 0\,.$$

To obtain the above relations, it has additionally been assumed that there is no mass production between the constituents ($\hat{\rho}^\alpha = 0$). Furthermore, the volume balance proceeds from the fact that both constituents, solid skeleton and pore-liquid, are materially incompressible ($\rho^{\alpha R} = \text{const.}$), such that $(110)_1$ is obtained by combination of the saturation condition (4) and the mass balance relations $(93)_1$, where, furthermore, use is made of the solid displacement function $\mathbf{u}_S$ and the seepage velocity $\mathbf{w}_F$ from (17). Concerning the momentum and moment of momentum balances, the momentum and angular momentum production constraints from $(92)_{1,2}$ have been used. Finally, the entropy inequality is a modification of (109), where, additionally, the volume balance equation multiplied by a *Lagrange*an multiplier p has been added to the thermodynamical process in the sense of an incompressibility constraint. As a result, $\mathbf{T}^\alpha$ and $\hat{\mathbf{p}}^F$ are substituted by the so-called "extra" terms

$$\begin{aligned} \mathbf{T}^\alpha_E &= \mathbf{T}^\alpha + n^\alpha p\,\mathbf{I}\,, \\ \hat{\mathbf{p}}^F_E &= \hat{\mathbf{p}}^F - p\,\operatorname{grad} n^F. \end{aligned} \tag{111}$$

Prescribing quasi-static conditions, the addition of the solid and fluid stresses yields

$$\mathbf{T} = -p\,\mathbf{I} + \mathbf{T}_E\,, \qquad \text{where} \qquad \mathbf{T}_E = \mathbf{T}^S_E + \mathbf{T}^F_E \tag{112}$$

is known as the *effective stress*, cf. e. g. Bishop [4] or Skempton [93]. Note in passing that, once (112) is given, p is easily identified as the unspecified pore pressure. Furthermore, the stress power terms included in the entropy inequality $(110)_6$ can be substituted by

$$\mathbf{T}^\alpha \cdot \mathbf{L}_\alpha = \mathbf{T}^\alpha_{\text{sym}} \cdot \mathbf{D}_\alpha + \mathbf{T}^\alpha_{\text{skw}} \cdot \mathbf{W}_\alpha\,, \tag{113}$$

where $\mathbf{T}^\alpha_{\text{skw}}$ can be obtained from $\hat{\mathbf{m}}^F$ via

$$\begin{aligned} \mathbf{T}^S_{E\,\text{skw}} &= -2\,(\hat{\mathbf{m}}^F \times \mathbf{I})\,, \\ \mathbf{T}^F_{E\,\text{skw}} &= 2\,(\hat{\mathbf{m}}^F \times \mathbf{I})\,. \end{aligned} \tag{114}$$

Finally, since the binary model under study is governed by an isothermal process, it is obvious that the energy balance relations can be dropped.

To close the model under consideration, constitutive equations must be found for the extra stresses $\mathbf{T}^\alpha_E$, the production terms $\hat{\mathbf{p}}^F_E$ and $\hat{\mathbf{m}}^F$ as well as for the free energy functions ψ^α such that the entropy principle $(110)_6$ is satisfied. Based on the fundamental principles of constitutive modelling, *determinism*, *local action*, *equipresence*, *frame indifference* and *dissipation* stemming from the work by Truesdell, Noll, Coleman and Noll and others, cf. e. g. Ehlers [28], the following response functions $\tilde{\mathcal{R}}$ have to be determined:

$$\tilde{\mathcal{R}} := \{\psi^S,\ \psi^F,\ \mathbf{T}^S_E,\ \mathbf{T}^F_E,\ \hat{\mathbf{p}}^F_E,\ \hat{\mathbf{m}}^F\}. \tag{115}$$

However, proceeding from the fact that both constituents of the binary aggregate, solid skeleton and pore-fluid, are intrinsically non-polar materials with symmetric Cauchy stresses on their microscales, a standard homogenization procedure reveals that, as a result, also the macroscopic stresses $\mathbf{T}^\alpha$ must be symmetric. Thus,

$$\mathbf{T}^\alpha = (\mathbf{T}^\alpha)^T \quad \longrightarrow \quad \hat{\mathbf{m}}^F \equiv \mathbf{0}\,. \tag{116}$$

Given (116), it is assumed on the basis of the principle of equipresence that the remaining elements of $\tilde{\mathcal{R}}$ summarized in $\mathcal{R}$ can depend on the same set $\mathcal{S}$ of constitutive variables:

$$\mathcal{R} := \mathcal{R}(\mathcal{S}). \tag{117}$$

Therein, $\mathcal{S}$ is a subset of the fundamental set $\mathcal{V}$ of constitutive variables. Proceeding from general porous media consisting of a single elastic porous skeleton φ^S and an arbitrary number of viscous pore-fluids φ^β, the fundamental set is given by [28]

$$\begin{aligned} \bar{\mathcal{V}} := \{\theta^\alpha,\ \operatorname{grad}\theta^\alpha,\ n^\beta,\ \operatorname{grad} n^\beta,\ \rho^{\alpha R},\ \operatorname{grad}\rho^{\alpha R},\\ \mathbf{F}_S,\ \operatorname{Grad}_S \mathbf{F}_S,\ \overset{\prime}{\mathbf{x}}_\beta,\ \operatorname{Grad}_\beta \overset{\prime}{\mathbf{x}}_\beta,\ \mathbf{X}_\alpha\}. \end{aligned} \tag{118}$$

However, if φ^S is a viscoelastic, an elasto-plastic or an elasto-viscoplastic skeleton material, $\mathcal{V}$ has still to be extended to describe this particular behaviour. The interested reader is referred, e. g., to the work by Ehlers [28–31] for elasto-plastic skeletons and to the work by Ehlers and Markert [44–47] for viscoelastic materials. Note in passing that $\mathcal{V}$ describes materials of second grade by the inclusion of the gradients of the respective basic variables.

In particular, $\mathcal{V}$ includes the individual temperature functions and temperature gradients of the constituents φ^α ($\alpha = S,\ \beta$). Furthermore, the *deformation* of an arbitrary constituent (solid, fluid crystal or fluid) is basically described by the variables $\mathbf{F}_\alpha$, n^α and $\rho^{\alpha R}$ in combination with the corresponding gradients. However, recall that the complete motion of a non-porous single-phasic material is defined by $\mathbf{F}_\alpha$ alone, since $\rho^{\alpha R}$ is uniquely determined by $\det \mathbf{F}_\alpha$ through the integration of the mass balance equation, whereas the volume fraction n^α looses the character of a variable ($n^\alpha \equiv 1$). On the other hand, if general multiphasic materials are concerned, the mass balance of a given constituent may be influenced by the density production term $\hat{\rho}^\alpha$. Following this, the partial density ρ^α cannot be obtained by integration of the mass balance only through $\det \mathbf{F}_\alpha$. Consequently, ρ^α is an independent variable governed by the product of n^α and $\rho^{\alpha R}$. In addition, since n^α is generally independent from $\mathbf{F}_\alpha$ and, therefore, also an independent variable, one has to add either n^α and $\rho^{\alpha R}$ or n^α and ρ^α to the deformation gradient for a full description of the local deformation. Concerning $\mathcal{V}$ from (118), n^S and $\operatorname{grad} n^S$ are dropped, since these terms are not considered as independent variables, since they are uniquely determined by the remainder of volume fractions and volume fraction gradients through the saturation condition (4).

In case that a given constituent φ^β is a fluid crystal or a real fluid, $\mathbf{F}_\beta$ has to be substituted by $\det \mathbf{F}_\beta$, cf. (23). Furthermore, it has been shown by Cross [21] that the deformation variables of real fluids (isotropic fluid crystals) are independent form the fact, whether or not the fluid constituent under consideration is influenced by a density production term. Obviously, considering, for example, the case without density productions, $\det \mathbf{F}_\beta$ can be substituted by ρ^β. Thus, $\mathbf{F}_\beta$ and, additionally, also $\operatorname{Grad}_\beta \mathbf{F}_\beta$ have not been included in $\mathcal{V}$, cf. [28].

Apart from the deformation variables, $\overset{\prime}{\mathbf{x}}_\beta$ and $\operatorname{Grad}_\beta \overset{\prime}{\mathbf{x}}_\beta$ are incorporated into $\mathcal{V}$ to describe fluid viscosity effects, whereas $\mathbf{X}_\alpha$ finally serves for the incorporation of possible inhomogeneities of the constituents φ^α in their respective reference configurations.

Given the fundamental set $\mathcal{V}$, some simplifications are necessary to describe the binary model under consideration. Firstly, the assumptions of an isothermal process and of homogeneous constituents φ^S and φ^F lead to

$$\begin{aligned} \tilde{\mathcal{V}} = \{n^F,\ \operatorname{grad} n^F,\ \rho^{SR},\ \operatorname{grad} \rho^{SR},\ \rho^{FR},\ \operatorname{grad} \rho^{FR}, \\ \mathbf{F}_S,\ \operatorname{Grad}_S \mathbf{F}_S,\ \overset{\prime}{\mathbf{x}}_F,\ \operatorname{Grad}_F \overset{\prime}{\mathbf{x}}_F\} \end{aligned} \tag{119}$$

substituting (118). In the next step, $\tilde{\mathcal{V}}$ is reduced with respect to the assumption of materially incompressible solid and liquid constituents. Following this, the effective densities are constant during deformation and, thus, loose the character of a variable. Furthermore, since the gradients of the constant effective densities vanish for homogeneous materials, ρ^{SR}, $\operatorname{grad} \rho^{SR}$, ρ^{FR} and $\operatorname{grad} \rho^{FR}$ have to be dropped from $\tilde{\mathcal{V}}$. In addition, since the mass balance equation of the materially incompressible solid constituent changes into a volume balance relation, n^F is uniquely defined through $\det \mathbf{F}_S$ by integration of this volume balance in combination with the saturation condition (4). Thus,

$$\begin{aligned} n^F &= 1 - n^S_{0S} \det \mathbf{F}_S^{-1} \\ \longrightarrow \quad \operatorname{grad} n^F &= -n^F\, \mathbf{F}_S^{T-1} (\mathbf{F}_S^{T-1} \operatorname{Grad}_S \mathbf{F}_S) \overset{1}{\cdot} . \end{aligned} \tag{120}$$

Since n^F and, as a consequence, $\operatorname{grad} n^F$ are uniquely determined by $\mathbf{F}_S$ and $\operatorname{Grad}_S \mathbf{F}_S$ through (120), these variables have to be crossed out additionally from $\tilde{\mathcal{V}}$. Following this, one finally obtains

$$\mathcal{S} = \{\mathbf{F}_S,\ \operatorname{Grad}_S \mathbf{F}_S,\ \mathbf{w}_F,\ \mathbf{D}_F\} \tag{121}$$

as the set of constitutive variables governing the binary model under consideration. Note in passing that the principle of frame indifference has been used to substitute $\overset{\prime}{\mathbf{x}}_F$ by the seepage velocity $\mathbf{w}_F$ and $\text{Grad}_F \overset{\prime}{\mathbf{x}}_F$ by the rate of deformation tensor $\mathbf{D}_F$. Combination of (115), (117) and (121) yields

$$\{\psi^S,\ \psi^F,\ \mathbf{T}^S_E,\ \mathbf{T}^F_E,\ \hat{\mathbf{p}}^F_E\} = \mathcal{R}\,(\mathbf{F}_S,\ \text{Grad}_S\,\mathbf{F}_S,\ \mathbf{w}_F,\ \mathbf{D}_F)\,. \qquad (122)$$

Proceeding from (122), the standard evaluation of the entropy principle $(110)_6$ can be carried out, e. g. on the basis of the *Coleman-Noll* [17] or the *Liu-Müller* procedure [75, 76]. Concerning the latter possibility, the interested reader is referred to the work by Svendsen and Hutter [96]. Note in passing that both methods generally lead to basically identical results except for a few particular problems, where the use of the Liu-Müller approach gives reason for more sophisticated results. In the present article, the method by Coleman and Noll is preferred because of its simpler handling without considerable disadvantages. Moreover, additional use is made of the ***principle of phase separation*** by the author [27]. This principle proceeds from the ***a priori*** constitutive assumption that the free energy ψ^α of a given constituent φ^α only depends of the variables included into the process by φ^α itself. Obviously, this principle is not in contradiction with the principle of equipresence, since the dependence of the remainder of $\mathcal{R}$ is not ***a priori*** restricted. Furthermore, following a statement by Bowen [15, 16], the second-grade character of the biphasic model only influences the production terms. Thus, the assumption is made that

$$\psi^S = \psi^S(\mathbf{F}_S)\,, \qquad \psi^F = \psi^F(n^F)\,, \qquad (123)$$

where $(120)_1$ has been used. Note in passing that a general exploitation of the entropy principle with $\psi^\alpha = \psi^\alpha(\mathcal{S})$, i. e. without the above assumption, would have recovered (123) as a natural result of the incompressible binary model under study, cf. e. g. [28, 31]. However, if more sophisticated models are concerned, the principle of phase separation yields a considerable facilitation when the entropy principle has to be analyzed. Furthermore, although a general evaluation of the entropy inequality may result in an admissibility of further variables like those having been assumed by application of the principle of phase separation, one finally has to choose the energy potentials ψ^α on the basis of admissible variables and experimental results. This procedure, how extended the number of admissible variables of ψ^α may be, generally reverts to (123) or even further simplifications.

However, proceeding from (123), the additional constitutive assumption is made that the energy potential of a fluid constituent does not recognize the domain, where the fluid exists on. In particular, since the volume fraction n^F is the local domain variable of φ^F in $\mathcal{B}$, it is assumed that n^F does not influence ψ^F. Thus, (123) is finally substituted by

$$\psi^S = \psi^S(\mathbf{F}_S), \qquad \psi^F = \psi^F(-). \qquad (124)$$

Once (124) is given, the entropy principle $(110)_6$ yields

$$(\mathbf{T}_E^S - \rho^S \frac{\partial \psi^S}{\partial \mathbf{F}_S} \mathbf{F}_S^T) \cdot \mathbf{L}_S + \mathbf{T}_E^F \cdot \mathbf{L}_F - \hat{\mathbf{p}}_E^F \cdot \mathbf{w}_F \geq 0 , \tag{125}$$

where

$$\begin{aligned} (\psi^S)'_S &= \frac{\partial \psi^S}{\partial \mathbf{F}_S} \cdot (\mathbf{F}_S)'_S = \frac{\partial \psi^S}{\partial \mathbf{F}_S} \mathbf{F}_S^T \cdot \mathbf{L}_S , \\ (\psi^F)'_F &\equiv 0 \end{aligned} \tag{126}$$

has been used. Proceeding from standard arguments, the exploitation of (125), in the state of thermodynamic equilibrium given by $\mathbf{w}_F \equiv \mathbf{0}$ and $\mathbf{D}_F \equiv \mathbf{0}$, yields

$$\mathbf{T}_E^S = \rho^S \frac{\partial \psi^S}{\partial \mathbf{F}_S} \mathbf{F}_S^T . \tag{127}$$

Given (127), the dissipative part of the entropy principle yields the dissipation inequality

$$\mathbf{T}_E^F \cdot \mathbf{L}_F - \hat{\mathbf{p}}_E^F \cdot \mathbf{w}_F \geq 0 . \tag{128}$$

Furthermore, the dissipation inequality is sufficiently satisfied with

$$\mathbf{T}_E^F = \overset{4}{\mathbf{D}} \, \mathbf{D}_F \quad \text{and} \quad \hat{\mathbf{p}}_E^F = -\mathbf{S}_v \mathbf{w}_F , \tag{129}$$

where $\overset{4}{\mathbf{D}}$ and $\mathbf{S}_v$ are the positive definite fourth order liquid viscosity tensor and the general permeability tensor of second order. Note that these tensors may depend on the full set $\mathcal{S}$, cf. e. g. [28, 31].

As was mentioned above, it can be shown by the arguments of dimensional analysis, that the friction force div $\mathbf{T}_E^F$ incorporated in the momentum balance of φ^F can be neglected in comparison with the extra momentum production term $\hat{\mathbf{p}}_E^F$, cf. [43]. Following this, it is assumed that

$$\mathbf{T}_E^F \approx \mathbf{0} . \tag{130}$$

Furthermore, to relate $\mathbf{S}_v$ to experimental data, the intrinsic permeability tensor $\mathbf{K}^S$ is introduced via

$$\mathbf{S}_v = (n^F)^2 \mu^{FR} (\mathbf{K}^S)^{-1} , \tag{131}$$

where μ^{FR} represents the effective shear viscosity parameter of the pore-liquid. Note in passing that, although the effective liquid stress or the frictional stress $\mathbf{T}_E^F$ has been neglected, the fluid viscosity is included via the production term $\hat{\mathbf{p}}_E^F$ also known as the effective drag force. In addition, the intrinsic permeability $\mathbf{K}^S$ not only provides information on the local porosity and, therefore, on the local permeability of the solid skeleton, but it also includes the possibility of providing directional information of anisotropic permeability values. Furthermore, the intrinsic permeability tensor is related to the Darcy permeability tensor $\mathbf{K}^F$ through

$$\mathbf{K}^F = \frac{\gamma^{FR}}{\mu^{FR}}\,\mathbf{K}^S, \tag{132}$$

where $\gamma^{FR} = \rho^{FR} g$ is the specific weight of the pore-liquid related to the effective density through the gravitation $g = |\,\mathbf{b}\,|$.

Incorporating $(129)_2$ and (131) in the fluid momentum balance $(110)_3$ yields the well-known Darcy law

$$n^F \mathbf{w}_F = -\frac{\mathbf{K}^S}{\mu^{FR}}\,(\operatorname{grad} p - \rho^{FR}\mathbf{b})\,. \tag{133}$$

Therein, $n^F \mathbf{w}_F$ is usually known as the filter velocity of the pore-fluid motion.

Finally, applying the principle of material frame indifference to both the solid free energy and the solid extra stress yields

$$\boldsymbol{\tau}_E^S = \mathbf{F}_S\,(\rho_{0S}^{SR}\,\frac{\partial \psi^S}{\partial \mathbf{E}_S})\,\mathbf{F}_S^T, \tag{134}$$

where $\boldsymbol{\tau}_E^S = (\det \mathbf{F}_S)\,\mathbf{T}_E^S$ is the Kirchhoff extra stress, ρ_{0S}^{SR} is the effective solid density in the solid reference configuration, and $\rho_{0S}^{SR}\,\psi^S$ is the elastic potential or the stored elastic energy, respectively. Furthermore, note that the principle of frame indifference led to the result that ψ^S depends on the deformation gradient $\mathbf{F}_S$ in such a way that this dependence can be expressed, e. g., by the Green-Lagrangean strain $\mathbf{E}_S = \frac{1}{2}(\mathbf{F}_S^T\mathbf{F}_S - \mathbf{I})$. Further possibilities to express $\boldsymbol{\tau}_E^S$ or $\mathbf{T}_E^S$ can be found in the relevant literature, cf. e. g. [8, 28, 31].

Given the above relations, the binary model under consideration is complete. Finally, note in passing that there has not been any restriction to small strains. Following this, one may include finite elastic materials in the frame of extended Ogden type models as well as geometrically linear ones like the well-known extended Hookean model.

5.3 Sophisticated models

Based on the general concepts of constitutive modelling, a variety of biphasic and triphasic models can be considered and has been investigated by the author and coworkers, cf. [23–26, 33, 34, 36, 38–42, 44–55, 77, 83]. In particular, saturated and partially saturated porous solids have been treated in the framework of both small and finite solid deformations based on elastic, plastic and viscous solid material behaviour. Note in passing that in technical applications like, e. g., soil mechanics, the notion "saturated" generally means a saturation by a single fluid (materially compressible or incompressible) and that the notion "partially saturated" means that the pore content consists of a binary fluid consisting of an incompressible pore-liquid and a compressible pore-gas. The partially saturated model, for example, is appropriate to describe phenomena like leaking and wetting problems.

A triphasic model. To set an example of a more sophisticated model like that of the preceding section, a triphasic material is considered consisting of a micropolar elasto-plastic or elasto-viscoplastic solid skeleton φ^S saturated by a binary pore content of a pore-liquid φ^L and a pore-gas φ^G, cf. [33]. Thus, fully saturated as well as partially saturated conditions can be described. Like in the preceding section, the solid and the pore-liquid are assumed to be materially incompressible (ρ^{SR} = const., ρ^{LR} = const.), whereas the pore-gas, of course, is materially compressible ($\rho^{GR} \neq$ const.). Proceeding from quasi-static problems ($\overset{\prime\prime}{\mathbf{x}}_\alpha = \mathbf{0}$, $(\bar{\boldsymbol{\omega}}_\alpha)'_\alpha = \mathbf{0}$) and isothermal conditions ($\theta^\alpha \equiv \theta \equiv 0$) without mass exchanges ($\hat{\rho}^\alpha = 0$), the governing equations can be given in analogy to (110) on the basis of the relations (93) and (94) as follows:

- pore-liquid volume balance:
$$(n^L)'_S + \operatorname{div}(n^L \mathbf{w}_L) + n^L \operatorname{div}(\mathbf{u}_S)'_S = 0\,,$$
- pore-gas mass balance:
$$(\rho^G)'_S + \operatorname{div}(\rho^G \mathbf{w}_G) + \rho^G \operatorname{div}(\mathbf{u}_S)'_S = 0\,,$$
- momentum balances:
$$\begin{aligned} &\operatorname{div}\mathbf{T}^S + \rho^S \mathbf{b} - \hat{\mathbf{p}}^F = \mathbf{0}\,, \\ &\left.\begin{aligned} &\operatorname{div}\mathbf{T}^L + \rho^L \mathbf{b} + \hat{\mathbf{p}}^L = \mathbf{0}\,, \\ &\operatorname{div}\mathbf{T}^G + \rho^G \mathbf{b} + \hat{\mathbf{p}}^G = \mathbf{0}\,, \end{aligned}\right\} \quad \hat{\mathbf{p}}^F = \hat{\mathbf{p}}^L + \hat{\mathbf{p}}^G \end{aligned} \tag{135}$$
- solid moment of momentum balance:
$$\mathbf{I} \times \mathbf{T}^S + \operatorname{div}\mathbf{M}^S + \rho^S \mathbf{c}^S = \mathbf{0}\,.$$

As was pointed in the preceding section, there is no production of angular momentum incorporated into the solid moment of momentum balance $(135)_6$, since the fluid constituents are microscopically non-polar, thus resulting in $\hat{\mathbf{m}}^L \equiv \mathbf{0}$ and $\hat{\mathbf{m}}^G \equiv \mathbf{0}$ and, hence, $\hat{\mathbf{m}}^S = \mathbf{0}$. Furthermore, to obtain $(135)_6$ from $(94)_3$, it has been assumed that

$$\rho^S (\bar{\boldsymbol{\Theta}}^S \bar{\boldsymbol{\omega}}_S)'_S = \rho^S \left[(\bar{\boldsymbol{\Theta}}^S)'_S \bar{\boldsymbol{\omega}}_S + \bar{\boldsymbol{\Theta}}^S (\bar{\boldsymbol{\omega}}_S)'_S \right] = \mathbf{0}\,. \tag{136}$$

Under quasi-static conditions, this assumption only holds, if not only $(\bar{\omega}_S)'_S$ but also $(\bar{\boldsymbol{\Theta}}^S)'_S$ vanishes. However, recall from Section 4.3 that, in a rotating frame fixed to the microparticles, the values of the tensor of microinertia are constants so that the *Green-Naghdi* derivative of $\bar{\boldsymbol{\Theta}}^S$ vanishes. Following this yields

$$(\bar{\boldsymbol{\Theta}}^S)'_S = 2\,(\bar{\boldsymbol{\Omega}}_S\,\bar{\boldsymbol{\Theta}}^S)_{\rm sym}\,. \tag{137}$$

Given the above result, it is immediately seen from the skew-symmetry of $\bar{\boldsymbol{\Omega}}_S$ that $(\bar{\boldsymbol{\Omega}}_S\,\bar{\boldsymbol{\Theta}}^S)_{\rm sym}$ vanishes if $\bar{\boldsymbol{\Theta}}^S = \Theta^S\,\mathbf{I}$. This result, however, does not only hold in case of spherical microparticles but it also generally holds as the result of an averaging process over an assembly of randomly shaped microparticles.

Furtheron, the entropy inequality (109) of the triphasic material under study is not considered in this section, since its evaluation leads to a quite lengthy procedure far beyond the scope of this article. However, the procedure can be carried out in analogy to the preceding section, where, additionally, internal variables like the inelastic strain have to be considered. The interested reader is referred to the work by de Boer [8], Diebels [23], Ehlers [28, 31] and Haupt [64].

The above equations (135) are governed by a set of primary variables given by the effective liquid and gas pressures, p^{LR} and p^{GR}, the solid displacement $\mathbf{u}_S$, the liquid and gas seepage velocities, $\mathbf{w}_L$ and $\mathbf{w}_G$, and the total average grain rotation $\bar{\varphi}_S$. Following this, the remainder of unspecified quantities included into (135) must be found by constitutive equations, basically, as functions of the above set of primary variables. Furthermore, it is convenient for a numerical treatment of the governing equations (135) to express the volume fractions n^L and n^G as well as the saturation s^G by the solid volume fraction n^S and the liquid saturation s^L. Thus, by use of (8) and (9),

$$n^L = (1-n^S)s^L, \qquad n^G = (1-n^S)(1-s^L), \qquad s^G = 1-s^L. \tag{138}$$

However, the above model can easily be simplified to non-polar solids by dropping $(135)_6$ or to binary media by either dropping $(135)_1$ and $(135)_4$ or by dropping $(135)_2$ and $(135)_5$. In case of binary media, the mass balance of the pore-liquid or of the pore-gas is usually extended by adding the solid mass balance equation. Thus, if the skeleton is fully liquid-saturated ($\varphi^F \equiv \varphi^L$), one obtains

$$\operatorname{div}\,[\,(\mathbf{u}_S\,)'_S + n^F\,\mathbf{w}_F\,] = 0 \tag{139}$$

recovering (110), whereas, if the skeleton is fully gas-saturated ($\varphi^F \equiv \varphi^G$), one has

$$n^F\,(\rho^{FR})'_S + \operatorname{div}\,(n^F\,\rho^{FR}\,\mathbf{w}_F) + \rho^{FR}\operatorname{div}\,(\mathbf{u}_S\,)'_S = 0\,. \tag{140}$$

To close the triphasic model under consideration, the set of equations (135) must be completed by constitutive assumptions for the solid and the fluid stresses, $\mathbf{T}^S$ and $\mathbf{T}^\beta$ ($\beta = \{L, G\}$), the solid couple stress $\mathbf{M}^S$ and the momentum production terms $\hat{\mathbf{p}}^L$ and $\hat{\mathbf{p}}^G$. Furthermore, the gravitation $\mathbf{b}$ and the body couple $\mathbf{c}^S$ are understood as prescribed quantities. Note in passing that the above model is downward compatible to simpler situations. In particular, by setting $\mathbf{M}^S = \mathbf{0}$ and $\mathbf{c}^S = \mathbf{0}$ to result in $\mathbf{T}^S = (\mathbf{T}^S)^T$ and by setting $\overset{*}{\boldsymbol{\varphi}}_S = \mathbf{0}$ to result in $\bar{\mathbf{R}}_S = \mathbf{R}_S$, the description of a standard non-polar skeleton material is naturally included.

As was seen in the preceding section, cf. (111) and (112), the solid and fluid stresses as well as the momentum production were split into "extra" quantities and an additional term governed by the pore pressure. Extending the *effective stress principle* to arbitrary multiphasic materials (Ehlers [31]), the following relations hold:

$$
\begin{aligned}
\mathbf{T}^S &= -n^S p\,\mathbf{I} + \mathbf{T}^S_E\,,\\
\mathbf{T}^\beta &= -n^\beta p^{\beta R}\,\mathbf{I} + \mathbf{T}^\beta_E\,,\\
\hat{\mathbf{p}}^\beta &= p^{\beta R}\,\operatorname{grad} n^\beta + \hat{\mathbf{p}}^\beta_E\,.
\end{aligned}
\tag{141}
$$

Therein, the pore pressure p is given through *Dalton*'s law

$$
p = s^L p^{LR} + s^G p^{GR},
\tag{142}
$$

where p^{LR} is the unspecified effective liquid pressure, whereas the effective gas pressure p^{GR} is related to the effective gas density ρ^{GR} through a constitutive relation. In the remainder of this article, the pressures p^{LR} and p^{GR} are understood as the "*effective excess pressures*" exceeding a typical surrounding pressure like, e. g., the atmospheric pressure p_0. Furthermore, it should be mentioned that *Dalton*'s law is more than only a convenient expression, since it can be recovered from thermodynamical considerations to satisfy the entropy principle. Note in passing that, in case of a binary model of a solid skeleton and a single pore-fluid, p either changes to a *Lagrange*an multiplier like in the preceding section (materially incompressible solid and materially incompressible pore-liquid) or to the gas-pressure (materially incompressible solid and materially compressible pore-gas).

The fluid constituents. Using the same arguments as in the preceding section, the fluid friction forces $\operatorname{div} \mathbf{T}^\beta_E$ are neglected in comparison to the viscous interaction terms $\hat{\mathbf{p}}^\beta_E$. Thus,

$$
\mathbf{T}^\beta_E \approx \mathbf{0}\,.
\tag{143}
$$

Furthermore, in analogy to $(129)_2$, (131) and (132), it is assumed that

$$
\hat{\mathbf{p}}^\beta_E \;=\; -\,(n^\beta)^2\,\gamma^{\beta R}(\mathbf{K}^\beta_r)^{-1}\,\mathbf{w}_\beta\,,
\tag{144}
$$

where $\mathbf{K}_r^\beta$ is the *relative* permeability tensor of φ^β related to the Darcy permeability tensor $\mathbf{K}^\beta$ via

$$\mathbf{K}_r^\beta = \kappa_r^\beta \, \mathbf{K}^\beta , \qquad \text{where} \qquad \kappa_r^\beta = (s^\beta)^{\lambda^\beta} . \tag{145}$$

Therein, κ_r^β is the relative permeability factor depending on the saturation of φ^β through the parameter λ^β, whereas $\mathbf{K}^\beta$ is understood as the permeability tensor of φ^β measured under fully saturated conditions ($s^\beta = 1$). Furthermore, $\mathbf{K}^\beta$ depends on the intrinsic permeability $\mathbf{K}^S$ through

$$\mathbf{K}^\beta = \frac{\gamma^{\beta R}}{\mu^{\beta R}} \, \mathbf{K}^S , \tag{146}$$

cf. (132). Note that this equation can be used to relate the Darcy permeability tensor $\mathbf{K}^\beta$ of various fluids through their specific weight $\gamma^{\beta R}$ and their effective shear viscosity $\mu^{\beta R}$ to the intrinsic permeability $\mathbf{K}^S$. In order to describe the deformation dependence of the intrinsic permeability, it is assumed [54] that

$$\mathbf{K}^S = \left(\frac{1 - n^S}{1 - n_{0S}^S} \right)^\pi (\overset{+}{\mathbf{B}}_S)^\xi \, \mathbf{K}_{0S}^S , \tag{147}$$

where $\overset{+}{\mathbf{B}}_S = (\det \mathbf{B}_S) \, \mathbf{B}_S^{-1}$ is the adjoint of the left Cauchy-Green tensor $\mathbf{B}_S = \mathbf{F}_S \mathbf{F}_S^T$ or the Finger tensor, respectively. Proceeding from geometrically linear conditions for the solid constituent, the volumetric solid deformation enters (147) through

$$n^S = n_{0S}^S \, (1 - \operatorname{div} \mathbf{u}_S) . \tag{148}$$

Furthermore, $\mathbf{K}_{0S}^S$ is the intrinsic permeability tensor of the undeformed skeleton, whereas π and ξ are material parameters. In particular, in case of an initially isotropic solid, $\mathbf{K}_{0S}^S$ reduces to

$$\mathbf{K}_{0S}^S = K_{0S}^S \, \mathbf{I} , \tag{149}$$

where K_{0S}^S is the initial intrinsic permeability coefficient related to the initial Darcy permeability coefficient k_{0S}^β in analogy to (146):

$$k_{0S}^\beta = \frac{\gamma^{\beta R}}{\mu^{\beta R}} \, K_{0S}^S . \tag{150}$$

Furthermore, it was shown by the author and coworkers [43, 54] that ξ governs the deformation induced part of the anisotropic permeability. Thus, if the material is fully isotropic, ξ reduces to 0 and (147) yields

$$\mathbf{K}^S = \left(\frac{1 - n^S}{1 - n_{0S}^S} \right)^\pi K_{0S}^S \, \mathbf{I} . \tag{151}$$

Inserting (144) into the quasi-static fluid momentum balance relations $(135)_4$ or $(135)_5$ leads to the Darcy equations

$$n^\beta \mathbf{w}_\beta = -\frac{\mathbf{K}_r^\beta}{\gamma^{\beta R}}\,(\operatorname{grad} p^{\beta R} - \rho^{\beta R}\mathbf{b}) \tag{152}$$

comparable to (133).

The effective density functions of the materially incompressible pore-liquid and the materially compressible pore-gas are given by

$$\rho^{LR} = \text{const.}\,, \qquad \rho^{GR} = \frac{p^{GR} + p_0}{\bar{R}^G\,\theta}\,. \tag{153}$$

Recall that $(153)_1$ stems from the property of liquid incompressibility, whereas $(153)_2$ is known as the ideal gas law (*Boyle-Mariotte*'s law). Therein, $\bar{R}^G$ denotes the specific gas constant of the pore-gas and θ the absolute *Kelvin*'s temperature. However, in the present investigations, it is assumed that the overall model can be described under isothermal conditions ($\theta = \text{const.}$).

The effective liquid and gas pressures, p^{LR} and p^{GR}, are coupled by the capillary pressure $p^C > 0$ through

$$p^{LR} = p^{GR} - p^C, \tag{154}$$

where p^C depends on the liquid saturation s^L or, vice versa, s^L depends on p^C. In particular, use is made of the *van Genuchten* model [105] given by

$$s^L = [\,1 + (\alpha\, p^C)^j\,]^{-h}. \tag{155}$$

Therein, α, j and h are material constants, where j and h are often used as coupled variables through $h = 1 - 1/j$.

The solid constituent. Following the geometrically linear approach to elasto-plasticity, both the *Cosserat* strain $\bar{\boldsymbol{\varepsilon}}_S$ and the curvature tensor $\bar{\boldsymbol{\kappa}}_S$ are additively decomposed into elastic and plastic parts:

$$\begin{aligned} \bar{\boldsymbol{\varepsilon}}_S &= \bar{\boldsymbol{\varepsilon}}_{Se} + \bar{\boldsymbol{\varepsilon}}_{Sp}\,,\\ \bar{\boldsymbol{\kappa}}_S &= \bar{\boldsymbol{\kappa}}_{Se} + \bar{\boldsymbol{\kappa}}_{Sp}\,. \end{aligned} \tag{156}$$

Note that, once $(156)_1$ is given, the decomposition $(156)_2$ is a natural consequence of the micropolar compatibility condition (74). This has been pointed out in more detail by Ehlers & Volk [52].

As was discussed in [52], the non-symmetric solid extra stress and the couple stress are given by

$$\begin{aligned} \mathbf{T}_E^S &= 2\,\mu^S\,\bar{\boldsymbol{\varepsilon}}_{Se\,\mathrm{sym}} + 2\,\mu_c^S\,\bar{\boldsymbol{\varepsilon}}_{Se\,\mathrm{skw}} + \lambda^S\,(\,\bar{\boldsymbol{\varepsilon}}_{Se}\cdot\mathbf{I}\,)\,\mathbf{I}\,,\\ \mathbf{M}^S &= 2\,\mu_c^S\,(l_c^S)^2\,\bar{\boldsymbol{\kappa}}_{Se}\,. \end{aligned} \tag{157}$$

In the above equations, μ^S and λ^S are the *Lamé* constants of the porous skeleton material, whereas μ_c^S is an additional parameter governing the influence of the skew-symmetric part of the elastic *Cosserat* strain on the effective stress of the skeleton material. Furthermore, the symmetric part of $\mathbf{T}_E^S$ is equivalent to the stress tensor of non-polar skeleton materials, whereas the skew-symmetric part is directly related to the independent micropolar rotation $\overset{*}{\varphi}_S$ through (65). Finally, as was pointed out, e. g. by de Borst [13], l_c^S represents an intrinsic length scale parameter relating the couple stress to the elastic curvature tensor.

In order to describe the plastic or the viscoplastic material properties of both non-polar and micropolar skeleton materials, one has to consider a convenient yield function to bound the elastic domain. In extension of the yield criterion by Ehlers [31] towards micropolar cohesive-frictional materials, it is assumed that

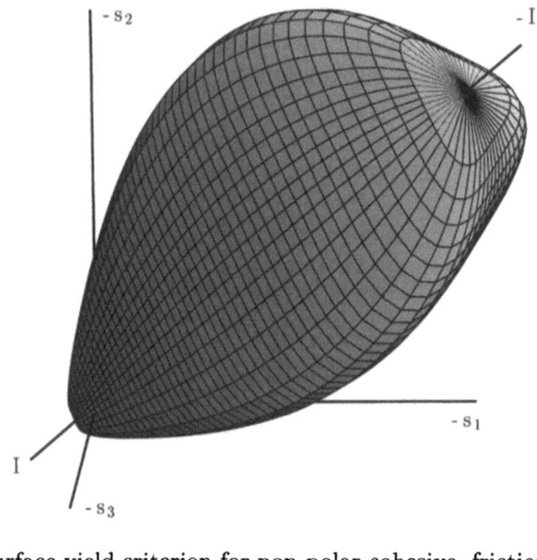

Fig. 7. Single-surface yield criterion for non-polar cohesive- frictional materials; s_1, s_2, s_3: principal stresses of $\mathbf{T}_E^S$ (tension positive).

$$\begin{aligned}
\bar{F} &= \bar{\Phi}^{1/2} + \beta\,\mathrm{I} + \epsilon\,\mathrm{I}^2 + \tfrac{1}{2}\,k_M\,(\mathbf{M}^S\cdot\mathbf{M}^S)^{1/2} - \kappa = 0\,,\\
\bar{\Phi} &= \mathrm{II}^D_{\mathrm{sym}}\,(1+\gamma\,\vartheta)^m + k_T\,\mathrm{II}_{\,\mathrm{skw}} + \tfrac{1}{2}\alpha\,\mathrm{I}^2 + \delta^2\,\mathrm{I}^4\,,\\
\vartheta &= \mathrm{III}^D_{\mathrm{sym}}/(\mathrm{II}^D_{\mathrm{sym}})^{3/2}
\end{aligned} \tag{158}$$

holds [52]. Therein, I, $\mathrm{II}^D_{\mathrm{sym}}$ and $\mathrm{III}^D_{\mathrm{sym}}$ are the first principal invariant of $\mathbf{T}_E^S$ and the (negative) second and third principal invariants of the symmetric part of the effective stress deviator $(\mathbf{T}_E^S)^D$, whereas $\mathrm{II}_{\,\mathrm{skw}}$ defines the second principal invariant of the skew-symmetric part of $\mathbf{T}_E^S$. Furthermore,

$\boldsymbol{p} = (\alpha, \beta, \gamma, \delta, \epsilon, m, \kappa)^T$ and $\boldsymbol{p}^c = (k_M, k_T)^T$ contain two sets of material parameters, where the parameters of $\boldsymbol{p}$ govern the non-polar part and the parameters of $\boldsymbol{p}^c$ the micropolar part of the yield function. In case of non-polar materials ($\boldsymbol{p}^c \equiv \mathbf{0}$), the yield criterion exhibits a closed single-surface yield function in the principal stress space, cf. Figure 7.

Proceeding either from the viscoplastic approach or from the ideal plasticity concept, $\boldsymbol{p}$ as well as $\boldsymbol{p}^c$ are constant during the deformation process. However, while $\boldsymbol{p}$ can be computed from standard experimental data by use of an optimization procedure [49], k_M and k_T have not been satisfactorily determined yet. This is due to the fact that these parameters as well as the internal length scale l_c^S and the micropolar shear modulus μ_c^S must be computed by a back analysis of a typical non-homogeneous boundary-value problem, e. g., including a shear band localization, since these parameters strongly depend on the micropolar rotation which, in case of shear banding phenomena, is only active in the localization zones.

Proceeding from the fact that the associated plasticity concept cannot be applied to frictional materials (cf. e. g. Ehlers & Volk [52]), the plastic potential

$$\begin{aligned} \bar{G} &= \bar{\Gamma}^{1/2} + \beta\,\mathrm{I} + \epsilon\,\mathrm{I}^2 - g(\mathrm{I})\,, \\ \bar{\Gamma} &= \mathrm{II}^D_{\mathrm{sym}} + k_T\,\mathrm{II}_{\mathrm{skw}} + \tfrac{1}{2}\alpha\,\mathrm{I}^2 + \delta^2\,\mathrm{I}^4 \end{aligned} \tag{159}$$

is considered, where $g(\mathrm{I})$ serves to relate the dilatation angle to experimental data. From the concept of a plastic potential, it is straight forward to obtain the evolution equation (flow rule) for the plastic *Cosserat* strain $\bar{\boldsymbol{\varepsilon}}_{Sp}$ via

$$(\bar{\boldsymbol{\varepsilon}}_{Sp})'_S = \Lambda\,\frac{\partial \bar{G}}{\partial \mathbf{T}^S_E}\,, \tag{160}$$

where Λ is the plastic multiplier. As was pointed out by Ehlers & Volk [52], there exists no evolution equation for the plastic rate of curvature tensor independent from both the evolution equation (160) and the micropolar compatibility condition (74). Thus, once (160) is given, the most convenient possibility to obtain an evolution equation for $\bar{\boldsymbol{\kappa}}_{Sp}$ directly results from (160) and (74). Thus,

$$(\bar{\boldsymbol{\kappa}}_{Sp})'_S = \tfrac{1}{2}\overset{3}{\mathbf{E}}\left(\mathrm{Grad}_S\,(\bar{\boldsymbol{\varepsilon}}_{Sp})'_S + \overset{13}{\mathrm{Grad}^T_S}\,(\bar{\boldsymbol{\varepsilon}}_{Sp})'_S - \overset{23}{\mathrm{Grad}^T_S}\,(\bar{\boldsymbol{\varepsilon}}_{Sp})'_S\right)\overset{2}{\cdot}\,. \tag{161}$$

In contrast to the investigation of elasto-plastic problems of non-polar materials, where the standard compatibility condition is generally fulfilled, one has to take into account the micropolar compatibility condition explicitly, if one considers elasto-plastic problems of micropolar materials [52]. Following this, one has two principally different choices to assume an evolution equation for the plastic rate of curvature tensor. The first and simplest one is given by the consideration of (161), where the micropolar compatibility condition is automatically fulfilled. The second one could proceed, on the one hand, from an evolution equation for $\bar{\boldsymbol{\kappa}}_{Sp}$ independent from (161) but must then take into account, on the other hand, the micropolar compatibility condition as a constraint to the computation of the elasto-plastic process under study. Using the second choice without taking into account the micropolar compatibility constraint is not only insufficient but also leads to an extremely unsatisfactory convergence of the numerical computations, whenever the micropolar rotation is active.

In the framework of viscoplasticity using the overstress concept of *Perzyna* type [85], the plastic multiplier included in (160) is given by

$$\Lambda = \frac{1}{\eta}\left\langle \frac{\bar{F}(\mathbf{T}_E^S)}{\sigma_0} \right\rangle^r . \tag{162}$$

Therein, $\langle\cdot\rangle$ are the *Macaulay* brackets, η is the relaxation time, σ_0 is the reference stress, and r is the viscoplastic exponent. However, in the framework of elasto-plasticity, where the plastic strains are rate-independent, Λ has to be computed from the *Kuhn-Tucker* conditions

$$\bar{F} \leq 0, \quad \Lambda \geq 0, \quad \Lambda \bar{F} = 0 \tag{163}$$

rather than from (162).

6 Numerical tools

6.1 Weak formulation of the governing field equations

Once a constitutive model is found following the procedure of the preceding sections, and the respective material parameters have been determined from experimental data, the numerical tools are based on weak formulations of the governing field equations together with discretization methods in the space and time domains. The following considerations proceed from the triphasic material that has been discussed in the preceding section, where the investigation of binary materials is naturally included as a side case.

Based on consideration of six independent fields, the solid displacement $\mathbf{u}_S$, the seepage velocities, $\mathbf{w}_L$ and $\mathbf{w}_G$, the effective liquid and gas pressures, p^{LR} and p^{GR}, and the total average grain rotation $\bar{\boldsymbol{\varphi}}_S$, the corresponding six equations of the weak formulation can be obtained from the kinematics, the balance relations (135) and the constitutive equations of the model. Concerning the quasi-static problem under study, the seepage velocities can be eliminated by use of the momentum balance equations of the pore-fluids φ^β, cf. (152). Thus, $\mathbf{w}_\beta$ looses the status of an independent field variable, and the number of equations of the weak formulation reduces to four. Following the work by Schrefler and coworker [87–90], Lewis & Schrefler [74] and Ehlers and coworkers [35, 41], these equations are given, in the framework of the standard *Galerkin* procedure (*Bubnov-Galerkin*), firstly by the sum of the solid and fluid momentum balance equations $(135)_{3-5}$ or the mixture momentum balance, respectively, multiplied by the test function $\delta\mathbf{u}_S$, secondly by the solid moment of momentum balance $(135)_6$ multiplied by the test function $\delta\bar{\boldsymbol{\varphi}}_S$, thirdly by the volume balance relation $(135)_1$ of the pore-liquid multiplied by the test function δp^{LR} and finally by the mass balance equation $(135)_2$ of the pore-gas multiplied by δp^{GR}. Thus, the weak formulation of the triphasic model reads:

- overall momentum balance:

$$\begin{aligned}\int_\Omega [\mathbf{T}_E^S - (s^L p^{LR} + s^G p^{GR})\,\mathbf{I}\,] \cdot \operatorname{grad}\delta\mathbf{u}_S\,\mathrm{d}v \\ = \int_\Omega (n^S\rho^{SR} + n^L\rho^{LR} + n^L\rho^{LR})\,\mathbf{b}\cdot\delta\mathbf{u}_S\,\mathrm{d}v + \int_{\Gamma_t} \bar{\mathbf{t}}\cdot\delta\mathbf{u}_S\,\mathrm{d}a\,,\end{aligned} \tag{164}$$

- solid angular momentum balance:

$$\int_\Omega \mathbf{M}^S \cdot \operatorname{grad}\delta\bar{\boldsymbol{\varphi}}_S\,\mathrm{d}v - \int_\Omega (\mathbf{I}\times\mathbf{T}^S)\cdot\delta\bar{\boldsymbol{\varphi}}_S\,\mathrm{d}v = 0\,, \tag{165}$$

- pore-liquid volume balance:

$$\begin{aligned}\int_\Omega \{\frac{k^L}{\gamma^{LR}}\operatorname{grad}p^{LR}\cdot\operatorname{grad}\delta p^{LR} + [\,(n^L)'_S + n^L\operatorname{div}(\mathbf{u}_S)'_S\,]\,\delta p^{LR}\}\,\mathrm{d}v \\ = \int_\Omega k^L\,\frac{\mathbf{b}}{g}\cdot\operatorname{grad}\delta p^{LR}\mathrm{d}v - \int_{\Gamma_v} \bar{v}^L\,\delta p^{LR}\,\mathrm{d}a\,,\end{aligned} \tag{166}$$

- pore-gas mass balance:

$$\begin{aligned}\int_\Omega \{\frac{k^G}{g}\operatorname{grad}p^{GR}\cdot\operatorname{grad}\delta p^{GR} + [\,(\rho^G)'_S + \rho^G\operatorname{div}(\mathbf{u}_S)'_S\,]\,\delta p^{GR}\}\,\mathrm{d}v \\ = \int_\Omega k^G\,\rho^{GR}\,\frac{\mathbf{b}}{g}\cdot\operatorname{grad}\delta p^{GR}\mathrm{d}v - \int_{\Gamma_q} \bar{q}^G\,\delta p^{GR}\,\mathrm{d}a\,.\end{aligned} \tag{167}$$

However, if the triphasic model reduces to a binary one, where the skeleton is a fully liquid-saturated binary medium ($\varphi^F = \varphi^L$), (166) reduces to

$$\begin{aligned}\int_\Omega \{\frac{k^F}{\gamma^{FR}}\operatorname{grad}p\cdot\operatorname{grad}\delta p + \operatorname{div}(\mathbf{u}_S)'_S\,\delta p\}\,\mathrm{d}v \\ = \int_\Omega k^F\,\frac{\mathbf{b}}{g}\cdot\operatorname{grad}\delta p\,\mathrm{d}v - \int_{\Gamma_v} \bar{v}^F\,\delta p\,\mathrm{d}a\end{aligned} \tag{168}$$

while (167) is dropped. On the other hand, if the triphasic model reduces to a binary one, where the skeleton is fully gas-saturated ($\varphi^F = \varphi^L$), (166) is dropped, while (167) reduces to

$$\begin{aligned} \int_\Omega \{ \frac{k^F}{g} \operatorname{grad} p \cdot \operatorname{grad} \delta p + [\, n^F (\rho^{FR})'_S + \rho^{FR} \operatorname{div} (\mathbf{u}_S)'_S \,] \, \delta p \} \, dv \\ = \int_\Omega k^F \rho^{FR} \frac{\mathbf{b}}{g} \cdot \operatorname{grad} \delta p \, dv - \int_{\Gamma_q} \bar{q}^G \, \delta p \, da \,, \end{aligned} \tag{169}$$

cf. (139) and (140). In the weak formulation of the problem under study given by the above equations (164)–(169), $\bar{\mathbf{t}}$ is the external load vector acting on the *Neumann* boundary Γ_t of the overall medium. Furthermore, $\bar{v}^L = n^L \mathbf{w}_L \cdot \mathbf{n}$ is the efflux of liquid volume through the *Neumann* boundary Γ_v, whereas $\bar{q}^G = \rho^G \mathbf{w}_G \cdot \mathbf{n}$ characterizes the efflux of gaseous mass through the *Neumann* boundary Γ_q; $\mathbf{n}$ is the outward oriented unit surface normal. To obtain the moment of momentum equation (165), it has been assumed that there is no external loading by volume couples $\mathbf{c}^S$ and by surface couples $\mathbf{m}^S = \mathbf{M}^S \mathbf{n}$.

The equations (164)–(167) represent the weak form of the so-called displacement-rotation-pressures formulation of the strongly coupled solid-fluid problem of triphasic media. In case of non-polar skeleton materials, these equations reduce to the well-known displacement-pressures formulation, where (165) is dropped. Furthermore, if binary media are concerned, (166) and (167) are substituted by (168) or by (169), respectively, thus resulting in either the displacement-rotation-pressure formulation or in the displacement-pressure formulation. In addition, it may be noted that the possibility to deal with an empty skeleton material is always included by simply disregarding the liquid and gas equations (166) and (167) together with $p^{LR} = p^{GR} \equiv 0$. Finally, in the framework of the standard *Galerkin* procedure, the included test functions $\delta \mathbf{u}_S$, $\delta \bar{\varphi}_S$, δp^{LR} and δp^{GR} correspond to the respective field quantities and, as a result, vanish at the *Dirichlet* boundaries with prescribed displacements, rotations and pressure values.

6.2 Discretization in space and time

Preliminaries. In the framework of the finite element method (FEM), the spatial discretization (semi-discretization with respect to the space variable $\mathbf{x}$) of the field equations (164)–(167) is based on quadratic shape functions for the solid displacement $\mathbf{u}_S$ and linear shape functions for the total average grain rotation $\bar{\varphi}^S$ and the fluid pressures p^{LR} and p^{GR} (external variables). Furthermore, both the evolution equation (160) for the plastic strain tensor $\bar{\varepsilon}_{Sp}$ (internal variable) and the plastic multiplier Λ are computed, in the sense of the collocation method, at the integration points of the numerical quadrature. Note again that there is no independent evolution equation for the plastic curvature tensor, since $(\bar{\kappa}_{Sp})'_S$ is obtained from $(\bar{\varepsilon}_{Sp})'_S$ by (161).

For a mesh of N_u nodes and N_q integration points, the space-discrete variables of the semi-discrete problem are collected in the vectors

$$\begin{aligned} \boldsymbol{u} &= ((\mathbf{u}_S^1, \bar{\varphi}_S^1, p_1^{LR}, p_1^{GR}), \ldots, (\mathbf{u}_S^{N_u}, \bar{\varphi}_S^{N_u}, p_{N_u}^{LR}, p_{N_u}^{GR}))^T, \\ \boldsymbol{q} &= ((\bar{\boldsymbol{\varepsilon}}_{Sp}^1, \Lambda^1), \qquad \ldots, (\bar{\boldsymbol{\varepsilon}}_{Sp}^{Nq}, \Lambda^{Nq}))^T. \end{aligned} \tag{170}$$

Using the abbreviation $(\,\cdot\,)' := (\,\cdot\,)'_S$ and the vector $\boldsymbol{y} := (\boldsymbol{u}^T, \boldsymbol{q}^T)^T$, one obtains the semi-discrete initial-value problem

$$\boldsymbol{F}(t, \boldsymbol{y}, \boldsymbol{y}') \equiv \begin{bmatrix} \boldsymbol{F}_1(t, \boldsymbol{u}, \boldsymbol{u}', \boldsymbol{q}) \\ \boldsymbol{F}_2(t, \boldsymbol{q}, \boldsymbol{q}', \boldsymbol{u}) \end{bmatrix} \equiv \begin{bmatrix} \boldsymbol{M}\boldsymbol{u}' + \boldsymbol{k}(\boldsymbol{u}, \boldsymbol{q}) - \boldsymbol{f} \\ \boldsymbol{A}\,\boldsymbol{q}' - \boldsymbol{g}(\boldsymbol{q}, \boldsymbol{u}) \end{bmatrix} \stackrel{!}{=} \mathbf{0} \tag{171}$$

of first order in the time variable t, where $t \geq t_0$ and $\boldsymbol{y}(t_0) = \boldsymbol{y}_0$ are the corresponding initial conditions [26, 55]. In (171), the first equation ($\boldsymbol{F}_1 = \mathbf{0}$) represents the discretization of the governing field equations, where $\boldsymbol{M}$ is the generalized mass matrix, $\boldsymbol{k}$ is the generalized stiffness vector, and $\boldsymbol{f}$ is the vector of the external forces. The second equation ($\boldsymbol{F}_2 = \mathbf{0}$) exhibits the plastic or the viscoplastic evolution equations together with the constraints resulting from the *Kuhn-Tucker* conditions of the elasto-plastic formulation. The introduction of the matrix $\boldsymbol{A}$ formally allows for a joint formulation of elasto-viscoplastic and elasto-plastic problems. Finally, $\boldsymbol{g}$ represents the right-hand side of the evolution equations and constraints, which are element-wise decoupled as a result of their evaluation at the integration points of the finite elements [55].

As a result of the quasi-static problem under consideration, it may occur that the generalized mass matrix $\boldsymbol{M}$ is not regular. Then, the system (171) turns out to be a system of differential-algebraic equations (DAE) of index one in the time variable. Furthermore, in case of elasto-plastic solid material behaviour, the matrix $\boldsymbol{A}$ is also not regular, thus additionally yielding a DAE system on the Gauß point level. Details on the solution of DAE systems can be taken from the literature [55, 59, 60].

Time adaptivity. The time integration as well as the following time-adaptive strategy are based on one-step methods with an embedded time step control, where the solution at time t_{n+1} only depends on the solution at time t_n. This choice is of essential importance with respect to space-adaptive methods (refinements as well as coarsenings), since the transfer of the numerical solution thus only includes two meshes, cf. Diebels et al. [26], Ellsiepen [55] Ehlers & Ellsiepen [41] and Ehlers et al. [42]. Based on the fact that it may occur that the system (171) is a DAE system of index one, it is convenient to apply diagonally implicit *Runge-Kutta* methods (DIRK) with suitable stability properties. With respect to both the size of the system and the treatment of elasto-plastic problems, DIRK methods yield the advantage of being able to solve the non-linear equation systems in a decoupled way.

In addition, embedded methods allow for an efficient estimation of the time error [26, 41, 55]. In particular, one obtains two numerical solutions of (171) at time t_{n+1}, namely $\boldsymbol{y}_{n+1}$ with the convergence order r and $\hat{\boldsymbol{y}}_{n+1}$ with the convergence order $\hat{r} \leq r$. As a result, an embedded error estimation is given by the difference of these solutions through

$$ERR \approx \|\boldsymbol{y}_{n+1} - \hat{\boldsymbol{y}}_{n+1}\| \,. \tag{172}$$

As was pointed out by Diebels et al. [26], this type of an error estimation is "cheap", since it does not require the additional solution of non-linear systems but only a weighted sum of already computed quantities. Following this, *Runge-Kutta* methods with embedded error estimators are well suited for large equation systems. In the present contribution, the numerical examples are carried out by use of a 2-stage singly diagonally implicit *Runge-Kutta* method (SDIRK) with order $r = 2$ and embedded order $\hat{r} = 1$ [55].

Using the relative and absolute tolerances ϵ_r and ϵ_a together with the weighted error measures

$$e_u := \left(\frac{1}{N} \sum_{k=1}^{N} \left[\frac{u_{n+1}^k - \hat{u}_{n+1}^k}{\epsilon_r \, |u_n^k| + \epsilon_a} \right]^2 \right)^{1/2}, \quad e_q := \max_k \left| \frac{q_{n+1}^k - \hat{q}_{n+1}^k}{\epsilon_r \, |q_n^k| + \epsilon_a} \right|, \tag{173}$$

where $N = \dim \boldsymbol{u}$, the time-step is accepted if $e_y := \max\{e_u, \, e_q\} \leq 1$ and rejected otherwise. In both cases, a new step size is predicted from the above error measures together with the order $\hat{r}$ of the embedded method by

$$\Delta t_{\text{new}} := \Delta t_{\text{old}} \, \min\Big\{ f_{\max}, \, \max\Big\{ f_{\min}, \, f_{\text{safety}} \; e_y^{-1/(\hat{p}+1)} \Big\}\Big\} \,. \tag{174}$$

Therein, $f_{\text{safety}} < 1$ is a safety factor, which prevents an oscillation of the time-step size, whereas $f_{\max} > 1$ and $f_{\min} < 1$ are used to limit the step size variation. Concerning further details of this procedure, the reader is referred to [55].

Space adaptivity. Concerning the model under consideration, no mathematically founded methods are known so far to estimate the spatial error [34]. Thus, the following procedure is applied, cf. Figure 8. A time step of the non-stationary problem is treated as a stationary problem, where the initial conditions are taken from the solution of the previous step. In order to estimate the spatial error of the discretized problem, the gradient-based error indicator of *Zienkiewicz-Zhu* type [108] is extended in such a way that all the driving quantities of saturated and unsaturated non-polar and micropolar elasto-plastic and elasto-viscoplastic materials are included.

Apart from the standard consideration of the effective solid stresses representing the elastic part of the problem, the error indicator is extended towards the plastic part of the strain state representing the accumulated plasticity and

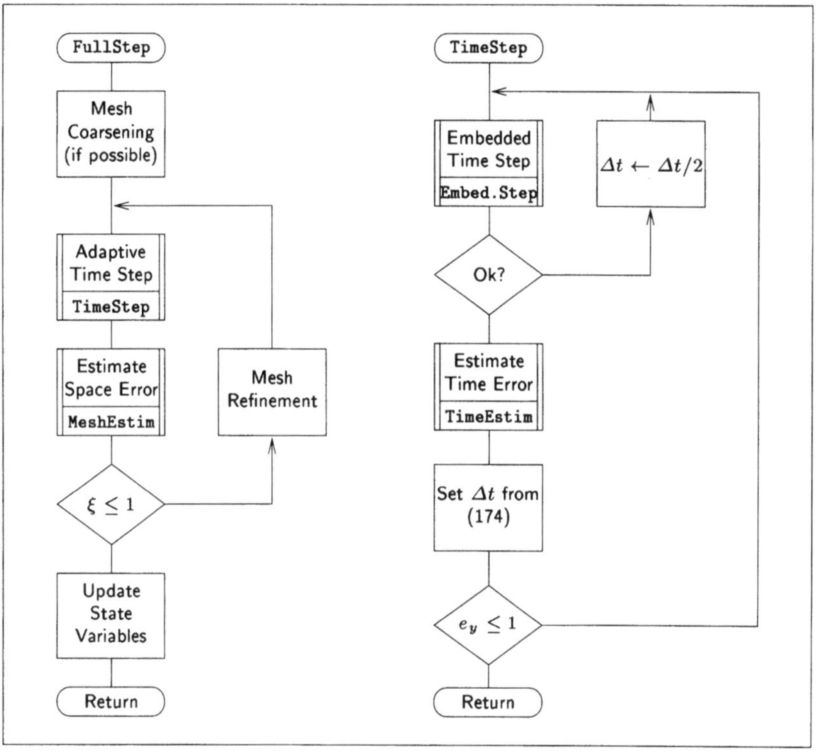

Fig. 8. Algorithm for a time- and space-adaptive step.

towards the seepage velocities representing the viscosities of the pore-fluids. As was pointed out by the author and coworkers [33, 41, 42], there is, in general, no need to either change or extend this set of mechanical quantities incorporated into the error indicator, even in case of micropolar problems. This statement, valid in a wide range of numerical applications like, e. g., localization phenomena, can be obtained as a result of the moment of momentum balance $(135)_6$ together with the micropolar compatibility condition (74). From $(135)_6$, the couple stress tensor $\mathbf{M}^S$ representing the elastic curvature is included through the skew-symmetric part of the effective stress $\mathbf{T}^S_E$, whereas the plastic curvature $\bar{\boldsymbol{\kappa}}_{Sp}$ is included through the plastic strain $\bar{\boldsymbol{\varepsilon}}_{Sp}$ by (74). However, if micropolar boundary layer effects are concerned, an explicit incorporation of $\mathbf{M}^S$ and $\bar{\boldsymbol{\kappa}}_{Sp}$ into the following error indicators is advantageous.

Proceeding from the L_2-norm $\|\cdot\|_2$ and the corresponding element-wise norm $\|\cdot\|_{2,e}$ (per element e), smoothened values $(\,\cdot\,)^*$ of all relevant quantities are computed on the basis of the FEM quantities $(\,\cdot\,)^h$. Following this, the error indicators

$$
\begin{aligned}
\eta_1^{(e)} &:= \|\mathbf{T}_E^{S*} - \mathbf{T}_E^{Sh}\|_{2,e}, & \eta_2^{(e)} &:= \|\bar{\varepsilon}_{Sp}^* - \bar{\varepsilon}_{Sp}^h\|_{2,e}, \\
\eta_3^{(e)} &:= \|\mathbf{M}^{S*} - \mathbf{M}^{Sh}\|_{2,e}, & \eta_4^{(e)} &:= \|\bar{\boldsymbol{\kappa}}_{Sp}^* - \bar{\boldsymbol{\kappa}}_{Sp}^h\|_{2,e}, \\
\eta_5^{(e)} &:= \|\mathbf{w}_L^* - \mathbf{w}_L^h\|_{2,e}, & \eta_6^{(e)} &:= \|\mathbf{w}_G^* - \mathbf{w}_G^h\|_{2,e}
\end{aligned}
\tag{175}
$$

can be applied, where η_1 considers the solid elasticity through the stresses and η_2 the accumulated solid plasticity or viscoplasticity through the plastic strains, whereas η_3 and η_4 explicitely include the couple stresses and plastic curvatures into the error measure. Finally, η_5 and η_6 consider the pore-liquid and the pore-gas flow processes through their seepage velocities. The domain integrals

$$
\begin{aligned}
W_1 &:= \|\mathbf{T}_E^{Sh}\|_2, & W_2 &:= \|\bar{\varepsilon}_{Sp}^h\|_2, \\
W_3 &:= \|\mathbf{M}^{Sh}\|_2, & W_4 &:= \|\bar{\boldsymbol{\kappa}}_{Sp}^h\|_2, \\
W_5 &:= \|\mathbf{w}_L^h\|_2, & W_6 &:= \|\mathbf{w}_G^h\|_2
\end{aligned}
\tag{176}
$$

serve as reference quantities of the respective error indicators. For practical reasons, the absolute errors $\eta_i^{(e)}$ are transferred into dimensionless (relative) errors by dividing through W_i. Consequently, tolerance-weighted error measures $\xi_{e,i}$ can be defined on the basis of user-specified relative and absolute tolerances, ϵ_r and $\epsilon_{a,i}$:

$$
\xi_{e,i} = \frac{\eta_i^{(e)}}{\epsilon_r\, W_i + \epsilon_{a,i}}, \qquad i = 1, 2, 3, 4, 5, 6. \tag{177}
$$

In contrast to the usual considerations on spatial error measures, where a user-specified combination of the absolute element-wise errors $\xi_{e,i}$ is taken to contribute to the global error measure ξ [42], it has been shown by Ehlers et al. [34] that the maximum error indicator ξ_e of each element is a very convenient measure to contribute to ξ. Thus,

$$
\xi_e = \max_{i=1,2,3,4} \left(\frac{\eta_i^{(e)}}{\epsilon_r\, W_i + \epsilon_{a,i}} \right), \qquad \xi = \left(\sum_{e=1}^{E} \xi_e^2 \right)^{1/2}. \tag{178}
$$

This criterion naturally includes $\eta_3^{(e)}$ and $\eta_4^{(e)}$ in case that micropolar boundary layer problems have to be considered. However, the solution on the actual mesh is accepted if $\xi \le 1$ and not accepted else. In order to refine ($\xi > 1$) or to coarsen the mesh ($\xi \le 1$), a new element radius h_{new} must be computed on the basis of a given density function. Concerning the choice of a convenient density function, it has been found by Ehlers et al. [34] that the function proposed by Ladevèse et al. [73] represents an excellent tool for mesh refinements and mesh coarsenings both in the framework of remeshing and hierarchical strategies. Thus, a new element radius h_{new} can be computed by the scaling factor $\mathcal{S}^{LPR}$ (*Ladevèse*, *Pelle* & *Rougeot*) via

$$\begin{aligned} h_{\text{new}} &:= \mathcal{S}^{LPR}\, h_{\text{old}}\,, \\ \mathcal{S}^{LPR} &= \xi_e^{-1/(r+1)} \left[\sum_{e=1}^{N_e} \xi_e^{2/(r+1)}\right]^{-1/(2r)}, \end{aligned} \tag{179}$$

where r is again the convergence order of the FEM discretization. Given (179), h_{new} represents the new element radius optimized per element with respect to the number of elements, cf. Ehlers & Ellsiepen [41], Ellsiepen [55] or the basic work by Gallimard et al. [57]. Concerning the following numerical examples, the present space-adaptive strategy proceeds from both remeshing and hierarchical h-adaptive schemes.

Remeshing strategy. Proceeding from a remeshing strategy means that a completely new mesh must be created, whenever a modification of the mesh is necessary. Therefore, after having evaluated the density function, the new element sizes h_{new} are written into a file, thus delivering the basic information for the mesh generator during the creation of the new mesh. In the present case, this procedure is based on a modified version of the triangular mesh generator `Triangle` presented by Shewchuk [92]. Computing time-dependent problems, the complete data of the current mesh has to be transferred to the new mesh in order to avoid a restart of the computation. When transferring data between FE meshes, two different data types have to be considered: data at nodal points and data at integration points. Concerning the transfer of nodal data, the first task is to find the specific element in the old mesh, wherein a given nodal point $\hat{P}$ of the new mesh is located, cf. Figure 9.

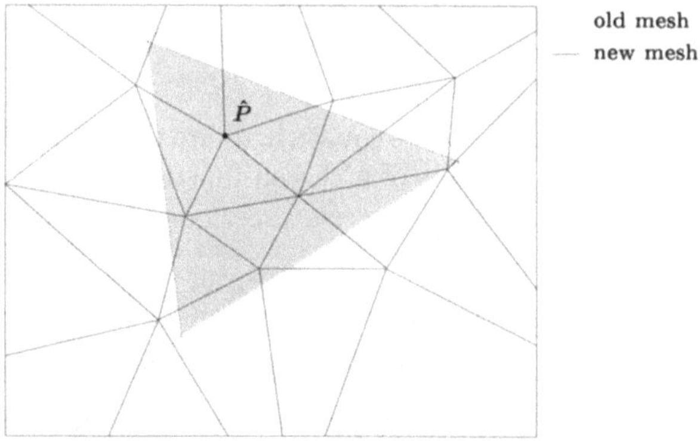

Fig. 9. Data transfer of nodal points.

The location of an element in a FE mesh plays a crucial role in the data transfer of remeshing h-adaptive methods. Following this, an efficient data transfer can only be realized, if an efficient algorithm for the element location is available. The algorithm used in this presentation consists of a combination of two methods: (1) a quadtree search [71] and (2) an inversion of the shape functions. In order to locate an element, one firstly uses the quadtree search to reduce the amount of the possible elements in the whole mesh. Secondly, the inversion of the shape functions is used to exactly locate the requested element. For example, proceeding from biquadratic shape functions as are widely used in the framework of multiphasic problems, an efficient method for inverting shape functions was shown by Crawford et al. [20]. Finally, the correct element is found, if certain conditions for the local coordinates ξ and η hold. For triangular elements, theses conditions are

$$\begin{aligned} 0 \;\le\; & \xi \le 1, \\ 0 \;\le\; & \eta \le 1, \\ & \xi + \eta \le 1. \end{aligned} \tag{180}$$

After having found the requested element, the local coordinates $(\xi_{\hat{P}},\, \eta_{\hat{P}})$ are computed, and the FE shape functions are evaluated to yield the transferred data at the nodal point $\hat{P}$ of the new mesh, e. g. the horizontal displacement $u_{\hat{P}}$ in a triangular element with quadratic shape functions (6 nodes):

$$u_{\hat{P}} = \sum_{i=1}^{6} N_i(\xi_{\hat{P}},\, \eta_{\hat{P}})\; u_i . \tag{181}$$

When transferring data at integration points, the element-wise data, in a first step, must be projected onto the nodal points. Therefore, a function $f(x_1,\, x_2)$ defined by

$$f(x_1,\, x_2) = \sum_i a_i\; \phi_i(x_1,\, x_2) \tag{182}$$

has to be created. In the above equation, a_i are the coefficients and $\phi_i(x_1,\, x_2)$ are the corresponding bases of a chosen function. For a quadratic function, these terms yield

$$\begin{aligned} \boldsymbol{a} &= (\; a_1, \; a_2, \; a_3, \; a_4, \; a_5, \; a_6 \;)^T, \\ \boldsymbol{\phi} &= (\; 1, \; x_1, \; x_2, \; x_1 x_2, \; x_1^2, \; x_2^2 \;)^T. \end{aligned} \tag{183}$$

By minimizing the sum of the quadratic difference between the value f_k of the data and the function $f(x_1,\, x_2)$ over all integration points K,

$$\Sigma(a_1, \ldots, a_n) := \sum_{k=1}^{K} \left(f_k - \sum_i a_i\; \phi_i(x_{1k},\, x_{2k}) \right)^2 \stackrel{!}{\longrightarrow} \text{min.}, \tag{184}$$

the coefficients a_i of the function $f(x_1, x_2)$ can be computed. After having evaluated this function at the nodal points, the same strategy as for the transfer of nodal data can be applied.

Hierarchical strategy based on bisection. In a hierarchical strategy, refinement or derefinement of meshes can be carried out by adding or removing FE edges. In addition, it is very important for the stability of the adaptive process that degenerated elements are avoided. This, however, strongly depends on the method how the refinement or the derefinement process is carried out. Using triangular elements, the *Newest Vertex Bisection* by Mitchell [79] in combination with the recursive algorithm by Kossaczký [70] was found very stable in a lot of adaptive computations [41, 55].

Using the *Newest Vertex Bisection* strategy, one firstly marks the edges of the initial FE mesh which have to be bisected during a first modification of the mesh. Basically, any edge can be chosen but it is obviously reasonable to mark the longest edge of each element. Subsequently, that edge of a triangle is marked for bisection which faces the most recently generated vertex: the *Newest Vertex*. This condition has to be accomplished by both neighbouring elements of the dividing edge. An example for this strategy is shown in Figure 10. Therein, the shaded triangle is the element to be refined. The marked edges are shown by the small arrows in each element.

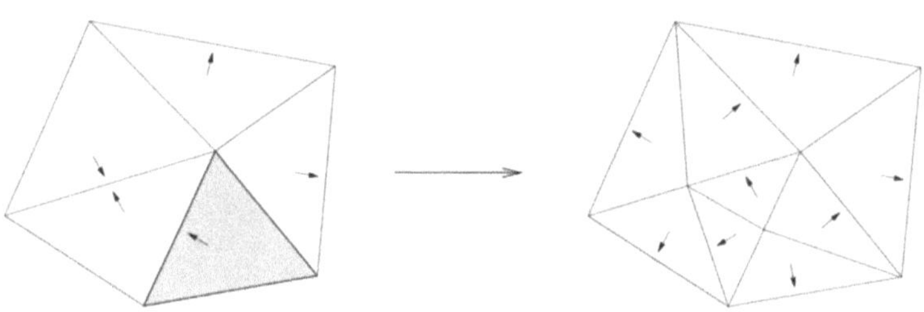

Fig. 10. Newest vertex bisection.

By use of the scaling factor $\mathcal{S}^{LPR}$ from (179), the element remains unchanged if $\mathcal{S}^{LPR} \geq 1$, the element is bisected in case of $1/2 \leq \mathcal{S}^{LPR} < 1$, and it is divided into three new elements if $1/3 \leq \mathcal{S}^{LPR} < 1/2$, and so forth. Derefinement is only possible, if all neighbouring elements having been refined by bisection of the same edge suffice the condition $\mathcal{S}^{LPR} \geq 2$.

The data transfer in the hierarchical strategy is totally different from the data transfer of the previous method. In the present case, only a local data transfer of the modified elements has to be carried out. Furthermore, an element location algorithm is not necessary while refining or derefining an

Parameter	Symbol	Value	Symbol	Value
Lamé constants	μ^S	$5\,583\,\mathrm{kN/m^2}$	λ^S	$8\,375\,\mathrm{kN/m^2}$
effective densities	ρ^{SR}	$2\,600\,\mathrm{kg/m^3}$	ρ^{LR}	$1\,000\,\mathrm{kg/m^3}$
	ρ_0^{GR}	$1.23\,\mathrm{kg/m^3}$		
gas constants	$\bar{R}^G$	$287.17\,\mathrm{J/(kg\,K)}$	θ	$283\,\mathrm{K} = \mathrm{const.}$
	p_0	$10^5\,\mathrm{N/m^2}$		
volume fractions	n_{0S}^S	0.67	n_{0S}^F	0.33
gravitation	g	$10\,\mathrm{m/s^2}$		
permeability	K_{01}^S	$1.2 \cdot 10^{-14}\,\mathrm{m^2}$	K_{02}^S	$10^{-9}\,\mathrm{m^2}$
parameters	μ^{LR}	$10^{-3}\,\mathrm{N\,s/m^2}$	μ^{GR}	$1.8 \cdot 10^{-5}\,\mathrm{N\,s/m^2}$
	λ^L	3.0	λ^G	3.0
	π	1.0	ξ	0
	α	$2 \cdot 10^{-4}$	j	2.3
	h	1.5		
parameters of the	α	$1.074\,0 \cdot 10^{-2}$	β	$0.119\,6$
single-surface	γ	1.555	δ	$1.377 \cdot 10^{-4}\,\mathrm{m^2/kN}$
yield criterion	ϵ	$4.330 \cdot 10^{-6}\,\mathrm{m^2/kN}$	κ	$10.27\,\mathrm{kN/m^2}$
	m	$0.593\,5$		
viscoplasticity	η	$2 \cdot 10^3\,\mathrm{s}$	σ_0	$10.27\,\mathrm{kN/m^2}$
	r	1		
Cosserat	l_c^S	$1 \cdot 10^{-3}\,\mathrm{m}$	μ_c^S	$4 \cdot 10^3\,\mathrm{kN/m^2}$
parameters	k_M	12	k_T	0

Table 3. Material parameters.

element by adding or removing FE edges, because the location of the element is obviously already known. The transfer of the nodal data even drops out in the case of derefinement. The actual transfer of the data, however, is handled as was shown above in the remeshing strategy by evaluation of the shape functions.

7 Numerical examples

7.1 Outline

The numerical examples presented here concern the wide range of applications of the Theory of Porous Media to engineering, especially, geomechanical problems. In particular, the leaking and wetting of a porous column is considered as well as the saturated and the unsaturated consolidation and two principally different localization phenomena like the well-known biaxial experiment and the base failure problem. Finally, to set an additional example for the treatment of unsaturated media, the flow of pore-water through an embankment is taken into consideration. The examples generally proceed from the material parameters included into Table 3, where isotropic permeability conditions ($\xi = 0$) are assumed. In particular, K_{0S}^S is generally taken as K_{01}^S leading to a Darcy permeability of $k_{0S}^L = 1.2 \cdot 10^{-7}$ m/s, whereas K_{02}^S leads to $k_{0S}^L = 10^{-2}$ m/s. Furthermore, the computations have been carried out by use of the FE package PANDAS [39], where time- and space-adaptive methods [33, 34, 41, 42] are widely applied. Concerning further examples, the reader is referred to [47, 48], where finite viscoelastic deformations of gas-saturated polyurethane foams are considered on the basis of an extended Ogden type formulation or to [45, 46], where the biphasic behaviour of biological soft tissues is treated.

Proceeding from the fact that localization phenomena like shear banding play an important role in geotechnical applications, it is also shown in some of the following examples, to what extend the inclusion of micropolarity, viscoplasticity or fluid viscosity yields a regularization of generally ill-posed localization problems. In particular, the first localization example (the biaxial experiment) concerns an empty elasto-plastic skeleton, where the shear band computation is regularized by the inclusion of micropolarity, the second localization example exhibits the same basic situation, however applied to a fluid-saturated elasto-viscoplastic material. The third localization example (the base failure problem) finally proceeds from the same material combination, the fluid-saturated elasto-viscoplastic solid skeleton.

7.2 Leaking and wetting of a porous column

(a) The leaking problem. The present example exhibits a rigid soil column of 1 m height which is fully liquid-saturated ($s^L = 1$) at time $t_0 = 0$. In order to correctly describe the leakage process, a triphasic medium is considered, where at times $t \geq t_0$ the values of the liquid saturation s^L characterize the distribution of the pore-fluids water and air throughout the soil column. In particular, the initial conditions are prescribed through $s^L = 0$ at the top and at the bottom of the sample. Figure 11 shows the progression of the leaking process driven by gravitation, whereas Figure 12 represents the efflux of liquid volume through the bottom of the soil column.

(b) The wetting problem. Basically, the wetting problem concerns the same soil column as before. However, the present sample is assumed to be fully gas-saturated ($s^L = 0$) at time $t_0 = 0$. Furthermore, the prescribed initial conditions are given by $s^L = 0$ at the top and $s^L = 1$ at the bottom of the column. Figure 13 shows the progression of the wetting process driven by capillary suction vs. gravitation, whereas Figure 14 represents the final distribution of the liquid-saturation vs. the column height.

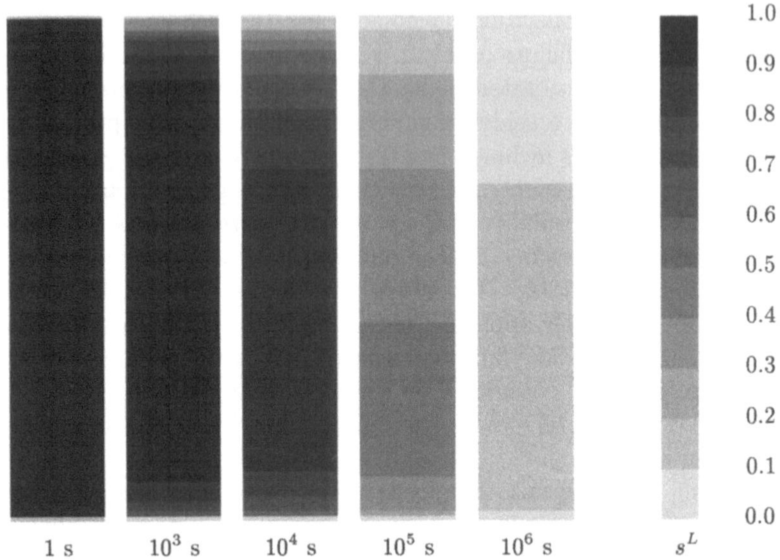

Fig. 11. Progression of liquid saturation s^L.

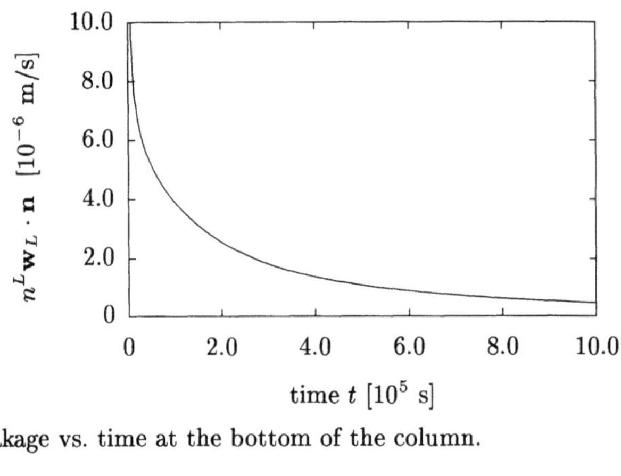

Fig. 12. Leakage vs. time at the bottom of the column.

7.3 Saturated and unsaturated consolidation

(a) The classical consolidation problem. In the classical literature, cf. e. g. Terzaghi [98], the consolidation problem is defined by the onset of an additional external load (e. g. a building) onto a fluid-saturated porous elastic solid (e. g. a soil), cf. Figure 15. As a result, a time-depending settlement process occurs which is accompanied by a drainage of the viscous pore-content. The present example concerns the well-known problem of a rigid strip footing on a soil half-space. Furthermore, the computations are based on the standard quasi-static formulation of a materially incompressible linear elas-

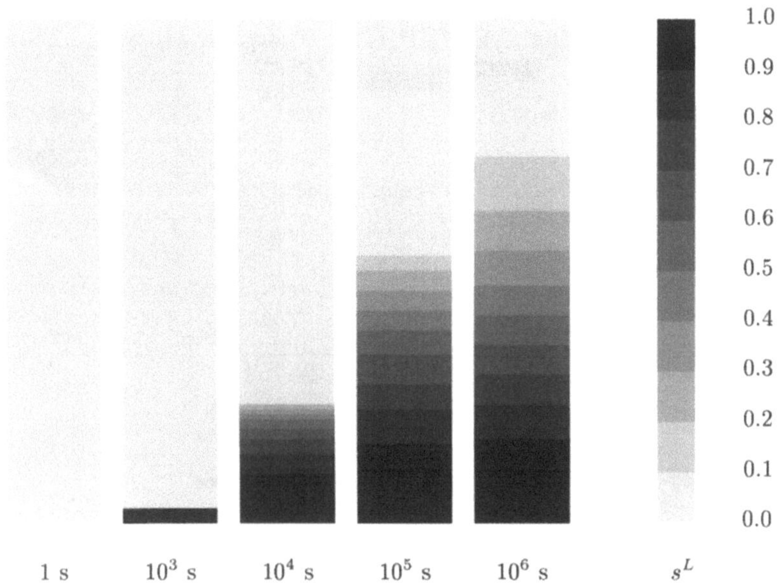

Fig. 13. Progression of liquid saturation s^L.

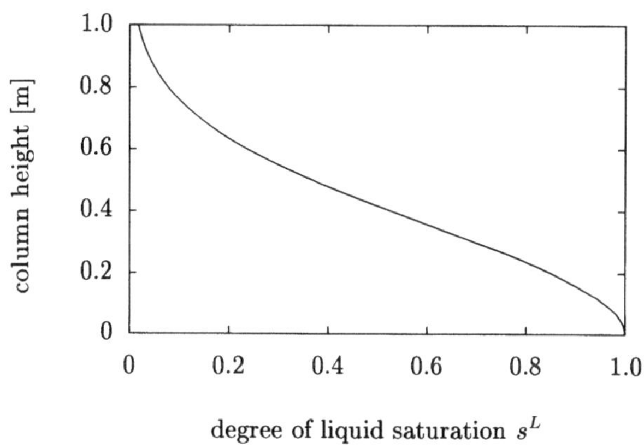

Fig. 14. Final distribution of the liquid saturation vs. column height.

tic (non-polar) skeleton which is fully saturated by a viscous and materially incompressible pore-water. In order to model the half-space, the dimension a is chosen large enough so that the impermeable boundaries at both sides and at the bottom of the sample do not influence the numerical solution.

In particular, the external load q is linearly increased from zero to its maximum value $q_{\max} = 300\,\mathrm{kN/m^2}$. As a reaction on the external load, there is not an instantaneous but a time-dependent deformation of the elastic skeleton due to the drainage process of the pore-water. Figure 16 exhibits the decrease of the excess pore-water pressure p (excess over the atmospheric

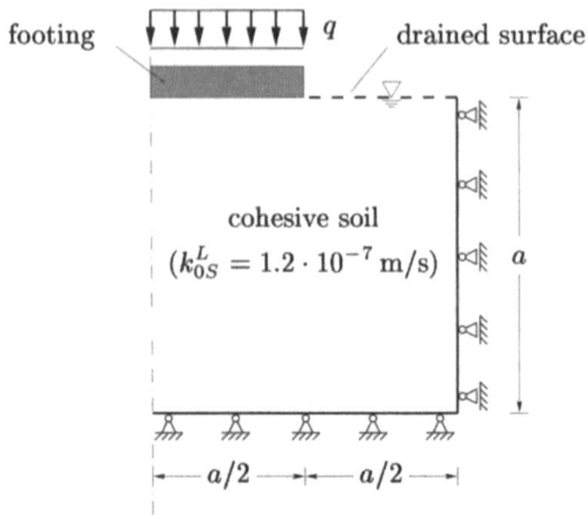

Fig. 15. Rigid strip footing on a water-saturated half-space.

pressure) throughout the consolidation process, where $t = 0$ indicates the time when $q_{\max}$ is reached and kept constant for $t > 0$. The development of the pressure isolines between $t = 0$ and $t = 10\,\text{h}$ shows that, firstly, the maximum pressure is directly under the footing, while $p = 0$ is prescribed at the drained surface. Secondly, the pore-water pressure decreases with time down to an overall value of $p = 0$ at $t \to \infty$.

(b) Soil subsidence by loss of ground water. Apart from the standard consolidation problem described above, is may occur under more realistic circumstances that the soil does not only behave purely elastically. Instead, it generally behaves elasto-plastically in a wide range of deformations. Moreover, in addition to an external loading process, there may be further reasons for a soil subsidence. Therefore, the present example concerns, firstly, the standard strip footing situation on a liquid-saturated half-space, where the soil is considered as a materially incompressible elasto-plastic (non-polar) skeleton. Secondly, when the maximum settlement of the elasto-plastic consolidation process is reached, it is assumed that there is a loss of ground-water through the bottom of the domain under consideration. This is modelled by switching the boundary condition at the bottom from impermeable to permeable, where, in addition, a fluid suction $p < 0$ is prescribed. Furthermore, $s^L = 0$ is prescribed at the top boundary, so that an unsaturated domain is generated in the sense of the general triphasic medium. In reality, these boundary condition correspond, e. g., to the case where the liquid pore-content in the area under the considered domain is pumped out. Problems like this occur, for example, in the surrounding of open coal mines or near oil and gas reservoirs under exploitation.

Figure 18 shows the time-settlement curve of the whole process including the pre-consolidation domain. In particular, point (1) indicates the end of

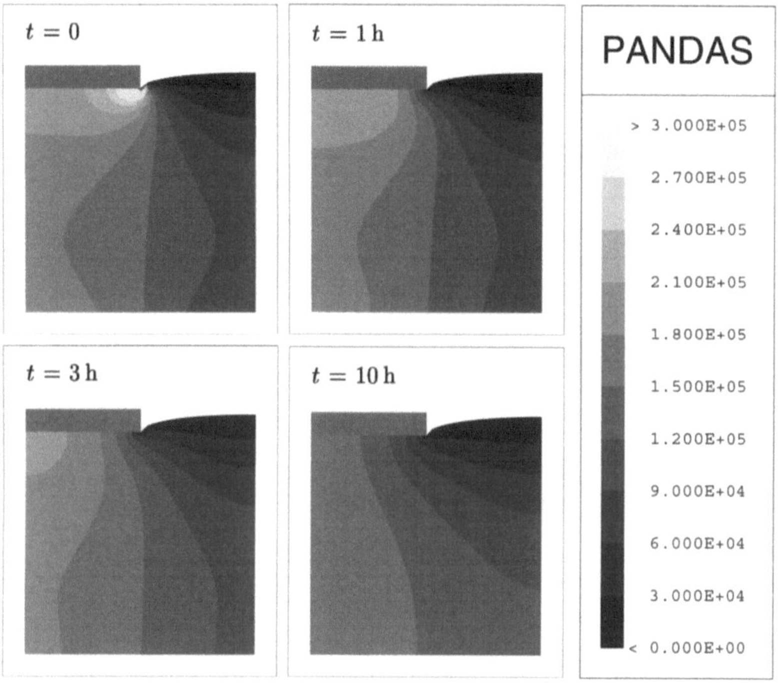

Fig. 16. Decrease of the pore-water pressure p [$\mathrm{N/m^2}$] at the deformed soil skeleton (scaling factor 10).

the pre-consolidation process driven by gravitation, where, in addition, the external load q is applied. Point (2) characterizes the end of the standard consolidation process under fully liquid-saturated conditions, where, at the same time, the loss of ground water is initiated. Finally, point (3) exhibits the final situation, when the whole process has come to an end. In addition to Figure 18, Figures 19 and 20 present the accumulated plastic strains after having reached the consolidation points (2) and (3). Note in passing that, obviously, the plastic strains localize in small bands, the so-called shear bands, thus giving a hint on a possible base failure problem.

7.4 The biaxial experiment

(a) The empty elasto-plastic micropolar solid (hierarchical h-adaptive scheme). The present example exhibits a simulation of a biaxial experiment, where plane strain conditions are prescribed, cf. Figure 21. In particular, the computations are carried out on the basis of time- and space-adaptive methods, where use is made of the hierarchical refinement and derefinement scheme described above and in more detail in [34, 41, 55]. Furthermore, an empty granular micropolar elasto-plastic solid skeleton is considered, where

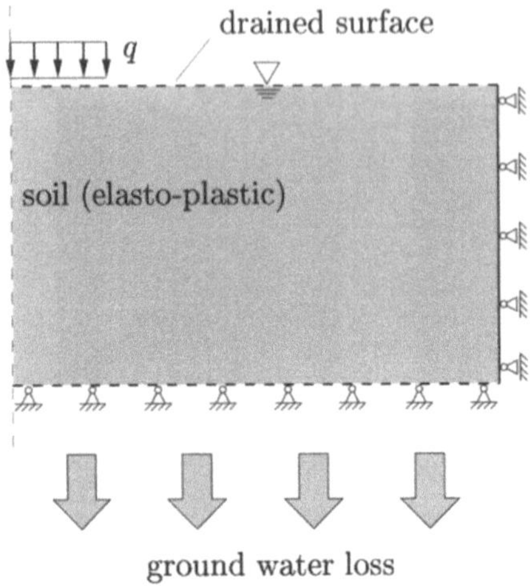

Fig. 17. Strip footing and loss of ground water.

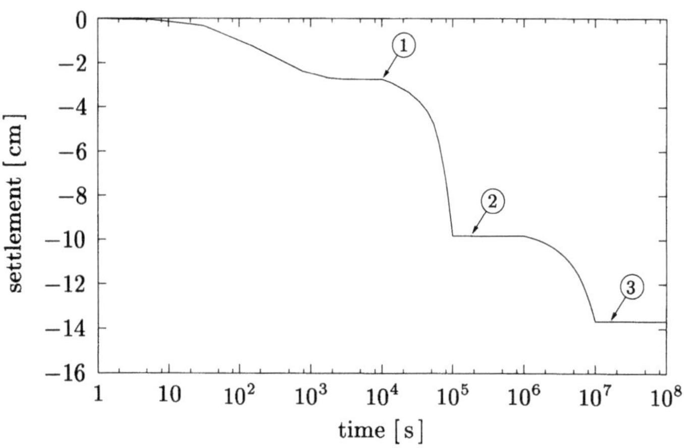

Fig. 18. Time-settlement curve.

at time $0 \leq t \leq \tau$ a linearly increasing horizontal stress σ_1 is applied, which is kept constant when $t > \tau$. Simultaneously, the specimen is loaded with a displacement-driven vertical stress such that $u_2 = (u_2)'_S\ (t - \tau)$. From the numerical point of view, Figure 21 characterizes a homogeneous problem. In order to obtain a shear band localization, an imperfection at the bottom of the material is included, thus initiating the shear band development.

Since there is no possibility to obtain an analytical solution of the problem, the adaptive computations are compared to a so-called overkill solution carried out on an extremely fine mesh which is kept constant during the numerical solution process. Figure 22 shows the load-deflection curves of the adaptive and the overkill solutions, where, as a result of the time- and space-adaptive strategy, both solutions are equivalent up to a great extent. It is

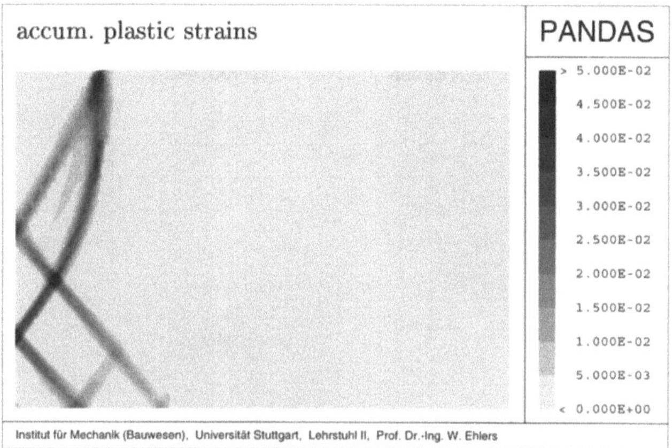

Fig. 19. Surface subsidence after elasto-plastic consolidation.

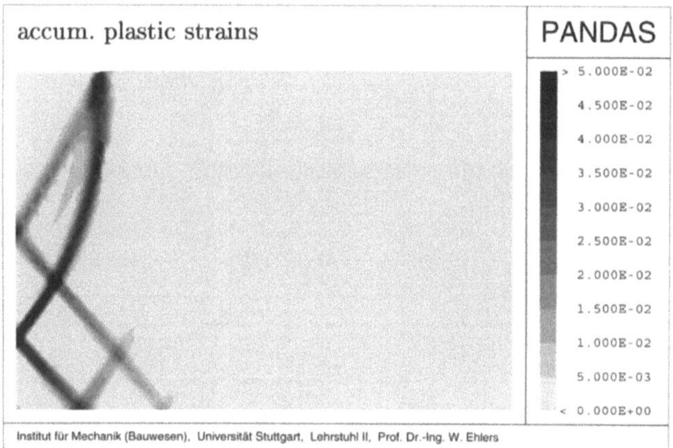

Fig. 20. Surface subsidence after loss of ground water.

furthermore seen by the crosses marking the varying time step sizes of the adaptive solution that very small time steps occur when the localization initiates and the load-defection curve decreases nearly vertically.

The computations start with an initial mesh of triangular elements with 2933 degrees of freedom (d. o. f.), cf. Figure 23. During the computational process, a mesh refinement takes place initiated by the onset of plastic yielding at the imperfection point. Based on prescribed values of the minimum element size, the final mesh results in 17808 d. o. f. To compare the solution of the adaptive computation with an overkill solution, which is assumed to be close to the "true" solution of the problem, a considerably fine mesh of 57177 d. o. f. is used. The regularization of the problem is seen from the fact that the distribution of the plastic strains, cf. Figure 24 (left), exhibits the shear band development which clearly spans several element layers.

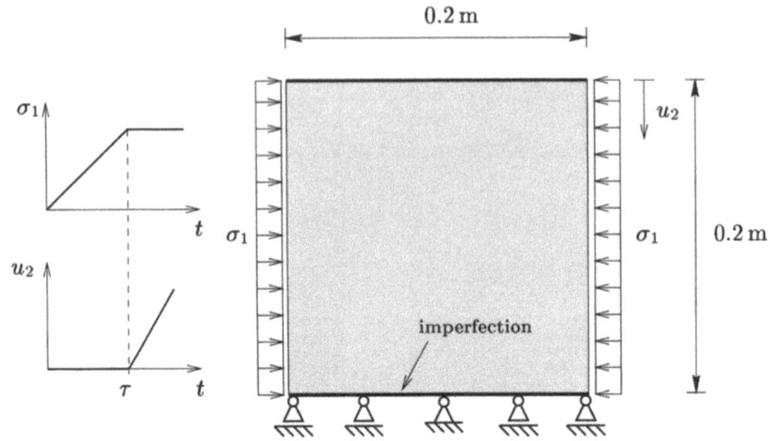

Fig. 21. Biaxial experiment: Initial boundary-value problem.

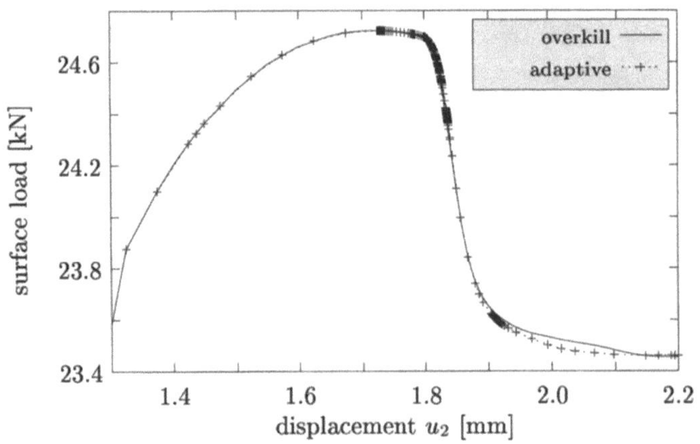

Fig. 22. Biaxial experiment: load-deflection curves.

Finally, Figure 24 (right) shows the distribution of the total rotation produced by micropolarity. Following this, it is clearly seen that there is approximately no rotation outside the shear bands. Inside the shear bands, the rotation is either positive or negative. As a result, the rotation is zero at the crossing point of positively and negatively rotating shear bands. In the present example, such a point is given at the imperfection which causes the shear band initiation.

(b) The liquid-saturated elasto-viscoplastic solid (hierarchical and remeshing h-adaptive schemes). The following considerations concern the same initial boundary-value problem as before, however, in contrast to the previous section, the present example makes use of a binary material, namely a liquid-saturated solid skeleton consisting of an incompressible pore-fluid

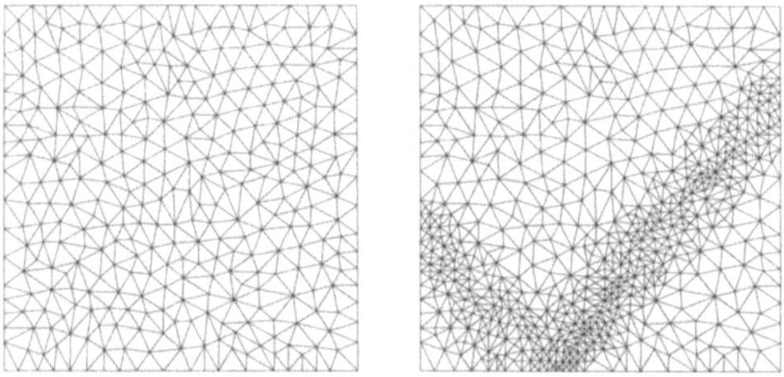

(a) Initial mesh: 2 933 d. o. f. (b) Refined mesh: 6 399 d. o. f.

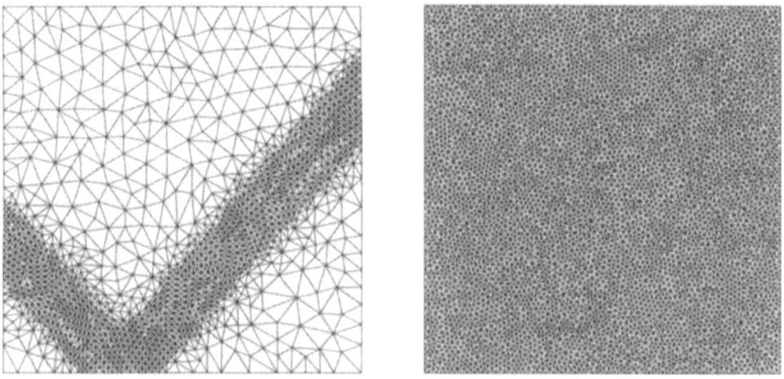

(c) Final mesh: 17 808 d. o. f. (d) Overkill mesh: 57 177 d. o. f.

Fig. 23. Hierarchical adaptive mesh refinement (micropolar formulation).

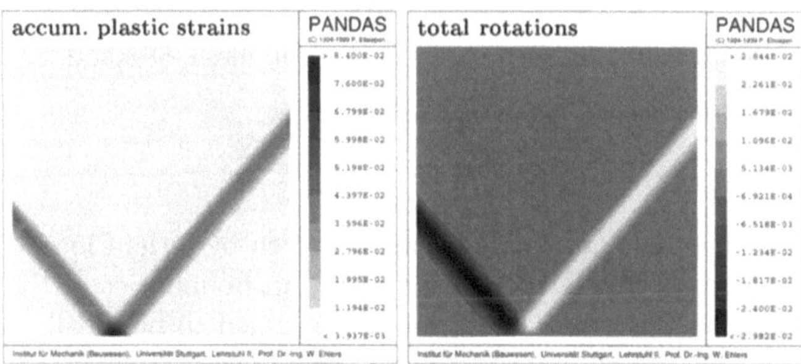

Fig. 24. Accumulated plastic strains and total rotations (micropolar formulation).

(the pore-water) and a materially incompressible elasto-viscoplastic skeleton in the framework of the standard non-polar formulation. Following this, the regularization of shear band phenomena is based on the included solid viscoplasticity and on the fluid viscosity. Furthermore, the boundary-value problem is defined as was described in the preceding example, except for the fact that here, after having applied the lateral load (phase 1), the rigid load platen is only driven into the specimen (phase 2) when the consolidation process resulting from σ_1 is finished. Furthermore, to compare the effort of hierarchical and remeshing strategies, the computations are carried out by use of both hierarchical and remeshing methods, cf. Ehlers et al. [34]. In partic-

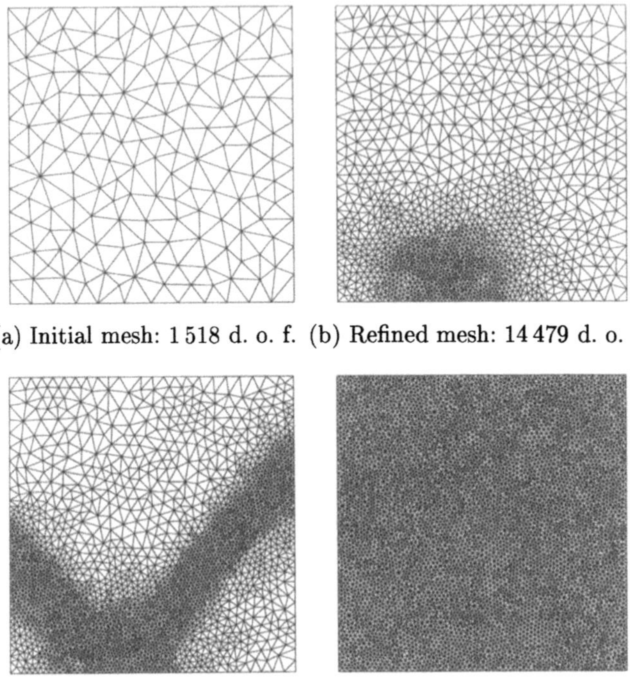

(a) Initial mesh: 1 518 d. o. f. (b) Refined mesh: 14 479 d. o. f.

(c) Final mesh: 26 947 d. o. f. (d) Overkill mesh: 58 504 d. o. f.

Fig. 25. Remeshing scheme (standard formulation).

ular, the specimen is loaded displacement driven by a rigid load platen with $\dot{u}_2 = 2 \cdot 10^{-7}$ m/s. While the top and bottom boundaries of the specimen are impermeable, the side boundaries are assumed to be ideally permeable. Furthermore, the side boundaries are stabilized by a linearly increasing stress of $\sigma_1 \leq 50$ kN/m^2 (Figure 21).

To compare the hierarchical (H) and the remeshing strategies (R), both the computing times and the accuracy of the solutions are considered. In particular, it turns out that there is a substantial difference between the computations based on the hierarchical and on the remeshing strategy. Proceeding from the same user-defined tolerances, the remeshing strategy, as far as this example is concerned, is more efficient than the hierarchical procedure. This is seen from the fact that the hierarchical scheme, on a SGI Power Challenge R 10000/195, needs a computing time of 6 h 56 min to result in 32 046 elements, whereas the remeshing strategy comes along with only 5 h 21 min and 26 947 elements. The mesh refinement of the remeshing process together with an overkill mesh can be taken from Figure 25.

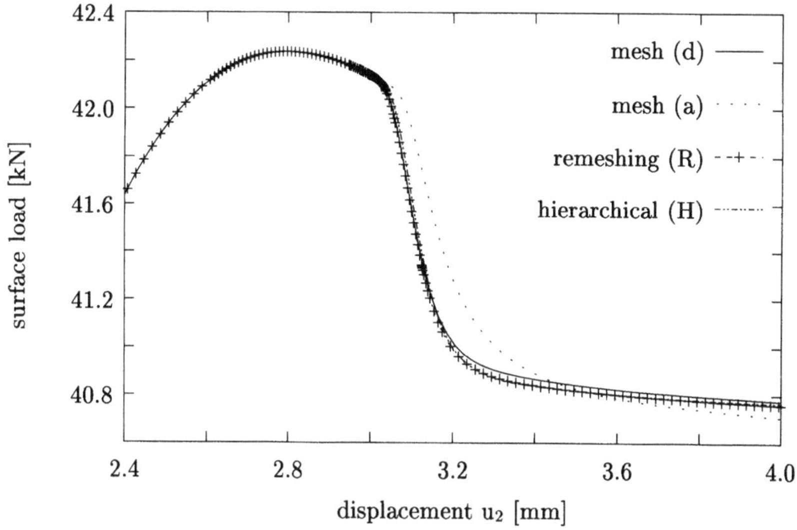

Fig. 26. Biaxial experiment: load-deflection curves.

However, although there were considerable differences in the performance of both strategies, there is no significant difference in the corresponding load-deflection curves, cf. Figure 26. Furthermore, it is seen from Figure 26 by the crosses marking the varying time step sizes of the adaptive solution obtained by use of the remeshing strategy that very small time steps occur when the localization is initiated.

(c) The liquid-saturated elasto-viscoplastic solid (compressive and dilatant shear bands). The following considerations concern the same example as before, namely the liquid-saturated solid skeleton consisting of an incompressible pore-fluid (the pore-water) and a materially incompressible elasto-viscoplastic skeleton. Following this, the regularization of shear band phenomena is again based on the included solid viscoplasticity and on the fluid viscosity. In particular, two different lateral loads are applied, thus leading to both contractant and dilatant shear bands, cf. Figures 27 and 28. The reason for the onset of either contractant or dilatant shear bands stems from the fact that, when viscoplastic yielding occurs, a lateral load of $\sigma_1 = 140\,\mathrm{kN/m^2}$ leads to a yield point in the ductile regime, whereas a lateral load of $\sigma_1 = 110\,\mathrm{kN/m^2}$ produces a yield point in the brittle regime. It is furthermore seen from Figure 27 that the final vertical displacement of $u_2 = 5\,\mathrm{mm}$ yields comparable shear band developments with slightly different inclinations ($43\,°$ vs. $47\,°$) in both cases the contractant and the dilatant one. However, as a result of dilatancy, the maximum value of the accumulated plastic strains is approximately 50 % higher in the dilatant case than in the contractant one. The reason for this fact mainly results from the incompressibility constraint of the pore-fluid. To explain this behaviour in more detail, the reader is referred

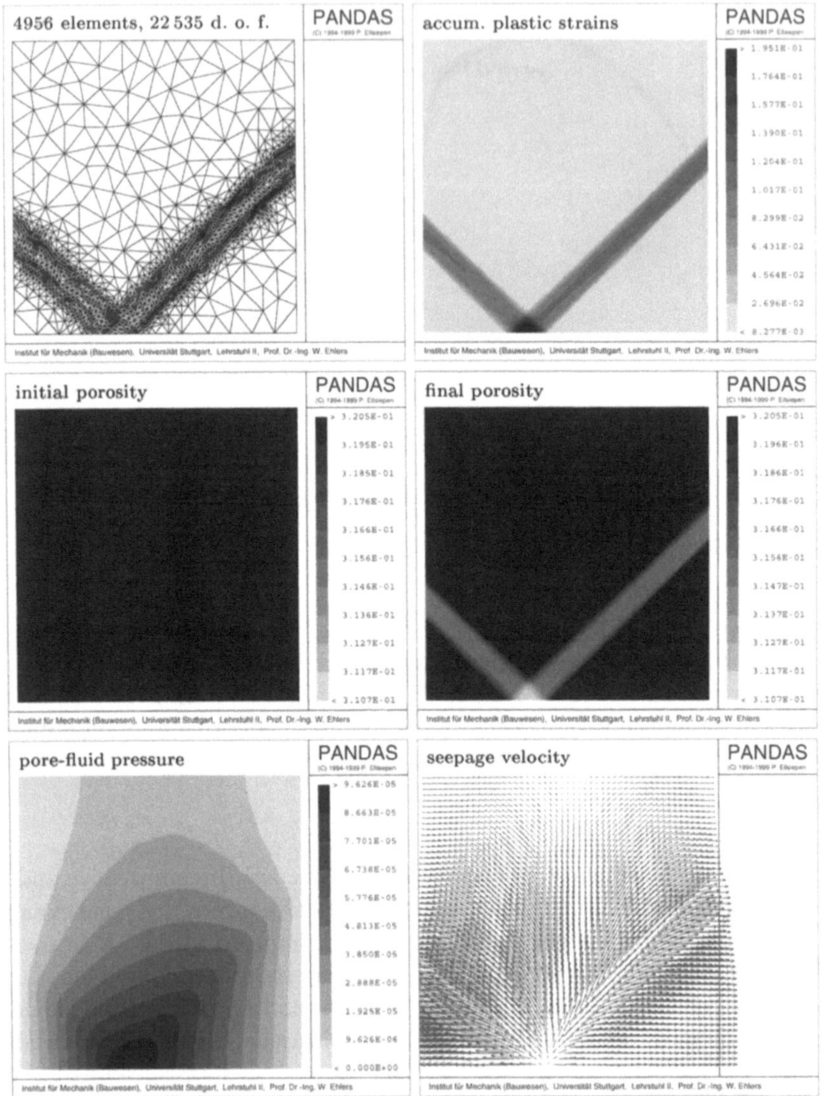

Fig. 27. Contractant shear band development ($\sigma_1 = 140\,\mathrm{kN/m^2}$).

to the pore-fluid pressure and to the seepage velocity developments of both computations. In the first case of Figure 27, one obtains a high pore-fluid pressure and, as a result of contractancy, a pore-fluid that is pressed out of the shear band, whereas, in the second case of Figure 28, one observes a considerable fluid suction and, as a result, a seepage flow into the shear band. Furthermore, as a result of contractancy, the skeleton parts separated by the shear band are gliding upon each other very easily, whereas, as a result of

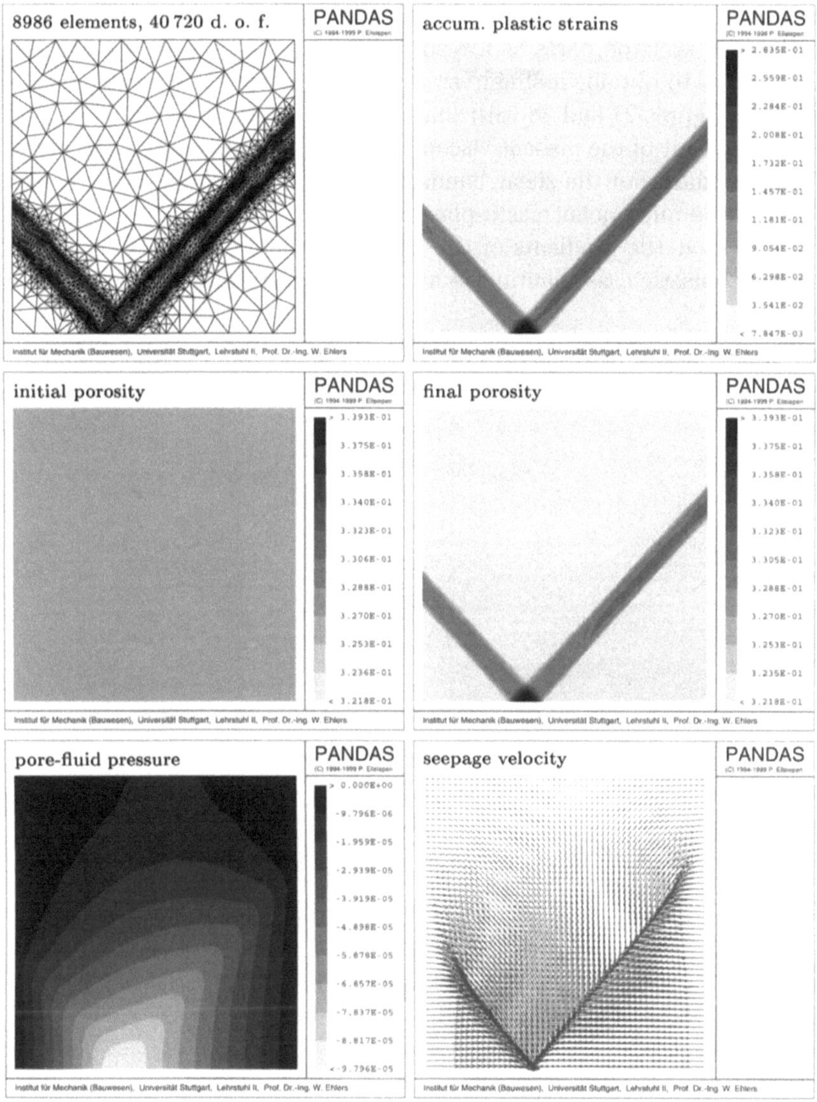

Fig. 28. Dilatant shear band development ($\sigma_1 = 110\,\mathrm{kN/m^2}$).

dilatancy, they are sucking onto each other, thus gluing the skeleton parts together.

Proceeding from the porosity of the specimen after phase 1 of the loading process is passed, the final porosity development in the shearing zones either exhibits an increase or a decrease of only 3 %. Note in passing that the final mesh obtained by use of the hierarchical procedure ranges from 22 535 d. o. f. in the contractant case to 40 720 in the dilatant one. Obviously, this is again a result of either contractancy or dilatancy, since, in the dilatant case, gluing together the skeleton parts as a result of fluid suction needs a much larger loading force to obtain the final vertical displacement. In comparison of the meshes of Figures 27 and 28 with that of Figure 23 (c), one observes that the mesh refinement of the present viscoplastic computation mainly concentrates at the boundaries of the shear bands, whereas it is approximately constant in case of the micropolar elasto-plastic material. Following this, it is finally concluded that the gradients of the plastic strains are much steeper in the elasto-viscoplastic case than in the micropolar elasto-plastic one.

7.5 The base failure problem

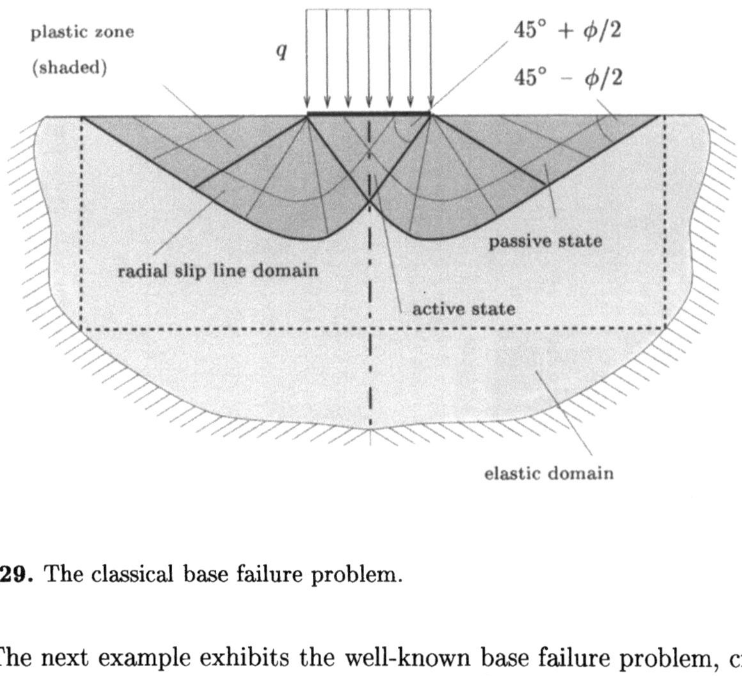

Fig. 29. The classical base failure problem.

The next example exhibits the well-known base failure problem, cf. Figure 29 and the textbook by Terzaghi & Jelinek [99]. Concerning the problem under study, an external load q of a rigid strip footing is applied onto the surface of a liquid-saturated half-space. As a result, a more or less undeformed wedge (the active state) is driven into the half-space, whereas the remainder of the shaded zone is either interspersed with plastic deformations and a field of shear bands (the radial slip line domain) or it is gliding upwards on the failure line (the passive state) which separates the plastic from the elastic domains of the overall problem. In order to solve this problem, two different strategies are applied proceeding from the assumption that the binary model either consists of a materially incompressible elasto-viscoplastic skeleton (standard formulation) or of a materially incompressible elasto-plastic micropolar skeleton (extended formulation) both saturated by an incompressible viscous pore-liquid. Assuming a rather large permeability through K_{02}^S from Table 3 leading to $k_{0S}^L = 10^{-2}\,\mathrm{m/s}$, the mathematical solution is only regularized by the included viscoplastic properties or the micropolarity of the solid skeleton.

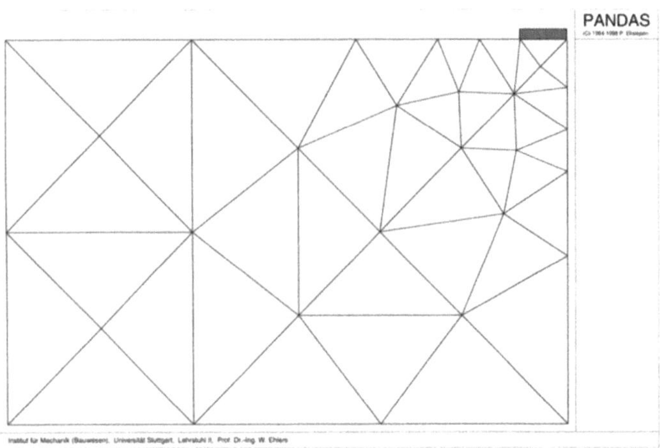

Fig. 30. Initial mesh (symmetric problem).

To get a high resolution result of this complex problem, time- and space-adaptive strategies must be applied. In both cases, the computations start with the initial mesh shown in Figure 30. Initiated by the singularity between the loaded and the non-loaded parts of the upper boundary, shear bands are initiated covering the whole area of the radial slip line domain. As a result of the symmetry condition included in the solution of the boundary-value problem, the shear bands are reflected at the symmetry line, thus delivering those shear band developments which include the final failure line, cf. e. g. Figures 31–35.

Concerning the regularization of the problem by the assumption of solid viscoplasticity, Figure 31 exhibits the final mesh development at a vertical displacement of the rigid strip footing of 48 cm. From this figure, it is furthermore not only seen that the mesh development excellently catches the shear band developments, cf. Figure 32, but also that the problem is well regularized, since the occurring shear bands are mesh-independent and the shear band widths exceed the mesh size. Concerning the quantification of the FE problem in the final state, it should be noted that the problem is characterized by 30 643 nodes, 15 254 elements, 68 981 d. o. f. and 228 810 internal variables.

On the other hand, proceeding from the regularization of the problem by the assumption of a micropolar elasto-plasticity, it is observed that the

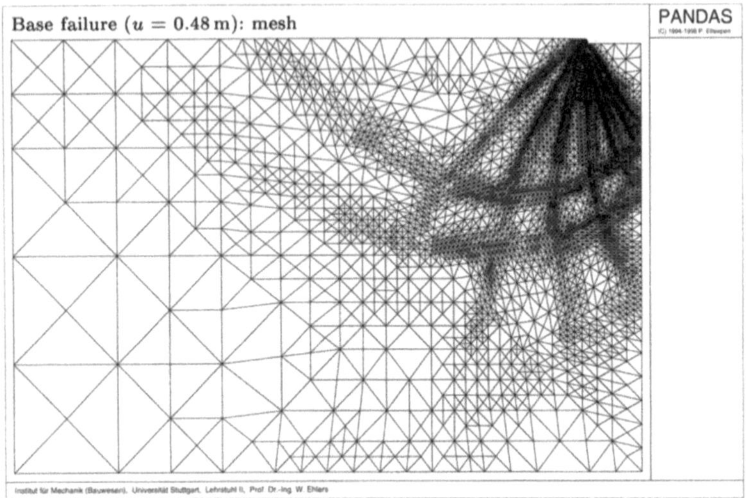

Fig. 31. Base failure problem: Final mesh (30 643 nodes, 15 254 elements, 68 981 d. o. f.).

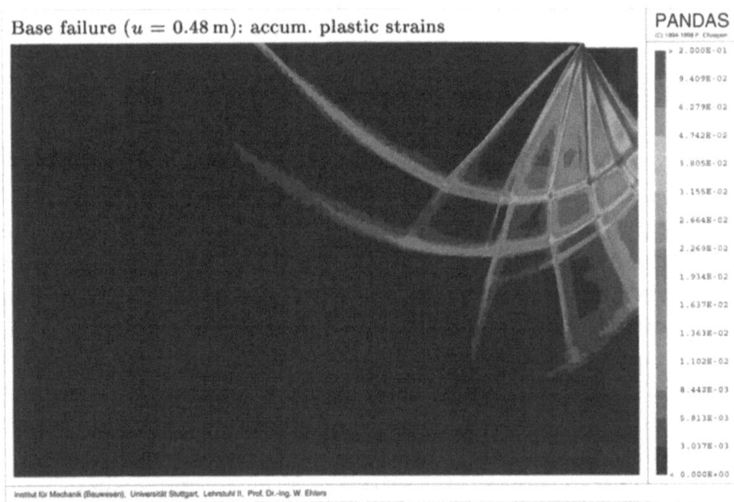

Fig. 32. Base failure problem: Accumulated plastic strains.

solution is qualitatively similar, although one obtains slightly different results. This difference, however, is mainly due to the assumption of different material properties of elasto-viscoplasticity, on the one hand, and of elasto-plasticity, on the other hand. Furthermore, according to the biaxial problem computed on the basis of micropolar material properties (cf. Figure 24), it is again seen from Figure 35 that the rotational degrees of freedom are only active in the localization zones, where the microparticles are rolling upon another, namely

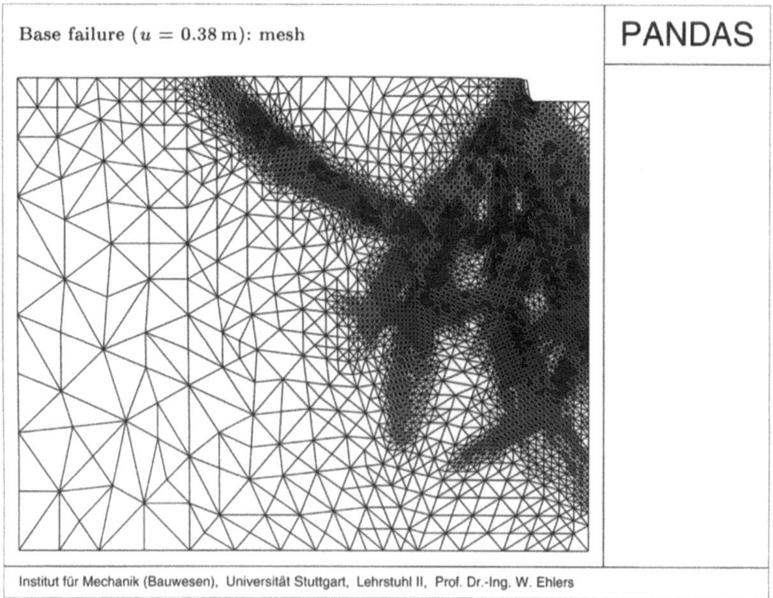

Fig. 33. Base failure problem (micropolar formulation): Final mesh (32 787 nodes, 16 306 elements, 82 056 d. o. f.).

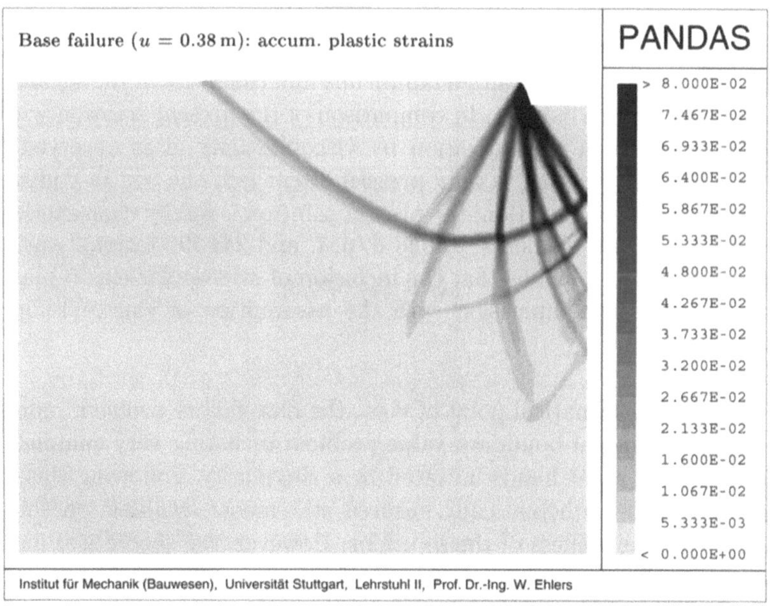

Fig. 34. Base failure problem (micropolar formulation): Accumulated plastic strains.

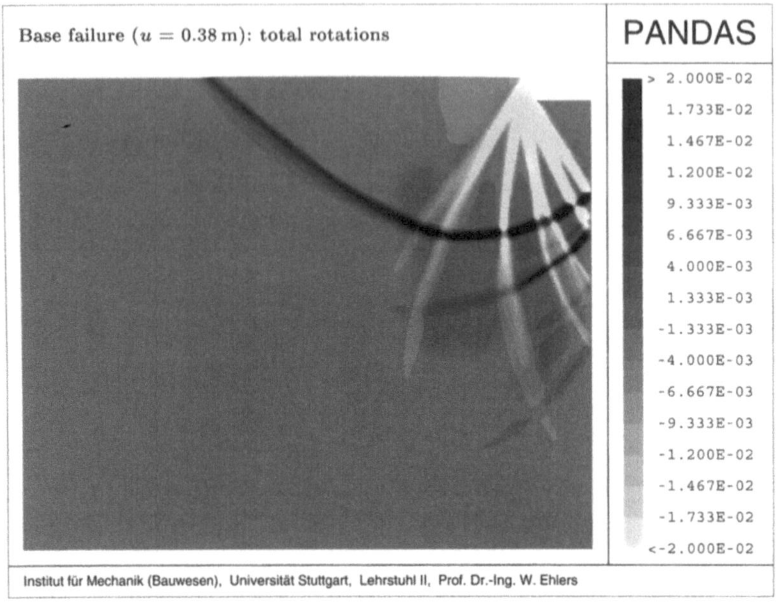

Fig. 35. Base failure problem (micropolar formulation): Total rotations.

counter-clockwise in the localized failure line and clockwise in the shear bands of the radial slip line domain. In comparison of the present solution with the first one based on a regularization by viscoplasticity, it is observed that, although the strip footing is only pressed 38 cm into the soil in contrast to 48 cm of the first computation, the present solution is finally characterized by 32 782 nodes, 16 306 elements, 82 056 d. o. f. and 244 590 internal variables. Following this, is is obvious that the inclusion of micropolar elasto-plasticity stiffens the soil in comparision with the assumption of elasto-viscoplastic material behaviour.

From the mathematical point of view, the base failure problem represents a very difficult initial boundary-value problem including very complex phenomena such as shear bands initiated by a singularity. Following this, there are generally no mathematically ensured statements available on the existence and the uniqueness of the solutions. However, the theoretical methods described in the present article nevertheless allow for a numerical computation of these problems without any theoretical *a priori* statement. Based on these remarks, the foregoing applications show that the continuum mechanical basis together with the time- and space-adaptive methods are not only convenient but also appropriate to solve a variety of practical problems occurring in the framework of geomechanical engineering.

7.6 Flow of pore-water through an embankment

The final example of this article concerns the study of a stationary pore-water flow through an embankment, cf. Figure 36, where use is made of the triphasic model consisting of an elasto-viscoplastic solid skeleton, a pore-liquid and a pore-gas. The embankment is loaded by gravitation and by a free water table of 8 m hight at the left side of the dam. Furthermore, the embankment consists of three parts, the ground level with an intrisic permeability of $K_0^S = 10^{-15}\,\mathrm{m}^2$ leading to the Darcy permeabilities $k_{0S}^L = 10^{-8}\,\mathrm{m/s}$ of the pore-water and $k_{0S}^G = 10^{-9}\,\mathrm{m/s}$ of the pore-gas. The dam itself is governed by $K_0^S = 10^{-12}\,\mathrm{m}^2$ thus leading to $k_{0S}^L = 10^{-5}\,\mathrm{m/s}$ and $k_{0S}^G = 10^{-6}\,\mathrm{m/s}$, whereas the filter at the right hand side of the embankment is governed by $K_{02}^S = 10^{-9}\,\mathrm{m}^2$, $k_{0S}^L = 10^{-2}\,\mathrm{m/s}$ and $k_{0S}^G = 10^{-6}\,\mathrm{m/s}$. The computations

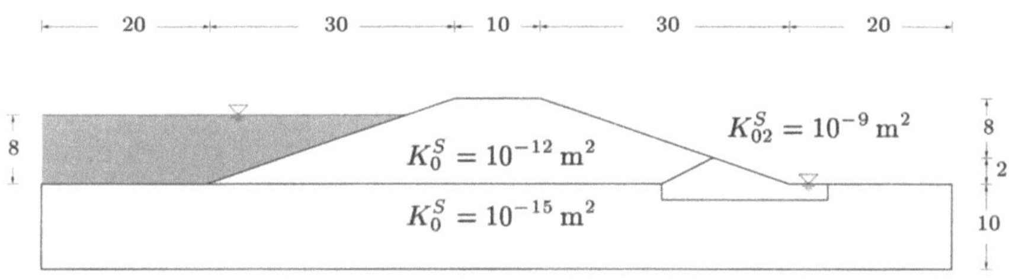

Fig. 36. Embankment with an impermeable sole [m].

have been started with the assumption of an initially empty embankment domain. Depending on the values of the intrinsic permeabilities in the different domains, one finally obtains the stationary state after a sufficient long computation time. Once this state is reached, the distribution of liquid saturation exhibits the values presented in Figure 37. Note in passing that the value of k_{0S}^L significantly influences the computation time to reach the stationary state rather than it influences the stationary state itself. Furthermore, as a result of the included capillary pressure-saturation relation, one observes a very narrow capillary zone. The values of liquid saturation are accompanied by a distribution of the excess liquid pressure ranging from $180\,\mathrm{kN/m}^2$ to $-40\,\mathrm{kN/m}^2$, cf. Figure 38. These values correspond to a water column of 18 meters at the water side of the embankment and a maximum suction head of 4 meters inside the embankment domain. It is observed that the fully saturated water table separating the fully and the partially saturated domains is clearly reproduced by the seepage velocity arrows.

Finally, Figure 39 shows the accumulated plastic strains under the external loads of gravitation and fluid flow. It is seen from this figure that the accumulated plastic strains are starting to localize and to form a shear band. Following this, it is obvious that a further increase of the water table could lead to a localized failure situation causing a destruction of the embankment corpus.

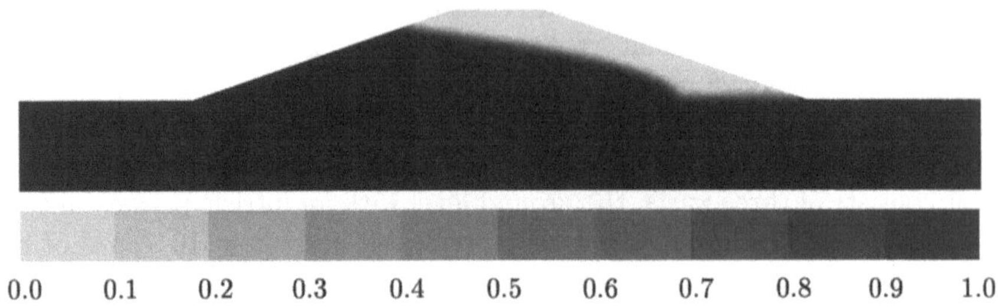

Fig. 37. Distribution of liquid saturation s^L [–].

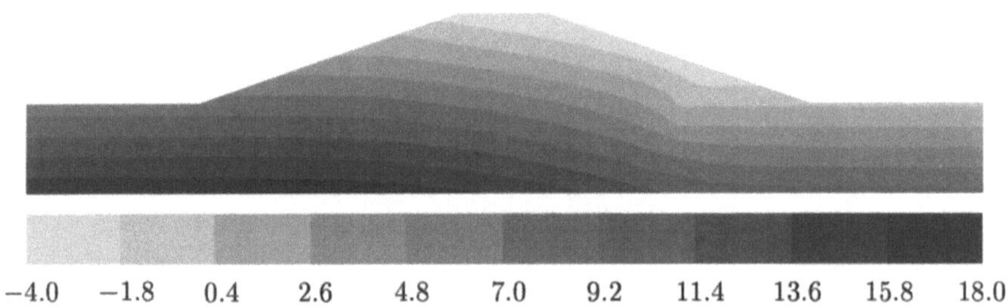

Fig. 38. Distribution of liquid pressure p^{LR} [10^4 N/m^2].

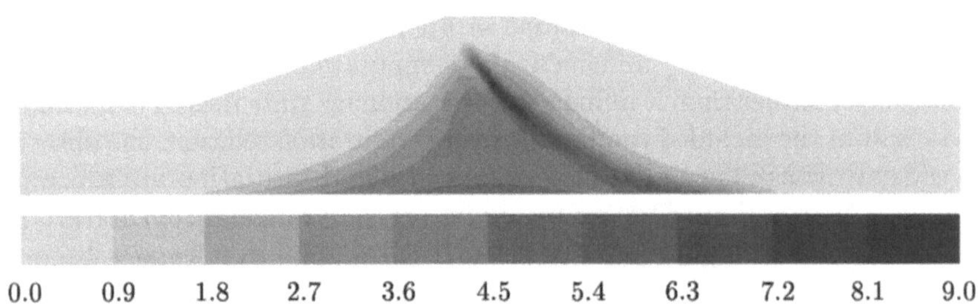

Fig. 39. Accumulated plastic strains [10^{-3}].

8 Conclusions

In the present article, the foundations of multiphasic and porous materials have been presented on the basis of the Theory of Porous Media including non-polar and micropolar constituents. In particular, fluid-saturated porous skeleton materials have been investigated in the framework of a biphasic and triphasic model including elastic, elasto-plastic and elasto-viscoplastic material properties of the solid skeleton and viscous material properties of the pore-fluids. Furthermore, the basic strategy of treating volumetrically coupled solid-fluid problems within the Finite Element Method was discussed and extended towards time- and space-adaptive methods.

Based on this approach, several numerical examples have been shown computed by use of the FE tool PANDAS. In particular, coupled solid-fluid problems have been treated including the mathematically difficult problem of shear band phenomena, where the regularization of a basically ill-posed problem, as for instance an empty skeleton material described within the standard elasto-plastic formulation, could be carried out by the inclusion of additional degrees of freedom in the sense of the Cosserat brothers, by the inclusion of elasto-viscoplastic skeleton behaviour or by viscous pore-fluid properties or by combinations of the above mechanisms.

The numerical examples demonstrate the efficiency of the proposed method. In particular, proceeding from the extended micropolar formulation opens the possibility to implicitly include the shear band width into the problem by proposing convenient numbers for the internal length scale parameter. Furthermore, since the micropolar rotations are only active in the localization zones, micropolarity is a very convenient regularization tool that does not affect the solution in the non-localizing zones. In contrast, although viscoplasticity is also very convenient to be applied as a regularization tool, is implies a principally different material behaviour and can thus only be used if the material also behaves viscoplastically in reality. On the other hand, since the inclusion of a viscous pore-fluid concerns a different mechanical problem than the description of an empty porous material, regularization by fluid viscosity is clearly restricted to volumetrically coupled saturated problems.

Furthermore, proceeding from triphasic materials consisting of a solid skeleton and two pore-fluids, a real pore-liquid and a pore-gas, opens the wide range of further realistic applications to engineering problems which could be additionally extended by the inclusion of positive and negative charges [80] to describe swelling and shrinking phenomena or by the inclusion of chemical properties [19, 87, 104] to describe, e. g., the aging of concrete.

The work presented here and the methods to describe coupled solid-fluid problems can of course not only be applied to civil engineering problems or to environmental mechanics, respectively, but also to mechanical engineering problems, as for example to porous polymeric and metallic foams as very promising new materials, cf. e. g. the work by Ehlers and Droste [36, 37] or by Ehlers and Markert [44, 47], or to biomechanical problems as for instance soft tissues or articular cartilage, cf. e. g. Mow et al. [80–82], Huyghe et al. [68] or Ehlers and Markert [45, 46]).

Acknowledgement

The work presented in this article is a result of several years of investigation at my institute. In particular, the numerical computations have been carried out by my coworkers or former coworkers Martin Ammann, Peter Blome, Peter Ellsiepen and Tobias Graf. Their work was essential for the preparation of this article and is herewith gratefully acknowledged.

References

1. Biot, M. A.: Le problème de la consolidation de matières argileuses sous une charge. *Ann. Soc. Sci. Bruxelles* B 55 (1935), 110–113.
2. Biot, M. A.: General theory of three-dimensional consolidation. *J. Appl. Phys.* **12** (1941), 155–164.
3. Biot, M. A.: Theory of propagation of elastic waves in a fluid-saturated porous solid, I. Low frequency range. *J. Acoust. Soc. Am.* **28** (1956), 168–178.
4. Bishop, A. W.: The effective stress principle. *Teknisk Ukeblad* **39** (1959), 859–863.
5. Bluhm, J.: *A consistent model for saturated and empty porous media.* Forschungsberichte aus dem Fachbereich Bauwesen, Heft 74, Universität-GH-Essen 1997.
6. de Boer, R.: *Vektor- und Tensorrechnung für Ingenieure.* Springer-Verlag, Berlin 1982.
7. de Boer, R.: Highlights in the historical development of porous media theory: toward a consistent macroscopic theory. *Applied Mechanics Review* **49** (1996), 201–262.
8. de Boer, R.: *Theory of Porous Media.* Springer-Verlag, Berlin 2000.
9. de Boer, R., Ehlers, W.: *Theorie der Mehrkomponentenkontinua mit Anwendung auf bodenmechanische Probleme.* Forschungsberichte aus dem Fachbereich Bauwesen, Heft 40, Universität-GH-Essen 1986.
10. de Boer, R., Ehlers, W.: The development of the concept of effective stresses. *Acta Mech.* **83** (1990), 77–92.
11. de Boer, R., Ehlers, W.: Uplift, friction and capillarity – three fundamental effects for liquid-saturated porous media. *Int. J. Solids Structures* **26** (1990), 43–57.
12. de Boer, R., Ehlers, W., Kowalski, S., Plischka, J.: *Porous media – a survey of different approaches.* Forschungsberichte aus dem Fachbereich Bauwesen, Heft 54, Universität-GH-Essen 1991.
13. de Borst, R.: Simulation of strain localization: A reappraisal of the Cosserat continuum. *Engineering Computations* **8** (1991), 317–332.
14. Bowen, R. M.: Theory of mixtures. In Eringen, A. C. (ed.): *Continuum Physics*, Vol. III, Academic Press, New York 1976, pp. 1–127.
15. Bowen, R. M.: Incompressible porous media models by use of the theory of mixtures. *Int. J. Engng. Sci.* **18** (1980), 1129–1148.
16. Bowen, R. M.: Compressible porous media models by use of the theory of mixtures. *Int. J. Engng. Sci.* **20** (1982), 697–735.
17. Coleman, B. D., Noll, W.: The thermodynamics of elastic materials with heat conduction and viscosity. *Arch. Rational Mech. Anal.* **13** (1963), 167–178.

18. Cosserat, E., Cosserat, F.: *Théorie des corps déformables.* A. Hermann et fils, Paris 1909. (Theory of Deformable Bodies, NASA TT F-11 561, 1968).
19. Coussy, O.: *Mechanics of Porous Continua.* Wiley, Chichester 1995.
20. Crawford, R. H., Anderson, D. C., Waggenspack, W. N.: Mesh rezoning of 2d isoparametric elements by inversion. *International Journal for Numerical Methods in Engineering* **28** (1989), 523–531.
21. Cross, J. J.: Mixtures of fluids and isotropic solids. *Arch. Mech.* **25** (1973), 1025–1039.
22. Delesse, M.: Pour déterminer la composition des roches. *Annales des mines, 4. séries* **13** (1848), 379–388.
23. Diebels, S.: *Mikropolare Zweiphasenmodelle: Formulierung auf der Basis der Theorie Poröser Medien.* Habilitation, Bericht Nr. II-4 aus dem Institut für Mechanik (Bauwesen), Universität Stuttgart 2000.
24. Diebels, S., Ehlers, W.: Dynamic analysis of a fully saturated porous medium accounting for geometrical and material non-linearities. *Int. J. Numer. Methods Engng.* **39** (1996), 81–97.
25. Diebels, S., Ehlers, W.: On fundamental concepts of multiphase micropolar materials. *Technische Mechanik*, 16:77–88 (1996).
26. Diebels, S., Ellsiepen, P., Ehlers, W.: Error-controlled *Runge-Kutta* time integration of a viscoplastic hybrid two-phase model. *Technische Mechanik* **19** (1999), 19–27.
27. Ehlers, W.: On thermodynamics of elasto-plastic porous media. *Arch. Mech.* **41** (1989), 73–93.
28. Ehlers, W.: *Poröse Medien – ein kontinuumsmechanisches Modell auf der Basis der Mischungstheorie.* Forschungsberichte aus dem Fachbereich Bauwesen, Heft 47, Universität-GH-Essen 1989.
29. Ehlers, W.: Toward finite theories of liquid-saturated elasto-plastic porous media. *Int. J. Plasticity* **7** (1991), 443–475.
30. Ehlers, W.: Compressible, incompressible and hybrid two-phase models in porous media theories. In Angel, Y. C. (ed.): *Anisotropy and Inhomogeneity in Elasticity and Plasticity*, AMD-Vol. 158. The American Society of Mechanical Engineers, New York 1993, pp. 25–38.
31. Ehlers, W.: Constitutive equations for granular materials in geomechanical context. In Hutter, K. (ed.): *Continuum Mechanics in Environmental Sciences and Geophysics*, CISM Courses and Lectures No. 337. Springer-Verlag, Wien 1993, pp. 313–402.
32. Ehlers, W.: Grundlegende Konzepte in der Theorie Poröser Medien. *Technische Mechanik* **16** (1996), 63–76.
33. Ehlers, W.: Continuum and numerical simulation of porous materials in science and technology. In Capriz, G., Ghionna, V. N., Giovine, G. (eds.): *Modelling and Mechanics of Granular and Porous Materials*, Modelling and Simulation in Science, Engineering and Technology Series, Birkhäuser Verlag, Basel 2002, pp. 243–289.
34. Ehlers, W., Ammann, M., Diebels, S.: *h*-Adaptive FE methods applied to single- and multiphase problems. *International Journal for Numerical Methods in Engineering* **54** (2002), 219–239.
35. Ehlers, W., Blome, P.: A triphasic model for unsaturated soils based on the theory of porous media. *Mathematical and Computer Modelling*, submitted.
36. Ehlers, W., Droste, A.: A continuum model for highly porous aluminium foam. *Technische Mechanik* **19** (1999), 341–350.

37. Ehlers, W., Droste, A.: FE simulations of metal foams based on the macroscopic approach of the Theory of Porous Media. In Banhart, J., Ashby, M., Fleck, N. (eds.): *Proceedings of the International Conference of Metal Foams and Porous Metal Structures*. Verlag MIT, Bremen 1999, pp. 299–302.
38. Ehlers, W., Ellsiepen, P.: Adaptive Zeitintegrations-Verfahren für ein elastisch-viskoplastisches Zweiphasenmodell. *ZAMM* **78** (1998), S361–S362.
39. Ehlers, W., Ellsiepen, P.: PANDAS: Ein FE-System zur Simulation von Sonderproblemen der Bodenmechanik. In Wriggers, P., Meißner, U., Stein, E., Wunderlich, W. (eds.): *Finite Elemente in der Baupraxis: Modellierung, Berechnung und Konstruktion*. Ernst & Sohn, Berlin 1998, pp. 391–400. Beiträge zur Tagung FEM '98 an der TU Darmstadt am 5. und 6. März 1998.
40. Ehlers, W., Ellsiepen, P.: Zeit- und ortsadaptive Verfahren zur Berechnung von Scherbändern in porösen Materialien. In Mahnken, R. (ed.): *Theoretische und Numerische Methoden in der Angewandten Mechanik mit Praxisbeispielen*. Forschungs- und Seminarberichte aus dem Bereich der Mechanik der Universität Hannover, Bericht-Nr. F98/4, 1998, pp. 99–106. Festschrift anläßlich der Emeritierung von Herrn Prof. Dr.-Ing. Dr.-Ing. E. h. E. Stein.
41. Ehlers, W., Ellsiepen, P.: Theoretical and numerical methods in environmental continuum mechanics based on the theory of porous media. In Schrefler, B. A. (ed.): *Environmental Geomechanics*, CISM Courses and Lectures No. 417. Springer-Verlag, Wien 2001, pp. 1–81.
42. Ehlers, W., Ellsiepen, P., Ammann, M.: Time- and space-adaptive methods applied to localization phenomena in empty and saturated micropolar and standard porous materials. *International Journal for Numerical Methods in Engineering* **52** (2001), 503–526.
43. Ehlers, W., Ellsiepen, P., Blome, P., Mahnkopf, D., Markert, B.: *Theoretische und numerische Studien zur Lösung von Rand- und Anfangswertproblemen in der Theorie Poröser Medien, Abschlußbericht zum DFG-Forschungsvorhaben Eh 107/6-2*. Bericht aus dem Institut für Mechanik (Bauwesen), Nr. 99-II-1, Universität Stuttgart 1999.
44. Ehlers, W., Markert, B.: On the viscoelastic behaviour of fluid-saturated materials. *Granular Matter* **2** (2000), 153–161.
45. Ehlers, W., Markert, B.: A linear viscoelastic biphasic model for soft tissues based on the theory of porous media. *ASME Journal of Biomechanical Engineering* **123** (2001), 418–424.
46. Ehlers, W., Markert, B.: A visco-elastic two-phase model for articular cartilage tissues. In Ehlers, W. (ed.): *IUTAM Symposium on Theoretical and Numerical Methods in Continuum Mechanics of Porous Materials*. Kluwer, Dordrecht 2001, pp. 87–92.
47. Ehlers, W., Markert, B.: A macroscopic finite strain model for cellular polymers. *International Journal of Plasticity*, in press.
48. Ehlers, W., Markert, B., Klar, O.: Biphasic description of viscoelastic foams by use of an extended Ogden-type formulation. In Ehlers, W., Bluhm, J. (eds.): *Porous Media: Theory, Experiments and Numerical Applications*. Springer-Verlag, Berlin 2002, pp. 275–294.
49. Ehlers, W., Müllerschön, H.: Stress-strain behaviour of cohesionless soils: Experiments, theory and numerical computations. In Cividini, A. (ed.): *Application of Numerical Methods to Geotechnical Problems*, CISM Courses and Lectures No. 397. Springer-Verlag, Wien 1998, pp. 675–684.

50. Ehlers, W., Volk, W.: On shear band localization phenomena induced by elasto-plastic consolidation of fluid-saturated soils. In Owen, D. R. J., Oñate, E., Hinton, E. (eds.): *Computational Plasticity – Fundamentals and Applications.* CIMNE, Barcelona 1997, pp. 1657–1664.
51. Ehlers, W., Volk, W.: On shear band localization phenomena of liquid-saturated granular elasto-plastic porous solid materials accounting for fluid viscosity and micropolar solid rotations. *Mech. Cohes.-Frict. Mater.* **2** (1997), 301–320.
52. Ehlers, W., Volk, W.: On theoretical and numerical methods in the theory of porous media based on polar and non-polar elasto-plastic solid materials. *Int. J. Solids Structures* **35** (1998), 4597–4617.
53. Ehlers, W., Volk, W.: Localization phenomena in liquid-saturated and empty porous solids. *Transport in Porous Media* **34** (1999), 159–177.
54. Eipper, G.: *Theorie und Numerik finiter elastischer Deformationen in fluidgesättigten porösen Medien.* Dissertation, Bericht Nr. II-1 aus dem Institut für Mechanik (Bauwesen), Universität Stuttgart 1998.
55. Ellsiepen, P.: *Zeit- und ortsadaptive Verfahren angewandt auf Mehrphasenprobleme poröser Medien.* Dissertation, Bericht Nr. II-3 aus dem Institut für Mechanik (Bauwesen), Universität Stuttgart 1999.
56. Eringen, A. C., Kafadar, C. B.: Polar field theories. In Eringen, A. C. (ed.): *Continuum Physics*, Vol. VI. Academic Press, New York 1976, pp. 1–73.
57. Gallimard, L., Ladevéze, P., Pelle, J. P.: Error estimation and adaptivity in elastoplasticity. *Int. J. Numer. Methods Engng.* **39** (1996), 129–217.
58. Günther, W.: Zur Statik und Kinematik des Cosseratschen Kontinuums. *Abh. Braunschweig. Wiss. Ges.* **10** (1958), 195–213.
59. Hairer, E., Lubich, C., Roche, M.: *The Numerical Solution of Differential-Algebraic Equations by Runge-Kutta Methods.* Springer-Verlag, Berlin 1989.
60. Hairer, E., Wanner, G.: *Solving Ordinary Differential Equations, Vol. 2: Stiff and Differential-Algebraic Problems.* Springer-Verlag, Berlin 1991.
61. Hassanizadeh, S. M., Gray, W. G.: General conservation equations for multiphase systems: 1. Averaging procedure. *Adv. Water Resources* **2** (1979), 131–144.
62. Hassanizadeh, S. M., Gray, W. G.: General conservation equations for multiphase systems: 2. Mass, momentum, energy and entropy equations. *Adv. Water Resources* **2** (1979), 191–203.
63. Haupt, P.: Foundation of continuum mechanics. In Hutter, K. (ed.): *Continuum Mechanics in Environmental Sciences and Geophysics*, CISM Courses and Lectures No. 337. Springer-Verlag, Wien 1993, pp. 1–77.
64. Haupt, P.: *Continuum Mechanics and Theory of Materials.* Springer-Verlag, Berlin 2000.
65. Heinrich, G., Desoyer, K.: Hydromechanische Grundlagen für die Behandlung von stationären und instationären Grundwasserströmungen. *Ing.-Archiv* **23** (1955), 182–185.
66. Heinrich, G., Desoyer, K.: Hydromechanische Grundlagen für die Behandlung von stationären und instationären Grundwasserströmungen, II. Mitteilung. *Ing.-Archiv* **24** (1956), 81–84.
67. Heinrich, G., Desoyer, K.: Theorie dreidimensionaler Setzungsvorgänge in Tonschichten. *Ing.-Archiv* **30** (1961), 225–253.

68. Huyghe, J. M., Jansson, C. F., Lanier, Y., von Donkelaar, C. C., Maroudas, A., van Campen, D. H.: Experimental measurement of electrical conductivity and electro-osmotic permeability of ionised porous media. In Ehlers, W., Bluhm, J. (eds.): *Porous Media: Theory, Experiments and Numerical Applications*. Springer-Verlag, Berlin 2002, pp. 295–313.
69. Kafadar, C. B., Eringen, A. C.: Micropolar media - I: the classical theory. *Int. J. Engng. Sci.* **9** (1971), 271–305.
70. Kossaczký, I.: A recursive approach to local mesh refinement in two and three dimensions. *J. Comp. Appl. Math.* **55** (1994), 275–288.
71. Krause, R., Rank, E.: A fast algorithm for point-location in a finite element mesh. *Computing* **57** (1996), 49–62.
72. Lade, P., de Boer, R.: The concept of effective stress for soil, concrete and rock. *Géotechnique* **47** (1997), 61–78.
73. Ladevèse, P., Pelle, J. P., Rougeot, P.: Error estimation and mesh optimization for classic finite elements. *Eng. Comp.* **8** (1991), 69–80.
74. Lewis, R. W., Schrefler, B. A.: *The Finite Element Method in the Static and Dynamic Deformation and Consolidation of Porous Media*, 2nd Edition. Wiley, Chichester 1998.
75. Liu, I-S.: Method of Lagrange multipliers for exploitation of the entropy principle. *Arch. Rational Mech. Anal.* **46** (1972), 131–148.
76. Liu, I-S., Müller, I.: Thermodynamics of mixtures of fluids. In Truesdell, C. (ed.): *Rational Thermodynamics*, 2nd Edition. Springer-Verlag, New York 1984, pp. 264–285.
77. Mahnkopf, D.: *Lokalisierung fluidgesättigter poröser Festkörper bei finiten elastoplastischen Deformationen*. Dissertation, Bericht Nr. II-5 aus dem Institut für Mechanik (Bauwesen), Universität Stuttgart 2000.
78. Miehe, C., Schröder, J., Schotte, J.: Computational homogenization analysis in finite plasticity. Simulation of texture development in polycrystalline materials. *Computer Methods in Applied Mechanics and Engineering* **171** (1999), 387–418.
79. Mitchell, W. F.: Adaptive refinement for arbitrary finite-element spaces with hierarchical bases. *J. Comp. Appl. Math.* **36** (1991), 65–78.
80. Mow, V. C., Ateshian, G. A., Lai, W. M., Gu, W. Y.: Effects on fixed charges on the stress-relaxation behavior of hydrated soft tissues in a confined compression problem. *Int. J. Solids Structures* **35** (1998), 4945–4962.
81. Mow, V. C., Gibbs, M. C., Lai, W. M., Zhu, W. B., Athanasiou, K. A.: Biphasic indentation of articular cartilage - II. *J. Biomechanics* **22** (1989), 853–861.
82. Mow, V. C., Sun, D. D., Guo, X. E., Likhitpanichkul, M., Lai, W. M.: Fixed negative charges modulate mechanical behaviours and electrical signals in articular cartilage under confined compression. In Ehlers, W., Bluhm, J. (eds.): *Porous Media: Theory, Experiments and Numerical Applications*. Springer-Verlag, Berlin 2002, pp. 227–247.
83. Müllerschön, H.: *Spannungs-Verzerrungsverhalten granularer Materialien am Beispiel von Berliner Sand*. Dissertation, Bericht Nr. II-6 aus dem Institut für Mechanik (Bauwesen), Universität Stuttgart 2000.
84. Nowacki, W.: *Theory of Asymmetric Elasticity*. Pergamon Press, Oxford 1986.
85. Perzyna, P.: Fundamental problems in viscoplasticity. *Adv. Appl. Mech. Eng.* **9** (1966), 243–377.

86. Plischka, J.: *Die Bedeutung der Durchschnittsbildungstheorie für die Theorie poröser Medien.* Dissertation, Fachbereich Bauwesen, Universität-GH-Essen 1992.
87. Schrefler, B. A.: Modelling of subsidence due to water or hydrocarbon withdraw from the subsoil. In Schrefler, B. A. (ed.): *Environmental Geomechanics,* CISM Courses and Lectures No. 417. Springer-Verlag, Wien 2001, pp. 235–301.
88. Schrefler, B. A., Scotta, R.: A fully coupled model for two-pase flow in deformable porous media. *Comp. Methods Appl. Mech. Engrg.* **190** (2001), 3223–3246.
89. Schrefler, B. A., Simoni, L., Xikui, L., Zienkiewicz, O. C.: Mechanics of partially saturated porous media. In Desai, C. S., Gioda, G. (eds.): *Numerical Methods and Constitutive Modelling in Geomechanics,* CISM Courses and Lectures No. 311. Springer-Verlag, Wien 1990, pp. 169–209.
90. Schrefler, B. A., Zhan, X.: A fully coupled model for water flow and air flow in deformable porous media. *Water Res. Research* **29** (1993), 155–167.
91. Schröder, J.: *Homogenisierungsmethoden der nichtlinearen Kontinuumsmechanik unter Beachtung von Stabilitätsproblemen.* Habilitation, Bericht Nr. I-7 aus dem Institut für Mechanik (Bauwesen), Universität Stuttgart 2000.
92. Shewchuk, J. R.: *Triangle: A Two-Dimensional Quality Mesh Generator and Delaunay Triangulator.* School of Computer Science, Carnegie Mellon University, Pittsburgh, Pensilvania 1996. `http://www.cs.cmu.edu/~quake/triangel.html`.
93. Skempton, A. W.: Significance of Terzaghi's concept of effective stress (Terzaghi's discovery of effective stress). In Bjerrum, L., Casagrande, A., Peck, R. B., Skempton, A. W. (eds.): *From Theory to Practice in Soil Mechanics.* Wiley, New York 1960, pp. 42–53.
94. Steinmann, P.: A micropolar theory of finite deformation and finite rotation multiplicative elasto-plasticity. *Int. J. Solids Structures* **31** (1994), 1063–1084.
95. Suquet, P. M.: Elements of homogenization for inelastic solid mechanics. In Sanches-Palencia, E., Zaoui, A. (eds.): *Homogenization techniques for composite media.* Lecture Notes in Physics, Springer-Verlag, Berlin 1987, pp. 193–277.
96. Svendsen, B., Hutter, K.: On the thermodynamics of a mixture of isotropic materials with constraints. *Int. J. Engng Sci.* **33** (1995), 2021–2054.
97. Terzaghi, K.: Die Berechnung der Durchlässigkeitsziffer des Tones aus dem Verlauf der hydrodynamischen Spannungserscheinungen. *Sitzungsber. Akad. Wiss. Wien, Math.-Naturwiss. Kl., Abt. IIa* **132** (1923), 125–138.
98. Terzaghi, K.: *Erdbaumechanik auf bodenphysikalischer Grundlage.* Franz Deuticke, Leipzig 1925.
99. Terzaghi, K., Jelinek, R.: *Theoretische Bodenmechanik.* Springer-Verlag, Berlin 1954.
100. Truesdell, C.: Sulle basi delle termomeccanica. *Rend. Lincei* **22** (1957), 158–166.
101. Truesdell, C.: *Rational Thermodynamics,* 2nd Edition. Springer-Verlag, New York 1984.
102. Truesdell, C.: Thermodynamics of diffusion. In Truesdell, C. (ed.): *Rational Thermodynamics.* 2nd Edition, Springer-Verlag, New York 1984, pp. 219–236.
103. Truesdell, C., Toupin, R. A.: The classical field theories. In Flügge, S. (ed.): *Handbuch der Physik,* Vol. III/1, Springer-Verlag, Berlin 1960, pp. 226–902.

104. Ulm, F.-J., Coussy, O.: Environmental chemomechanics of concrete. In Schrefler, B. A. (ed.): *Environmental Geomechanics*, CISM Courses and Lectures No. 417. Springer-Verlag, Wien 2001, pp. 301–350.
105. van Genuchten, M. T.: A closed-form equation for predicting the hydraulic conductivity of unsaturated soils. *Soil Science Society of America Journal* **44** (1980), 892–898.
106. Volk, W.: *Untersuchung des Lokalisierungsverhaltens mikropolarer poröser Medien mit Hilfe der Cosserat-Theorie*. Dissertation, Bericht Nr. II-2 aus dem Institut für Mechanik (Bauwesen), Universität Stuttgart 1999.
107. Woltman, R.: *Beyträge zur Hydraulischen Architektur*. Dritter Band, Johann Christian Dietrich, Göttingen 1794.
108. Zienkiewicz, O. C., Zhu, J. Z.: A simple error estimator and adaptive procedure for practical engineering analysis. *Int. J. Numer. Methods Engng.* **24** (1987), 337–357.

Modelling of saturated thermo-elastic porous solids with different phase temperatures

Joachim Bluhm

University of Essen, Institute of Mechanics, 45117 Essen, Germany

Abstract. Based on the Theory of Porous Media (TPM), a binary model for the description of saturated thermo-elastic porous solids with different phase temperatures will be presented. The constituents solid and fluid can be compressible or incompressible, i. e. the binary model discussed here includes the compressible model, the hybrid models of first and second type and the incompressible model. For the four different binary models the field equations, the constitutive relations and the dissipation mechanism will be developed and discussed.

1 Introduction

In view of engineering problems, fluid-saturated porous media can be described within the framework of a macroscopic point of view with the well-known and well-founded Theory of Porous Media (TPM), e. g. see Bowen [8, 9], Ehlers [12–14] and de Boer [4]. In this article, a consistent binary model consisting of a thermo-elastic porous solid and a non-viscous fluid will be presented, where different temperatures of the constituents will be considered. Mass exchange between the phases will be neglected. The constituents can be compressible or incompressible. Thus, the binary model includes the compressible model (compressible solid and compressible fluid), the hybrid models of first and second type (compressible solid, incompressible fluid and incompressible solid, compressible fluid) and the incompressible model (incompressible solid, incompressible fluid).

After a short discussion of the describing field equations of the binary model (concept of volume fraction, saturation condition and balance equations) the entropy inequality for the porous medium will be evaluated, i. e. restrictions concerning the constitutive relations and the dissipation mechanism of the model will be developed and discussed. The main part of the article works on the formulation of constitutive relations for the free *Helmholtz* energy functions and the stresses of the solid and fluid phase.

With respect to the description of compressible and incompressible porous solids, a constitutive relation will be developed. This law is based on the introduction of a material parameter which describes the compression point of the porous solid. For empty compressible porous solids the law presented here can be converted into the well-known constitutive relation for compressible one-component materials of Simo and Pister [20].

Concerning the compressible fluid phase it will be assumed that the behaviour fluid is like an ideal gas. In this case the pore fluid pressure depends on the free *Helmholtz* energy of the fluid. If the fluid is incompressible, the model shows that the fluid pressure is determined by the real and partial volume deformations of the porous solids. For the incompressible model the pore fluid pressure is an indeterminate reaction force in the model.

2 Basics of the porous media theory

Today, the most widely used theory for describing the thermodynamic behaviour of empty or saturated porous solids is the Theory of Porous Media, i. e. mixture theory in consideration of the volume fraction concept. This concept is understood as the determination of the fraction of a body occupied by a constituent with the local ratio of the partial volume (volume of the real material of the corresponding constituent) in relation to the total volume. Within the framework of the volume fraction concept it will be assumed that the porous solid always models a control space in which the pores are filled with liquids and/or gases. Furthermore, it is assumed that the pores are statistically distributed.

The basis of the description of porous media, using the theory of mixtures restricted by the volume fraction concept, is the model of a macroscopic body, where neither a geometrical interpretation of the pore structure nor the exact location of the individual components of the body (constituents) are considered. Thus, one proceeds on the assumption that the constituents, which are bound together, are smeared over the control space, which the porous solid spans, i. e. each substitute constituent occupies the total volume of the control space simultaneously with the other constituents. This approach is based on the concept that the constituents are statistically distributed over the control space, see Figure 1. Thus, all geometrical and physical quantities of the constituents φ^α such as motion, deformation and stress are defined in the total control space and they can be interpreted as the statistically average values of the true quantities of φ^α.

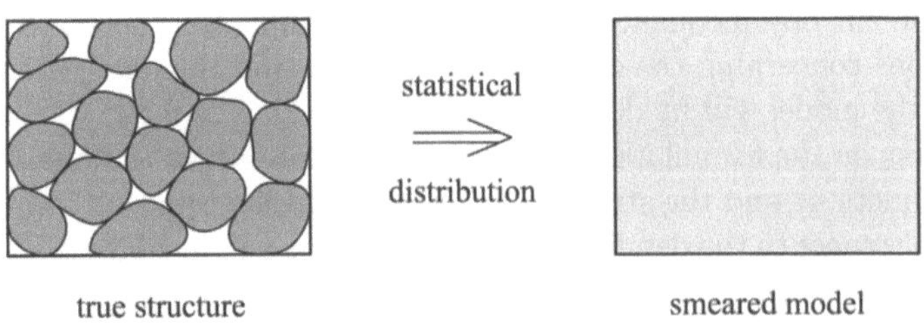

Fig. 1. Illustration of the statistical distribution of a binary porous medium consisting of a granular solid phase and a gas phase.

2.1 The concept of volume fractions

In the actual placement a porous medium occupying the deformed control space of the porous solid B_S with the boundary ∂B_S at time t consists of κ constituents φ^α with the partial volumes V^α. The boundary ∂B_S is a material surface for the solid phase and a non-material surface for the fluid phases (liquid and/or gas phases). In order to formulate the volume fraction concept, the existence of the volume fraction

$$n^\alpha = n^\alpha(\mathbf{x}, t) = \frac{\mathrm{d}v^\alpha}{\mathrm{d}v} \tag{1}$$

is assumed, where $\mathbf{x}$ is the position vector to the spatial point x. This point is simultaneously occupied with material points X_α of all κ constituents φ^α. The quantities $\mathrm{d}v = \mathrm{d}v(\mathbf{x}, t)$ and $\mathrm{d}v^\alpha = \mathrm{d}v^\alpha(\mathbf{x}, t) = n^\alpha\,\mathrm{d}v$ denote the volume element and the partial volume element of the constituent φ^α referred to the actual placement. The volume fraction concept can be founded in micromechanics, see e. g. de Boer and Didwania [6].

The partial volume of φ^α in the region B_S is defined by

$$V^\alpha = \int_{B_S} n^\alpha\,\mathrm{d}v = \int_{B_S} \mathrm{d}v^\alpha\,. \tag{2}$$

Thus, the total volume V of B_S with the boundary ∂B_S can be expressed as

$$V = \int_{B_S} \mathrm{d}v = \sum_{\alpha=1}^{\kappa} V^\alpha = \int_{B_S} \sum_{\alpha=1}^{\kappa} \mathrm{d}v^\alpha = \int_{B_S} \sum_{\alpha=1}^{\kappa} n^\alpha\,\mathrm{d}v\,, \tag{3}$$

and, corresponding to the aforementioned derivations, the volume fraction condition

$$\sum_{\alpha=1}^{\kappa} n^\alpha = 1 \tag{4}$$

in the actual placement is gained.

The consideration of the volume fraction concept in the mixture theory is connected with the introduction of the so called real densities of the constituents φ^α. Therefore, the total mass of the porous medium of actual placement of the solid at time t will be introduced:

$$M = \sum_{\alpha=1}^{\kappa} M^\alpha = \int_{B_S} \sum_{\alpha=1}^{\kappa} \rho^\alpha\,\mathrm{d}v = \int_{B_S} \sum_{\alpha=1}^{\kappa} \frac{\rho^\alpha}{n^\alpha}\,\mathrm{d}v^\alpha = \int_{B_S} \sum_{\alpha=1}^{\kappa} \rho^{\alpha R}\,\mathrm{d}v^\alpha\,, \tag{5}$$

where $\rho^\alpha = \rho^\alpha(\mathbf{x}, t)$ and $\rho^{\alpha R} = \rho^{\alpha R}(\mathbf{x}, t)$ denote the bulk or partial density and the real density of the constituent φ^α at the position $\mathbf{x}$ at time t. The partial density is related to the real density of φ^α via the volume fraction n^α:

$$\rho^\alpha = n^\alpha\,\rho^{\alpha R}\,. \tag{6}$$

It is worth mentioning that the volume fraction condition (4) is not only a mathematical expression with no special physical meaning but also a significant statement, if the porous solid is saturated. In this case the volume fraction condition represents the so-called saturation condition and must be considered as a constraint concerning the motion of the individual constituents φ^α, e. g. see Bluhm [3] and de Boer [4].

2.2 Kinematics

Within the framework of the general Porous Media Theory, a fluid saturated porous solid will be treated as an immiscible mixture of the constituents φ^α with particles X_α, where at any time t each spatial point of the current placement of the solid phase is simultaneously occupied by particles X_α of the constituents φ^α. These particles proceed from different reference positions $\mathbf{X}_\alpha$ at time $t = t_0$. Thus, each constituent is assigned its own independent motion function, e. g. see de Boer and Ehlers [7] and Ehlers [12]:

$$\mathbf{x} = \chi_\alpha(\mathbf{X}_\alpha, t)\,, \tag{7}$$

where $\mathbf{x}$ is element of the control space of the porous solid at time t ($\mathbf{x} \in B_S$). In general, the reference position $\mathbf{X}_F$ of the fluid phase must not be element of the reference placement of the solid at time $t = t_0$. Only for deformation processes in which the fluid phase leaves the control space of the solid, is the reference position $\mathbf{X}_F$ element of B_{0S}. A geometrical interpretation of the function of motion (7) concerning the motion of a solid and a fluid particle is shown in Figure 2.

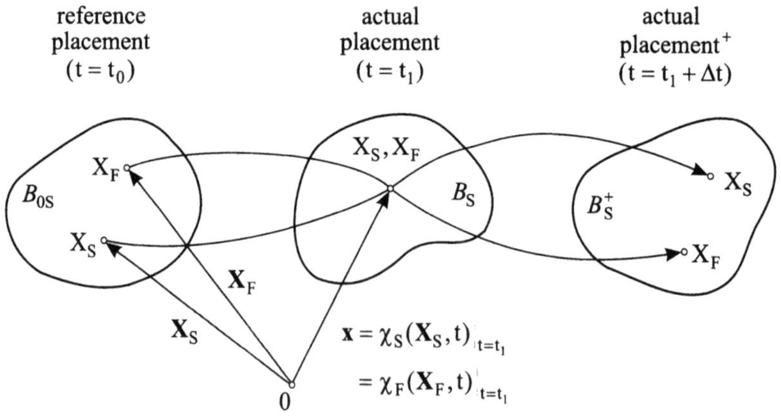

Fig. 2. Illustration of the motion of a solid and a fluid particle in a fluid saturated porous solid.

By using the *Lagrange*an description of motion, see (7), the velocity and the acceleration field of the constituent φ^α are defined as

$$\overset{\prime}{\mathbf{x}}_\alpha = \frac{\partial \chi_\alpha(\mathbf{X}_\alpha, t)}{\partial t}, \quad \overset{\prime\prime}{\mathbf{x}}_\alpha = \frac{\partial^2 \chi_\alpha(\mathbf{X}_\alpha, t)}{\partial t^2}. \tag{8}$$

The function χ_α is postulated to be unique and uniquely invertible at any time t. The existence of a function inverse to (7) leads to the *Euler*ian description of motion, viz.

$$\mathbf{X}_\alpha = \chi_\alpha^{-1}(\mathbf{x}, t). \tag{9}$$

With the help of (9) one obtains an alternative representation of the velocity and acceleration field:

$$\begin{aligned} \overset{\prime}{\mathbf{x}}_\alpha &= \overset{\prime}{\mathbf{x}}_\alpha(\mathbf{x}, t) = \overset{\prime}{\mathbf{x}}_\alpha[\chi_\alpha^{-1}(\mathbf{x}, t), t], \\ \overset{\prime\prime}{\mathbf{x}}_\alpha &= \overset{\prime\prime}{\mathbf{x}}_\alpha(\mathbf{x}, t) = \overset{\prime\prime}{\mathbf{x}}_\alpha[\chi_\alpha^{-1}(\mathbf{x}, t), t]. \end{aligned} \tag{10}$$

A mathematically sufficient condition for the existence of (9) is given, if the *Jacobi*an

$$J_\alpha = \det \mathbf{F}_\alpha \tag{11}$$

differs from zero. In (11) $\mathbf{F}_\alpha$ is the deformation gradient of the constituent φ^α, which is defined as

$$\mathbf{F}_\alpha = \frac{\partial \chi_\alpha(\mathbf{X}_\alpha, t)}{\partial \mathbf{X}_\alpha} = \operatorname{Grad}_\alpha \chi_\alpha. \tag{12}$$

The inverse of the deformation gradient $\mathbf{F}_\alpha$ is given by

$$\mathbf{F}_\alpha^{-1} = \frac{\partial \chi_\alpha^{-1}(\mathbf{x}, t)}{\partial \mathbf{x}} = \frac{\partial \mathbf{X}_\alpha}{\partial \mathbf{x}} = \operatorname{grad} \mathbf{X}_\alpha. \tag{13}$$

The material and spatial velocity gradient of the $\mathbf{F}_\alpha$ read as follows:

$$\begin{aligned} (\mathbf{F}_\alpha)'_\alpha &= \frac{\partial \overset{\prime}{\mathbf{x}}_\alpha}{\partial \mathbf{X}_\alpha} = \operatorname{Grad}_\alpha \overset{\prime}{\mathbf{x}}_\alpha, \\ \mathbf{L}_\alpha &= (\operatorname{Grad}_\alpha \overset{\prime}{\mathbf{x}}_\alpha)\, \mathbf{F}_\alpha^{-1} = \operatorname{grad} \overset{\prime}{\mathbf{x}}_\alpha. \end{aligned} \tag{14}$$

The spatial velocity gradient can be additively decomposed into a symmetric and a skew-symmetric part,

$$\mathbf{L}_\alpha = \mathbf{D}_\alpha + \mathbf{W}_\alpha, \tag{15}$$

where

$$\mathbf{D}_\alpha = \frac{1}{2}(\mathbf{L}_\alpha + \mathbf{L}_\alpha^T), \quad \mathbf{W}_\alpha = \frac{1}{2}(\mathbf{L}_\alpha - \mathbf{L}_\alpha^T) \tag{16}$$

denote the symmetric and the skew-symmetric part of the spatial velocity gradient $\mathbf{L}_\alpha$.

2.3 Balance equations and entropy inequality

The balance equations of mass, momentum, moment of momentum, and energy for a mixture consisting of κ constituents in connection with the corresponding balance equations for each constituent have been discussed at length, e. g. by Truesdell and Toupin [24], Green and Naghdi [15], Müller [18], Bowen and Wiese [10], and de Boer and Ehlers [7].

In order to derive the balance equations of the mixture theory, balance equations for each constituent have to be established, where interaction or supply terms, respectively, are used for the description of chemical and physical exchange processes between the different phases. Then, the sum of the balance equations of mass, momentum, moment of momentum and energy of the κ constituents φ^α results in the corresponding balance laws of the mixture, which, formally, must be equivalent to the conservation laws (balance laws without supply terms) of a one-component material (*Truesdell*'s third "metaphysical principle", see Truesdell [23]).

Balance of mass. Excluding mass exchanges between the phases, the balance of mass for the individual constituent φ^α requires that the material time derivative of the mass M^α following the motion of φ^α be equal to zero,

$$(M^\alpha)'_\alpha = (\int_{B_\alpha} \rho^\alpha \,\mathrm{d}v)'_\alpha = \int_{B_\alpha} [(\rho^\alpha)'_\alpha + \rho^\alpha \operatorname{div} \overset{\prime}{\mathbf{x}}_\alpha] \,\mathrm{d}v = 0\,. \tag{17}$$

With respect to the actual placement of φ^α the local statement of the balance of mass reads

$$(\rho^\alpha)'_\alpha + \rho^\alpha \operatorname{div} \overset{\prime}{\mathbf{x}}_\alpha = 0\,. \tag{18}$$

The material time derivative of a scalar quantity depending on $\mathbf{x}$ and t is defined as $(\,\cdot\,)'_\alpha = \partial(\,\cdot\,)/\partial t + \operatorname{grad}(\,\cdot\,)\cdot \overset{\prime}{\mathbf{x}}_\alpha$. The expression "div" denotes the divergence operator.

Balance of momentum and moment of momentum. The balance equation of momentum for the constituent φ^α states that the material time derivative of the momentum $\mathbf{l}^\alpha$ is equal to the sum of external forces and the supply term of momentum:

$$(\mathbf{l}^\alpha)'_\alpha = (\int_{B_\alpha} \rho^\alpha \overset{\prime}{\mathbf{x}}_\alpha \,\mathrm{d}v)'_\alpha = \int_{B_\alpha} \rho^\alpha \overset{\prime\prime}{\mathbf{x}}_\alpha \,\mathrm{d}v = \mathbf{k}^\alpha + \int_{B_\alpha} \hat{\mathbf{p}}^\alpha \,\mathrm{d}v\,, \tag{19}$$

where

$$\mathbf{k}^\alpha = \int_{B_\alpha} \rho^\alpha \mathbf{b}^\alpha \,\mathrm{d}v + \int_{\partial B_\alpha} \mathbf{t}^\alpha \,\mathrm{da} \tag{20}$$

is the vector of external forces, which is to be split into two parts, namely into a part which is caused by the local external body force $\rho^\alpha \mathbf{b}^\alpha \, dv$ and a part caused by the local external contact force $\mathbf{t}^\alpha$ da. The quantity $\hat{\mathbf{p}}^\alpha = \hat{\mathbf{p}}^\alpha(\mathbf{x}, t)$ in (19) denotes the local supply term of momentum of φ^α. With the theorem of *Cauchy*, $\mathbf{t}^\alpha = \mathbf{T}^\alpha \, \mathbf{n}$, where $\mathbf{T}^\alpha$ is the partial *Cauchy* stress tensor of φ^α and $\mathbf{n}$ the unit normal vector at the surface, the local statement

$$\operatorname{div} \mathbf{T}^\alpha + \rho^\alpha (\mathbf{b}^\alpha - \overset{\prime\prime}{\mathbf{x}}_\alpha) = -\hat{\mathbf{p}}^\alpha \tag{21}$$

of the balance of momentum for the constituent φ^α is obtained from (19). The local supply terms of momentum are restricted by

$$\sum_{\alpha=1}^{\kappa} \hat{\mathbf{p}}^\alpha = \mathbf{0} \, . \tag{22}$$

The statement of the balance of moment of momentum for non-polar materials reads that the material time derivative of the moment of momentum of φ^α along the trajectory of the mentioned phase is equal to the moment of external forces and the moment of the supply term of momentum. The moments are referred to a fixed reference point of the configuration space. For non-polar materials supply terms of moment of momentum can not be considered because these materials have no rotational degrees of freedom for the material point. Thus, the evaluation of this balance equation of moment of momentum yields the local statement

$$\mathbf{T}^\alpha = (\mathbf{T}^\alpha)^T \, , \tag{23}$$

i. e. the partial *Cauchy* stress tensor of the constituent φ^α is symmetric.

Balance of energy. The first law of thermodynamics (balance of energy) for the constituent φ^α states that the material time derivative of the sum of the internal energy E^α and kinetic energy K^α equals the sum of the increment of the mechanical work of the external forces W^α, the increment of the non-mechanical work Q^α and the supply term of energy of φ^α:

$$\begin{aligned} (E^\alpha)'_\alpha + (K^\alpha)'_\alpha &= \Big(\int_{B_\alpha} \rho^\alpha \, \varepsilon^\alpha \, dv\Big)'_\alpha + \Big(\int_{B_\alpha} \frac{1}{2} \rho^\alpha \, \overset{\prime}{\mathbf{x}}_\alpha \cdot \overset{\prime}{\mathbf{x}}_\alpha \, dv\Big)'_\alpha \\ &= \int_{B_\alpha} [\rho^\alpha \, (\varepsilon^\alpha)'_\alpha + \rho^\alpha \, \overset{\prime\prime}{\mathbf{x}}_\alpha \cdot \overset{\prime}{\mathbf{x}}_\alpha] \, dv \\ &= W^\alpha + Q^\alpha + \int_{B_\alpha} \hat{\mathrm{e}}^\alpha \, dv \, , \end{aligned} \tag{24}$$

where the scalar quantities $\varepsilon^\alpha = \varepsilon^\alpha(\mathbf{x}, t)$ and $\hat{\mathrm{e}}^\alpha = \hat{\mathrm{e}}^\alpha(\mathbf{x}, t)$ are the specific internal energy and the local supply term of energy of φ^α. The increments of the mechanical work of the external forces and of the non-mechanical work are given by

$$
\begin{aligned}
W^\alpha &= \int\limits_{B_\alpha} \overset{\prime}{\mathbf{x}}_\alpha \cdot \rho^\alpha \, \mathbf{b}^\alpha \, dv + \int\limits_{\partial B_\alpha} \overset{\prime}{\mathbf{x}}_\alpha \, \mathbf{T}^\alpha \, \mathbf{da} \\
&= \int\limits_{B_\alpha} [\overset{\prime}{\mathbf{x}}_\alpha \cdot (\rho^\alpha \, \mathbf{b}^\alpha + \operatorname{div} \mathbf{T}^\alpha) + \mathbf{T}^\alpha \cdot \mathbf{L}_\alpha] \, dv \,, \\
Q^\alpha &= \int\limits_{B_\alpha} \rho^\alpha \, \mathrm{r}^\alpha \, dv - \int\limits_{\partial B_\alpha} \mathbf{q}^\alpha \cdot \mathbf{da} = \int\limits_{B_\alpha} (\rho^\alpha \, \mathrm{r}^\alpha - \operatorname{div} \mathbf{q}^\alpha) \, dv \,.
\end{aligned}
\tag{25}
$$

The non-mechanical work Q^α consists of two parts caused by the local external heat supply $\mathrm{r}^\alpha = \mathrm{r}^\alpha(\mathbf{x}, t)$ and the heat influx vector $\mathbf{q}^\alpha = \mathbf{q}^\alpha(\mathbf{x}, t)$.

Taking into consideration the balance of momentum (21), the insertion of (25) into (24) yields the following local statement of the balance of energy concerning the constituent φ^α:

$$
\rho^\alpha \, (\varepsilon^\alpha)'_\alpha - \mathbf{T}^\alpha \cdot \mathbf{D}_\alpha - \rho^\alpha \, \mathrm{r}^\alpha + \operatorname{div} \mathbf{q}^\alpha = \hat{\mathrm{e}}^\alpha - \hat{\mathbf{p}}^\alpha \cdot \overset{\prime}{\mathbf{x}}_\alpha \,. \tag{26}
$$

On account of $\mathbf{T}^\alpha = (\mathbf{T}^\alpha)^T$ in (26) the scalar product $\mathbf{T}^\alpha \cdot \mathbf{L}_\alpha$ has been replaced by $\mathbf{T}^\alpha \cdot \mathbf{D}_\alpha$. The condition

$$
\sum_{\alpha=1}^{\kappa} \hat{\mathrm{e}}^\alpha = 0 \tag{27}
$$

restricts the local supply terms of energy.

Entropy inequality for the mixture. In order to gain restrictions for constitutive relations, the second law of thermodynamics (entropy principle) is a helpful tool. The entropy inequality for the mixture with different temperatures of the constituents is defined as

$$
\begin{aligned}
\sum_{\alpha=1}^{\kappa} (H^\alpha)'_\alpha &= \sum_{\alpha=1}^{\kappa} (\int\limits_{B_\alpha} \rho^\alpha \, \eta^\alpha \, dv)'_\alpha = \sum_{\alpha=1}^{\kappa} \int\limits_{B_\alpha} \rho^\alpha \, (\eta^\alpha)'_\alpha \, dv \\
&\geq \sum_{\alpha=1}^{\kappa} \int\limits_{B_\alpha} \frac{1}{\Theta^\alpha} \, \rho^\alpha \mathrm{r}^\alpha \, dv - \sum_{\alpha=1}^{\kappa} \int\limits_{\partial B_\alpha} \frac{1}{\Theta^\alpha} \, \mathbf{q}^\alpha \cdot \mathbf{da} \\
&\geq \sum_{\alpha=1}^{\kappa} \int\limits_{B_\alpha} [\frac{1}{\Theta^\alpha} \, \rho^\alpha \mathrm{r}^\alpha - \operatorname{div} (\frac{1}{\Theta^\alpha} \, \mathbf{q}^\alpha)] \, dv \,,
\end{aligned}
\tag{28}
$$

see Ehlers [12], where H^α and $\eta^\alpha = \eta^\alpha(\mathbf{x}, t)$ denote the entropy and the specific entropy of the constituent φ^α. The quantity $\Theta^\alpha = \Theta^\alpha(\mathbf{x}, t)$ is the local absolute temperature of the corresponding constituent.

With the free *Helmholtz* energy function

$$\psi^\alpha = \varepsilon^\alpha - \Theta^\alpha \, \eta^\alpha \tag{29}$$

and the balance of energy (26), the local statement

$$\begin{aligned} \sum_{\alpha=1}^{\kappa} \frac{1}{\Theta^\alpha} \Big\{ & -\rho^\alpha \, [(\psi^\alpha)'_\alpha + (\Theta^\alpha)'_\alpha \, \eta^\alpha] - \hat{\rho}^\alpha \, (\psi^\alpha - \frac{1}{2} \, \overset{\prime}{\mathbf{x}}_\alpha \cdot \overset{\prime}{\mathbf{x}}_\alpha) + \\ & + \mathbf{T}^\alpha \cdot \mathbf{D}_\alpha - \hat{\mathbf{p}}^\alpha \cdot \overset{\prime}{\mathbf{x}}_\alpha - \frac{1}{\Theta^\alpha} \, \mathbf{q}^\alpha \cdot \operatorname{grad} \Theta^\alpha + \hat{\mathrm{e}}^\alpha \Big\} \geq 0 \end{aligned} \tag{30}$$

of the inequality (28) is obtained.

3 Field equations and constitutive quantities

The field equations for porous media consist of the balance equations of the constituents taken from the mixture theory and the saturation condition. Thus, the local field equations for the description of the thermodynamic behaviour of binary porous media, consisting of a compressible and/or an incompressible solid and fluid phase, are given by the local statements of the balance equations of mass (18),

$$(\rho^S)'_S + \rho^S \operatorname{div} \overset{\prime}{\mathbf{x}}_S = 0 \, , \quad (\rho^F)'_F + \rho^F \operatorname{div} \overset{\prime}{\mathbf{x}}_F = 0 \, , \tag{31}$$

the balance equations of momentum (21),

$$\operatorname{div} \mathbf{T}^S + \rho^S \, (\mathbf{b} - \overset{\prime\prime}{\mathbf{x}}_S) = \hat{\mathbf{p}}^F \, , \quad \operatorname{div} \mathbf{T}^F + \rho^F \, (\mathbf{b} - \overset{\prime\prime}{\mathbf{x}}_F) = -\hat{\mathbf{p}}^F \, , \tag{32}$$

where it has been assumed that all constituents undergo the same external acceleration ($\mathbf{b}^\alpha = \mathbf{b}$), the balance equations of energy (26) with consideration of (29),

$$\begin{aligned} & \rho^S [(\psi^S)'_S + (\Theta^S)'_S \, \eta^S + \Theta^S \, (\eta^S)'_S] - \mathbf{T}^S \cdot \mathbf{D}_S - \rho^S \, \mathrm{r}^S + \operatorname{div} \mathbf{q}^S = \\ & \quad = -\hat{\mathrm{e}}^F + \hat{\mathbf{p}}^F \cdot \overset{\prime}{\mathbf{x}}_S \, , \\ & \rho^F [(\psi^F)'_F + (\Theta^F)'_F \, \eta^F + \Theta^F \, (\eta^F)'_F] - \mathbf{T}^F \cdot \mathbf{D}_F - \rho^F \, \mathrm{r}^F + \operatorname{div} \mathbf{q}^F = \\ & \quad = \hat{\mathrm{e}}^F - \hat{\mathbf{p}}^F \cdot \overset{\prime}{\mathbf{x}}_F \, , \end{aligned} \tag{33}$$

and the saturation condition (4),

$$n^S + n^F = \frac{\rho^S}{\rho^{SR}} + \frac{\rho^F}{\rho^{FR}} = \frac{J_{SR}}{\rho_{0S}^{SR}}\,\rho^S + \frac{\rho^F}{\rho^{FR}} = 1\,. \tag{34}$$

In the field equations $(32)_1$ and $(33)_1$ the relations $\hat{\mathbf{p}}^S = -\hat{\mathbf{p}}^F$ and $\hat{\mathrm{e}}^S = -\hat{\mathrm{e}}^F$ concerning the supply terms of momentum and energy have been considered, see (22) and (27). Furthermore, in (34) the expressions $n^\alpha = \rho^\alpha/\rho^{\alpha R}$, see (6), and $\rho^{SR} = \rho_{0S}^{SR}/J_{SR}$ have been used, where ρ_{0S}^{SR} is the real solid density at the position $\mathbf{X}_S$ at time $t = t_0$ (lower index $(\cdot)_{0S}$). The quantity $J_{SR} = \det \mathbf{F}_{SR}$ is the determinant of the deformation part $\mathbf{F}_{SR}$, which results from the multiplicative decomposition of the deformation gradient $\mathbf{F}_S = \mathbf{F}_{SN}\,\mathbf{F}_{SR}$ of the solid phase. The deformation parts $\mathbf{F}_{SN}$ and $\mathbf{F}_{SR}$ describe the change of the pores in size and shape and the deformation of the real solid material. A detailed discussion of the multiplicative decomposition of $\mathbf{F}_S$ and the corresponding expression of the determinants can be found in Bluhm [3].

Taking into account (31) – (34), the binary model presented here will be described by 11 field equations, viz., the balance equations of mass, momentum and energy ($2 + 6 + 2 = 10$ equations) and the saturation condition (1 equation). In the field equations mentioned previously, there exists a total of 44 field variables:

$$\begin{aligned} \mathcal{F} = \{&\mathbf{T}^\alpha,\, \chi_\alpha,\, \mathbf{q}^\alpha,\, \hat{\mathbf{p}}^F,\, \mathbf{b}, \\ &\psi^\alpha,\, \eta^\alpha,\, \rho^\alpha,\, J_{SR},\, \rho^{FR},\, \rho_{0S}^{SR},\, \Theta^\alpha,\, \mathrm{r}^\alpha,\, \hat{\mathrm{e}}^F\} \implies \mathcal{V}^{44}\,. \end{aligned} \tag{35}$$

In (35), the velocity $\overset{\prime}{\mathbf{x}}_\alpha$ and the acceleration $\overset{\prime\prime}{\mathbf{x}}_\alpha$ of the constituent φ^α are represented by the motion function χ_α, see (7). Furthermore, the local statement $\mathbf{T}^\alpha = (\mathbf{T}^\alpha)^T$ of the balance of moment of momentum has been considered in the form that the dimension of the partial *Cauchy* stress tensor $\mathbf{T}^\alpha$ in the vector space is equal to six ($\mathbf{T}^\alpha \to \mathcal{V}^6$).

In the following, the closure problem will be discussed for four different binary models, namely, for a compressible model (compressible solid, compressible fluid), for hybrid models (compressible solid, incompressible fluid and incompressible solid, compressible fluid), and for an incompressible model (incompressible solid, incompressible fluid). In order to close the system of equations for the model of binary porous media consisting of compressible and/or incompressible phases (i. e. the number of field equations must be equal to the number of unknown field quantities) additional equations must be formulated. In the present work, only constitutive relations will be formulated to close the system of field equations. The known, constitutive and unknown field quantities concerning the different binary models will be summarized in the lists $\mathcal{K}_{(\dots)}$, $\mathcal{C}_{(\dots)}$, and $\mathcal{U}_{(\dots)}$.

– Model A: Compressible model (compressible solid, compressible fluid)

$$
\begin{aligned}
\mathcal{K}_{\mathrm{A}} &= \{\mathbf{b},\, \rho_{0S}^{SR},\, \mathrm{r}^\alpha\} && \Longrightarrow && \mathcal{V}^6\,,\\
\mathcal{C}_{\mathrm{A}} &= \{\mathbf{T}^\alpha,\, \mathbf{q}^\alpha,\, \hat{\mathbf{p}}^F,\, \psi^\alpha,\, \eta^\alpha,\, J_{SR},\, \hat{\mathrm{e}}^F\} && \Longrightarrow && \mathcal{V}^{27},\\
\mathcal{U}_{\mathrm{A}} &= \{\boldsymbol{\chi}_\alpha,\, \rho^\alpha,\, \rho^{FR},\, \Theta^\alpha\} && \Longrightarrow && \mathcal{V}^{11}.
\end{aligned}
\tag{36}
$$

– Model B: Hybrid model of first type
(compressible solid, incompressible fluid)

$$
\begin{aligned}
\mathcal{K}_{\mathrm{B}} &= \{\mathbf{b},\, \rho^{FR} = \text{const.},\, \rho_{0S}^{SR},\, \mathrm{r}^\alpha\} && \Longrightarrow && \mathcal{V}^7\,,\\
\mathcal{C}_{\mathrm{B}} &= \{\mathbf{T}^\alpha,\, \mathbf{q}^\alpha,\, \hat{\mathbf{p}}^F,\, \psi^\alpha,\, \eta^\alpha,\, \hat{\mathrm{e}}^F\} && \Longrightarrow && \mathcal{V}^{26},\\
\mathcal{U}_{\mathrm{B}} &= \{\boldsymbol{\chi}_\alpha,\, \rho^\alpha,\, J_{SR},\, \Theta^\alpha\} && \Longrightarrow && \mathcal{V}^{11}.
\end{aligned}
\tag{37}
$$

– Model C: Hybrid model of second type
(incompressible solid, compressible fluid)

$$
\begin{aligned}
\mathcal{K}_{\mathrm{C}} &= \{\mathbf{b},\, J_{SR} = 1,\, \rho_{0S}^{SR},\, \mathrm{r}^\alpha\} && \Longrightarrow && \mathcal{V}^7\,,\\
\mathcal{C}_{\mathrm{C}} &= \{\mathbf{T}^\alpha,\, \mathbf{q}^\alpha,\, \hat{\mathbf{p}}^F,\, \psi^\alpha,\, \eta^\alpha,\, \hat{\mathrm{e}}^F\} && \Longrightarrow && \mathcal{V}^{26},\\
\mathcal{U}_{\mathrm{C}} &= \{\boldsymbol{\chi}_\alpha,\, \rho^\alpha,\, \rho^{FR},\, \Theta^\alpha\} && \Longrightarrow && \mathcal{V}^{11}.
\end{aligned}
\tag{38}
$$

– Model D: Incompressible model (incompressible solid, incompressible fluid)

$$
\begin{aligned}
\mathcal{K}_{\mathrm{D}} &= \{\mathbf{b},\, J_{SR} = 1,\, \rho^{FR} = \text{const.},\, \rho_{0S}^{SR},\, \mathrm{r}^\alpha\} && \Longrightarrow && \mathcal{V}^8\,,\\
\mathcal{C}_{\mathrm{D}} &= \{\mathbf{T}^\alpha,\, \mathbf{q}^\alpha,\, \hat{\mathbf{p}}^F,\, \psi^\alpha,\, \eta^\alpha,\, \hat{\mathrm{e}}^F\} && \Longrightarrow && \mathcal{V}^{26},\\
\mathcal{U}_{\mathrm{D}} &= \{\boldsymbol{\chi}_\alpha,\, \rho^\alpha,\, \Theta^\alpha,\, \tilde{\kappa}\} && \Longrightarrow && \mathcal{V}^{11}.
\end{aligned}
\tag{39}
$$

One sees that for all four models the number of field equations is equal to the number of unknown field quantities. For the incompressible model (Model D) an additional unknown field quantity ($\tilde{\kappa}$) has been introduced in $(39)_3$ to close the system of field equations. From the mathematical point of view, this quantity is to be understood as an indeterminate reaction force assigned to the saturation condition which is an equation in excess for the incompressible model.

4 Evaluation of the entropy inequality of binary porous media

The difference between mixture theory and the Porous Media Theory is that within the framework of the Porous Media Theory the volume fractions and the saturation condition are considered. Therefore, the saturation condition $n^S + n^F = 1$, which is understood as a constraint, i. e. as a restriction of the motion of the solid and the fluid phase, must be accounted for in the entropy inequality. The influence of the saturation constraint on the inequality (30) will be taken into account by using the material time derivative of the saturation condition following the motion of the solid or the fluid phase, together with the concept of *Lagrange* multipliers. Here, the material time derivative of the saturation condition following the motion of the solid phase will be used:

$$(n^S)'_S + (n^F)'_S = (n^S)'_S + (n^F)'_F - \operatorname{grad} n^F \cdot \mathbf{w}_{FS} = 0\,, \tag{40}$$

where

$$\mathbf{w}_{FS} = \overset{\prime}{\mathbf{x}}_F - \overset{\prime}{\mathbf{x}}_S \tag{41}$$

is the difference velocity of the fluid and the solid phase. With

$$\begin{aligned} (n^\alpha)'_\alpha &= \frac{n^\alpha}{\rho^\alpha}\,(\rho^\alpha)'_\alpha - \frac{n^\alpha}{\rho^{\alpha R}}\,(\rho^{\alpha R})'_\alpha = -n^\alpha\,(\mathbf{D}_\alpha \cdot \mathbf{I}) - \frac{n^\alpha}{\rho^{\alpha R}}\,, \\ (\rho^{SR})'_S &= (\frac{1}{J_{SR}}\,\rho^{SR}_{0S})'_S = -\frac{\rho^{SR}_{0S}}{J^2_{SR}}\,(J_{SR})'_S = -\rho^{SR}\,\frac{1}{J_{SR}}\,(J_{SR})'_S\,, \end{aligned} \tag{42}$$

the rate of the saturation condition (40) can be written as

$$\begin{aligned} & n^S\,(\mathbf{D}_S \cdot \mathbf{I}) + n^F\,(\mathbf{D}_F \cdot \mathbf{I}) - \\ & \quad - n^S\,\frac{1}{J_{SR}}\,(J_{SR})'_S + \frac{n^F}{\rho^{FR}}\,(\rho^{FR})'_F + \operatorname{grad} n^F \cdot \mathbf{w}_{FS} = 0\,. \end{aligned} \tag{43}$$

The multiplication of (43) by the *Lagrange* multiplier λ and the positive scalar value

$$\frac{1}{\Theta_{\text{ref}}} = \frac{1}{2}\,\frac{\Theta^S + \Theta^F}{\Theta^S\,\Theta^F} \tag{44}$$

yields the constraint

$$\frac{1}{\Theta_{\text{ref}}}\,\lambda\,\{\text{left-hand side of (43)}\} = 0\,, \tag{45}$$

which will be added to the entropy inequality (30) of the mixture:

$$
\begin{aligned}
& -\frac{1}{\Theta^S}\,\rho^S\,[(\psi^S)'_S + \eta^S\,(\Theta^S)'_S] - \frac{1}{\Theta^F}\,\rho^F\,[(\psi^F)'_F + \eta^F\,(\Theta^F)'_F] + \\
& + \frac{1}{\Theta^S}\,\mathbf{D}_S \cdot (\mathbf{T}^S + n^S\,\lambda\,\frac{\Theta^S}{\Theta_{\text{ref}}}\,\mathbf{I}) + \frac{1}{\Theta^F}\,\mathbf{D}_F \cdot (\mathbf{T}^F + n^F\,\lambda\,\frac{\Theta^F}{\Theta_{\text{ref}}}\,\mathbf{I}) - \\
& - n^S\,\lambda\,\frac{1}{\Theta_{\text{ref}}}\,\frac{1}{J_{SR}}\,(J_{SR})'_S + n^F\,\lambda\,\frac{1}{\Theta_{\text{ref}}}\,\frac{1}{\rho^{FR}}\,(\rho^{FR})'_F - \\
& - \frac{1}{\Theta^F}(\hat{\mathbf{p}}^F - \lambda\,\frac{\Theta^F}{\Theta_{\text{ref}}}\,\operatorname{grad} n^F) \cdot \mathbf{w}_{FS} + \frac{\Theta^F - \Theta^S}{\Theta^S\,\Theta^F}\,\hat{\mathbf{p}}^F \cdot \overset{\prime}{\mathbf{x}}_S - \\
& - \frac{\Theta^F - \Theta^S}{\Theta^S\,\Theta^F}\,\hat{\mathrm{e}}^F - \frac{1}{(\Theta^S)^2}\,\mathbf{q}^S \cdot \operatorname{grad}\Theta^S - \frac{1}{(\Theta^F)^2}\,\mathbf{q}^F \cdot \operatorname{grad}\Theta^F \geq 0\,,
\end{aligned}
\tag{46}
$$

where the relations $\hat{\mathbf{p}}^S = -\,\hat{\mathbf{p}}^F$ and $\hat{\mathrm{e}}^S = -\,\hat{\mathrm{e}}^F$ with respect to the supply terms of momentum and energy have been considered, see (22) and (27). The inequality (46) represents the entropy inequality of the porous medium. The *Lagrange* multiplier λ is understood as the reaction force of the saturation condition, weighted with $1/\Theta_{\text{ref}}$. The introduction of the aforementioned value is not necessary, but it simplifies the evaluation of the entropy inequality of a binary model with different temperatures and the quantity λ has the dimension of a stress.

In the following the entropy inequality will be evaluated for the four different binary models, namely, for the compressible model, for the hybrid models and for the incompressible model. With respect to the aforementioned models, the below listed constitutive assumptions for ψ^S, ψ^F, and J_{SR} and the restrictions for the real densities of φ^S and φ^F if the constituents are incompressible must be considered in the entropy inequality (46):

- Model A: Compressible model (compressible solid, compressible fluid)

$$\psi^S = \psi^S(\Theta^S,\,\mathbf{C}_S)\,,\ \psi^F = \psi^F(\Theta^F,\,\rho^{FR})\,,\ J_{SR} = J_{SR}(J_S)\,. \tag{47}$$

- Model B: Hybrid model of first type
  (compressible solid, incompressible fluid)

$$\psi^S = \psi^S(\Theta^S,\,\mathbf{C}_S,\,J_{SR})\,,\ \psi^F = \psi^F(\Theta^F)\,,\ \rho^{FR} = \text{const.} \tag{48}$$

- Model C: Hybrid model of second type
  (incompressible solid, compressible fluid)

$$\psi^S = \psi^S(\Theta^S,\,\mathbf{C}_S)\,,\ \psi^F = \psi^F(\Theta^F,\,\rho^{FR})\,,\ \rho^{SR} = \text{const.} \tag{49}$$

- Model D: Incompressible model (incompressible solid, incompressible fluid)

$$\psi^S = \psi^S(\Theta^S,\,\mathbf{C}_S)\,,\ \psi^F = \psi^F(\Theta^F)\,,\ \rho^{SR} = \text{const.}\,,\ \rho^{FR} = \text{const.} \tag{50}$$

Considering $(47)_{1,2}$, $(48)_{1,2}$, $(49)_{1,2}$, and $(50)_{1,2}$ the expressions $\rho^S(\psi^S)'_S$ and $\rho^F(\psi^F)'_F$ in (46) can be replaced by

$$\begin{aligned} \rho^S\,(\psi^S)'_S &= \rho^S\,\frac{\partial\psi^S}{\partial\Theta^S}\,(\Theta^S)'_S + 2\,\rho^S\,\mathbf{F}_S\,\frac{\partial\psi^S}{\partial\mathbf{C}_S}\,\mathbf{F}_S^T\cdot\mathbf{D}_S + \underbrace{\rho^S\,\frac{\partial\psi^S}{\partial J_{SR}}\,(J_{SR})'_S}_{\text{n. e. (A, C, D)}}\,, \\ \rho^F\,(\psi^F)'_F &= \rho^F\,\frac{\partial\psi^F}{\partial\Theta^F}\,(\Theta^F)'_F + \underbrace{\rho^F\,\frac{\partial\psi^F}{\partial\rho^{FR}}\,(\rho^{FR})'_F}_{\text{n. e. (B, D)}}\,. \end{aligned} \tag{51}$$

The constitutive ansatz for J_{SR}, see $(47)_3$, and the material time derivative

$$(J_{SR})'_S = \frac{\partial J_{SR}}{\partial J_S}\,(J_S)'_S = J_S\,\frac{\partial J_{SR}}{\partial J_S}(\mathbf{D}_S\cdot\mathbf{I})\,, \tag{52}$$

respectively, will be considered in the inequality (46) with the help of the concept of *Lagrange* multipliers, i. e. the relation

$$\frac{1}{\Theta_{\text{ref}}}\,\lambda^{SR}[(J_{SR})'_S - J_S\,\frac{\partial J_{SR}}{\partial J_S}\,\mathbf{I}\cdot\mathbf{D}_S] = 0 \tag{53}$$

will be added to (46). The influence of the incompressibility of the real materials concerning the entropy inequality will be considered with the conditions

$$\begin{aligned} \frac{1}{\Theta_{\text{ref}}}\,\kappa^{SR}\,\frac{1}{\rho^{SR}}\,(\rho^{SR})'_S &= -\frac{1}{\Theta_{\text{ref}}}\,\kappa^{SR}\,\frac{1}{J_{SR}}\,(J_{SR})'_S = 0\,, \\ \frac{1}{\Theta_{\text{ref}}}\,\kappa^{FR}\,\frac{1}{\rho^{FR}}\,(\rho^{FR})'_F &= 0\,. \end{aligned} \tag{54}$$

With (51), (53), and (54) the entropy inequality for the four different binary models reads

$$\begin{aligned} &-\frac{1}{\Theta^S}\,(\Theta^S)'_S \quad \left(\rho^S\,\eta^S + \rho^S\,\frac{\partial\psi^S}{\partial\Theta^S}\right) - \\ &-\frac{1}{\Theta^F}\,(\Theta^F)'_F \quad \left(\rho^F\,\eta^F + \rho^F\,\frac{\partial\psi^F}{\partial\Theta^F}\right) + \\ &+\frac{1}{\Theta^S}\,\mathbf{D}_S \quad \cdot\left(\mathbf{T}^S + n^S\,\lambda\,\frac{\Theta^S}{\Theta_{\text{ref}}}\,\mathbf{I} - 2\,\rho^S\,\mathbf{F}_S\,\frac{\partial\psi^S}{\partial\mathbf{C}_S}\,\mathbf{F}_S^T - \right. \\ &\qquad\qquad \left. \underbrace{-\,\lambda^{SR}\,\frac{\Theta^S}{\Theta_{\text{ref}}}\,J_S\,\frac{\partial J_{SR}}{\partial J_S}\,\mathbf{I}}_{\text{n. e. (B, C, D)}}\right) + \\ &+\frac{1}{\Theta^F}\,\mathbf{D}_F \quad \cdot\left(\mathbf{T}^F + n^F\,\lambda\,\frac{\Theta^F}{\Theta_{\text{ref}}}\,\mathbf{I}\right) - \end{aligned} \tag{55}$$

$$
\begin{aligned}
-\frac{1}{\Theta^S}(J_{SR})'_S \quad & \cdot \Big(n^S \lambda \frac{1}{J_{SR}} \frac{\Theta^S}{\Theta_{\text{ref}}} + \underbrace{\rho^S \frac{\partial \psi^S}{\partial J_{SR}}}_{\text{n. e. (A, C, D)}} - \underbrace{\lambda^{SR} \frac{\Theta^S}{\Theta_{\text{ref}}}}_{\text{n. e. (B, C, D)}} + \\
& \quad + \underbrace{\frac{\Theta^S}{\Theta_{\text{ref}}} \frac{1}{J_{SR}} \kappa^{SR}}_{\text{n. e. (A, B)}} \Big) + \\
+\frac{1}{\Theta^F}(\rho^{FR})'_F \quad & \cdot \Big(n^F \lambda \frac{1}{\rho^{FR}} \frac{\Theta^F}{\Theta_{\text{ref}}} - \underbrace{\rho^F \frac{\partial \psi^F}{\partial \rho^{FR}}}_{\text{n. e. (B, D)}} + \underbrace{\frac{\Theta^F}{\Theta_{\text{ref}}} \frac{1}{\rho^{FR}} \kappa^{FR}}_{\text{n. e. (A, C)}} \Big) - \\
-\frac{1}{\Theta^F} \mathbf{w}_{FS} \quad & \cdot \Big(\hat{\mathbf{p}}^F - \lambda \frac{\Theta^F}{\Theta_{\text{ref}}} \operatorname{grad} n^F \Big) + \\
+\frac{\Theta^F - \Theta^S}{\Theta^S \Theta^F} \quad & \Big(\hat{\mathbf{p}}^F \cdot \overset{\prime}{\mathbf{x}}_S - \hat{e}^F \Big) - \\
-\frac{1}{(\Theta^S)^2} \operatorname{grad} \Theta^S \quad & \cdot \Big(\mathbf{q}^S \Big) - \\
-\frac{1}{(\Theta^F)^2} \operatorname{grad} \Theta^F \quad & \cdot \Big(\mathbf{q}^F \Big) \geq 0 ,
\end{aligned}
$$

where the marked terms are not existent (n. e.) for the models which are listed in the brackets.

Following the argumentation of Coleman and Noll [11], the entropy inequality must hold for fixed values of the process variables

$$
\{\Theta^S, \Theta^F, \operatorname{grad}\Theta^S, \operatorname{grad}\Theta^F, \mathbf{C}_S, J_{SR}, \rho^{FR}, \operatorname{grad} n^F, \mathbf{w}_{FS}\} \tag{56}
$$

and for arbitrary values of the so-called freely available quantities, i. e. for the derivations of the process variables in time and space,

$$
\{(\Theta^S)'_S, (\Theta^F)'_F, \mathbf{D}_S, (J_{SR})'_S, \mathbf{D}_F, (\rho^{FR})'_F\} , \tag{57}
$$

where only the freely available quantities appearing in (55) are listed. In consideration of the aforementioned remarks, one obtains the following necessary and sufficient conditions for the unrestricted validity of the second law of thermodynamics for the here presented binary models:

$$
\begin{aligned}
&\rho^S \eta^S + \rho^S \frac{\partial \psi^S}{\partial \Theta^S} = 0\,, \quad \rho^F \eta^F + \rho^F \frac{\partial \psi^F}{\partial \Theta^F} = 0\,,\\
&\mathbf{T}^S + n^S \lambda \frac{\Theta^S}{\Theta_{\text{ref}}} \mathbf{I} - 2\,\rho^S \mathbf{F}_S \frac{\partial \psi^S}{\partial \mathbf{C}_S} \mathbf{F}_S^T \underbrace{- \lambda^{SR} \frac{\Theta^S}{\Theta_{\text{ref}}} J_S \frac{\partial J_{SR}}{\partial J_S} \mathbf{I}}_{\text{n. e. (B, C, D)}} = \mathbf{0}\,,\\
&\mathbf{T}^F + n^F \lambda \frac{\Theta^F}{\Theta_{\text{ref}}} \mathbf{I} = \mathbf{0}\,,\\
&n^S \lambda \frac{1}{J_{SR}} \frac{\Theta^S}{\Theta_{\text{ref}}} + \underbrace{\rho^S \frac{\partial \psi^S}{\partial J_{SR}}}_{\text{n. e. (A, C, D)}} \underbrace{- \lambda^{SR} \frac{\Theta^S}{\Theta_{\text{ref}}}}_{\text{n. e. (B, C, D)}} + \underbrace{\frac{\Theta^S}{\Theta_{\text{ref}}} \frac{1}{J_{SR}} \kappa^{SR}}_{\text{n. e. (A, B)}} = 0\,,\\
&n^F \lambda \frac{1}{\rho^{FR}} \frac{\Theta^F}{\Theta_{\text{ref}}} \underbrace{- \rho^F \frac{\partial \psi^F}{\partial \rho^{FR}}}_{\text{n. e. (B, D)}} + \underbrace{\frac{\Theta^F}{\Theta_{\text{ref}}} \frac{1}{\rho^{FR}} \kappa^{FR}}_{\text{n. e. (A, C)}} = 0
\end{aligned}
\tag{58}
$$

and

$$
\mathcal{D} \geq 0\,, \tag{59}
$$

where

$$
\begin{aligned}
\mathcal{D} = &-\frac{1}{\Theta^F} \mathbf{w}_{FS} \cdot (\hat{\mathbf{p}}^F - \lambda \frac{\Theta^F}{\Theta_{\text{ref}}} \operatorname{grad} n^F) + \\
&+ \frac{\Theta^F - \Theta^S}{\Theta^S \Theta^F} (\hat{\mathbf{p}}^F \cdot \overset{\prime}{\mathbf{x}}_S - \hat{\text{e}}^F) - \\
&- \frac{1}{(\Theta^S)^2} \operatorname{grad} \Theta^S \cdot \mathbf{q}^S - \frac{1}{(\Theta^F)^2} \operatorname{grad} \Theta^F \cdot \mathbf{q}^F
\end{aligned}
\tag{60}
$$

is the dissipation of the model.

From (58) one obtains the following constitutive relations for the specific entropies η^S and η^F, for the *Cauchy* stress tensors $\mathbf{T}^S$ and $\mathbf{T}^F$ of the solid and fluid phases and the quantity λ, where for the incompressible model (Model D) λ is an indeterminate reaction force of the saturation condition, which is an equation in excess for the model in question:

– Model A: Compressible model (compressible solid, compressible fluid)

$$
\eta^S = -\frac{\partial \psi^S(\Theta^S, \mathbf{C}_S)}{\partial \Theta^S}\,, \; \eta^F = -\frac{\partial \psi^F(\Theta^F, \rho^{FR})}{\partial \Theta^F} \tag{61}
$$

and

$$
\begin{aligned}
\mathbf{T}^S &= -n^S\, p\, \frac{\Theta^S}{\Theta^F} \left(1 - \frac{J_S}{J_{SR}}\, \frac{\partial J_{SR}(J_S)}{\partial J_S}\right) \mathbf{I} + 2\, \rho^S\, \mathbf{F}_S\, \frac{\partial \psi^S(\Theta^S,\, \mathbf{C}_S)}{\partial \mathbf{C}_S}\, \mathbf{F}_S^T, \\
\mathbf{T}^F &= -n^F\, p\, \mathbf{I}\,,\; p = \frac{\Theta^F}{\Theta_{\text{ref}}}\, \lambda = (\rho^{FR})^2\, \frac{\partial \psi^F(\Theta^F,\, \rho^{FR})}{\partial \rho^{FR}}\,.
\end{aligned}
\tag{62}
$$

– Model B: Hybrid model of first type
(compressible solid, incompressible fluid)

$$
\eta^S = -\frac{\partial \psi^S(\Theta^S,\, \mathbf{C}_S,\, J_{SR})}{\partial \Theta^S}\,, \quad \eta^F = -\frac{\partial \psi^F(\Theta^F)}{\partial \Theta^F}
\tag{63}
$$

and

$$
\begin{aligned}
\mathbf{T}^S &= -n^S\, p\, \frac{\Theta^S}{\Theta^F}\, \mathbf{I} + 2\, \rho^S\, \mathbf{F}_S\, \frac{\partial \psi^S(\Theta^S,\, \mathbf{C}_S,\, J_{SR})}{\partial \mathbf{C}_S}\, \mathbf{F}_S^T, \\
\mathbf{T}^F &= -n^F\, p\, \mathbf{I}\,, \\
p &= \frac{\Theta^F}{\Theta_{\text{ref}}}\, \lambda = -\frac{\Theta^F}{\Theta^S}\, \rho^{SR}\, J_{SR}\, \frac{\partial \psi^S(\Theta^S,\, \mathbf{C}_S,\, J_{SR})}{\partial J_{SR}} = -\frac{\Theta^F}{\Theta_{\text{ref}}}\, \frac{1}{n^F}\, \kappa^{FR}\,.
\end{aligned}
\tag{64}
$$

– Model C: Hybrid model of second type
(incompressible solid, compressible fluid)

$$
\eta^S = -\frac{\partial \psi^S(\Theta^S,\, \mathbf{C}_S)}{\partial \Theta^S}\,, \quad \eta^F = -\frac{\partial \psi^F(\Theta^F,\, \rho^{FR})}{\partial \Theta^F}
\tag{65}
$$

and

$$
\begin{aligned}
\mathbf{T}^S &= -n^S\, p\, \frac{\Theta^S}{\Theta^F}\, \mathbf{I} + 2\, \rho^S\, \mathbf{F}_S\, \frac{\partial \psi^S(\Theta^S,\, \mathbf{C}_S)}{\partial \mathbf{C}_S}\, \mathbf{F}_S^T, \\
\mathbf{T}^F &= -n^F\, p\, \mathbf{I}\,, \\
p &= \frac{\Theta^F}{\Theta_{\text{ref}}}\, \lambda = (\rho^{FR})^2\, \frac{\partial \psi^F(\Theta^F,\, \rho^{FR})}{\partial \rho^{FR}} = -\frac{\Theta^F}{\Theta_{\text{ref}}}\, \frac{1}{n^S}\, \kappa^{SR}\,.
\end{aligned}
\tag{66}
$$

– Model D: Incompressible model (incompressible solid, incompressible fluid)

$$
\eta^S = -\frac{\partial \psi^S(\Theta^S,\, \mathbf{C}_S)}{\partial \Theta^S}\,, \quad \eta^F = -\frac{\partial \psi^F(\Theta^F)}{\partial \Theta^F}
\tag{67}
$$

and

$$
\begin{aligned}
\mathbf{T}^S &= -n^S p \frac{\Theta^S}{\Theta^F} \mathbf{I} + 2 \rho^S \mathbf{F}_S \frac{\partial \psi^S(\Theta^S, \mathbf{C}_S)}{\partial \mathbf{C}_S} \mathbf{F}_S^T, \\
\mathbf{T}^F &= -n^F p \mathbf{I}, \\
p &= \frac{\Theta^F}{\Theta_{\text{ref}}} \lambda = -\frac{\Theta^F}{\Theta_{\text{ref}}} \frac{1}{n^F} \kappa^{FR} = -\frac{\Theta^F}{\Theta_{\text{ref}}} \frac{1}{n^S} \kappa^{SR}.
\end{aligned}
\tag{68}
$$

For the heat flux vectors $\mathbf{q}^S$ and $\mathbf{q}^F$, the supply term of momentum $\hat{\mathbf{p}}^F = -\hat{\mathbf{p}}^S$ and the supply term of energy $\hat{\mathrm{e}}^F = -\hat{\mathrm{e}}^S$ of the liquid phase for all four models will be postulated:

$$
\begin{aligned}
\frac{1}{(\Theta^S)^2} \mathbf{q}^S &= -\alpha^S_{\partial\Theta^S} \operatorname{grad}\Theta^S - \alpha^S_{\partial\Theta^F} \operatorname{grad}\Theta^F - \alpha^S_{\mathbf{w}_{FS}} \mathbf{w}_{FS}, \\
\frac{1}{(\Theta^F)^2} \mathbf{q}^F &= -\alpha^F_{\partial\Theta^S} \operatorname{grad}\Theta^S - \alpha^F_{\partial\Theta^F} \operatorname{grad}\Theta^F - \alpha^F_{\mathbf{w}_{FS}} \mathbf{w}_{FS},
\end{aligned}
\tag{69}
$$

and

$$
\begin{aligned}
\frac{1}{\Theta^F} (\hat{\mathbf{p}}^F - \lambda \frac{\Theta^F}{\Theta_{\text{ref}}} \operatorname{grad} n^F) &= -\beta^F_{\partial\Theta^S} \operatorname{grad}\Theta^S - \beta^F_{\partial\Theta^F} \operatorname{grad}\Theta^F - \\
&\quad -\beta^F_{\mathbf{w}_{FS}} \mathbf{w}_{FS}, \\
\frac{1}{\Theta^S\Theta^F} (\hat{\mathrm{e}}^F - \hat{\mathbf{p}}^F \cdot \overset{\prime}{\mathbf{x}}_S) &= -\gamma^F_{\Theta^{FS}} (\Theta^F - \Theta^S).
\end{aligned}
\tag{70}
$$

Considering (60), the insertion of (69) and (70) into (59) yields

$$
\begin{aligned}
\mathcal{D} = \quad & \alpha^S_{\partial\Theta^S} \operatorname{grad}\Theta^S \cdot \operatorname{grad}\Theta^S + \alpha^F_{\partial\Theta^F} \operatorname{grad}\Theta^F \cdot \operatorname{grad}\Theta^F + \\
& + (\alpha^S_{\partial\Theta^F} + \alpha^F_{\partial\Theta^S}) \operatorname{grad}\Theta^S \cdot \operatorname{grad}\Theta^F + \\
& + (\alpha^S_{\mathbf{w}_{FS}} + \beta^F_{\partial\Theta^S}) \operatorname{grad}\Theta^S \cdot \mathbf{w}_{FS} + \\
& + (\alpha^F_{\mathbf{w}_{FS}} + \beta^F_{\partial\Theta^F}) \operatorname{grad}\Theta^F \cdot \mathbf{w}_{FS} + \\
& + \beta^F_{\mathbf{w}_{FS}} \mathbf{w}_{FS} \cdot \mathbf{w}_{FS} + \gamma^F_{\Theta^{FS}} (\Theta^F - \Theta^S)^2 \geq 0.
\end{aligned}
\tag{71}
$$

The dissipation mechanism is fulfilled if the restrictions

$$
\alpha^S_{\partial\Theta^S} \geq 0, \ \alpha^F_{\partial\Theta^F} \geq 0, \ \beta^F_{\mathbf{w}_{FS}} \geq 0, \ \gamma^F_{\Theta^{FS}} \geq 0 \tag{72}
$$

and

$$
\alpha^S_{\partial\Theta^F} + \alpha^F_{\partial\Theta^S} = 0, \ \alpha^S_{\mathbf{w}_{FS}} + \beta^F_{\partial\Theta^S} = 0, \ \alpha^F_{\mathbf{w}_{FS}} + \beta^F_{\partial\Theta^F} = 0 \tag{73}
$$

for the parameters in (71) are considered.

5 Constitutive relations for the stresses

In this section, constitutive relations for the stresses of the solid and fluid phase will be formulated with respect to the compressible model, the hybrid models, and the incompressible model. The development of the equations for the stress tensors is connected with the formulation of the free *Helmholtz* energy functions for both constituents and, in addition, for the compressible model (Model A), a constitutive relation for the real part of the volume deformation of the solid.

5.1 Constitutive relations for the stresses of the solid phase

For the description of the so-called effective stresses of compressible solids within the framework of finite deformation processes, well-known constitutive laws for compressible one-component materials can be used. The reason for this is the fact that compressible one-component materials and compressible empty porous solids show the same macroscopic deformation behaviour in the compression phase, namely, that no further volume compression can occur if the *Jacobian* of the solid is equal to zero. From the microscopic point of view, the deformation behaviour of the aforementioned solids is different. At the compression point the density of the one-component material is infinity; for compressible porous materials the so-called real density is infinity and in addition all pores are closed at the compression point. For incompressible porous solids, i. e. for those whose real density is constant, one has to consider that no further volume compression can occur if all pores are closed. Thus, the point of compaction or the compression point depends on the volume fraction of the reference placement of the solid phase, i. e. the compression point of an incompressible porous solid is reached if the *Jacobian* is equal to the aforementioned volume fraction.

The evaluation of the entropy inequality of the binary model presented here consisting of compressible and/or incompressible phases yields the following structure for the stresses of the porous solid:

$$\mathbf{T}^S = -n^S p \frac{\Theta^S}{\Theta^F} \Bigg(1 - \underbrace{\frac{J_S}{J_{SR}} \frac{\partial J_{SR}(J_S)}{\partial J_S}}_{\text{n. e. (B, C, D)}} \Bigg) \mathbf{I} + \mathbf{T}^S_C \, , \tag{74}$$

compare $(62)_1$, $(64)_1$, $(66)_1$, and $(68)_1$. The so-called classic *Cauchy* stress tensor

$$\mathbf{T}^S_C = 2 \rho^S \mathbf{F}_S \frac{\partial \psi^S}{\partial \mathbf{C}_S} \mathbf{F}^T_S \tag{75}$$

of the solid phase is defined as the part of $\mathbf{T}^S$ which depends on the derivative of the free *Helmholtz* energy ψ^S referring to the right *Cauchy-Green* tensor

$\mathbf{C}_S = \mathbf{F}_S^T \mathbf{F}_S$. For the compressible model, the hybrid model of second type and the incompressible model (Model A, C, and D) the free *Helmholtz* energy of the solid phase is a function of the absolute temperature Θ^S and the deformation tensor $\mathbf{C}_S$,

$$\psi^S = \hat{\psi}^S_{(\mathrm{A,C,D})}(\Theta^S, \mathbf{C}_S)\,, \tag{76}$$

and for the hybrid model of first type (Model B), in comparison with (76) the additional dependence on the free *Helmholtz* energy on the real part J_{SR} of the *Jacobian* $J_S = \det \mathbf{F}_S$ must be considered, i. e.

$$\psi^S = \hat{\psi}^S_{(B)}(\Theta^S, \mathbf{C}_S, J_{SR})\,. \tag{77}$$

In the following, only isotropic solid materials will be considered. For these kinds of materials the condition

$$\hat{\psi}^S(\mathbf{C}_S, \dots) = \mathring{\psi}^S(\mathbf{B}_S, \dots) \tag{78}$$

concerning the free *Helmholtz* energy for φ^S must hold, see Beatty [1], where the quantity $\mathbf{B}_S = \mathbf{F}_S \mathbf{F}_S^T$ denotes the left *Cauchy-Green* tensor. The symbol $(\dots)$ represents the quantity Θ^S (Model A, C, and D) and the quantities Θ^S and J_{SR} (Model B), respectively. The condition (78) is fulfilled if the free *Helmholtz* energy $\hat{\psi}^S$ is an isotropic scalar valued function of the principal invariants of $\mathbf{C}_S$ or $\mathbf{B}_S$ and other scalar quantities. Thus, for the description of isotropic thermo-elastic materials, the quantity $\mathbf{C}_S$ in (76) and (77), respectively, must be replaced by the invariants of $\mathbf{C}_S$, i. e.

$$\begin{aligned} \psi^S = \hat{\psi}^S(\mathbf{C}_S, \dots) \implies \psi^S &= \hat{\hat{\psi}}^S(\mathrm{I}_{\mathbf{C}_S}, \mathrm{II}_{\mathbf{C}_S}, \mathrm{III}_{\mathbf{C}_S}, \dots) \\ &= \hat{\hat{\psi}}^S(\mathrm{I}_{\mathbf{C}_S}, \mathrm{II}_{\mathbf{C}_S}, J_S, \dots)\,, \end{aligned} \tag{79}$$

where

$$\begin{aligned} \mathrm{I}_{\mathbf{C}_S} &= \mathbf{C}_S \cdot \mathbf{I}\,, \\ \mathrm{II}_{\mathbf{C}_S} &= \frac{1}{2}\left[(\mathbf{C}_S \cdot \mathbf{I})^2 - \mathbf{C}_S \cdot \mathbf{C}_S\right], \\ \mathrm{III}_{\mathbf{C}_S} &= \det \mathbf{C}_S = J_S^2 \end{aligned} \tag{80}$$

are the three invariants of the right *Cauchy-Green* tensor $\mathbf{C}_S$.

With (79), the classic stress of the solid phase can be written as

$$\begin{aligned} \mathbf{T}_C^S &= 2\,\rho^S\,\mathbf{F}_S \left(\frac{\partial \hat{\hat{\psi}}^S}{\partial \mathrm{I}_{\mathbf{C}_S}} \frac{\partial \mathrm{I}_{\mathbf{C}_S}}{\partial \mathbf{C}_S} + \frac{\partial \hat{\hat{\psi}}^S}{\partial \mathrm{II}_{\mathbf{C}_S}} \frac{\partial \mathrm{II}_{\mathbf{C}_S}}{\partial \mathbf{C}_S} + \frac{\partial \hat{\hat{\psi}}^S}{\partial J_S} \frac{\partial J_S}{\partial \mathbf{C}_S} \right) \mathbf{F}_S^T \\ &= \alpha_0\,\mathbf{I} + \alpha_1\,\mathbf{B}_S + \alpha_2\,\mathbf{B}_S\,\mathbf{B}_S\,. \end{aligned} \tag{81}$$

The quantities

$$\begin{aligned} \alpha_0 &= \rho^S J_S \frac{\partial \hat{\hat{\psi}}^S}{\partial J_S}, \\ \alpha_1 &= 2\rho^S \left(\frac{\partial \hat{\hat{\psi}}^S}{\partial \mathrm{I}_{\mathbf{C}_S}} + \mathrm{I}_{\mathbf{C}_S} \frac{\partial \hat{\hat{\psi}}^S}{\partial \mathrm{II}_{\mathbf{C}_S}} \right), \\ \alpha_2 &= -2\rho^S \frac{\partial \hat{\hat{\psi}}^S}{\partial \mathrm{II}_{\mathbf{C}_S}} \end{aligned} \tag{82}$$

are the material or elastic response functions of the solid. The form (81) of the relation for $\mathbf{T}_C^S$ is called the general constitutive equation for isotropic hyperelastic materials, compare Beatty [1].

It will be postulated that with respect to the description of isotropic thermo-hyperelastic compressible and incompressible saturated porous solids the free *Helmholtz* energy function is independent on the second invariant of $\mathbf{C}_S$, i. e. ψ^S is a function of Θ^S, $\mathrm{I}_{\mathbf{C}_S}$ and J_S for the Models A, C, and D,

$$\psi^S = \hat{\hat{\psi}}^S = \hat{\hat{\psi}}^S_{(\mathrm{A,C,D})} = \hat{\hat{\psi}}^S_{(\mathrm{A,C,D})}(\Theta^S, \mathrm{I}_{\mathbf{C}_S}, J_S), \tag{83}$$

and of Θ^S, $\mathrm{I}_{\mathbf{C}_S}$, J_S, and J_{SR} for the Model B,

$$\psi^S = \hat{\hat{\psi}}^S = \hat{\hat{\psi}}^S_{(\mathrm{B})} = \hat{\hat{\psi}}^S_{(\mathrm{B})}(\Theta^S, \mathrm{I}_{\mathbf{C}_S}, J_S, J_{SR}). \tag{84}$$

Following the structure of the well-known constitutive relations for the description of finite deformation problems of elastic and thermo-elastic one-component materials of Simo and Pister [20] and Miehe [17], the following ansatz for the free *Helmholtz* energy ψ^S for thermo-elastic porous solids will be postulated:

$$\psi^S = \hat{\hat{\psi}}^S = \hat{\hat{\psi}}^S_E(\Theta^S, \mathrm{I}_{\mathbf{C}_S}, J_S) + \underbrace{\hat{\hat{\psi}}^S_p(J_S, J_{SR})}_{\text{n. e. (A, C, D)}}, \tag{85}$$

where

$$\begin{aligned} \hat{\hat{\psi}}^S_E &= \frac{1}{\rho^S_{0S}} \Big\{ \lambda^S_{cp} [\tfrac{1}{2} (\log J_S)^2 + \xi^S] - \mu^S \log J_S + \frac{1}{2} \mu^S (\mathrm{I}_{\mathbf{C}_S} - 3) - \\ &\quad - 3\alpha^S k^S (\log J_S)(\Theta^S - \Theta^S_{0S}) - \\ &\quad - \rho^S_{0S} c^S (\Theta^S \log \frac{\Theta^S}{\Theta^S_{0S}} - \Theta^S + \Theta^S_{0S}) \Big\}, \qquad (86) \\ \hat{\hat{\psi}}^S_p &= \frac{1}{2\rho^S_{0S}} n^S_{0S} \lambda^S_{cp} k_p \left(\log \frac{J_{SR}}{J_S^{k^S/k^{SR}}} \right)^2 . \end{aligned}$$

The first part of the free *Helmholtz* energy is assigned to the so-called effective stresses of the solid, and the second part, which only exists for the Model B, is related to the pore pressure p (this fact will be shown later). In $(86)_1$ the abbreviation ξ^S stands for

$$\xi^S = J^S_{cp} \log J_S + \frac{1 - J^S_{cp}}{J^S_{cp} - 2} \left[\log \frac{J^S_{cp} - J_S}{J_S\,(J^S_{cp} - 1) - J^S_{cp}} - \log(1 - J^S_{cp}) \right] . \tag{87}$$

In (86), $k^S = 2/3\mu^S + \lambda^S$ and k^{SR} are the macroscopic compression modulus and the compression modulus of the real solid material, μ^S and λ^S are the macroscopic *Lamé* constants and α^S and c^S are the macroscopic thermal expansion coefficient and the specific heat of φ^S. The macroscopic material parameter λ^S_{cp} is proportional to the *Lamé* constant λ^S, and k_p is a dimensionless parameter which controls in connection with λ^S_{cp} the pore pressure and the volume deformations J_S and J_{SR} of the solid for the hybrid model of first type (Model B). Furthermore, the quantity J^S_{cp} $(0 \leq J^S_{cp} \leq 1)$ in (87) is assigned to the compression point of the solid phase ($J^S_{cp} = 0$ for compressible solids, $J^S_{cp} = n^S_{0S}$ for incompressible solids).

It is worth mentioning that for the compressible model (Model A), i. e. $J^S_{cp} = 0$ and $\xi^S = 0$, the structure of the ansatz (85) and $(86)_1$, respectively, is identical with the ansatz of Simo and Pister [20] for compressible elastic one-component materials, where they have not considered thermal effects. The structure of the terms in $(86)_1$, which are connected with the temperature, is based on the work of Miehe [17].

With the additive decomposition of the free *Helmholtz* energy of ψ^S, see (85), the classic *Cauchy* stress tensor of the solid can be split into two parts, viz.

$$\mathbf{T}^S_C = 2\,\rho^S\,\mathbf{F}_S\,\frac{\partial\hat{\psi}^S}{\partial\mathbf{C}_S}\,\mathbf{F}^T_S = 2\,\rho^S\,\mathbf{F}_S\,\frac{\partial\hat{\psi}^S_E}{\partial\mathbf{C}_S}\,\mathbf{F}^T_S + \underbrace{2\,\rho^S\,\mathbf{F}_S\,\frac{\partial\hat{\psi}^S_p}{\partial\mathbf{C}_S}\,\mathbf{F}^T_S}_{\text{n. e. (A, C, D)}} . \tag{88}$$

The first part

$$\mathbf{T}^S_E = 2\,\rho^S\,\mathbf{F}_S\,\frac{\partial\hat{\psi}^S_E}{\partial\mathbf{C}_S}\,\mathbf{F}^T_S \tag{89}$$

is the effective *Cauchy* stress tensor and the second part

$$\mathbf{T}^S_p = 2\,\rho^S\,\mathbf{F}_S\,\frac{\partial\hat{\psi}^S_p}{\partial\mathbf{C}_S}\,\mathbf{F}^T_S \tag{90}$$

is the pore pressure part of the classic *Cauchy* stress for the hybrid model of first type (Model B).

Using the ansatz (85), the material functions (82) for the representation form (81) of $\mathbf{T}_C^S$ read

$$
\begin{aligned}
\alpha_0 &= \rho^S J_S \frac{\partial \hat{\hat{\psi}}_E^S}{\partial J_S} + \rho^S J_S \frac{\partial \hat{\hat{\psi}}_p^S}{\partial J_S} = \\
&= \frac{1}{J_S}[\lambda_{cp}^S (\log J_S + \zeta^S) - \mu^S - 3\,\alpha^S k^S (\Theta^S - \Theta_{0S}^S)] - \\
&\quad - \frac{1}{J_S} n_{0S}^S \lambda_{cp}^S k_p \frac{k^S}{k^{SR}} \log \frac{J_{SR}}{J_S^{k^S/k^{SR}}}, \\
\alpha_1 &= 2\,\rho^S \frac{\partial \hat{\hat{\psi}}_E^S}{\partial \mathrm{I}_{\mathbf{C}_S}} = \frac{1}{J_S}\mu^S, \\
\alpha_2 &= 0,
\end{aligned}
\tag{91}
$$

where the relation $J_S = \rho_{0S}^S/\rho^S$ has been used. The quantity ζ^S in $(91)_1$ is the abbreviation for

$$
\zeta^S = J_S \frac{\partial \xi^S}{\partial J_S} = J_{cp}^S \left[1 - \frac{J_S}{J_S^2 + \dfrac{(J_{cp}^S)^2}{1 - J_{cp}^S}(J_S - 1)} \right]. \tag{92}
$$

The insertion of (91) into (81) yields the following constitutive relations for the effective *Cauchy* stress tensor and the pore pressure part for Model B of the classic *Cauchy* stress of the solid phase:

$$
\begin{aligned}
\mathbf{T}_E^S &= \frac{1}{J_S}[2\,\mu^S \mathbf{K}_S + \lambda_{cp}^S (\log J_S + \zeta^S)\,\mathbf{I} - 3\,\alpha^S k^S (\Theta^S - \Theta_{0S}^S)\,\mathbf{I}], \\
\mathbf{T}_p^S &= -\frac{1}{J_S} n_{0S}^S \lambda_{cp}^S k_p \frac{k^S}{k^{SR}} \log \frac{J_{SR}}{J_S^{k^S/k^{SR}}}\,\mathbf{I}.
\end{aligned}
\tag{93}
$$

In $(93)_1$, the quantity

$$
\mathbf{K}_S = \frac{1}{2}(\mathbf{B}_S - \mathbf{I}) \tag{94}
$$

is the *Karni-Reiner* strain tensor referred to the actual placement of the constituent φ^S.

It is to be bear in mind that for Model B the pore pressure is proportional to the derivative of the free *Helmholtz* energy of the solid phase referring to the real part J_{SR} of the *Jacobian* of φ^S, see $(64)_3$. With the ansatz (85) in consideration of $(86)_2$, the pore pressure of Model B can be expressed as

$$p = -\frac{\Theta^F}{\Theta^S}\,\rho^{SR}\,J_{SR}\,\frac{\partial \hat{\hat{\psi}}^S}{\partial J_{SR}} = -\frac{\Theta^F}{\Theta^S}\,\rho^{SR}\,J_{SR}\,\frac{\partial \hat{\hat{\psi}}^S_p}{\partial J_{SR}} = \\ = -\frac{\Theta^F}{\Theta^S}\,\frac{1}{J_{SR}}\,\lambda^S_{cp}\,k_p\,\log\frac{J_{SR}}{J_S^{k^S/k^{SR}}}\,, \tag{95}$$

where the relations $J_{SR} = \rho^{SR}_{0S}/\rho^{SR}$ and $\rho^S_{0S} = n^S_{0S}\,\rho^{SR}_{0S}$ have been used. Using the expression (95) for the pore pressure, the pore pressure part of the classic *Cauchy* stress tensor $(93)_2$ can be replaced by

$$\mathbf{T}^S_p = \frac{\Theta^S}{\Theta^F}\,\frac{J_{SR}}{J_S}\,n^S_{0S}\,\frac{k^S}{k^{SR}}\,p\,\mathbf{I} = \frac{\Theta^S}{\Theta^F}\,n^S\,\frac{k^S}{k^{SR}}\,p\,\mathbf{I}\,. \tag{96}$$

For the last expression of (96) the identity $J_{SR}/J_S = J_{SN}^{-1} = n^S/n^S_{0S}$ has been considered.

Summarizing the aforementioned implementations with respect to the pore pressure part of $\mathbf{T}^S_C$, the classic *Cauchy* stress of the solid can be written in the form

$$\mathbf{T}^S_C = \underbrace{\frac{\Theta^S}{\Theta^F}\,n^S\,\frac{k^S}{k^{SR}}\,p\,\mathbf{I}}_{\text{n. e. (A, C, D)}} + \mathbf{T}^S_E\,. \tag{97}$$

The constitutive relation for the effective *Cauchy* stress tensor of the solid is given by $(93)_1$. One sees that, for the Models A, C, and D, i. e. for the compressible model, the hybrid model of second type (incompressible solid, compressible liquid) and the incompressible model the effective *Cauchy* stress tensor is identical with the classic *Cauchy* stress of the constituent φ^S.

With respect to the constitutive relation of the *Cauchy* stress tensor of the solid for the compressible model (Model A), one has to formulate an ansatz for the real part of the volume deformation J_{SR} in dependence on the *Jacobian* J_S, see $(62)_1$ and (74), respectively. With the ansatz

$$J_{SR} = (J_S)^m\,, \quad m = \frac{k^S}{k^{SR}}\,, \tag{98}$$

see de Boer and Bluhm [5], and the relation

$$J_S\,J_{SR}^{-1}\,\frac{\partial J_{SR}}{\partial J_S} = \frac{k^S}{k^{SR}} \tag{99}$$

one obtains in connection with (97) the constitutive equation

$$\mathbf{T}^S = -\frac{\Theta^S}{\Theta^F}\,n^S\,p\,(1 - \frac{k^S}{k^{SR}})\,\mathbf{I} + \mathbf{T}^S_E \tag{100}$$

for the models with compressible solid phases (Model A and B). The ansatz (98) results from (95) for $p = 0$. The reduction of the influence of the pore pressure on $\mathbf{T}^S$ due to the compressibility of the real solid material results on the one hand from the ansatz for J_{SR} (Model A) and on the other hand from the consideration of J_{SR} as process variable (Model B). For incompressible porous solids (Model C and D), i. e. $k^{SR} = \infty$, the expression (100) turns into

$$\mathbf{T}^S = -\frac{\Theta^S}{\Theta^F} n^S p \mathbf{I} + \mathbf{T}^S_E . \tag{101}$$

If the constituents have the same temperatures ($\Theta^S = \Theta^F$), the relations (100) and (101) simplify to

$$\mathbf{T}^S = -n^S p \left(1 - \frac{k^S}{k^{SR}}\right) \mathbf{I} + \mathbf{T}^S_E , \quad \mathbf{T}^S = -n^S p \mathbf{I} + \mathbf{T}^S_E . \tag{102}$$

$(102)_1$ represents a version of the effective stress principle for saturated compressible porous solids originally proposed by Šuklje [21]. The validity of this formula for hydrostatic stress states has been recently proved experimentally by Lade and de Boer [16]. The relation $(102)_2$ is *Terzaghi*'s well-known statement of the effective stress principle for incompressible saturated porous solids, see Terzaghi [22].

As already mentioned, the material parameter λ^S_{cp} is proportional to the *Lamé* constant λ^S. In order to show the structure of λ^S_{cp}, the effective stresses of the solid will be derived for the description of small deformation processes. In analogy with the well-known relation $\mathbf{S}^S = J_S \mathbf{F}_S^{-1} \mathbf{T}^S \mathbf{F}_S^{T-1}$ for the second (symmetric) *Piola-Kirchhoff* stress tensor $\mathbf{S}^S$ referred to the reference placement of φ^S, the effective stress

$$\begin{aligned} \mathbf{S}^S_E &= J_S \mathbf{F}_S^{-1} \mathbf{T}^S_E \mathbf{F}_S^{T-1} = \\ &= 2\mu^S \overset{R}{\mathbf{K}}_S + \lambda^S_{cp} (\log J_S + \zeta^S) \mathbf{C}_S^{-1} - 3\,\alpha^S k^S (\Theta^S - \Theta^S_{0S}) \mathbf{C}_S^{-1} \end{aligned} \tag{103}$$

concerning the reference placement of the solid phase will be introduced. The tensor

$$\overset{R}{\mathbf{K}}_S = \frac{1}{2} (\mathbf{I} - \mathbf{C}_S^{-1}) \tag{104}$$

denotes the *Karni-Reiner* strain tensor referred to the reference placement of the constituent φ^S.

In order to obtain a *Hooke*an-type constitutive law, the linearization

$$\mathbf{S}^S_{E,\text{lin}} = \mathbf{S}^S_E \Big|_{\mathcal{P}_0} + \frac{\partial \mathbf{S}^S_E}{\partial \mathbf{C}_S}\Big|_{\mathcal{P}_0} \Delta \mathbf{C}_S + \frac{\partial \mathbf{S}^S_E}{\partial \Theta^S}\Big|_{\mathcal{P}_0} \Delta \Theta^S , \tag{105}$$

where

$$(\cdot)\Big|_{\mathcal{P}_0} = (\cdot)\Big|_{\mathbf{C}_S=\mathbf{I},\,\Theta^S=\Theta^S_{0S}} \tag{106}$$

of the stress tensor $\mathbf{S}^S_E$ at the position $\mathcal{P}_0$ is introduced, i. e. the stresses (103) will be evaluated in a *Taylor* series for $\mathbf{C}_S = \mathbf{I}$ and $\Theta^S = \Theta^S_{0S}$ by neglecting higher-order terms.

Considering (103), it can be concluded that the stress tensor $\mathbf{S}^S_E$ at the position $\mathcal{P}_0$ vanishes:

$$\mathbf{S}^S_E\Big|_{\mathcal{P}_0} = \mathbf{0}\,. \tag{107}$$

The elastic and the thermal tangent operator concerning the reference placement at the position $\mathcal{P}_0$ are given by

$$\frac{\partial \mathbf{S}^S_E}{\partial \mathbf{C}_S}\Big|_{\mathcal{P}_0} = \mu^S \overset{4}{\mathbf{I}} + \frac{1}{2}\,\lambda^S \overset{4}{\bar{\mathbf{I}}}\,,\quad \frac{\partial \mathbf{S}^S_E}{\partial \Theta^S}\Big|_{\mathcal{P}_0} = -3\,\alpha^S\,k^S\,\mathbf{I}\,. \tag{108}$$

In $(108)_1$ the abbreviation

$$\lambda^S = \lambda^S_{cp}\left[1 + J^S_{cp}\left(1 + \frac{(J^S_{cp})^2}{1 - J^S_{cp}}\right)\right] \tag{109}$$

has been used. Thus, the linearized effective stress tensor $\mathbf{S}^S_{E,\mathrm{lin}}$ reads

$$\mathbf{S}^S_{E,\mathrm{lin}} = \mu^S\,\Delta\mathbf{C}_S + \frac{1}{2}\,\lambda^S(\Delta\mathbf{C}_S\cdot\mathbf{I})\,\mathbf{I} - 3\,\alpha^S\,k^S\,\Delta\Theta^S\,\mathbf{I}\,. \tag{110}$$

Within the framework of the linear theory the rate of the right *Cauchy-Green* tensor can be expressed with help of the *Lagrange*an strain tensor

$$\mathbf{E}_S = \frac{1}{2}\,(\mathbf{C}_S - \mathbf{I})\,, \tag{111}$$

viz.

$$\Delta\mathbf{C}_S = \mathbf{C}_S - \mathbf{C}_S\Big|_{\mathcal{P}_0} = \mathbf{C}_S - \mathbf{I} = 2\,\mathbf{E}_S\,. \tag{112}$$

The rate of the temperature within the framework of the linear theory is given by

$$\Delta\Theta^S = \Theta^S - \Theta^S\Big|_{\mathcal{P}_0} = \Theta^S - \Theta^S\Big|_{\hat{\mathcal{P}}_0} = \Theta^S - \Theta^S_{0S}\,. \tag{113}$$

Thus, the insertion of (112) and (113) into (110) yields the following form of the linearized law of *Hooke*an type, with respect to the reference placement:

$$\begin{aligned} \mathbf{S}^S_{E,\mathrm{lin}} &= 2\,\mu^S\,\mathbf{E}_S + \lambda^S(\mathbf{E}_S\cdot\mathbf{I})\,\mathbf{I} - 3\,\alpha^S\,k^S\,(\Theta^S - \Theta^S_{0S})\,\mathbf{I} \\ &= 2\,\mu^S\,\mathbf{E}^D_S + k^S(\mathbf{E}_S\cdot\mathbf{I})\,\mathbf{I} - 3\,\alpha^S\,k^S\,(\Theta^S - \Theta^S_{0S})\,\mathbf{I}\,, \end{aligned} \tag{114}$$

where the symbol $(\cdot)^D$ denotes the deviatorical part of the tensor quantity $(\cdot)$. The determination of the strain tensor

$$\mathbf{E}_S = \frac{1}{2\,\mu^S}\,\mathbf{S}^S_{E,\mathrm{lin}} - \frac{\lambda^S}{2\,\mu^S\,(2\,\mu^S + 3\,\lambda^S)}\,(\mathbf{S}^S_{E,\mathrm{lin}}\cdot\mathbf{I})\,\mathbf{I} + \alpha^S(\Theta^S - \Theta^S_{0S})\,\mathbf{I} \tag{115}$$

from (114) reveals that the response equation for $\mathbf{E}_S$ can be compared with the well-known strain relation of the linear theory of one-component thermo-elastic materials.

(109) shows that the material parameter λ^S_{cp} of the free *Helmholtz* energy ψ^S, see (85) and (86), respectively, can be expressed with help of λ^S and the quantity J^S_{cp}, which is related to the compression point of the porous solid. For compressible materials, i. e. $J^S_{cp} = 0$, the *Lamé* constant λ^S is equal to λ^S_{cp}.

In the following, the constitutive law for $\mathbf{T}^S_E$, see $(93)_1$, for compressible and incompressible empty elastic porous solids ($p = 0$, $\mathbf{T}^S = \mathbf{T}^S_E$) within the framework of finite deformation processes will be discussed with the boundary problem shown in Figure 3 a. Thermal effects will be neglected, i. e. $\Theta^S = \Theta^S_{0S}$. In Figure 3 b the results (displacements versus load referring to μ^S) of the FEM-simulation of the boundary problem for the plain strain state for a compressible porous solid and for two different incompressible porous solids are presented. For a given *Lamé* constant μ^S the other material parameters are taken as $\lambda^S = 0.6667\,\mu^S$ and $k^S = 1.3333\,\mu^S$ (*Poisson*'s ratio is 0.2), i. e. stresses perpendicular to the load direction are considered. The parameter λ^S_{cp} can be calculated with help of (109) for a given J^S_{cp}. For the compressible empty porous solid the real volume deformation versus the determinant of the deformation gradient $\mathbf{F}_S$ is shown in Figure 3 c, where the ratio $k^S/k^{SR} = 1/3$ concerning the partial and the real compression moduli has been used.

As mentioned, the dimensionless parameter k_p controls, in connection with λ^S_{cp}, the pore pressure and the volume deformations J_S and J_{SR} of the compressible solid for the hybrid model of first type (Model B), see (95). The influence of k_p for different pore fluid pressures is shown in Figure 4, where it has been considered that for the hybrid model of first type the parameter which is related to the compression point of the solid is equal to zero ($J^S_{cp} = 0$) and therefore $\lambda^S_{cp} = \lambda^S$. Furthermore, the same ratio of the compression moduli of the partial and the real solid phase has been used as

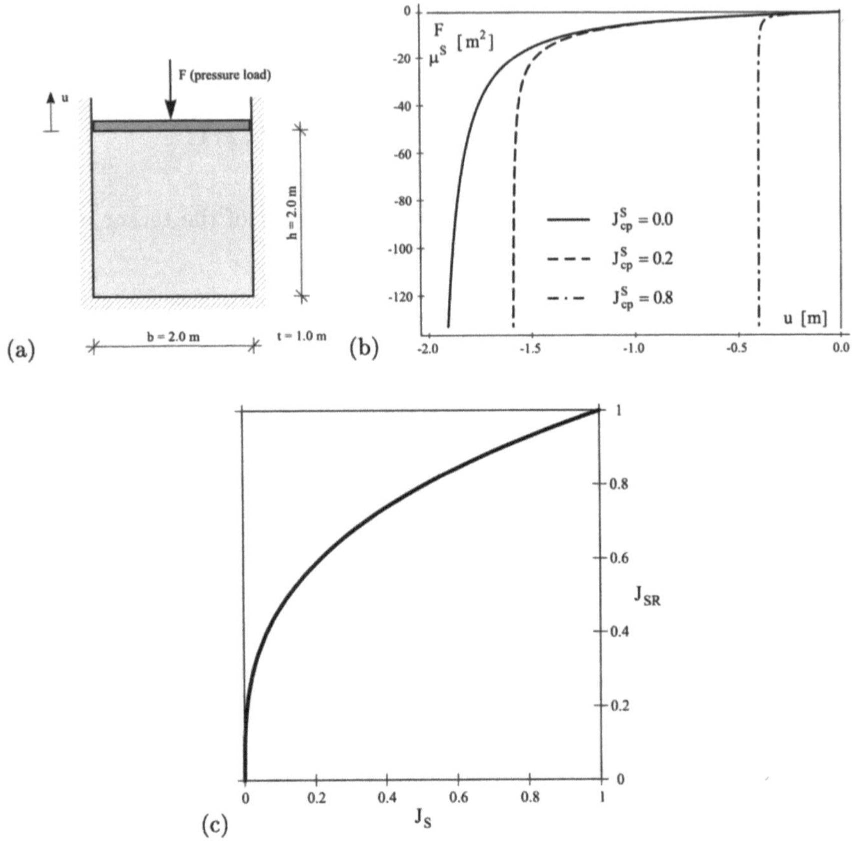

Fig. 3. (a) Boundary problem and load, (b) Displacements versus load referring to μ^S for empty porous solids (compression), (c) J_{SR} versus J_S for $k^S/k^{SR} = 1/3$ (compressible porous solid, $p = 0$).

for Model A. Thus, the curve for $p = 0$ is identical with the representation of J_{SR} versus J_S of the curve of the compressible model.

It is worth mentioning that the principle of effective stresses which has been formulated by Biot and Willis [2] for the theory of consolidation can also be derived within the framework of the model. In this case, the constitutive assumptions

$$J_{SR} = \frac{1}{n_{0S}^S\, k^{SR}} \left[k^S\,(J_S - 1) + n_{0S}^S\, k^{SR}\right] \tag{116}$$

and

$$\hat{\psi}_p^S = \frac{1}{2\,\rho_{0S}^S}\, n_{0S}^S\, \lambda_{cp}^S\, k_p \log\left[J_{SR} - \frac{k^S}{n_{0S}^S\, k^{SR}}\,(J_S - 1)\right]^2 \tag{117}$$

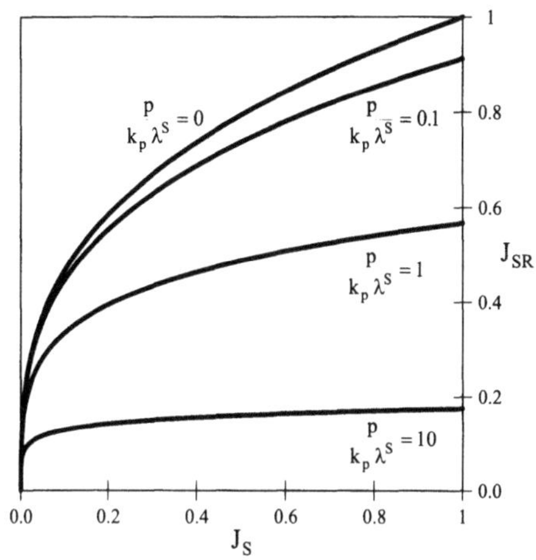

Fig. 4. Illustration of the with $1/k_p\lambda^S$ weighted pore pressure of the incompressible fluid in dependence on the partial and real volume deformations of the compressible solid (Model B).

for the real volume deformation of the solid (Model A) and the pressure part of the free *Helmholtz* energy of φ^S (Model B) instead of $(98)_1$ and $(86)_2$ must be used. The relations (116) and (117) are only valid for small deformation processes, because J_{SR} is a linear function of J_S. With (116) and (117) one obtains for $\Theta^S = \Theta^F$ the well-known formula

$$\mathbf{T}^S = -n^S p \left(1 - \frac{1}{n^S} \frac{k^S}{k^{SR}}\right) \mathbf{I} + \mathbf{T}^S_E \tag{118}$$

of Biot and Willis [2] for the stress tensor of the solid. This relation of the concept of effective stresses has also been formulated by Nur and Byerlee [19] for wet rocks under high hydrostatic pressure. In order to demonstrate the physical validity of (118), Nur and Byerlee [19] made several simple compression and pore pressure tests on *Weber* sandstone, for example. It is worth noting that the porosity of the tested *Weber* sandstone is nearly 6 %, i. e. $n^S = 0.94$. Thus, the calculation of the effective hydrostatic pressure of the solid phase using (118) or $(102)_1$ yields nearly the same results so that, in the opinion of the author, the experiments of Nur and Byerlee [19] do not clearly show whether the ratio of the compression moduli is connected with the volume fraction of the solid or not.

5.2 Constitutive relations for the stresses of the fluid phase

For the compressible model and the hybrid model of second type (Model A and C), a constitutive relation for the pore fluid pressure p of the compressible fluid phase must be formulated. Concerning the aforementioned models the fluid pressure is proportional to the derivative of the free *Helmholtz* energy function referring to the real density of the fluid, see $(62)_3$ and $(66)_3$. With respect to the compressible fluid, an ideal gas will be taken into consideration. An ideal gas is defined as an inviscid fluid, the pressure is related to the temperature Θ^F and the real density ρ^{FR}. With the ansatz

$$\psi^F = \psi^F_{(\mathrm{A,C})} = -\frac{R}{M}\,\Theta^F \ln\frac{1}{\rho^{FR}} - c^F\left(\Theta^F \log\frac{\Theta^F}{\Theta^F_{0F}} - \Theta^F + \Theta^F_{0F}\right) \tag{119}$$

one obtains the well-known constitutive relation

$$p = \frac{R}{M}\,\Theta^F \rho^{FR} \tag{120}$$

for an ideal gas, see de Boer [4]. In (119) and (120), respectively, the quantities M and R are the molecular weight of the fluid (gas) and the gas constant, which is the same for all ideal gases. Furthermore, the quantities c^F and Θ^F_{0F} denote the macroscopic thermal expansion coefficient of the fluid (gas) phase and the fluid temperature at the position $\mathbf{X}_F$ at time $t = t_0$.

If the pores are filled with water, the pore fluid can be considered in a certain range as incompressible (Model B and D). In this case, the first expression in (119) does not exist, i. e. the ansatz for ψ^F reduces to

$$\psi^F = \psi^F_{(\mathrm{B,D})} = -c^F\left(\Theta^F \log\frac{\Theta^F}{\Theta^F_{0F}} - \Theta^F + \Theta^F_{0F}\right). \tag{121}$$

As shown before, for an incompressible fluid, the pore pressure is determined by real and partial volume deformations of the porous solid (Model A), or the pressure is an indeterminate reaction force in the model (Model D).

6 Conclusions

In the present paper, a consistent binary model consisting of a thermo-elastic porous solid and a non-viscous fluid with different temperatures of the constituents has been developed. The binary model includes the compressible model, the hybrid models of first and second type and the incompressible model.

The investigations show that the pore pressure part of the solid stress tensor is connected with the temperatures of both phases. For fluid-saturated compressible porous solids, the effective stresses of the solid phase are influenced by the reduced pore fluid pressure. The reduction factor depends on the real material property of the solid. Furthermore, if the temperatures of the phases are equal, it has been shown that the results of the constitutive equations for the stresses of the binary model can be transferred to the corresponding constitutive assumptions of the models of Biot and Willis [2], Šuklje [21] and Nur and Byerlee [19]. Also, *Terzaghi*'s statement of the effective stress principle (Terzaghi [22]) can be derived.

The results show that further experimental and numerical investigations are needed to clarify the influence of the constitutive relations for the heat flux vectors and the supply terms of momentum and energy. Consequently the aim of further investigations will be to simulate initial and boundary value problems numerically in connection with test observations in order to identify the parameters of the aforementioned constitutive equations.

References

1. Beatty, M. F.: Topics in finite elasticity: Hyperelasticty of rubber, elastomers, and biological tissues – with examples. *Appl. Mech. Rev.* **40** (1987), 1699–1734.
2. Biot, M. A., Willis, D. G.: The elastic coefficients of the theory of consolidation. *J. Appl. Mech.* **24** (1957), 594–601.
3. Bluhm, J.: *A consistent model for saturated and empty porous media.* Forschungsberichte aus dem Fachbereich Bauwesen 74, Universität-GH Essen 1997.
4. de Boer, R.: *Theory of porous media – highlights in the historical development and current state.* Springer-Verlag, Berlin 2000.
5. de Boer, R., Bluhm, J.: The influence of compressibility on the stresses of elastic porous solids – semimicroscopic investigations. *Int. J. Solids Structures* **36** (1999), 4805–4819.
6. de Boer, R., Didwania, A. K.: The effect of uplift in liquid-saturated porous solids – Karl von Terzaghi's contributions and recent findings. *Géotechnique* **47**(2) (1997), 289–298.
7. de Boer, R., Ehlers, W.: *Theorie der Mehrkomponentenkontinua mit Anwendung auf bodenmechanische Probleme*, Teil I. Forschungsberichte aus dem Fachbereich Bauwesen 40, Universität-GH Essen 1986.
8. Bowen, R. M.: Incompressible porous media models by use of the theory of mixtures. *Int. J. Eng. Sci.* **18** (1980), 1129–1148.
9. Bowen, R. M.: Compressible porous media models by use of the theory of mixtures. *Int. J. Eng. Sci.* **20** (1982), 697–735.
10. Bowen, R. M., Wiese, J. C.: Diffusion in mixtures of elastic materials. *Int. J. Eng. Sci.* **7** (1969), 689–722.
11. Coleman, B. D., Noll, W.: The thermodynamics of elastic materials with heat conduction and viscosity. *Arch. Rational Mech. Anal.* **13** (1963), 167–178.
12. Ehlers, W.: *Poröse Medien – ein kontinuumsmechanisches Modell auf der Basis der Mischungstheorie.* Forschungsberichte aus dem Fachbereich Bauwesen 47, Universität-GH Essen 1988.
13. Ehlers, W.: On the thermodynamics of elasto-plastic porous media. *Arch. Mech.* **41** (1989), 73–93.
14. Ehlers, W.: Constitutive equations for granular materials in geomechanical contexts. In Hutter, H. (ed.): *Continuum mechanics in environmental sciences and geophysics*, CISM Courses and Lecture Notes No. 337, Springer-Verlag, Wien 1993, 313–402.
15. Green, A. E., Naghdi, P. M.: A dynamical theory of interacting continua. *Int. J. Engng. Sci.* **3** (1965), 231–241.
16. Lade, P. V., de Boer, R.: The concept of effective stress for soil, concrete and rock. *Géotechnique* **47**(1) (1997), 61–78.

17. Miehe, C.: *Zur numerischen Behandlung thermomechanischer Prozesse.* Forschungs- und Seminarberichte aus dem Bereich der Mechanik der Universität Hannover, Bericht-Nr. F88/6, Universität Hannover 1988.
18. Müller, I.: A thermodynamic theory of mixtures of fluids. *Arch. Rational Mech. Anal.* **28** (1968), 1–39.
19. Nur, A., Byerlee, J. D.: An exact effective stress law for elastic deformation of rock with fluids. *J. Geophys. Res.* **76** (1971), 6414–6419.
20. Simo, J. C., Pister, K. S.: Remarks on rate constitutive equations for finite deformation problems: Computational implications. *Comput. Meth. Appl. Mech. Eng.* **46** (1984), 201–215.
21. Šuklje, L.: *Rheological aspects of soil mechanics.* Wiley Interscience, New York 1969.
22. von Terzaghi, K.: The shearing resistance of saturated soils and the angle between the planes of shear. In Casagrande, A. *et al.* (eds.): *Proceedings of the Internationale Conference on Soil Mechanics and Foundation Engineering,* Vol. I, Harvard University 1936, pp. 54–56.
23. Truesdell, C.: *Rational thermodynamics.* 2nd. ed., Springer-Verlag, New York 1984.
24. Truesdell, C., Toupin, R. A.: The classical field theories. In Flügge, S. (ed.): *Handbuch der Physik,* Band III/1, Springer-Verlag, Berlin 1960.

II. Theory of Porous Media

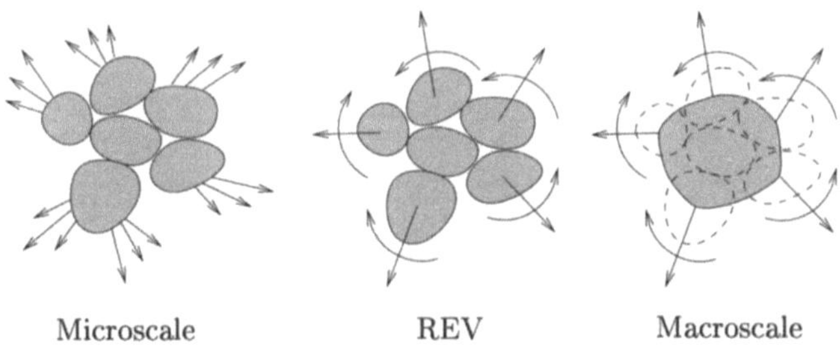

Micropolar mixture models on the basis of the Theory of Porous Media

Stefan Diebels

University of Stuttgart, Institute of Applied Mechanics (CE),
70550 Stuttgart, Germany

Abstract. The behaviour of porous media can be described in a continuum mechanical setting by the Theory of Porous Media, i. e. by a mixture theory extended by the concept of volume fractions. In addition to the volume fractions, micropolarity is taken into account to model the internal structure of porous media on the macroscopic scale. After a microscopic motivation of the approach, which shows that it is physically motivated to deal with micropolar mixture models, the kinematics, the balance relations, and the constitutive framing of such a theory are discussed. A set of model equations is formulated within the presented frame and applied to some boundary value problems showing the evidence of the theoretical approach.

1 Motivation

The theoretical description of porous media plays an important role in several branches of engineering such as geomechanics, petrol engineering, and ground water flow. On the other hand, new man-made porous materials such as polymer foams or metal foams became important in the production and in the applications during the last years. Therefore, theoretical tools must be developed, which lead to thermodynamically consistent models of porous materials on a macroscopic scale and which describe the mechanical behaviour of these materials possessing a clearly visible microstructure.

During the last decades, the Theory of Porous Media (TPM) was developed and successfully applied to a wide range of engineering problems, cf. de Boer [4], Ehlers and Ellsiepen [17], and Lewis and Schrefler [26]. In addition, the TPM was extended in the sense of the *Cosserat* theory, i. e. rotational degrees of freedom of the material points are taken into account within the continuum description [10, 11, 39]. In this context, the advantage of extended continuum theories is twofold: On the one hand, the micropolar extention to the TPM is motivated from the present microstructure. It arises due to the internal structure of the grains in geomaterials like soils or due to the structure of cells in metal foams or polymer foams. Furthermore, the extended theory inherently describes the size effect which can be observed in these materials. On the other hand, such an extended theory is able to deal in a realistic fashion with softening material behaviour. In numerical examples it leads to shearbands of finite width, independent of the spatial discretization and, therefore, regularizes the otherwise ill-posed problem [32, 39].

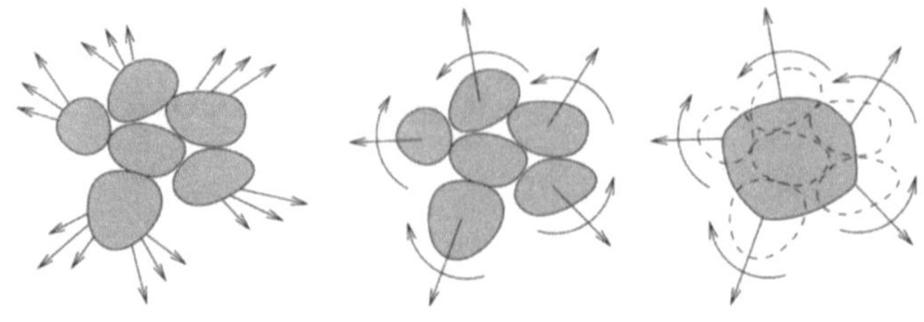

Fig. 1. Homogenization of granular materials.

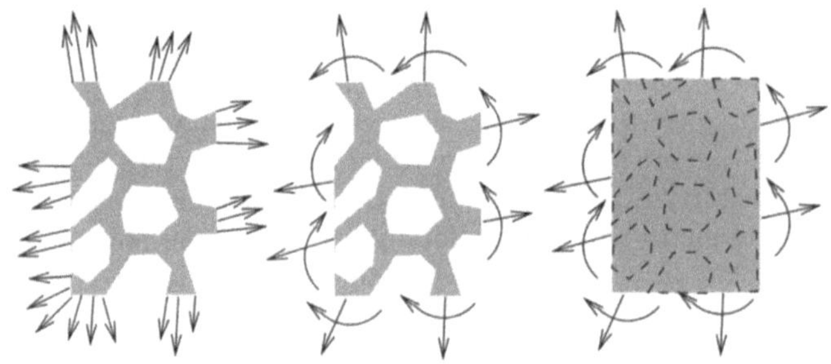

Fig. 2. Homogenization of foams.

Figures 1 and 2 show principle sketches of a representative elementary volume of a granular medium and of a foam, respectively. Both materials show a strongly fluctuating stress distribution on the microscale. If a homogenization procedure is applied to these materials, the fluctuating stresses are replaced by mean stresses on a mesoscale and by resulting couple stresses. In the case of the grains, the mean stresses represent the resultant contact forces with respect to the centre of gravity of the individual grains and the related moments. The corresponding micromechanical model is based on particle dynamics or on the Discrete Element Method [8]. The equivalence between rigid body dynamics on the microscale and the *Cosserat* theory on the macroscale is evident [12, 16]. For foams, beam models as proposed by Gibson and Ashby [22] or Onck [30] are suitable on the microscale. The inhomogeneous stress distribution within the beams again leads to resultant forces and moments on the mesoscale, and, therefore, these models may also be homogenized to micropolar media on the macroscale.

The mentioned microscopic models are quite useful to get insight into the physical processes governing the behaviour of the porous medium, but in order to model real engineering problems, millions of degrees of freedom are necessary. Therefore, the computational cost is extremely high. Furthermore, this kind of models becomes rather complicated, if pore fluids are taken into account, i. e. they are not well suited to model the coupled solid-fluid problem, which often arises in the description of saturated porous materials.

According to these examples, it becomes evident that a macroscopic multiphase approach is necessary and that the micropolar extension of the TPM is physically well motivated if granular media or foams have to be modelled.

Section 2 gives a brief outline of the extended kinematics of a micropolar twophase model starting from the concept of superimposed continua. Section 3 collects the governing balance equations and in Section 4 the constitutive framing is discussed. In Section 5, the consequences of material incompressibility are resumed, before an example is presented in Section 6.

2 Kinematics

In the following sections, a binary mixture is taken into account consisting of a solid matrix and a viscous pore fluid. The macroscopic TPM approach to model such a mixture is based on the concept of superimposed continua [38] and can be applied to an arbitrary number of constituents. After a real or a virtual homogenization procedure of the microstructure, the material points of the solid matrix φ^S and of the pore fluid φ^F occupy the same spatial position $\mathbf{x}$, even if they follow their own function of motion starting from different positions $\mathbf{X}_S$ and $\mathbf{X}_F$ in the reference configuration, i. e.

$$\mathbf{x} = \chi_S(\mathbf{X}_S, t) = \chi_F(\mathbf{X}_F, t)\,. \tag{1}$$

Material line elements of the individual constituents are transformed from the reference configuration into the actual configuration by the deformation gradient:

$$\mathrm{d}\mathbf{x}_\alpha = \mathbf{F}_\alpha\,\mathrm{d}\mathbf{X}_\alpha \qquad \text{with} \qquad \mathbf{F}_\alpha = \frac{\partial \mathbf{x}}{\partial \mathbf{X}_\alpha}\,, \qquad \alpha = S, F\,. \tag{2}$$

The velocity of the material points is obtained by taking the derivative of (1) with respect to time t:

$$\mathbf{v}_\alpha = \frac{\partial \chi_\alpha(\mathbf{X}_\alpha, t)}{\partial t} = \overset{\prime}{\mathbf{x}}_\alpha\,. \tag{3}$$

The *Euler*ian representation with respect to the current configuration is obtained by use of the inverse function of motion:

$$\mathbf{X}_S = \chi_S^{-1}(\mathbf{x}, t)\,, \qquad \mathbf{X}_F = \chi_F^{-1}(\mathbf{x}, t)\,. \tag{4}$$

As motivated in Section 1, the porous skeleton is assumed to be micropolar. Material points of the continuous model are able to transfer couple stresses, and, as a consequence, they possess an orientation and may undergo rotations independently of the macroscopic displacement field. In order to describe these rotations of the material points of the skeleton, a director Ξ_S

is attached to each material point of the solid phase in its reference configuration. This director is rotated by the micromotion $\bar{\mathbf{R}}_S$ into the director $\boldsymbol{\xi}_S$ of the actual configuration [21]:

$$\boldsymbol{\xi}_S = \bar{\mathbf{R}}_S \, \Xi_S \qquad \text{with} \qquad \bar{\mathbf{R}}_S \bar{\mathbf{R}}_S^T = \mathbf{I}, \qquad \det \bar{\mathbf{R}}_S = 1 . \tag{5}$$

For simplicity, micropolarity of the fluid phase is excluded, but the presented concepts also hold in the case of micropolar fluids as discussed in [10].

The deformation tensors of the micropolar matrix can be derived in the usual way [10, 21, 39]. If the scalar product between the directors $\boldsymbol{\xi}_S$ and line elements $d\mathbf{x}_S$ is taken into account, the following non-symmetric deformation tensors of the reference configuration and of the actual configuration, respectively, may be derived:

$$\bar{\mathbf{U}}_S = \bar{\mathbf{R}}_S^T \mathbf{F}_S , \qquad \bar{\mathbf{V}}_S = \mathbf{F}_S \bar{\mathbf{R}}_S^T . \tag{6}$$

The corresponding micropolar *Lagrange* and *Almansi* strain tensors read

$$\bar{\mathbf{E}}_S = \bar{\mathbf{U}}_S - \mathbf{I}, \qquad \bar{\mathbf{A}}_S = \mathbf{I} - \bar{\mathbf{V}}_S^{-1} . \tag{7}$$

Beside the deformation and strain tensors, the micropolar continuum requires additional deformation measures related to the gradient $\text{Grad}_S \, \bar{\mathbf{R}}_S$ of the micromotion. The so-called curvature tensor is defined as [21, 33]

$$^R\overset{3}{\mathcal{K}}_S = (\bar{\mathbf{R}}_S^T \, \text{Grad}_S \, \bar{\mathbf{R}}_S)^{\underline{3}} . \tag{8}$$

The underlined superscript represents the order of the resulting tensorial object. The third order curvature tensor (8) is skew symmetric with respect to the first and second basis system and can be contracted to a second order tensor without loss of information by application of the *Ricci* tensor $\overset{3}{\mathbf{E}}$:

$$^R\bar{\mathcal{K}}_S = -\tfrac{1}{2} (\overset{3}{\mathbf{E}} \, {}^R\overset{3}{\mathcal{K}}_S)^{\underline{2}} . \tag{9}$$

As for the deformation tensor $\bar{\mathbf{U}}_S$, it can be shown that $^R\bar{\mathcal{K}}_S$ is a tensor of the reference configuration [10]. Push forward of (9) with the micromotion and the inverse deformation gradient yields the second order curvature tensor of the current configuration:

$$\bar{\mathcal{K}}_S = \bar{\mathbf{R}}_S \, {}^R\bar{\mathcal{K}}_S \, \mathbf{F}_S^{-1} . \tag{10}$$

Following [39], the linearization of both finite strain tensors $\bar{\mathbf{E}}_S$ and $\bar{\mathbf{A}}_S$ according to (7) leads to the *Cosserat* strain tensor

$$\bar{\varepsilon}_S = \text{Grad}_S \, \mathbf{u}_S + \overset{3}{\mathbf{E}} \, \bar{\varphi}_S = \text{sym} \, \bar{\varepsilon}_S + \text{skw} \, \bar{\varepsilon}_S \tag{11}$$

in the case of small deformations. In (11), $\mathbf{u}_S$ is the displacement field of the solid skeleton and $\bar{\varphi}_S$ is the angle of rotation of the solid skeleton related to the orthogonal tensor $\bar{\mathbf{R}}_S$. If the linearized strain tensor is split into a symmetric and a skew symmetric part, it is found that the symmetric part is equal to the usual linearized strain tensor of the classical continuum theory while the skew symmetric part represents the so-called extra rotation, i. e. the difference between the continuum rotation $\boldsymbol{\varphi}_S = -\operatorname{skw}\operatorname{Grad}_S \mathbf{u}_S$ and the total or *Cosserat* rotation $\bar{\boldsymbol{\varphi}}_S$. Linearization of the curvature tensor yields the gradient of the *Cosserat* rotation [39]:

$$\bar{\boldsymbol{\kappa}}_S = \operatorname{Grad}_S \bar{\boldsymbol{\varphi}}_S \,. \tag{12}$$

If elastic behaviour of the skeleton is assumed, the stresses and couple stresses are functions of the deformation tensors according to (6) or of the strain tensors according to (7), respectively, and of the curvature (9).

The constitutive behaviour of the non-polar viscous fluid can be taken into account by the strain rate or stretching defined in the usual way as symmetric part of the velocity gradient $\mathbf{L}_F$:

$$\mathbf{D}_F = \operatorname{sym}\mathbf{L}_F \qquad \text{with} \qquad \mathbf{L}_F = (\mathbf{F}_F)'_F \, \mathbf{F}_F^{-1} \,. \tag{13}$$

The material time derivative following the motion of phase $\alpha = S,\, F$ is given by

$$(\,\cdot\,)'_\alpha = \frac{\partial(\,\cdot\,)}{\partial t} + \operatorname{grad}(\,\cdot\,) \cdot \mathbf{v}_\alpha \tag{14}$$

for a scalar quantity. It represents the change of the physical quantity under study that is realized by an observer moving with the velocity $\overset{\prime}{\mathbf{x}}_\alpha$ of the constituent φ^α.

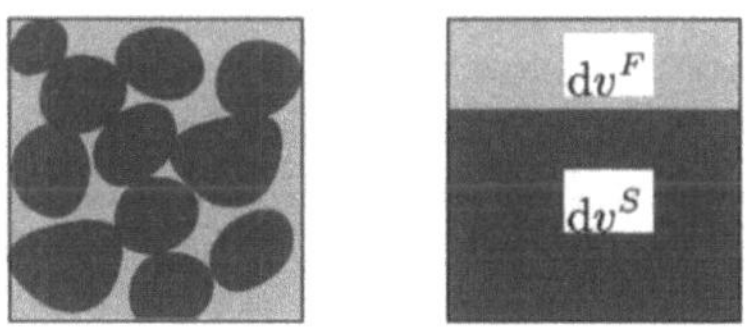

Fig. 3. Volume element with microstructure and with partial volume elements.

Due to the fact, that both phases occupy the same spatial point in the actual configuration according to the concept of superimposed continua, the local composition of the mixture is described by the volume fractions n^α. They relate the partial volume element $\mathrm{d}v^\alpha$ occupied by the phase α on the microscale to the volume element $\mathrm{d}v$ occupied by the whole mixture (Figure 3):

$$n^\alpha = \frac{\mathrm{d}v^\alpha}{\mathrm{d}v} \qquad \text{with} \qquad n^S + n^F = 1 \,. \tag{15}$$

According to the saturation constraint $(15)_2$, the whole volume element $\mathrm{d}v$ is nothing more than the sum of the partial volume elements occupied by the individual phases. Depending on the choice of the volume element $\mathrm{d}v$ and $\mathrm{d}v^\alpha$, respectively, two different densities may be introduced. If the element of mass $\mathrm{d}m^\alpha$ is related to the volume element $\mathrm{d}v^\alpha$ occupied by the constituent φ^α, the so-called effective density is obtained,

$$\rho^{\alpha R} = \frac{\mathrm{d}m^\alpha}{\mathrm{d}v^\alpha}, \tag{16}$$

and, otherwise, if $\mathrm{d}m^\alpha$ is related to dv, the partial density is obtained,

$$\rho^{\alpha} = \frac{\mathrm{d}m^\alpha}{\mathrm{d}v}. \tag{17}$$

Both density functions are connected by the volume fraction according to $(15)_1$

$$\rho^\alpha = n^\alpha \rho^{\alpha R}. \tag{18}$$

Note that incompressibility of a constituent φ^α is understood in a material way on the microscale. In this case, the effective density is assumed to be constant, $\rho^{\alpha R} = \text{const}$. The partial density ρ^α may still vary due to changes in the volume fractions n^α.

3 Balance equations

According to *Truesdell*'s metaphysical principles [37], the behaviour of the constituents of mixtures is governed by balance equations of the same structure as is the behaviour of a single phase continuum. The interactions between the constituents are taken into account by additional production terms, cf. [4–6, 10, 11, 13, 38]. In these references, detailed discussions of the derivation of the balance equations and of the resultant consequences for the interactions between the constituents are given. In the present article, only a brief summary is presented.

3.1 Balance of mass

If phase changes are not taken into account, the mass of an individual constituent is conserved. Following the usual argumentation, the local form of the balance of mass for the individual constituent is obtained as

$$(\rho^\alpha)'_\alpha + \rho^\alpha \operatorname{div} \overset{\prime}{\mathbf{x}}_\alpha = 0. \tag{19}$$

Summation over both constituents φ^S and φ^F yields the balance of mass of the mixture as a whole:

$$\dot{\rho} + \rho \operatorname{div} \dot{\mathbf{x}} = 0. \tag{20}$$

In (20), the definitions of the density

$$\rho = \sum_\alpha \rho^\alpha \tag{21}$$

of the mixture and of the barycentric velocity

$$\dot{\mathbf{x}} = \frac{1}{\rho}\,(\rho^S \overset{\prime}{\mathbf{x}}_S + \rho^F \overset{\prime}{\mathbf{x}}_F) \tag{22}$$

are used.

3.2 Balance of momentum

The balance of momentum relates the change of momentum of the constituent φ^α to the forces acting on it:

$$\rho^\alpha \overset{\prime\prime}{\mathbf{x}}_\alpha = \operatorname{div}\mathbf{T}^\alpha + \rho^\alpha\,\mathbf{b}^\alpha + \hat{\mathbf{p}}^\alpha\,. \tag{23}$$

The forces are given by the divergence of the *Cauchy* stress tensor $\mathbf{T}^\alpha$, by the body forces $\rho^\alpha\,\mathbf{b}^\alpha$, and by the momentum exchange $\hat{\mathbf{p}}^\alpha$. In the two phase model under study, the momentum exchange may be interpreted as the interaction force density between both constituents. Again, the balance of momentum for the whole mixture

$$\rho\,\ddot{\mathbf{x}} = \operatorname{div}\mathbf{T} + \rho\,\mathbf{b} \tag{24}$$

is obtained by summing up over both constituents. Due to the fact, that the momentum of the mixture is conserved, the sum of the production terms is zero. Therefore, it follows that

$$\hat{\mathbf{p}}^S + \hat{\mathbf{p}}^F = \mathbf{0}\,. \tag{25}$$

3.3 Balance of moment of momentum

It was shown by Hassanizadeh and Gray [23] that the classical result obtained from the balance of moment of momentum for non-polar continua also holds in the case of non-polar constituents of mixtures: The *Cauchy* stress tensor is symmetric. This is valid for the fluid phase:

$$\mathbf{T}^F = (\mathbf{T}^F)^T\,. \tag{26}$$

On the other hand, the balance of moment of momentum for the micropolar solid skeleton yields a more complicated result. Following e. g. Eringen and Kafadar [21], micropolar media transfer couple stresses $\mathbf{M}$ and body couples $\rho\,\mathbf{c}$. The balance of moment of momentum for φ^S may be given in the following form:

$$\rho^S\,\mathbf{\Theta}^S\,(\bar{\omega}_S)'_S + 2\,\mathrm{sym}(\rho^S\,\bar{\mathbf{\Omega}}_S\,\mathbf{\Theta}^S)\,\bar{\omega}_S = \operatorname{div}\mathbf{M}^S + \mathbf{I}\times\mathbf{T}^S + \rho^S\,\mathbf{c}^S. \tag{27}$$

Therein, $\rho^S \, \mathbf{\Theta}^S$ is the tensor of microinertia, $\bar{\mathbf{\Omega}}_S = (\bar{\mathbf{R}}_S)'_S \, \bar{\mathbf{R}}_S^T$ is the gyration tensor related to the micromotion $\bar{\mathbf{R}}_S$, and $\bar{\boldsymbol{\omega}} = \frac{1}{2} \overset{3}{\mathbf{E}} \, \bar{\mathbf{\Omega}}_S^T$ is the corresponding angular velocity. The *balance of microinertia* as introduced by Eringen [20] was taken into account in (27). From the physical point of view, the *balance of microinertia* is a kinematical condition derived from the assumption of rigid bodies on the microscale. The structure of equation (27) is exactly the same as found in *Euler*'s gyroscopic equation. As a consequence of the balance of moment of momentum (27), the stress tensor $\mathbf{T}^S$ of the micropolar matrix material is not necessarily symmetric.

Due to the fact that the fluid constituent is assumed to be non-polar, a production $\hat{\mathbf{h}}^S$ of moment of momentum is *a priori* set to zero in (27).

3.4 Balance of energy

The change of the internal energy ε^F of the non-polar fluid is given by

$$\rho^F \left(\varepsilon^F\right)'_F = \mathbf{T}^F \cdot \mathbf{D}^F - \operatorname{div} \mathbf{q}^F + \rho^F \, r^F + \hat{\varepsilon}^F \tag{28}$$

and is caused by the stress power $\mathbf{T}^F \cdot \mathbf{D}^F$, by the heat flux $\mathbf{q}^F$, by the heat supply $\rho^F \, r^F$, and by the energy production $\hat{\varepsilon}^F$, which describes the exchange of energy between the fluid and the solid constituent.

The balance of energy reads for the micropolar solid skeleton

$$\rho^S \left(\varepsilon^S\right)'_S = \mathbf{T}^S \cdot \bar{\mathbf{\Delta}}_S + \mathbf{M}^S \cdot \operatorname{grad} \bar{\boldsymbol{\omega}}_S + \operatorname{div} \mathbf{q}^S + \rho^S \, r^S + \hat{\varepsilon}^S \, . \tag{29}$$

The stress power for the micropolar medium is computed with the micropolar rate of deformation [10, 21]

$$\bar{\mathbf{\Delta}}_S = \mathbf{L}_S - \bar{\mathbf{\Omega}}_S \, , \tag{30}$$

and, in addition, the spent power of the couple stresses $\mathbf{M}^S \cdot \operatorname{grad} \bar{\boldsymbol{\omega}}_S$ is taken into account.

Note in passing if both constituents possess the same temperature, only the energy balance of the mixture has to be taken into account. As a consequence, the exchange of energy $\hat{\varepsilon}^\alpha$ has not to be modelled.

3.5 Entropy balance

Finally, the entropy balance for each constituent reads

$$(\rho^\alpha \, \eta^\alpha)'_\alpha + \rho^\alpha \, \eta^\alpha \operatorname{div} \overset{\prime}{\mathbf{x}}_\alpha = - \operatorname{div} \boldsymbol{\varphi}^\alpha_\eta + \sigma^\alpha_\eta + \hat{\eta}^\alpha \, . \tag{31}$$

The evolution of the entropy is driven by the entropy flux

$$\boldsymbol{\varphi}^\alpha_\eta = \frac{\mathbf{q}^\alpha}{\vartheta} \, , \tag{32}$$

by the entropy supply

$$\sigma_\eta^\alpha = \frac{\rho^\alpha r}{\vartheta}, \tag{33}$$

and by an entropy production $\hat{\eta}^\alpha$. The entropy flux and its supply are chosen in form of *a priori* constitutive assumptions according to (32) and (33). This is usually done in classical thermodynamics and may be extended within the framework of rational thermodynamics as proposed by Liu and Müller [24, 27–29].

Because it would be too restrictive to assume that each of the production terms $\hat{\eta}^\alpha$ should be non-negative, the second law of thermodynamics is postulated for the mixture as a whole and may be stated in the form

$$\hat{\eta} = \hat{\eta}^S + \hat{\eta}^F \geq 0. \tag{34}$$

The second law of thermodynamics expresses the evidence of the irreversibility of the processes under study. It is applied to find restrictions for the formulation of constitutive equations. The consequences of the second law of thermodynamics for a micropolar binary mixture are briefly discussed in the following section without going into details. The interested reader is referred to Diebels [10] for details of this model. Further information may also be found in the work by Ehlers [13], Svendsen and Hutter [35], Bauer [1], and many others.

4 Constitutive modelling

4.1 Evolution of the volume fractions

The volume fractions were introduced in (16) as internal variables describing the local composition of the mixture under study. The evolution of internal variables is governed by evolution equations, i. e. by constitutively given relations.

Starting point for the motivation of an appropriate evolution equation for the solid volume fraction is the definition of the partial density $\rho^S = n^S \rho^{SR}$ and the balance of mass (14). Due to the fact, that changes of the partial density are governed by changes in the volume fraction n^S and by the effective density ρ^{SR}, respectively, the balance of mass can be re-written in the following form:

$$\begin{aligned} &\rho^{SR}\big((n^S)'_S + z^S n^S \operatorname{div} \overset{\prime}{\mathbf{x}}_S\big) + \\ &\quad + n^S\big((\rho^{SR})'_S + (1 - z^S)\,\rho^{SR} \operatorname{div} \overset{\prime}{\mathbf{x}}_S\big) = 0 . \end{aligned} \tag{35}$$

The function z^S is introduced in (35) as a function of the process parameters. It describes the relation of density changes due to changes in the volume fraction n^S and due to changes in the effective density ρ^{SR}. The range of values of z^S is limited by

$$0 \leq z^S \leq 1 . \tag{36}$$

The limiting case $z^S = 0$ describes the behaviour at constant volume fraction, while $z^S = 1$ represents the incompressible limit, i. e. the special case of $\rho^{SR} = \text{const}$. From the thermodynamical point of view, the volume fraction n^S and the effective density are independent process variables. Therefore, (35) must hold for any values of n^S and ρ^{SR}. As a consequence, both parts of the sum have to be zero independently [10]. This, on the one hand, yields an evolution equation for the volume fraction, namely

$$(n^S)'_S + z^S\, n^S \operatorname{div} \overset{\prime}{\mathbf{x}}_S = 0 , \tag{37}$$

and, on the other hand, a balance equation for the effective density is obtained:

$$(\rho^{SR})'_S + (1 - z^S)\, \rho^{SR} \operatorname{div} \overset{\prime}{\mathbf{x}}_S = 0 . \tag{38}$$

Finally, the evolution equation for the volume fraction n^F of the fluid phase follows from the saturation constraint. This requires the assumption, that the fluid fills all available pore space without cavitation or condensation. Taking the material time derivative of the saturation constraint with respect to the motion of the fluid and substituting (37) yields the evolution equation

$$(n^F)'_F = z^S\, n^S \operatorname{div} \overset{\prime}{\mathbf{x}}_S - \operatorname{grad} n^S \cdot \mathbf{w}_F . \tag{39}$$

Following the presented argumentation, the porosity n^F is determined by the saturation constraint, i. e. by the deformation of the solid constituent via the evolution equation (37). Both equations, (37) and (39), have the character of *a priori* constitutive equations like the entropy flux (33), and, therefore, must be taken into account if the entropy principle is investigated.

4.2 Process variables

The first step towards the evaluation of the entropy principle is the choice of process variables. According to this choice, the class of the material under study is restricted. In the present contribution, elastic behaviour for the solid skeleton is assumed, while the fluid is assumed to be viscous. This requires the following set of independent process variables:

$$\begin{aligned} \mathcal{S} = \{ &\bar{\mathbf{U}}_S,\, \nabla\bar{\mathbf{U}}_S,\, {}^R\bar{\mathcal{K}}_S,\, \nabla{}^R\bar{\mathcal{K}}_S,\, \rho^{FR},\, \operatorname{grad}\rho^{FR}, \\ &\mathbf{D}_F,\, \mathbf{w}_F,\, n^S,\, \operatorname{grad} n^S,\, \vartheta,\, \operatorname{grad}\vartheta \} . \end{aligned} \tag{40}$$

The elasticity of the skeleton is governed by the micropolar deformation tensor $\bar{\mathbf{U}}_S$ and by the curvature tensor ${}^R\bar{\mathcal{K}}_S$, the corresponding gradients are included into the set $\mathcal{S}$ due to the fact that mixtures have to be modelled as continua of degree two. Here, the gradient is taken with respect to the reference configuration:

$$\nabla \bar{\mathbf{U}}_S = \text{Grad}_S \, \bar{\mathbf{U}}_S, \qquad \nabla {}^R\bar{\mathcal{K}}_S = \text{Grad}_S {}^R\bar{\mathcal{K}}_S \,. \tag{41}$$

The properties of the fluid phase are taken into account by the real density ρ^{FR}, by its gradient grad ρ^{FR}, and by the rate of deformation $\mathbf{D}_F$. Furthermore, the volume fraction n^S and its gradient grad n^S are included as structural variables describing the local composition of the mixture. The seepage velocity $\mathbf{w}_F$ is taken into account to model the momentum exchange between both constituents and, finally, the common temperature of both constituents and its gradient are included in order to capture thermal effects [9, 10].

If the free *Helmholtz* energy

$$\psi^\alpha := \varepsilon^\alpha + \vartheta \, \eta^\alpha \tag{42}$$

is introduced, constitutive equations must be developed for the following set of quantities:

$$\mathcal{R} = \{\eta^\alpha, \, \varphi^\alpha_\eta, \, \psi^\alpha, \, \mathbf{T}^\alpha, \, \mathbf{M}^S, \, \hat{\mathbf{p}}^F, \, (n^S)'_S\} \,. \tag{43}$$

Following the standard argumentation, the functions of the set $\mathcal{R}$ explicitly depend on the process variables $\mathcal{S}$ only. The dependence on space and time is given implicitly via the process variables. Therefore, material time derivatives entering the entropy balance may be expressed via the chain rule of differentiation [7]. In a second step, the terms in the resulting inequality are collected, either if they are linear expressions in derivatives of the process variables $\mathcal{S}$ with respect to time or space on the one hand or, on the other hand, if they depend on the values of the process variables directly. The corresponding formalism and its mathematical consequences are discussed by Liu [27] for one phase materials and for mixtures e. g. by Bauer [1]. The results, presented in the following subsection, are derived and discussed in [9, 10].

4.3 Results from the entropy principle

The evaluation of the entropy principle is quite formal. The interested reader is referred to [1, 2, 13] and others. Therefore, this section briefly summarizes the results obtained from the entropy principle for the binary mixture under study [9, 10]. The presented results are valid close to the equilibrium state.

- Free energy function ψ^α:

$$\psi^\alpha = \psi^\alpha(\bar{\mathbf{U}}_S, \, {}^R\bar{\mathcal{K}}_S, \, \rho^{FR}, \, \text{grad}\,\rho^{FR}, \, n^S, \, \text{grad}\, n^S, \, \vartheta, \, \text{grad}\,\vartheta) \,. \tag{44}$$

  The equilibrium state is characterized by

$$\psi^\alpha_0 = \psi^\alpha_0(\bar{\mathbf{U}}_S, \, {}^R\bar{\mathcal{K}}_S, \, \rho^{FR}, \, n^S, \, \vartheta) \,. \tag{45}$$

The free energy at the equilibrium does not depend on all variables included in the set $\mathcal{S}$. For the following discussion, the reduced dependence according to (45) is assumed, $\psi^\alpha \equiv \psi_0^\alpha$.

- Entropy:

$$\rho^\alpha \eta^\alpha = -\rho^\alpha \frac{\partial \psi^\alpha}{\partial \vartheta} \,. \tag{46}$$

This result is well known from the *Legendre* transformation which is usually introduced in classical thermodynamics.

- Stress tensor $\mathbf{T}^S$ of the solid skeleton:

$$\mathbf{T}^S = \bar{\mathbf{R}}_S \left[\rho^S \frac{\partial \psi^S}{\partial \bar{\mathbf{U}}_S} + \rho^F \frac{\partial \psi^F}{\partial \bar{\mathbf{U}}_S} \right] \mathbf{F}_S^T - z^S n^S \left[p^{FR} + p^{KR} \right] \mathbf{I} \,. \tag{47}$$

Again, the free energy function serves as a potential for the so-called extra stresses, while the pressure terms arising in (47) are identified as the effective thermodynamic fluid pressure

$$p^{FR} = \left(\rho^{FR}\right)^2 \frac{\partial \psi^F}{\partial \rho^{FR}} + \frac{\rho^S \rho^{FR}}{n^F} \frac{\partial \psi^S}{\partial \rho^{FR}} \tag{48}$$

and the configurational pressure

$$p^{KR} = \rho^F \frac{\partial \psi^F}{\partial n^S} + \rho^S \frac{\partial \psi^S}{\partial n^S} \,. \tag{49}$$

The additional term $\rho^F \bar{\mathbf{R}}_S (\partial \psi^F / \partial \bar{\mathbf{U}}_S) \mathbf{F}_S^T$ can be interpreted as hydration stress, related to the swelling phenomena which can be observed in clay or in some kinds of polymers. This term only arises in (47) if the principle of equipresence is taken into account at the very beginning and if the mixture under study is modelled as continuum of second order.
The formal equivalence of (47) with the effective stress principle [36] is evident. This principle is well known in geomechanics and details are discussed by Bluhm and de Boer [3], de Boer [4], and many others. This principle allows for the determination of the function z^S which was introduced in the evolution equation (37) [10]. This will be discussed in the example below.

- Stress tensor $\mathbf{T}^F$ of the fluid phase:

$$\mathbf{T}^F = -n^F p^{FR} \mathbf{I} + \overset{4}{\mathbf{T}}_1 \mathbf{D}_F \,. \tag{50}$$

The effective fluid pressure is defined in (48). The fourth order material tensor

$$\overset{4}{\mathbf{T}}_1 = \left(\frac{\partial \mathbf{T}^F}{\partial \bar{\mathbf{D}}_F} \right)_0 \tag{51}$$

is obtained from the linearization of the fluid extra stresses close to the equilibrium as indicated by the subscript $(\,\cdot\,)_0$. It should be positive definite and it describes the viscous properties of φ^F.

- Couple stress tensor $\mathbf{M}^S$ of the solid skeleton:

$$\mathbf{M}^S = \bar{\mathbf{R}}_S \left[\rho^S \frac{\partial \psi^S}{\partial\, {}^R\bar{\mathcal{K}}_S} + \rho^F \frac{\partial \psi^F}{\partial\, {}^R\bar{\mathcal{K}}_S} \right] \mathbf{F}_S^T . \tag{52}$$

While the first term is known from single phase micropolar materials, the second one arises due to the principle of equipresence.

- Interaction force between the solid and the fluid phase:

$$\begin{aligned} \hat{\mathbf{p}}^F = &- \left[p^{FR} + \rho^F \frac{\partial \psi^F}{\partial n^S} \right] \operatorname{grad} n^S + \rho^S \frac{\partial \psi^S}{\partial \rho^{FR}} \operatorname{grad} \rho^{FR} - \\ &- \rho^F \mathbf{F}_S^{T-1} \left[\left(\frac{\partial \psi^F}{\partial \bar{\mathbf{U}}_S} \nabla \bar{\mathbf{U}}_S \right)^{\frac{1}{}} + \left(\frac{\partial \psi^F}{\partial\, {}^R\bar{\mathcal{K}}_S} \nabla\, {}^R\bar{\mathcal{K}}_S \right)^{\frac{1}{}} \right] + \mathbf{P}\,\mathbf{w}_F . \end{aligned} \tag{53}$$

The momentum exchange is governed by an equilibrium part depending on the gradients of the volume fraction n^S, the effective fluid density ρ^{FR}, the deformation tensor $\bar{\mathbf{U}}_S$, and the curvature ${}^R\bar{\mathcal{K}}_S$ and by a non-equilibrium part depending on the seepage velocity $\mathbf{w}_F$. The tensor $\mathbf{P}$ represents the permeability of the material under study. On the one hand, it is a function of the size and the shape of the pores, on the other hand, it is influenced by the viscosity of the pore fluid. It should be negative definite. In the isotropic case, it is usually chosen as

$$\mathbf{P} = -\frac{(n^F)^2\, \gamma^{FR}}{k^F} , \tag{54}$$

where γ^{FR} is the effective weight of the fluid and k^F is the so-called *Darcy* permeability.
Further terms may be taken into account in the presented relations [9, 10], which become necessary if the full thermo-mechanical coupling is described. In this case, the interaction may additionally depend on the deformation rate and on the heat flux. For simplicity, these effects are neglected in the representation (53). Generally, the second, third and fourth term appear in (53) due to the principle of equipresence. In applications, these terms are usually chosen to be zero.

- Internal part of the heat flux of the mixture:

$$\mathbf{q}^I = \mathbf{Q} \operatorname{grad} \vartheta \tag{55}$$

Fourier's law is included in (55), if the tensor $\mathbf{Q}$ is chosen as

$$\mathbf{Q} = -\,\kappa\, \mathbf{I} , \tag{56}$$

where κ is the positive coefficient of heat conduction. An additional part of the heat flux is due to diffusive processes. This diffusive parts arise in the flux quantities of the mixture in a natural way by the formalism connecting the partial balances of the individual constituents and the balance equations of the mixture as a whole.

5 Simplified models

In the previous section, the results of the entropy principle are collected in equations (44) – (55) for a binary mixture of an elastic porous skeleton saturated by a compressible pore fluid. Details of the derivation are given in [9, 10]. Even if more general results may be obtained within the framework of extended thermodynamics, the presented formulae require further simplifications before the resulting model is concrete enough for a numerical treatment. For this reason, the dependence of the free energy functions ψ^α is restricted to primary variables related to φ^α, i. e. the principle of equipresence is revoked such that

$$\begin{aligned} \psi^S &= \psi^S(\bar{\mathbf{U}}_S,\, {}^R\bar{\mathcal{K}}_S,\, n^S,\, \vartheta)\,, \\ \psi^F &= \psi^F(\rho^{FR},\, n^S,\, \vartheta)\,. \end{aligned} \tag{57}$$

This rules out the effect of hydration stresses as remarked above. In this case, the stresses of the skeleton are governed by

$$\mathbf{T}^S = \rho^S \bar{\mathbf{R}}_S \frac{\partial \psi^S}{\partial \bar{\mathbf{U}}_S} \mathbf{F}_S^T - z^S n^S \left[p^{FR} + p^{KR}\right] \mathbf{I}, \tag{58}$$

with the configuration pressure

$$p^{KR} = \rho^F \frac{\partial \psi^F}{\partial n^S} + \rho^S \frac{\partial \psi^S}{\partial n^S}, \tag{59}$$

while the equilibrium part of the fluid stress is given by the effective fluid pressure

$$\mathbf{T}_0^F = -n^F (\rho^{FR})^2 \frac{\partial \psi^F}{\partial \rho^{FR}} \mathbf{I} = -n^F p^{FR} \mathbf{I}. \tag{60}$$

Within the frame work of the simplified model, the couple stresses are derived from the free energy function by

$$\mathbf{M}^S = \rho^S \bar{\mathbf{R}}_S \frac{\partial \psi^S}{\partial\, {}^R\bar{\mathcal{K}}_S} \mathbf{F}_S^T. \tag{61}$$

Finally, the momentum production in the equilibrium state takes the reduced form

$$\hat{\mathbf{p}}_0^F = -\left(p^{FR} + \rho^F \frac{\partial \psi^F}{\partial n^S}\right) \operatorname{grad} n^S. \tag{62}$$

Beside the equilibrium parts the fluid extra stress

$$\mathbf{T}_E^F = \overset{4}{\mathbf{D}}\,\mathbf{D}^F \tag{63}$$

and the extra momentum exchange

$$\hat{\mathbf{p}}_E^F = \mathbf{P}\,\mathbf{w}_F \tag{64}$$

are not influenced by the mentioned simplification. Furthermore, they will not be effected, if the incompressibility of the constituents is taken into account.

Up to now, both phases are assumed to be compressible, but in several applications the material incompressibility of one or of both phases may be taken into account. Specially in the case of highly porous solids, the volumetric strain due to the deformation of the pores is much larger than the volumetric strain due to the deformation of the skeleton material itself. Therefore, it is justified to neglect the material compressibility of the skeleton compared to the compressibility of the pore space. The effective density of the solid phase is constant under this assumption,

$$\rho^{SR} = \text{const.} \tag{65}$$

Models of this type are called hybrid models of the second kind [14, 15]. The argumentation leading to the evolution equation (37) of the volume fraction n^S requires that the function z^S is chosen as

$$z^S \equiv 1 \tag{66}$$

in this case. Therefore, the balance of mass is reduced to a balance of volume for φ^S which can be integrated immediately yielding

$$n^S = n_{0S}^S \det \mathbf{F}_S^{-1} = n_{0S}^S \det \bar{\mathbf{U}}_S^{-1}\,. \tag{67}$$

The second part of the equation is obtained by the definition of the deformation tensor $\bar{\mathbf{U}}_S$ according to (6) and to the orthogonality of the micromotion $\bar{\mathbf{R}}_S$. Note in passing, that due to (67) the volume fraction n^S drops out of the list of independent variables of ψ^S because it can be directly expressed by the deformation tensor $\bar{\mathbf{U}}_S$. This leads to the following dependencies for the free energy functions:

$$\begin{aligned} \psi^S &= \psi^S(\bar{\mathbf{U}}_S,\, {}^R\bar{\mathcal{K}}_S,\, \vartheta)\,, \\ \psi^F &= \psi^F(n^S,\, \rho^{FR},\, \vartheta)\,. \end{aligned} \tag{68}$$

The stress tensors $\mathbf{T}^\alpha$ for both constituents, the couple stress tensor $\mathbf{M}^S$ and the momentum exchange $\hat{\mathbf{p}}^F$ follow immediately from the relations (58) – (62) if the reduced dependencies are taken into account as indicated in (68).

If, in addition, the pressure is of moderate height or the permeability of the matrix material is sufficiently large, the fluid undergoes very small volumetrical deformations only. In this case, both constituents can be assumed to be incompressible leading to the so-called incompressible model [2, 14, 15]. The assumption

$$\rho^{\alpha R} \; = \; \text{const.} \tag{69}$$

transforms both balances of mass into volume balances of the form

$$(n^\alpha)'_\alpha \; + n^\alpha \operatorname{div} \overset{\prime}{\mathbf{x}}_\alpha = 0 \,, \tag{70}$$

i. e. both volume fractions follow their own evolution equation. The saturation condition becomes a true constraint in this situation and has to be taken into account in the entropy principle by the introduction of a *Lagrange* multiplier P [6, 13–15]. On the other hand, the effective fluid density is no longer suitable as process variable. Without going into details, the simplified incompressible model is governed by the following relations:

- Stresses of the skeleton:

$$\mathbf{T}^S \; = \bar{\mathbf{R}}_S \left[\rho^S \, \frac{\partial \psi^S}{\partial \bar{\mathbf{U}}_S} \right] \mathbf{F}_S^T \; - \, n^S \left[P \, + \, \rho^F \, \frac{\partial \psi^F}{\partial n^S} \right] \mathbf{I} \tag{71}$$

  and equilibrium stresses of the fluid

$$\mathbf{T}_0^F \; = \; - \, n^F \, P \, \mathbf{I} \,. \tag{72}$$

  It can immediately be seen that the *Lagrange* multiplier P may be interpreted as an independent fluid pressure (pore pressure), which is governed by the boundary conditions.

- Momentum exchange:

$$\hat{\mathbf{p}}_0^F \; = \; - \, P \operatorname{grad} n^S \; + \, \rho^F \, \mathbf{F}_S^{T-1} \left[n^F \, \frac{\partial \psi^F}{\partial n^S} \, \bar{\mathbf{U}}_S^{T-1} \, \nabla \bar{\mathbf{U}}_S \right]^{\frac{1}{\cdot}} . \tag{73}$$

The other quantities, namely the couple stresses, the fluid extra stresses, the extra momentum supply and the head flux, are not influenced by these considerations concerning the material incompressibility of the individual constituents. Furthermore, it is found, that the micropolar properties of the skeleton material are not influenced by volumetric properties like the fluid or the configuration pressure, respectively, or by the assumption of compressibility or incompressibility of the constituents. This result is not surprizing because the micropolar properties are related to the orthogonal tensor $\bar{\mathbf{R}}_S$ which represents an isochoric motion of the directors.

6 Example

6.1 Choice of the constitutive equations

The constitutive equations must explicitly be given to describe the material behaviour under study. Especially, the function z^S in the evolution equation for the volume fraction n^S and the free energy functions ψ^α must be specified. This section offers a special choice which fits into the general framework discussed above. For simplicity, isothermal conditions are assumed.

The function z^S can be determined from an extended version of the effective stress principle as given by Šuklje [34] for compressible porous media. According to the principle of effective stress, the total hydrostatic stress σ of a binary mixture is a weighted sum of the effective (hydrostatic) stress σ' depending on the solid deformation and of the pore pressure p^{FR}:

$$\sigma = \sigma' + \left[1 - (1 - n^F)\,\frac{k^S}{k^{SR}}\right] p^{FR} . \tag{74}$$

Therein, k^S is the bulk modulus of compression of the porous skeleton and k^{SR} is the effective modulus of compression of the matrix material without pores. The relation (74) was experimentally verified on wooden bricks by Lade and de Boer [25]. Combining the hydrostatic part of the stress tensors $\mathbf{T}^S$ and $\mathbf{T}^F$ according to (47) and to (50) and comparison with (74) yields

$$1 - n^S\,\frac{k^S}{k^{SR}} = n^F + z^S n^S \quad \Rightarrow \quad z^S = 1 - \frac{k^S}{k^{SR}} . \tag{75}$$

The function z^S splitting the balance of mass into an evolution equation of the volume fraction and into an effective density balance is therefore determined by the ratio of the bulk modulus of compression k^S to the effective modulus of compression k^{SR}. This includes the limiting case of material incompressible matrix materials where $k^{SR} \to \infty$, i. e. $z^S \to 1$. Applying the *Voigt* bound in the form $k^S = n^S\,k^{SR}$ yields

$$z^S = 1 - n^S = n^F . \tag{76}$$

In this case the evolution equation becomes an equation of state after integration with respect to time. The current value of the volume fraction depends on the deformation of the matrix material only:

$$n^S = \frac{n^S_{0S}}{n^S_{0S}(1 - \det \mathbf{F}_S) + \det \mathbf{F}_S} . \tag{77}$$

Furthermore, the free energy function ψ^S must be specified as a function of the deformation tensor $\bar{\mathbf{U}}_S$, of the curvature tensor ${}^R\bar{\mathcal{K}}_S$ and of the volume fraction n^S. In extension to the Neo-*Hooke* material law, the following form is chosen:

$$\begin{aligned} \rho_{0S}^S \psi^S &= \mu^S(\mathrm{II}_U - \ln J_S - 3/2) + \mu_c^S(\mathrm{II}_U - \mathrm{IV}_U) + \\ &+ \lambda^S \left[\alpha W^S(J_S) + (1-\alpha) W^S(n_{0S}^S/n^S)\right] + \\ &+ \mu_c^S (l_c^S)^2 f(J_S)\, \mathrm{II}_K + P_0 J_S \,. \end{aligned} \tag{78}$$

The invariants are defined as

$$\begin{aligned} \mathrm{II}_U &:= \tfrac{1}{2} \bar{\mathbf{U}}_S \cdot \bar{\mathbf{U}}_S \,, \\ \mathrm{III}_U &:= \det \bar{\mathbf{U}}_S = \det \mathbf{F}_S = J_S \,, \\ \mathrm{IV}_U &:= \tfrac{1}{2} \bar{\mathbf{U}}_S \cdot \bar{\mathbf{U}}_S^T \,, \end{aligned} \tag{79}$$

and

$$\mathrm{II}_K := \tfrac{1}{2}\, {}^R\bar{\mathcal{K}}_S \cdot {}^R\bar{\mathcal{K}}_S \,. \tag{80}$$

In (78), W describes the volumetric extension of the free energy function depending on the volumetric strain $J_S = \det \mathbf{F}_S$ and on the volume fraction n^S. The function

$$f(J_S) = J_S^{-m} \qquad \text{with} \qquad m \geq 0 \tag{81}$$

allows for a strong coupling between strain and curvature in the relation for the stresses and for the couple stresses. This term is justified from a micromechanical point of view, e. g. a deformation of the matrix material leads to bending of individual beams and, therefore, to an increase of the arising moments. Finally, P_0 takes into account an initial pressure related to the initial density of the compressible fluid constituent.

Taking the derivative of (78) with respect to the deformation tensor $\bar{\mathbf{U}}_S$ yields the weighted *Cauchy* stress:

$$\begin{aligned} J_S \mathbf{T}_E^S &= \mu^S (\mathbf{B}_S - \mathbf{I}) + \mu_c^S (\mathbf{B}_S - (\bar{\mathbf{V}}_S^T)^2) + \\ &+ \alpha \lambda^S J_S \frac{\mathrm{d}W^S(J_S)}{\mathrm{d}J_S} \mathbf{I} + \mu_c^S (l_c^S)^2 J_S \frac{\mathrm{d}f(J_S)}{\mathrm{d}J_S} \mathrm{II}_K \mathbf{I} + J_S P_0 \mathbf{I} \end{aligned} \tag{82}$$

and the derivative with respect to ${}^R\bar{\mathcal{K}}_S$ yields the weighted couple stress with respect to the current configuration:

$$J_S \mathbf{M}^S = \mu_c^S (l_c^S)^2 f(J_S)\, \mathcal{K}_S \mathbf{B}_S \,. \tag{83}$$

Therein, the transport properties of the kinematic quantities are applied [10].

The volumetric part of the free energy for a compressible skeleton may be chosen as

$$W^S(J_S) = \frac{1}{4}\left[(J_S - 1)^2 + (\ln J_S)^2\right], \tag{84}$$

which is well known in solid mechanics, cf. [31], if, on the other hand, the matrix material is materially incompressible, the volumetric part of the free energy has to take care of the point of compaction. In this case, the hydrostatic stresses must go to infinity if the pores close under compression. This is guaranteed by the volumetric part of the energy function as proposed by Eipper [18]:

$$W^S(J_S) = (1 - n_{0S}^S)^2 \left[\frac{J_S - 1}{1 - n_{0S}^S} - \ln \frac{J_S - n_{0S}^S}{1 - n_{0S}^S} \right]. \tag{85}$$

For simplicity, the fluid is modelled as ideal gas if material compressibility is taken into account. In this case, the fluid pressure is directly proportional to the density

$$p^{FR} = \frac{p_0^{FR}}{\rho_0^{FR}} \rho^{FR}. \tag{86}$$

Finally, the interaction force is chosen as

$$\hat{\mathbf{p}}^F = p^{FR} \operatorname{grad} n^F - \frac{(n^F)^2 \gamma^{FR}}{k^F} \mathbf{w}_F. \tag{87}$$

Combining the constitutive equation (87) with the assumption, that the influence of the fluid extra stress $\mathbf{T}_E^F$ and of inertia effects are negligible with respect to the interaction force and inserting these assumption into the balance of momentum of the fluid phase, *Darcy*'s law is obtained in the well-known form

$$n^F \mathbf{w}_F = \frac{k^F}{\gamma^{FR}} (\rho^{FR} \mathbf{b} - \operatorname{grad} p^{FR}). \tag{88}$$

6.2 Applications of the model

The presented model is implemented into the finite element code *PANDAS* [19] and applied to some boundary value problems to show its evidence. The first example is a one-dimensional consolidation problem of a column of height h and an initial porosity of $n_0^F = 30\,\%$ as shown in Figure 4.

The computations are performed for the compressible model (both constituents compressible), the hybrid model (materially incompressible skeleton and compressible fluid), and the incompressible model (both constituents materially incompressible) and for two load cases. In the first case, the load $f(t)$ is increased in one second to its maximum value and then kept constant. The load-deflection curve of the surface is shown in Figure 5. In the second case, the load is increased linearly in time without limitation. The corresponding load-deflection of the surface is shown in Figure 6.

The examples show, that the three different model types lead to different responses: In the range of small load values, the compressible and the hybrid model show nearly the same more or less time independent behaviour

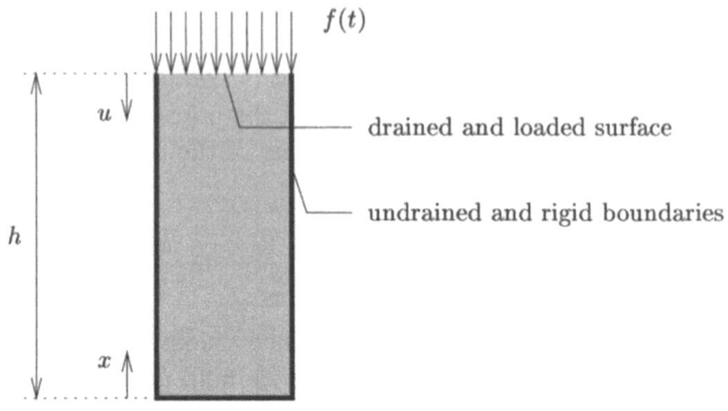

Fig. 4. Consolidation problem and boundary conditions.

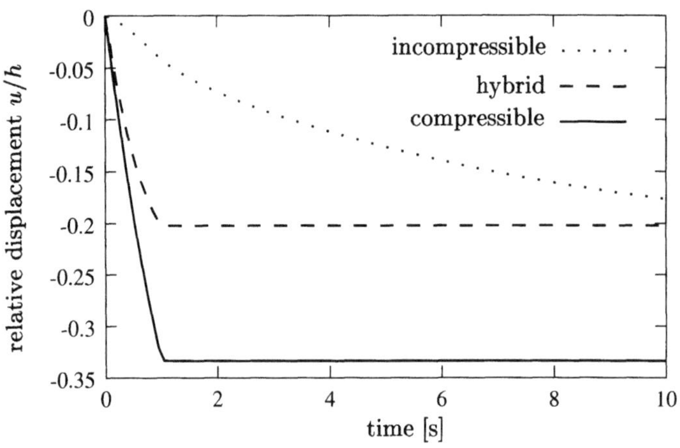

Fig. 5. Load-deflection curve for load case 1.

while the incompressible model shows a strong time dependence. The reason for this difference is found in the volumetric behaviour of the fluid: If it is incompressible, any deformation of the column is only possible if the fluid drains out. This process is time dependent. If, on the other hand, the fluid is compressible, it may immediately undergo volumetric deformations and a deformation of the column is possible before the fluid starts draining. For increasing loads the compressible model shows larger deformation than the hybrid and the incompressible model, i. e. the compressible model behaves not as stiff as the other two models due to the compressibility of the skeleton. If in the second load case the load becomes large enough the pores close in the hybrid and in the incompressible model and the point of compaction will be reached. As a consequence no further deformation is possible even if the load is further increased. Note in passing, that the additional degrees of

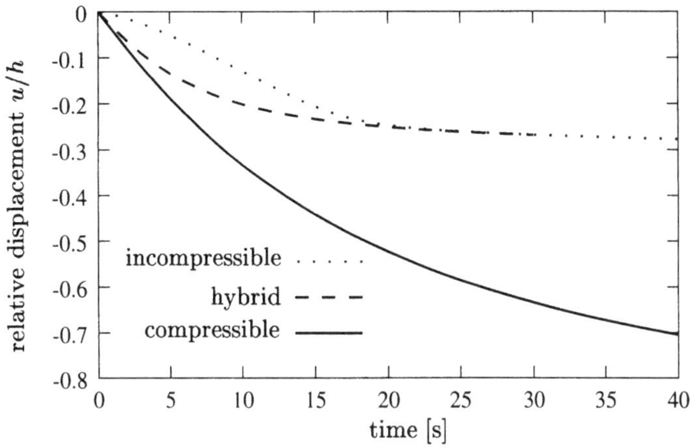

Fig. 6. Load-deflection curve for load case 2.

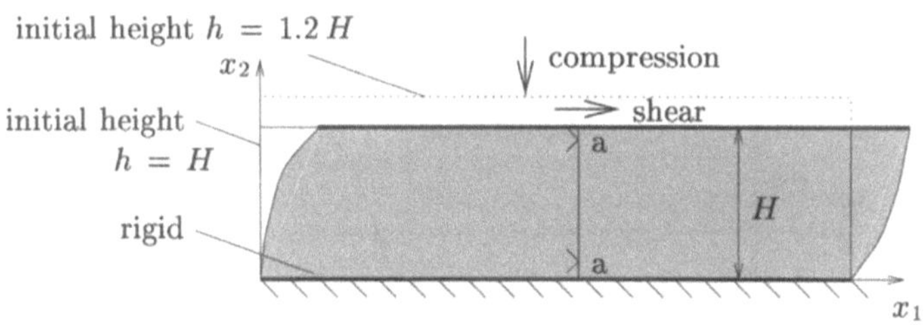

Fig. 7. Shear problem and boundary conditions.

freedom related to the *Cosserat* extension do not become active in this one dimensional problem.

The second example is a shear test with compression. In this case the rotations become active. The boundary value problem is sketched in Figure 7. Two initial geometries of the specimen are investigated. The first one is of height h. It is sheared on its surface by a horizontal displacement of $u_1 = 0.5\,H$ leading to a shear angle of $\phi = \arctan 0.5$. The second specimen has an initial height that is 1.2 times larger, i. e. $h = 1.2\,H$. It is first compressed to the height h and in a second load step it is sheared to $\phi = \arctan 0.5$.

The total *Cosserat* rotation in the cut a-a is shown in Figure 8. It can be seen that the coupling between the second invariant II_K of the curvature and the volumetric deformation J_S in the free energy function leads to an increase of the *effective* internal length, i. e. the boundary layer becomes becomes thicker where the boundary conditions show a significant influence. The same is observed for the couple stresses shown in Figure 9. In the case with superimposed compression, the gradient of the micromotion becomes

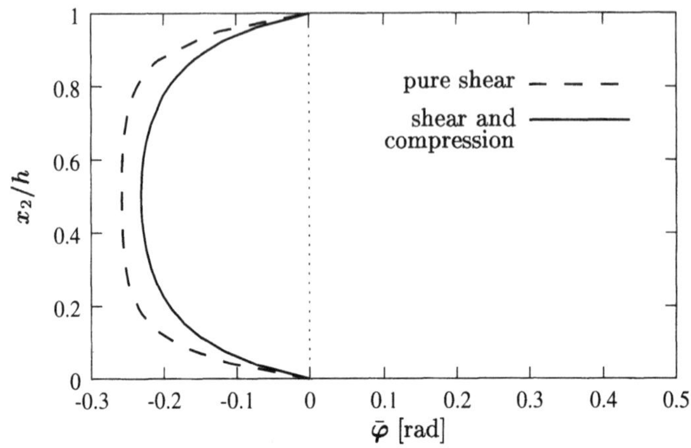

Fig. 8. Total rotation under shear loading.

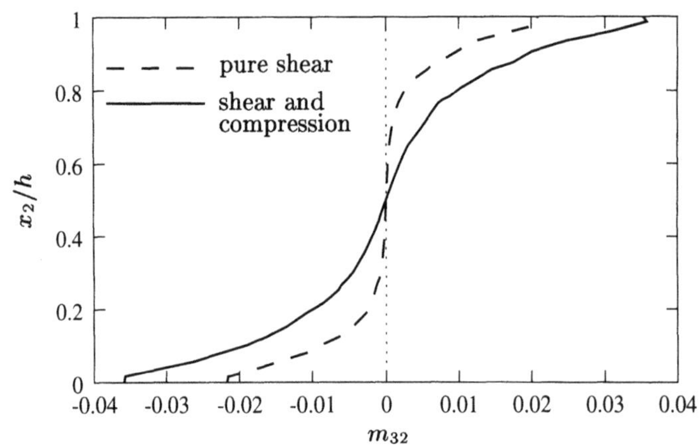

Fig. 9. Couple stress distribution under shear loading.

larger in the region close to the boundaries. Therefore, the curvature and the couple stresses increase.

Due to the fact that the thickness of the boundary layer is mainly governed by the internal length and not be the absolute thickness of the shear layer, the *Cosserat* approach is able to capture the size effect.

7 Conclusions

Based on a microscopic motivation, a micropolar mixture theory is developed. The kinematics and the balance equations for a binary mixture consisting of a micropolar porous skeleton and a viscous pore fluid are discussed and the general framing of the constitutive theory is presented.

It is found from the entropy principle that the free energy functions serve a potential for the equilibrium part of the stresses. Furthermore, it is found that the principle of effective stress arises in a natural way in mixture theories even if the constituents are assumed to be compressible. Finally, this principle allows to determine the constitutive functions arising in the evolution equation of the volume fractions.

Taking into account the material incompressibility either of the solid matrix material or of both constituents, the general results may be simplified. It is found in these cases, that only volumetric properties are influenced by the simplifications and that the micropolar properties are not. This results from the fact that micropolarity is related to rotational degrees of freedom only.

An appropriate choice of the free energy functions is presented and the model is applied to different boundary value problems. It is found that the model shows a physically evident behaviour.

Remark

This article covers the basic theoretical results of the habilitation thesis of the author [10] originally published in German as report No. II-4 of the Institute of Applied Mechanics at the University of Stuttgart.

References

1. Bauer, G.: *Thermodynamische Betrachtung einer gesättigten Mischung.* Dissertation, Technische Hochschule Darmstadt 1997.
2. Bluhm, J.: *A consistent model for saturated and empty porous media.* Forschungsberichte aus dem Fachbereich Bauwesen, 74, Universität-GH-Essen 1997.
3. Bluhm, J., de Boer, R.: Effective stress – a clarification. *Arch. Appl. Mech.* **66** (1996), 479–492.
4. de Boer, R.: *Theory of porous media – highlights in the historical development and current state.* Springer-Verlag, Berlin 2000.
5. de Boer, R., Ehlers, W.: *Theorie der Mehrkomponentenkontinua mit Anwendung auf bodenmechanische Probleme*, Teil I. Forschungsberichte aus dem Fachbereich Bauwesen, 40, Universität-GH-Essen 1986.
6. Bowen, R. M.: Incompressible porous media models by use of the theory of mixtures. *Int. J. Engng. Sci.* **18** (1980), 1129–1148.
7. Coleman, B. D., Noll, W.: The thermodynamics of elastic materials with heat conduction and viscosity. *Arch. Rat. Mech. Anal.* **13** (1963), 167–178.
8. Cundall, P. A, Strack, O. D. L.: A discrete numerical model for granular assemblies. *Géotechnique* **29** (1979), 47–65.
9. Diebels, S.: A micropolar theory of porous media: Constitutive modelling. *Transport in Porous Media* **34** (1999), 193–208.
10. Diebels, S.: *Mikropolare Zweiphasenmodelle: Formulierung auf der Basis der Theorie Poröser Medien.* Habilitationsschrift, Bericht Nr. II-4, Institut für Mechanik (Bauwesen), Lehrstuhl II, Universität Stuttgart 2000.

11. Diebels, S., Ehlers, W.: On basic equations of multiphase micropolar materials. *Technische Mechanik* **16** (1996), 77–88.
12. Diebels, S., Ehlers, W., Michelitsch, T.: Particle simulations as a microscopic approach to a Cosserat continuum. *J. Phys. IV France* (2001), submitted.
13. Ehlers, W.: *Poröse Medien – ein kontinuumsmechanisches Modell auf der Basis der Mischungstheorie.* Forschungsberichte aus dem Fachbereich Bauwesen, 47, Universität-GH-Essen 1988.
14. Ehlers, W.: Compressible, incompressible and hybrid two-phase models in porous media theories. In Angel, Y. C. (ed.): *Anisotropy and Inhomogeneity in Elasticity and Plasticity*, AMD-Vol. 158, ASME 1993, pp. 25–38.
15. Ehlers, W.: Constitutive equations for granular materials in geomechanical context. In Hutter, K. (ed.): *Continuum mechanics in environmental sciences and geophysics*, CISM Courses and Lectures No. 337, Springer-Verlag, Wien 1993, pp. 313–402.
16. Ehlers, W., Diebels, S., Michelitsch, T.: Microscopic modeling of granular materials taking into account particle rotations. In Vermeer, P. A. et al. (eds.): *Continuous and discontinuous modelling of cohesive frictional materials*, Springer-Verlag, Berlin 2001, pp. 259–274.
17. Ehlers, W., Ellsiepen, P.: Theoretical and numerical methods in environmental continuum mechanics based on the theorie of porous media. In Schrefler, B. A. (ed.): *Environmental mechanics*, CISM Courses and Lectures No. 417, Springer-Verlag, Wien 2001.
18. Eipper, G.: *Theorie und Numerik finiter elastischer Deformationen in fluidgesättigten porösen Festkörpern.* Dissertation, Bericht Nr. II-1, Institut für Mechanik (Bauwesen), Lehrstuhl II, Universität Stuttgart 1998.
19. Ellsiepen, P. (ed.): *PANDAS – Benutzer- und Referenzhandbuch.* Bericht Nr. 97-II-9, Institut für Mechanik (Bauwesen), Lehrstuhl II, Universität Stuttgart 1997.
20. Eringen, A. C.: Simple microfluids. *Int. J. Engng. Sci.* **2** (1964), 205–217.
21. Eringen, A. C., Kafadar, C. B.: Polar field theories. In Eringen, A. C. (ed.): *Continuum Physics, Vol IV – Polar and nonlocal field theories*, Academic Press, New York 1976, pp. 1–73.
22. Gibson, L. J., Ashby, M. F.: *Cellular solids – structure and properties.* 2nd ed., Cambridge University Press, Cambridge 1997.
23. Hassanizadeh, S. M., Gray, W. G.: General conservation equations for multiphase systems: 2. Mass, momentena, energy, and entropy equations. *Advances in Water Resources* **2** (1979), 191–208.
24. Hutter, K.: The foundations of thermodynamics, its basic postulates and implications. A review of modern thermodynamics. *Acta Mech.* **27** (1977), 1–54.
25. Lade, P., de Boer, R.: The concept of effective stress for soil, concrete and rock. *Géotechnique* **47** (1997), 61–78.
26. Lewis, R. W., Schrefler, B. A.: *The Finite Element Method in the Static and Dynamic Deformation and Consolidation of Porous Media.* 2nd ed., John Wiley & Sons, Chichester 1998.
27. Liu, I-S.: Method of Lagrange multipliers for exploitation of the entropy principle. *Arch. Rat. Mech. Anal.* **46** (1972), 131–148.
28. Liu, I-S., Müller, I.: Thermodynamics of mixtures of fluids. In Truesdell, C. (ed.): *Rational Thermodynamics*, 2nd ed., Springer-Verlag, New York 1984, pp. 264–285.

29. Müller, I.: *Thermodynamik – Grundlagen der Materialtheorie.* Bertelsmann Universitätsverlag, Düsseldorf 1973.
30. Onck, P. O.: Notch-strengthening in two-dimensional foams. *J. Phys IV France* (2001), submitted.
31. Simio, J. C., Taylor, R. L.: Penalty function formulations for incompressible nonlinear elastostatics. *Comp. Meth. Appl. Mech. Engng.* **35** (1982), 107–118.
32. Steinmann, P.: *Lokalisierungsprobleme in der Plasto-Mechanik.* Dissertation, Universität Karlsruhe 1992.
33. Steinmann, P.: A micropolar theory of finite deformation and finite rotation multiplicative elastoplasticity. *Int. J. Solids Struct.* **31** (1994), 1063–1084.
34. Šuklje, L.: *Rheological Aspects of Soil Mechanics.* Wiley Interscience, London 1969.
35. Svendsen, B., Hutter, K.: On the thermodynamics of a mixture of isotropic materials with constraints. *Int. J. Engng. Sci.* **33** (1995), 2021–2054.
36. von Terzaghi, K.: Zur Berechnung der Durchlässigkeitsziffer des Tones aus dem Verlauf der hydrodynamischen Spannungserscheinungen. *Sitzungsber. Akad. Wiss. Wien* **132** (1923), 125–138.
37. Truesdell, C.: Thermodynamics of diffusion. In Truesdell, C. (ed.): *Rational Thermodynamics*, 2nd ed., Springer-Verlag, New York 1984, pp. 219–236.
38. Truesdell, C., Toupin, R. A.: The classical field theories. In Flügge, S. (ed.): *Handbuch der Physik*, III/1, Springer-Verlag, Berlin 1960, pp. 226–793.
39. Volk, W.: *Untersuchung des Lokalisierungsverhaltens mikropolarer poröser Medien.* Dissertation, Bericht Nr. II-2, Institut für Mechanik (Bauwesen), Lehrstuhl II, Universität Stuttgart 1999.

Elasto-plastic behaviour of a granular material with an additional scalar degree of freedom

Nina P. Kirchner and Kolumban Hutter

Darmstadt University of Technology, Institute of Mechanics III,
64283 Darmstadt, Germany

Abstract. In granular material theories the introduction of internal variables is often useful or even necessary to adequately capture the material behaviour. Yet no common "rules" exist according to which equations governing the evolution of these internal variables should be formulated. In this article, 3 different approaches to model the evolution of such scalar valued internal variables are investigated in a continuum-thermodynamical framework, exploiting the entropy principle according to *Müller* and *Liu*. For all three models, a so-called generalized *Gibbs* equation is obtained, relating the differential of the entropy (which is a constitutive quantity) to that of the internal energy and an additional (model-specific) contribution. The main focus is then on the *Poincaré* conditions, the satisfaction of which provides a powerful tool to restrict the constitutive quantities in such a way that entropy is in fact a well defined scalar potential. The results emerging from this analysis performed for all three modelling approaches shed light on their ability and limitation, respectively. We arrive at reduced or even explicit forms of the constitutive equations and the *Lagrange* multipliers introduced to exploit the entropy principle.

1 Motivation, introductory remarks and outline of the present work

The continuum-thermodynamical modelling of many materials requires the consideration of (scalar) internal degrees of freedom entering the field problem in the form of so-called internal variables, see [8]. In the following, an example from soil mechanics illustrates the necessity of introducing such internal variables:

In a soil mechanics context, observations in-situ and triaxial experiments with dry soil specimens under quasi-static and dynamic loads indicate that a failure of the probes is often initiated by the formation of shear bands. It is suspected that a mechanism called abrasion is at least partly responsible for the initiation of such localizations. Abrasion is a consequence of the surfacial rubbing activity between individual particles of the geomaterial and describes the "peeling off" of small roughness elements from the grain surfaces such that these surfaces may become considerably smoother. According to the internal state of stress and deformation, the process of abrasion is introduced inhomogeneously such that the polishing of particles – leading e. g. to a reduction of the internal angle of friction – may (in certain regions of the material) act as a natural defect accomplishing the mentioned localization. Abrasion would certainly not be observed if the granular material was not rough, roughness thus being treated as a material property which will enter as an internal variable whose time rate of change is in fact abrasion.

If in the above example the abraded mass is neglected, the rough granular material can in a first approximation be treated as a single-constituent material. Although the investigations performed in this article can easily be extended also to apply to multi-constituent systems (needed e. g. to describe the phenomenon of particle size segregation in granular mixtures), we treat here – for simplicity of notation – a single constituent material only.

This article is organized as follows: Sections 2 and 3 serve to introduce different approaches to model scalar valued internal variables and to provide – by summarizing the balance equations and defining the constitutive class – the general background for the investigations to follow. Section 4 briefly reviews the entropy principle and its method of exploitation as suggested by [7] and [5]. The so-called extended entropy inequality is given explicitly for the various proposed modellings of the internal scalar valued variable, but a detailed investigation (which can however be found in [3]) is here not carried out. Selected results in thermodynamic equilibrium, presented in Section 5, will be of importance when it comes to the exploitation of the so-called *Poincaré* conditions. If fulfilled, the latter guarantee that the constitutive quantity "entropy" is in fact a well defined scalar potential, and this issue is addressed in Section 6. Fulfilling the *Poincaré* conditions is at the same time a means to decrease the complexity of the constitutive functionals, and the reduced forms of the latter are derived – using a few additional simplifying assumptions – in Section 7. Finally, a brief summary is given in Section 8, complemented by conclusions.

2 Modelling approaches for the internal variables

This section is devoted to the presentation of various modelling approaches that can be pursued when dealing with scalar internal variables. The additional internal scalar variable will throughout this article be denoted by κ and we will refrain from assigning a special interpretation to κ. Aiming at presenting a rather general way of handling such internal variables, we start from the *Euler*ian description of a balance law for a physical quantity Ψ:

$$\frac{\mathrm{d}}{\mathrm{d}t}\int_V \Psi \,\mathrm{d}v = -\oint_{\partial V} \mathbf{h}^{(\Psi)} \cdot \mathbf{n}\,\mathrm{d}a + \int_V \pi^{(\Psi)}\mathrm{d}v\,, \tag{1}$$

in which $\mathbf{h}^{(\Psi)}$ is the flux of Ψ across the surface ∂V (with unit normal vector $\mathbf{n}$) of the control volume V and $\pi^{(\Psi)}$ is the production of Ψ in V, respectively. For simplicity, we will restrict considerations to sufficiently smooth fields only, so that in localizing (1) discontinuities need not be accounted for.

Ψ	$\mathbf{h}^{(\Psi)}$	$\pi^{(\Psi)}$	Abbreviation	Reference
κ	–	$\pi^{(\kappa)}$	(I)	[9]
κ	$\mathbf{h}^{(\kappa)}$	$\pi^{(\kappa)}$	(II)	[12]
$\rho k\dot{\kappa}$	$\mathbf{h}^{(\rho k\dot{\kappa})} \backsimeq \mathbf{h}^{(\dot{\kappa})}$	$\pi^{(\rho k\dot{\kappa})} \backsimeq \pi^{(\dot{\kappa})}$	(III)	[2]

Table 1. Densities for the three proposed modelling approaches.

Three models designed to model the evolution of κ are suggested by choosing the assignments listed in Table 1 (note that in the fourth column, we have added references of researchers who used such postulates for specific scalar valued internal variables such as e. g. the volume fraction or the porosity): Models (I) and (II) regard κ as the basic variable, and constitutive relations will have to be formulated for the flux and production terms. Model (III) however follows Goodman's and Cowin's [2] concept of an equilibrated force balance and establishes a balance for the time rate of change of κ by choosing $\Psi = \rho\, k\, \dot{\kappa}$ with $k = \text{const}$. Compared to models (I) and (II), model (III) obviously treats $\dot{\kappa}$ as the basic variable (which is why we will for brevity henceforth add only $\dot{\kappa}$ as a superscript to the respective quantities, see also Table 1), but this is not the only difference. More significant is the fact that in the latter model kinetic energy and powers of working associated with $\dot{\kappa}$ are postulated as follows:

$$K^{(\dot{\kappa})} := \int_V \rho\, k\, \frac{\dot{\kappa}^2}{2}\, \mathrm{d}v, \qquad P^{(\dot{\kappa})} := \oint_{\partial V} \dot{\kappa}\, \mathbf{h}^{(\dot{\kappa})} \cdot \mathbf{n}\, \mathrm{d}a\,.$$

This changes the (localized) form of the energy balance the explicit form of which is given below in $(2)_4$. Localization of (1) with the specifications indicated in Table 1 yields the local balances for the internal variable and are given in $(2)_3$. For convenience, we have complemented those equations by the balance of mass $(2)_1$, (linear) momentum $(2)_2$, and entropy $(2)_5$:

$$0 = \dot{\rho} + \rho\, \mathrm{div}\, \mathbf{v},$$

$$\mathbf{0} = \rho\, \dot{\mathbf{v}} - \mathrm{div}\, \mathbf{T} - \rho\, \mathbf{b},$$

$$0 = \begin{cases} \text{(I)} & \dot{\kappa} + \kappa\, \mathrm{div}\, \mathbf{v} - \pi^{(\kappa)} \\ \text{(II)} & \dot{\kappa} + \kappa\, \mathrm{div}\, \mathbf{v} - \mathrm{div}\, \mathbf{h}^{(\kappa)} - \pi^{(\kappa)} \\ \text{(III)} & \rho\, k\, \ddot{\kappa} - \mathrm{div}\, \mathbf{h}^{(\dot{\kappa})} - \rho\, \pi^{(\dot{\kappa})}\,, \end{cases} \tag{2}$$

$$0 = \begin{cases} \text{(I), (II)} & \rho\dot{\varepsilon} + \operatorname{div} \mathbf{q} - \mathbf{T}\cdot\mathbf{D} - \rho r \\ \text{(III)} & \rho\dot{\varepsilon} + \operatorname{div} \mathbf{q} - \mathbf{T}\cdot\mathbf{D} - \rho r + \dot{\kappa}\,\rho\,\pi^{(\dot{\kappa})} - \mathbf{h}^{(\dot{\kappa})}\cdot\nabla\dot{\kappa}\,, \end{cases}$$

$$\pi = \rho\dot{\eta} + \operatorname{div} \boldsymbol{\phi} - \rho\sigma\,.$$

Here, $\mathbf{T}$ and $\mathbf{D}$ are the *Cauchy* stress and the stretching tensor, respectively, $\mathbf{b}$, $\mathbf{q}$, and $\boldsymbol{\phi}$ are an external body force (such as e. g. gravity), the heat flux and the entropy flux, and ε is the internal energy. Finally, r is an external radiation, and η and σ are the entropy and its supply. Complementing equations (2) by constitutive relations will yield a continuum mechanical model for the material under consideration.

3 Constitutive equations

As mentioned in Section 2, equations (2) have to be complemented by constitutive equations. To do so, we have to specify the state space $\mathbb{S}$, which comprises the independent variables of the theory and which is the domain of the constitutive equations. For the present modelling, the state space is given by[1]

$$\mathbb{S} := \{\kappa_0,\, \kappa,\, \nabla\kappa,\, \dot{\kappa},\, \theta,\, \nabla\theta, \dot{\theta}, \mathbf{D},\, \mathbf{B},\, \mathbf{Z}\}\,, \tag{3}$$

where $\mathbf{D}$ has been defined above, $\mathbf{B}$ is the left *Cauchy-Green* tensor, θ is an empirical temperature and $\mathbf{Z}$ is a symmetric internal tensorial (spatial) variable. $\mathbf{Z}$ is postulated to be an objective tensor of rank two and is introduced to model the plastic behaviour of the granular material. κ, $\dot{\kappa}$, and an initial value κ_0 are postulated to be objective scalar quantities, and with these assumptions all above quantities fulfill *Euclid*ean objectivity. A spatial variation $\nabla\kappa$ of the internal variable is accounted for, a spatial variation of $\dot{\kappa}$ is for reasons of simplicity not considered. For the internal variable $\mathbf{Z}$, an evolution equation is postulated such that its time-derivative, denoted by an overhead circle, is objective:

$$\overset{\circ}{\mathbf{Z}} := \dot{\mathbf{Z}} - [\mathbf{W}, \mathbf{Z}] := \dot{\mathbf{Z}} - \mathbf{W}\,\mathbf{Z} + \mathbf{Z}\,\mathbf{W} = \boldsymbol{\Phi}\,. \tag{4}$$

According to this definition, the overhead circle denotes in fact the *Jaumann* derivative of $\mathbf{Z}$, and $\boldsymbol{\Phi}$ is the constitutive part of (4) modelled e. g. in a hypoplastic framework, see e. g. [10]. The constitutive quantities depend on the modelling approach which is pursued and are gathered in the set

$$\mathbb{C}^{(\cdot)} := \{\mathbf{T},\, \varepsilon,\, \mathbf{q},\, \eta,\, \mathbf{h}^{(\cdot)},\, \boldsymbol{\phi},\, \boldsymbol{\Phi},\, \pi^{(\cdot)}\}.$$

On the right-hand side of this equation, the quantities marked by $(\cdot)$ have to be replaced according to Table 1 depending on whether $\mathbb{C}^{(\cdot)}$ stands for $\mathbb{C}^{(\mathrm{I})}$,

[1] The choice of the state space $\mathbb{S}$ can in essence be found in [2] and [11].

$\mathbb{C}^{(\mathrm{II})}$, or $\mathbb{C}^{(\mathrm{III})}$. For each $\mathcal{C} \in \mathbb{C}^{(\cdot)}$ we have a functional relation of the form $\mathcal{C} = \hat{\mathcal{C}}(\mathbb{S})$, assumed to be smooth and differentiable at least once with respect to its variables. Upon inserting the constitutive equations into the balance equations, the latter become the so-called field equations, and any solution to those is called a thermodynamic process.

4 The entropy principle

According to [5–7], the exploitation of the entropy principle starts from a modified entropy inequality in which all balance equations are considered as constraining equations and are thus subtracted from the entropy inequality $(2)_5$. Note that the balance of mass does not appear as a constraint since ρ is treated as a function of the independent constitutive field $\mathbf{B}$ via

$$\rho_r = \rho \det \mathbf{F} = \rho \sqrt{\det \mathbf{B}}\,,$$

where ρ_r is the mass density in the reference configuration. The extended entropy inequality thus reads as follows:

$$\begin{aligned}
\pi = {} & \rho\dot{\eta} + \operatorname{div} \boldsymbol{\phi} + \rho\sigma \\
& - \boldsymbol{\lambda}^{\mathrm{mom}} \cdot (\rho\dot{\mathbf{v}} - \operatorname{div} \mathbf{T} - \rho\mathbf{b}) \\
& - \begin{cases} \lambda^{\mathrm{I}}\,(\dot{\kappa} + \kappa \operatorname{div} \mathbf{v} - \pi^{(\kappa)}) & \text{for model (I)} \\ \lambda^{\mathrm{II}}\,(\dot{\kappa} + \kappa \operatorname{div} \mathbf{v} - \operatorname{div} \mathbf{h}^{(\kappa)} - \pi^{(\kappa)}) & \text{for model (II)} \\ \lambda^{\mathrm{III}}\,(\rho\, k\, \ddot{\kappa} - \operatorname{div} \mathbf{h}^{(\dot{\kappa})} - \rho\,\pi^{(\dot{\kappa})}) & \text{for model (III)} \end{cases} \\
& - \lambda^{\varepsilon} \begin{cases} \rho\dot{\varepsilon} + \operatorname{div}\mathbf{q} - \mathbf{T}\cdot\mathbf{D} + \rho r & \text{for models (I, II)} \\ \rho\dot{\varepsilon} + \operatorname{div}\mathbf{q} - \mathbf{T}\cdot\mathbf{D} + \rho r - \mathbf{h}^{(\dot{\kappa})} \cdot \nabla\dot{\kappa} + \dot{\kappa}\,\rho\,\pi^{(\dot{\kappa})} & \text{for model (III)} \end{cases} \\
& - \boldsymbol{\Lambda}^{\mathrm{plast}} \cdot \left(\dot{\mathbf{Z}} - [\mathbf{W}, \mathbf{Z}] - \boldsymbol{\Phi}\right) \\
\geq {} & 0\,.
\end{aligned} \tag{5}$$

Here, λ^{I}, λ^{II}, λ^{III}, and λ^{ε} are scalar valued, $\boldsymbol{\lambda}^{\mathrm{mom}}$ is a vector valued, and $\boldsymbol{\Lambda}^{\mathrm{plast}}$ is a second order tensor valued symmetric[2] *Lagrange* multiplier. In general, these *Lagrange* multipliers may depend on all independent variables collected in the set $\mathbb{S}$, and possibly on $\mathbf{v}$, r, $\mathbf{b}$, and σ.
Inequality (5) is a requirement aiming at restrictions of constitutive relations. It is physically reasonable to suppose that this constitutive behaviour cannot depend on the source terms. Thus, all source terms arising in (5) must add up to zero, implying $\sigma = \boldsymbol{\lambda}^{\mathrm{mom}} \cdot \mathbf{b} + \lambda^{\varepsilon}\, r$, which says that the entropy source is known in terms of all other sources if the *Lagrange* multipliers are known.

[2] Since $\dot{\mathbf{Z}} - [\mathbf{W}, \mathbf{Z}] - \boldsymbol{\Phi}$ is symmetric, only the symmetric part of $\boldsymbol{\Lambda}^{\mathrm{plast}}$ is relevant.

If the differentiations in inequality (5) are carried out with regard to the state space $\mathbb{S}$ using the chain rule of differentiation, one arrives after lengthy but straightforward calculations at an expression for π which reads in principle as follows:

$$\pi = \sum_{\alpha} f_{\alpha}^{s}(\mathbb{S})\, s_{\alpha} + \sum_{\beta} f_{\beta}^{n}(\mathbb{S})\, n_{\beta} + f_r \geq 0\,. \tag{6}$$

Here, f_{α}^{s} and f_{β}^{n} are coefficient functions preceding variables s_{α} which are *contained* in $\mathbb{S}$ and variables n_{β} *exceeding* the state space $\mathbb{S}$, respectively. The latter are often simply called the ***higher derivatives***. f_r is a function collecting those terms associated with variables that belong to none of the just described classes, examples are the terms $\lambda^{\mathrm{III}}\,\rho\,\pi^{(\dot{\kappa})}$ or $\boldsymbol{\Lambda}^{\mathrm{plast}} \cdot \boldsymbol{\Phi}$. It should be noted that s_{α} and n_{β} may be scalar, vector or tensor valued (and that therefore, the same holds true for the coefficient functions), but for simplicity of notation, this is not reflected in the symbols chosen in (6).

[5] has shown that in order to satisfy the entropy principle, the coefficient functions preceding the higher derivatives n_{β} are each required to vanish identically. This provides us with two kinds of useful information: First, the vanishing of the coefficient functions f_{β}^{n} can be exploited – it gives rise to the so-called *Liu*-equations which can be cast into a compact form and then yield the so-called generalized *Gibbs*-relations. The implications of the latter equations are studied in detail in Section 6. Second, inequality (6) simplifies considerably due to the vanishing coefficient functions and becomes the so-called ***residual inequality*** which is in essence of the form

$$\pi = \sum_{\alpha} f_{\alpha}^{s}(\mathbb{S})\, s_{\alpha} + \tilde{f}_r \geq 0\,, \tag{7}$$

where a modified function $\tilde{f}_r$ accounts for possible changes in f_r due to the vanishing coefficient functions f_{β}^{n}.

An evaluation of (7) in thermodynamic equilibrium is a powerful tool to constrain the materials constitutive response (in equilibrium), and as examples we will present the expressions for the equilibrium *Cauchy* stress tensor, the entropy and the partial derivative of the *Helmholtz* energy with respect to spatial variations of the internal variable κ resulting from all three modelling approaches in the next Section 5.

5 Selected results in thermodynamic equilibrium

Thermodynamic equilibrium is a state in which the thermodynamic processes are time-independent and characterized by a homogeneous temperature- and vanishing velocity field: $\nabla\theta = \mathbf{0}$ and $\mathbf{v} = \mathbf{0}$. The evaluation of a quantity in equilibrium is denoted by $(\cdot)\big|_{\mathrm{E}}$. It is further convenient to introduce the *Helmholtz* free energy ψ defined by $\psi := \varepsilon - \theta\,\eta$, and with this, the equilibrium

Cauchy stress reads as follows:

$$\mathbf{T}_{\mathrm{E}} = \begin{cases} \rho \dfrac{\partial \boldsymbol{\Phi}}{\partial \mathbf{D}}\Big|_{\mathrm{E}} \dfrac{\partial \psi}{\partial \mathbf{Z}}\Big|_{\mathrm{E}} + 2\,\rho\,\mathrm{sym}\Big(\dfrac{\partial \psi}{\partial \mathbf{B}}\Big|_{\mathrm{E}} \mathbf{B}\Big) + \theta\,\lambda^{\mathrm{I}}\Big|_{\mathrm{E}} \kappa\,\mathbf{I} & \text{(I)}, \\[2ex] \rho \dfrac{\partial \boldsymbol{\Phi}}{\partial \mathbf{D}}\Big|_{\mathrm{E}} \dfrac{\partial \psi}{\partial \mathbf{Z}}\Big|_{\mathrm{E}} + 2\,\rho\,\mathrm{sym}\Big(\dfrac{\partial \psi}{\partial \mathbf{B}}\Big|_{\mathrm{E}} \mathbf{B}\Big) + \\ \qquad + \theta\,\dfrac{\partial \lambda^{\mathrm{II}}}{\partial \dot{\kappa}}\Big|_{\mathrm{E}} \mathrm{sym}\Big(\nabla\kappa \otimes \mathbf{h}^{(\kappa)}\Big)\Big|_{\mathrm{E}} - \\ \qquad - \theta\,\lambda^{\mathrm{II}}\Big|_{\mathrm{E}} \Big(\dfrac{\partial \pi^{(\kappa)}}{\partial \mathbf{D}}\Big|_{\mathrm{E}} - \kappa\,\mathbf{I}\Big|_{\mathrm{E}}\Big) & \text{(II)}, \\[2ex] \rho \dfrac{\partial \boldsymbol{\Phi}}{\partial \mathbf{D}}\Big|_{\mathrm{E}} \dfrac{\partial \psi}{\partial \mathbf{Z}}\Big|_{\mathrm{E}} + 2\,\rho\,\mathrm{sym}\Big(\dfrac{\partial \psi}{\partial \mathbf{B}}\Big|_{\mathrm{E}} \mathbf{B}\Big) + \\ \qquad + \theta\,\dfrac{\partial \lambda^{\mathrm{III}}}{\partial \dot{\kappa}}\Big|_{\mathrm{E}} \mathrm{sym}\Big(\nabla\kappa \otimes \mathbf{h}^{(\dot{\kappa})}\Big|_{\mathrm{E}}\Big) - \\ \qquad - \mathrm{sym}\Big(\nabla\kappa \otimes \mathbf{h}^{(\dot{\kappa})}\Big|_{\mathrm{E}}\Big) & \text{(III)}. \end{cases} \tag{8}$$

Similarly, the equilibrium entropy is given by

$$\eta_{\mathrm{E}} = \begin{cases} -\dfrac{\partial \psi}{\partial \mathbf{Z}}\Big|_{\mathrm{E}} \cdot \dfrac{\partial \boldsymbol{\Phi}}{\partial \dot{\theta}}\Big|_{\mathrm{E}} - \dfrac{\partial \psi}{\partial \theta}\Big|_{\mathrm{E}} & \text{(I)}, \\[2ex] -\dfrac{\partial \psi}{\partial \mathbf{Z}}\Big|_{\mathrm{E}} \cdot \dfrac{\partial \boldsymbol{\Phi}}{\partial \dot{\theta}}\Big|_{\mathrm{E}} - \dfrac{\partial \psi}{\partial \theta}\Big|_{\mathrm{E}} + \\ \qquad + \dfrac{\theta}{\rho}\Big(\lambda^{\mathrm{II}}\Big|_{\mathrm{E}} \dfrac{\partial \pi^{(\kappa)}}{\partial \dot{\theta}}\Big|_{\mathrm{E}} - \dfrac{\partial^2 \lambda^{\mathrm{II}}}{\partial \kappa\, \partial \dot{\theta}}\Big|_{\mathrm{E}} \mathbf{h}^{(\kappa)}\Big|_{\mathrm{E}} \cdot \nabla\kappa\Big) & \text{(II)}, \\[2ex] -\dfrac{\partial \psi}{\partial \mathbf{Z}}\Big|_{\mathrm{E}} \cdot \dfrac{\partial \boldsymbol{\Phi}}{\partial \dot{\theta}}\Big|_{\mathrm{E}} - \dfrac{\partial \psi}{\partial \theta}\Big|_{\mathrm{E}} + \theta \dfrac{\partial \lambda^{\mathrm{III}}}{\partial \dot{\theta}}\Big|_{\mathrm{E}} \pi^{(\dot{\kappa})}\Big|_{\mathrm{E}} - \\ \qquad - \dfrac{\theta}{\rho} \dfrac{\partial^2 \lambda^{\mathrm{III}}}{\partial \kappa\, \partial \dot{\theta}}\Big|_{\mathrm{E}} \mathbf{h}^{(\dot{\kappa})}\Big|_{\mathrm{E}} \cdot \nabla\kappa & \text{(III)}, \end{cases} \tag{9}$$

and the response of the *Helmholtz* free energy to spatial changes of κ reads as follows:

$$\rho\,\dfrac{\partial \psi}{\partial \nabla\kappa}\Big|_{\mathrm{E}} = \begin{cases} \mathbf{0} & \text{(I)}, \\[2ex] -\theta\,\dfrac{\partial \lambda^{\mathrm{II}}}{\partial \dot{\kappa}}\Big|_{\mathrm{E}} \mathbf{h}^{(\kappa)}\Big|_{\mathrm{E}} & \text{(II)}, \\[2ex] \Big(1 - \theta\,\dfrac{\partial \lambda^{\mathrm{III}}}{\partial \dot{\kappa}}\Big|_{\mathrm{E}}\Big)\,\mathbf{h}^{(\dot{\kappa})}\Big|_{\mathrm{E}} & \text{(III)}. \end{cases} \tag{10}$$

From these expressions alone (which are in detail discussed in [3]) it is obvious that at this stage, it is difficult to favour one model over another. Thus, the effects which simplifying assumptions have on the models (and results) presented so far have to be investigated to be eventually able to classify the models.

Considering the above equations (8) – (10), it is tempting to invoke the following postulates:

1. Guided by the "classical" form of the equilibrium entropy, we postulate

$$\begin{aligned}
&\left.\frac{\partial \boldsymbol{\Phi}}{\partial \dot{\theta}}\right|_{\mathrm{E}} := \mathbf{0} && && && \text{for model (I),} \\
&\left.\frac{\partial \boldsymbol{\Phi}}{\partial \dot{\theta}}\right|_{\mathrm{E}} := \mathbf{0}, && \left.\frac{\partial \lambda^{\mathrm{II}}}{\partial \dot{\theta}}\right|_{\mathrm{E}} := 0, && \left.\frac{\partial \pi^{(\kappa)}}{\partial \dot{\theta}}\right|_{\mathrm{E}} := 0 && \text{for model (II),} \\
&\left.\frac{\partial \boldsymbol{\Phi}}{\partial \dot{\theta}}\right|_{\mathrm{E}} := \mathbf{0}, && \left.\frac{\partial \lambda^{\mathrm{III}}}{\partial \dot{\theta}}\right|_{\mathrm{E}} := 0 && && \text{for model (III),}
\end{aligned} \tag{11}$$

due to which $\eta\big|_{\mathrm{E}} = -(\partial\psi/\partial\theta)\big|_{\mathrm{E}}$ is obtained from (9) for all modelling approaches since one can show that e. g. $(\partial^2\lambda^{\mathrm{III}}/(\partial\kappa\partial\dot{\theta}))\big|_{\mathrm{E}} =$
$= \partial/\partial\kappa((\partial\lambda^{\mathrm{III}}/\partial\dot{\theta})\big|_{\mathrm{E}})$ holds.

2. Focussing on the equilibrium *Cauchy* stress, we first observe that it consists essentially of three contributions: In each line on the right-hand side of (8), the first term is related to the (hypo)plastic behaviour, the second one is capturing elastic effects, and the remaining terms are generated by the internal scalar variable κ, its spatial variation $\nabla\kappa$, and possibly its flux $\mathbf{h}^{(\cdot)}$. Driven by the desire not to "extinct" one of these types of contributions, we are forced to constrain the *Lagrange* multipliers associated with the evolution of κ and $\dot{\kappa}$, respectively, in the following way:
As far as model (I) is concerned, no simplifications can be imposed at all. Models (II) and (III) however enable us to postulate reduced dependencies of the *Lagrange* multipliers λ^{II} and λ^{III} without eliminating the *entire* contributions due to the modelling of κ and $\dot{\kappa}$, respectively. To be precise, we can postulate that

$$\left.\frac{\partial \lambda^{\mathrm{II}}}{\partial \dot{\kappa}}\right|_{\mathrm{E}} := 0 \quad \text{and} \quad \left.\frac{\partial \lambda^{\mathrm{III}}}{\partial \dot{\kappa}}\right|_{\mathrm{E}} := 0 \tag{12}$$

hold. It should be noted that with this, model (III) is the only one which is capable of capturing anisotropic (i. e. deviatoric) stress contributions which are caused by spatial inhomogeneities of the internal variable κ and the flux $\mathbf{h}^{(\dot{\kappa})}$ (via the term $\operatorname{sym}(\nabla\kappa \otimes \mathbf{h}^{(\dot{\kappa})})$), see (8). Moreover, the material response evoked by changes in the spatial variation of κ can thermodynamically – namely by means of the *Helmholtz* free energy ψ – only be detected with model (III): As is seen from (10), $(\partial\psi/\partial\nabla\kappa)\big|_{\mathrm{E}}$ vanishes identically for models (I), (II) if (12) is postulated to be valid.

For the time being, assumption (12) is hence considered a too severe restriction. However, before imposing assumption (11), it should be emphasized that considerations have here been restricted to thermodynamic equilibrium only which gives rise to two important questions: First, would it be beneficial to extend these assumptions to hold also in non-equilibrium? And secondly, would such an extension violate other results derived in non-equilibrium? The next Section 6 will deal with these issues.

6 The *Poincaré* conditions – integrability conditions in non-equilibrium

The central focus in this section is on the so-called *Poincaré* conditions, which – if fulfilled – guarantee that entropy η is in fact a well defined scalar potential.
To embark on the investigations, we take the so-called generalized *Gibbs*-equation as a starting point, which is given by

$$\mathrm{d}\eta = \lambda^{\varepsilon}\mathrm{d}\varepsilon + \frac{1}{\rho}\,\mathrm{d}\mathcal{P} =: R\,. \tag{13}$$

The scalar valued quantity $\mathrm{d}\mathcal{P}$ is a sum of differentials of variables s_α belonging to the state space $\mathbb{S}$, preceded by coefficient functions $\mathcal{P}_\alpha(\cdot)$:

$$\mathrm{d}\mathcal{P} = \sum_\alpha \mathcal{P}_\alpha(\cdot)\,\mathrm{d}s_\alpha\,, \tag{14}$$

and R has been introduced as a short hand for $\lambda^{\varepsilon}\mathrm{d}\varepsilon + \mathrm{d}\mathcal{P}/\rho$. Accounting for the dependencies of ε and using (14) , it is observed that R can be rewritten as

$$\begin{aligned} R = \Big(\lambda^{\varepsilon}\frac{\partial\varepsilon}{\partial\kappa_0} + \frac{1}{\rho}\mathcal{P}_{\kappa_0}\Big)\,\mathrm{d}\kappa_0 + \Big(\lambda^{\varepsilon}\frac{\partial\varepsilon}{\partial\kappa} + \frac{1}{\rho}\mathcal{P}_{\kappa}\Big)\,\mathrm{d}\kappa + \ldots + \\ + \Big(\lambda^{\varepsilon}\frac{\partial\varepsilon}{\partial\mathbf{Z}} + \frac{1}{\rho}\boldsymbol{\Lambda}^{\mathrm{plast}}\Big)\cdot\mathrm{d}\mathbf{Z} =: \sum_\alpha R_\alpha\,\mathrm{d}s_\alpha\,, \end{aligned}$$

where again $s_\alpha \in \mathbb{S}$, that is $s_1 = \kappa_0$, $s_2 = \kappa$, $\ldots$, $s_{10} = \mathbf{Z}$. Viewing R as a 1-*form* places us in the setting of *Poincaré*'s theorem, stating that the 1-*form* R is *exact* if and only if R is *closed*. Here, R being *exact* means that R is the exterior derivative (denoted by d) of a scalar field B, that is $R = \mathrm{d}B$, and R is called *closed* if its exterior derivative vanishes, $\mathrm{d}R = \mathbf{0}$. The exterior derivative of R is defined as (see e. g. [1])

$$\mathrm{d}R := \frac{\partial R_\beta}{\partial s_\alpha}\,\mathbf{b}^\alpha \wedge \mathbf{b}^\beta\,,$$

in which $\mathbf{b}^{(\cdot)}$ are the base vectors of the *Euclid*ean space and the symbol "$\wedge$" denotes the wedge product defined by $\mathbf{b}^\alpha \wedge \mathbf{b}^\beta := \mathbf{b}^\alpha \otimes \mathbf{b}^\beta - \mathbf{b}^\beta \otimes \mathbf{b}^\alpha$.

The important implication of *Poincaré*'s theorem is now the following: If one can show that $\mathrm{d}R = \mathbf{0}$ holds, R is exact – implying that the scalar potential from which it can be derived can be identified with the entropy η, which is then well defined modulo additive constants. In other words: To arrive at a well defined scalar potential which is called entropy we fulfill the requirement $\mathrm{d}R = \mathbf{0}$, which provides us with a considerable number of restrictions to be placed on the constitutive quantities:

$$\mathrm{d}R = \frac{\partial R_\beta}{\partial s_\alpha}\,\mathbf{b}^\alpha \wedge \mathbf{b}^\beta = \left(\frac{\partial R_\beta}{\partial s_\alpha} - \frac{\partial R_\alpha}{\partial s_\beta}\right)\mathbf{b}^\alpha \otimes \mathbf{b}^\beta \stackrel{!}{=} \mathbf{0}$$

implies that,

$$\frac{\partial R_\beta}{\partial s_\alpha} \stackrel{!}{=} \frac{\partial R_\alpha}{\partial s_\beta} \tag{15}$$

must hold for all $s_\alpha, s_\beta \in \mathbb{S}$ for R to be exact. The equations (15) are known as the *Poincaré* conditions.

At this stage, it is however not clear in which way the constitutive quantities are restricted by the conditions (15). To gain deeper insight into (15), one has thus to take a closer look at the term $\mathrm{d}\mathcal{P}$ in (13), which "perturbs" the collinearity of $\mathrm{d}\eta$ and $\mathrm{d}\varepsilon$. The general structure of $\mathrm{d}\mathcal{P}$ has been given above in (14), but it should be noted that irrespective of the modelling approach chosen, certain coefficient functions $\mathcal{P}_\alpha(\cdot)$ vanish due to *Liu*'s lemma [5]. To be precise, one finds that $\mathcal{P}_{\kappa_0}$, $\mathcal{P}_{\dot\theta}$, and $\mathcal{P}_{\mathbf{D}}$ vanish identically, reducing $\mathrm{d}\mathcal{P}$ to

$$\begin{aligned}\mathrm{d}\mathcal{P} = {} & \mathcal{P}_\kappa\,\mathrm{d}\kappa + \mathcal{P}_{\dot\kappa}\,\mathrm{d}\dot\kappa + \mathcal{P}_{\nabla\kappa}\cdot\mathrm{d}\nabla\kappa + \mathcal{P}_\theta\,\mathrm{d}\theta + \mathcal{P}_{\nabla\theta}\cdot\mathrm{d}\nabla\theta + \\ & + \mathcal{P}_{\mathbf{B}}\cdot\mathrm{d}\mathbf{B} + \mathcal{P}_{\mathbf{Z}}\cdot\mathrm{d}\mathbf{Z}\end{aligned}$$

with model specific expressions for $\mathcal{P}_{\nabla\kappa}$, $\mathcal{P}_{\dot\kappa}$, and $\mathcal{P}_{\nabla\theta}$ as indicated below:

$$\begin{aligned}\mathcal{P}_{\nabla\kappa} &= \begin{cases}\mathbf{0} & \text{(I)}\\ \dfrac{\partial\lambda^{\mathrm{II}}}{\partial\dot\kappa}\,\mathbf{h}^{(\kappa)} & \text{(II)}\\ \left(\dfrac{\partial\lambda^{\mathrm{III}}}{\partial\dot\kappa} - \lambda^\varepsilon\right)\mathbf{h}^{(\dot\kappa)} & \text{(III)}\ ,\end{cases}\\ \mathcal{P}_{\nabla\theta} &= \begin{cases}-\dfrac{\partial\lambda^\varepsilon}{\partial\dot\theta}\,\mathbf{q} & \text{(I)}\\ \dfrac{\partial\lambda^{\mathrm{II}}}{\partial\dot\theta}\,\mathbf{h}^{(\kappa)} - \dfrac{\partial\lambda^\varepsilon}{\partial\dot\theta}\,\mathbf{q} & \text{(II)}\\ \dfrac{\partial\lambda^{\mathrm{III}}}{\partial\dot\theta}\,\mathbf{h}^{(\dot\kappa)} - \dfrac{\partial\lambda^\varepsilon}{\partial\dot\theta}\,\mathbf{q} & \text{(III)}\ ,\end{cases}\end{aligned} \tag{16}$$

$\mathcal{P}_{\dot\kappa} = 0$ for models (I), (II) and $\mathcal{P}_{\dot\kappa} = \rho\,k\,\lambda^{\mathrm{III}}$ for model (III).

Examples for conditions imposed by (15) are hence

$$\frac{\partial R_{\nabla\theta}}{\partial\kappa} \overset{!}{=} \frac{\partial R_{\kappa}}{\partial\nabla\theta} \implies \frac{\partial}{\partial\kappa}\Big(\lambda^{\varepsilon}\frac{\partial\varepsilon}{\partial\nabla\theta} + \frac{1}{\rho}\mathcal{P}^{(\cdot)}_{\nabla\theta}\Big) \overset{!}{=} \frac{\partial}{\partial\nabla\theta}\Big(\lambda^{\varepsilon}\frac{\partial\varepsilon}{\partial\kappa} + \frac{1}{\rho}\mathcal{P}_{\kappa}\Big) \quad (17)$$

or

$$\frac{\partial R_{\dot\kappa}}{\partial\nabla\kappa} \overset{!}{=} \frac{\partial R_{\nabla\kappa}}{\partial\dot\kappa} \implies \frac{\partial}{\partial\nabla\kappa}\Big(\lambda^{\varepsilon}\frac{\partial\varepsilon}{\partial\dot\kappa} + \frac{1}{\rho}\mathcal{P}^{(\cdot)}_{\dot\kappa}\Big) \overset{!}{=} \frac{\partial}{\partial\dot\kappa}\Big(\lambda^{\varepsilon}\frac{\partial\varepsilon}{\partial\nabla\kappa} + \frac{1}{\rho}\mathcal{P}^{(\cdot)}_{\nabla\kappa}\Big), \quad (18)$$

where $\mathcal{P}^{(\cdot)}_{\nabla\theta}$, $\mathcal{P}^{(\cdot)}_{\dot\kappa}$, and $\mathcal{P}^{(\cdot)}_{\nabla\kappa}$ have to be replaced according to (16).

It is now obvious that a (partial) extension of (11) to non-equilibrium, namely the postulation that

$$\frac{\partial\lambda^{\mathrm{II}}}{\partial\dot\theta} := 0 \quad \text{and} \quad \frac{\partial\lambda^{\mathrm{III}}}{\partial\dot\theta} := 0 \quad (19)$$

holds, reduces the complexity of (17) and (18) for models (II) and (III) since $\mathcal{P}_{\nabla\theta}$ is then collinear to $\mathbf{q}$ for each modelling approach. Such reduced complexity is very welcome, since for the given state space $\mathbb{S}$, the number of *Poincaré* conditions to be evaluated amounts (provided the fields are sufficiently smooth so that the order of differentiation can be interchanged) to 45 in each model. Before embarking on the explicit evaluation of the former, let us hence compile additional pieces of information that are available (for details, see [3] or [4]) and might be useful to simplify certain expressions arising in (15) even further:

- The evaluation of isotropy conditions results in a significantly reduced complexity of the *Lagrange* multipliers λ^{ε}, λ^{II} and λ^{III}, which depend in general on the whole state space $\mathbb{S}$. In fact, one can show that

$$\begin{aligned} \lambda^{\varepsilon} &= \hat\lambda^{\varepsilon}(\kappa,\dot\kappa,\theta,\dot\theta), \\ \lambda^{\mathrm{II}} &= \hat\lambda^{(\mathrm{II})}(\kappa,\dot\kappa,\theta,\dot\theta), \\ \lambda^{\mathrm{III}} &= \hat\lambda^{(\mathrm{III})}(\kappa,\dot\kappa,\theta,\dot\theta) \end{aligned} \quad (20)$$

  holds; for λ^{I} however, no corresponding result can be achieved. Introducing a so-called *ideal wall postulate* [3] it is further possible to show that λ^{ε} depends on θ and $\dot\theta$ only, which is indeed a welcome simplification in the evaluation of the *Poincaré* conditions.

- In order not to exclude the possibility of the (linearized) heat conduction equation of becoming hyperbolic, it can be shown that one has to request that

$$\left.\frac{\partial\lambda^{\varepsilon}}{\partial\dot\theta}\right|_{\mathrm{E}} \neq 0 \quad (21)$$

[3] For general information see [7], for the case where κ denotes the roughness of a granular material see [3].

holds. $(20)_1$ encourages us to postulate that this statement is valid in non-equilibrium as well, and we will make frequent use of the fact that $\partial\lambda^\varepsilon/\partial\dot{\theta} :\neq 0$ in deriving explicit restrictions upon the constitutive functionals from (15).

In concluding this section, the postulates imposed to simplify the complex expressions arising from the exploitation of the entropy principle are synoptically summarized in Table 2: Column two lists those terms which motivate the final postulates, complemented by an equation number locating the context in which these "guidelines" have been identified. The final postulates themselves are given in the last column of Table 2. It should be noted that a reduction of the dependence of $\boldsymbol{\Phi}$ as suggested for all models in column three of Table 2 has not yet been invoked: For the evaluation of the *Poincaré* conditions (15), this has not been considered necessary. However, as soon as one turns to a specification of the modelling of the (hypo)plastic material behaviour, such simplifications will become an important issue.

Model	"Guideline"	Context	Postulate
(I)	$\left.\frac{\partial\lambda^\varepsilon}{\partial\dot{\theta}}\right\|_E \neq 0, \quad \left.\frac{\partial\boldsymbol{\Phi}}{\partial\dot{\theta}}\right\|_E = 0$	(12), (21)	$\frac{\partial\lambda^\varepsilon}{\partial\dot{\theta}} :\neq 0$
(II)	$\left.\frac{\partial\lambda^\varepsilon}{\partial\dot{\theta}}\right\|_E \neq 0, \quad \left.\frac{\partial\boldsymbol{\Phi}}{\partial\dot{\theta}}\right\|_E = 0$	(12), (21)	$\frac{\partial\lambda^\varepsilon}{\partial\dot{\theta}} :\neq 0$
	$\left.\frac{\partial\lambda^{II}}{\partial\dot{\theta}}\right\|_E = 0, \quad \left.\frac{\partial\pi^{(\kappa)}}{\partial\dot{\theta}}\right\|_E = 0$	(12)	$\frac{\partial\lambda^{II}}{\partial\dot{\theta}} := 0$
(III)	$\left.\frac{\partial\lambda^\varepsilon}{\partial\dot{\theta}}\right\|_E \neq 0, \quad \left.\frac{\partial\boldsymbol{\Phi}}{\partial\dot{\theta}}\right\|_E = 0$	(12), (21)	$\frac{\partial\lambda^\varepsilon}{\partial\dot{\theta}} :\neq 0$
	$\left.\frac{\partial\lambda^{III}}{\partial\dot{\theta}}\right\|_E = 0$	(12)	$\frac{\partial\lambda^{III}}{\partial\dot{\theta}} := 0$

Table 2. Postulates invoked for the different modelling approaches.

The next Section 7 will now deal with the main results emerging from the evaluation of the conditions (15) using the (model-specific) postulates indicated in the last column of Table 2.

7 The reduced form of the constitutive quantities and *Lagrange* multipliers

In this section, all *Poincaré* conditions resulting from model (III) will be stated explicitly, whereas we will refrain from doing so for models (I) and (II): For the latter, we restrict ourselves to a comparative presentation of the most important results. In Section 6 we have already mentioned that the number of *Poincaré* conditions amounts to a total of 45, and it turns out that these 45 equations can be grouped according to certain features: For all models investigated, five classes of results emerge, which are labelled "class 0" to "class 5" and which will now be discussed in detail.

7.1 Class 0 results

For all models investigated, class 0 comprises those *Poincaré* conditions which reduce to empty statements. For model (I), twelve conditions in (15) collapse to the identity $0 = 0$, whereas for models (II) and (III), only nine do. Why this difference in numbers occurs, will become clear at a later stage of the investigation.

7.2 Class 1 results

Let us now present the class 1 results for model (III) in detail: The results in this class are of such a form that the independence of a constitutive quantity of an element of the state space $\mathbb{S}$ can immediately be derived. Note that the only assumptions entering (15) are those stated in the last column of Table 2.

$$\text{Class 1:} \quad \begin{cases} \dfrac{\partial \varepsilon}{\partial \kappa_0} = 0\,, & \dfrac{\partial \varepsilon}{\partial \dot{\kappa}} = 0\,, & \dfrac{\partial \varepsilon}{\partial \mathbf{D}} = \mathbf{0}\,, \\[2ex] \dfrac{\partial \mathbf{q}}{\partial \kappa_0} = \mathbf{0}\,, & \dfrac{\partial \mathbf{q}}{\partial \dot{\kappa}} = \mathbf{0}\,, & \dfrac{\partial \mathbf{q}}{\partial \mathbf{D}} = \mathbf{0}\,, \\[2ex] \dfrac{\partial \mathbf{h}}{\partial \kappa_0} = \mathbf{0}\,, & \dfrac{\partial \mathbf{h}}{\partial \mathbf{D}} = \mathbf{0}\,. & \end{cases} \tag{22}$$

As is immediately seen, this reduces the complexity of the constitutive quantities ε, $\mathbf{q}$, and $\mathbf{h}$ to the following dependencies:

$$\begin{aligned} \varepsilon &= \hat{\varepsilon}(\mathbb{S}) = \hat{\varepsilon}(\kappa, \nabla\kappa, \theta, \dot{\theta}, \nabla\theta, \mathbf{B}, \mathbf{Z}) \\ \mathbf{q} &= \hat{\mathbf{q}}(\mathbb{S}) = \hat{\mathbf{q}}(\kappa, \nabla\kappa, \theta, \dot{\theta}, \nabla\theta, \mathbf{B}, \mathbf{Z}) \\ \mathbf{h} &= \hat{\mathbf{h}}(\mathbb{S}) = \hat{\mathbf{h}}(\kappa, \dot{\kappa}, \nabla\kappa, \theta, \dot{\theta}, \nabla\theta, \mathbf{B}, \mathbf{Z})\,. \end{aligned} \tag{23}$$

For completeness, it should be mentioned that the independence of $\mathbf{h}$ on κ_0, see the first expression in the third line of (22), requires an additional assumption: Whereas the remaining results in (22) follow immediately from evaluating equations (15) for suitable α and β, the choice $\alpha = \nabla\kappa$, $\beta = \kappa_0$ leads to the expression

$$\frac{\partial \mathbf{h}}{\partial \kappa_0}\left(\frac{\partial \lambda^{\mathrm{III}}}{\partial \dot{\kappa}} - \lambda^{\varepsilon}\right) = 0\,.$$

Since the assumption that the terms in brackets vanish is considered a too severe restriction of the *Lagrange* multipliers λ^{ε} and λ^{III}, we thus conclude that $\mathbf{h}$ is independent of κ_0.

Let us take a closer look at the first line of (22). In combination with the vanishing of the expressions $\mathcal{P}_{\kappa_0}$ and $\mathcal{P}_{\mathbf{D}}$ (which has been addressed in the context of (16)), they play a key role in reducing the dependencies of η: We find

$$\begin{aligned}
\mathcal{P}_{\kappa_0} &:= \rho\left(\frac{\partial \eta}{\partial \kappa_0} - \lambda^{\varepsilon}\frac{\partial \varepsilon}{\partial \kappa_0}\right) = 0 \quad \overset{(22)}{\Longrightarrow} \quad \frac{\partial \eta}{\partial \kappa_0} = 0\,,\\
\mathcal{P}_{\mathbf{D}} &:= \rho\left(\frac{\partial \eta}{\partial \mathbf{D}} - \lambda^{\varepsilon}\frac{\partial \varepsilon}{\partial \mathbf{D}}\right) = 0 \quad \overset{(22)}{\Longrightarrow} \quad \frac{\partial \eta}{\partial \mathbf{D}} = 0\,,
\end{aligned} \tag{24}$$

leading to the reduced dependency

$$\eta = \hat{\eta}(\mathbb{S}) = \hat{\eta}(\kappa, \dot{\kappa}, \nabla\kappa, \theta, \dot{\theta}\,, \nabla\theta, \mathbf{B}, \mathbf{Z}). \tag{25}$$

Note that the vanishing of $\mathcal{P}_{\dot{\theta}}$ cannot be exploited in this manner, since (22) contains no information about the dependency of ε on $\dot{\theta}$.
Let us now briefly address the corresponding results for models (I) and (II):

- As far as model (II) is concerned, the conditions emerging from (15) are exactly those given in (22) (and hence (23)) and need thus not be repeated. However, since $\mathcal{P}_{\dot{\kappa}}$ vanishes, see (16), a similar argumentation to the one given in (24) enables us to conclude that for model (II), η is a function of seven variables only:

$$\eta = \hat{\eta}(\mathbb{S}) = \hat{\eta}(\kappa,\ \nabla\kappa,\ \theta,\ \dot{\theta}\,, \nabla\theta,\ \mathbf{B},\ \mathbf{Z})\,. \tag{26}$$

- For model (I), class 1 comprises again 8 conditions, which is a somewhat unexpected result since for models (II) and (III), two conditions in (22) are related to the flux $\mathbf{h}$. However, for model (I) two additional restrictions concerning the heat flux vector $\mathbf{q}$ and the internal energy ε can be derived, compensating thus the "loss" of the two conditions related to $\mathbf{h}$.
  In particular, we find for model (I) the following reduced dependencies of the constitutive quantities ε and $\mathbf{q}$ from an investigation of the class 1 results:

$$\varepsilon = \hat{\varepsilon}(\mathbb{S}) = \hat{\varepsilon}(\kappa,\ \theta,\ \dot{\theta},\ \nabla\theta,\ \mathbf{B},\ \mathbf{Z}), \quad \mathbf{q} = \hat{\mathbf{q}}(\mathbb{S}) = \hat{\mathbf{q}}(\kappa,\ \theta,\ \dot{\theta},\ \nabla\theta,\ \mathbf{B},\ \mathbf{Z})\,.$$

  Compared to models (II) and (III), the dependence of ε and $\mathbf{q}$ on $\nabla\kappa$ has thus been eliminated, see (23). As far as entropy is concerned, the vanishing of $\mathcal{P}_{\nabla\kappa}$ (defined in complete analogy to e. g. $\mathcal{P}_{\kappa_0}$ in (24), see also

(16)) combined with the independence of ε on $\nabla\kappa$ implies that entropy in model (I) is hence not a function $\nabla\kappa$:

$$\eta = \hat{\eta}(\mathbb{S}) = \hat{\eta}(\kappa,\, \theta,\, \dot{\theta}\,, \nabla\theta,\, \mathbf{B},\, \mathbf{Z})\,. \tag{27}$$

7.3 Class 2 results

For model (III), the results of class 2 (which are eleven in total) are such that the vanishing of second derivatives of η is observed; in detail, these read as follows:

$$\text{Class 2:}\quad \begin{cases} \dfrac{\partial}{\partial\kappa_0}\left(\dfrac{\partial\eta}{\partial\kappa}\right) = 0\,, \quad \dfrac{\partial}{\partial\kappa_0}\left(\dfrac{\partial\eta}{\partial\mathbf{B}}\right) = 0\,, \quad \dfrac{\partial}{\partial\kappa_0}\left(\dfrac{\partial\eta}{\partial\mathbf{Z}}\right) = 0\,, \\ \qquad\qquad \dfrac{\partial}{\partial\kappa_0}\left(\dfrac{\partial\eta}{\partial\theta}\right) = 0\,, \\ \dfrac{\partial}{\partial\mathbf{D}}\left(\dfrac{\partial\eta}{\partial\kappa}\right) = 0\,, \quad \dfrac{\partial}{\partial\mathbf{D}}\left(\dfrac{\partial\eta}{\partial\mathbf{B}}\right) = 0\,, \quad \dfrac{\partial}{\partial\mathbf{D}}\left(\dfrac{\partial\eta}{\partial\mathbf{Z}}\right) = 0\,, \\ \qquad\qquad \dfrac{\partial}{\partial\mathbf{D}}\left(\dfrac{\partial\eta}{\partial\theta}\right) = 0\,, \\ \dfrac{\partial}{\partial\dot{\kappa}}\left(\dfrac{\partial\eta}{\partial\kappa}\right) = 0\,, \quad \dfrac{\partial}{\partial\dot{\kappa}}\left(\dfrac{\partial\eta}{\partial\mathbf{B}}\right) = 0\,, \quad \dfrac{\partial}{\partial\dot{\kappa}}\left(\dfrac{\partial\eta}{\partial\mathbf{Z}}\right) = 0\,. \end{cases} \tag{28}$$

Starting from (25), the exploitation of (28) shows us that necessarily, η must be of the form

$$\eta = \eta_1(\kappa, \nabla\kappa, \theta, \dot{\theta}, \nabla\theta, \mathbf{B}, \mathbf{Z}) + \eta_2(\dot{\kappa}, \nabla\kappa, \theta, \dot{\theta}, \nabla\theta)\,. \tag{29}$$

Let us now take a closer look at the very first equation in the third line of (28) – it will provide us with important information upon combining it with one of the *Liu*-equations (which have been briefly addressed in Section 4 in a general context).

Recall that the *Liu*-equations are obtained from the vanishing of coefficient functions preceding derivatives which exceed the state space $\mathbb{S}$ – in the particular case of $\mathbb{S}$ being given by (3), $\ddot{\kappa}$ is certainly one of those "higher" derivatives whose coefficient function, given by (see e. g. [3]) $\partial\eta/\partial\dot{\kappa} - \lambda^{\text{III}}\, k$, is required to vanish. With this, we obtain

$$0 = \underbrace{\frac{\partial}{\partial\kappa}\left(\frac{\partial\eta}{\partial\dot{\kappa}} - \lambda^{\text{III}}\, k\right)}_{=0} = \underbrace{\frac{\partial^2\eta}{\partial\kappa\,\partial\dot{\kappa}}}_{=0,\,(28)} - k\,\frac{\partial\lambda^{\text{III}}}{\partial\kappa}\,,$$

implying that λ^{III} is independent of κ provided that $k \neq 0$ holds, which is tacitly assumed. Combining this important result with $(20)_3$ and the postulate invoked for model (III), see Table 2, enables us to conclude that

$$\lambda^{\text{III}} = \hat{\lambda}^{\text{III}}(\dot{\kappa}, \theta) \tag{30}$$

holds, which depicts an enormous reduction of complexity for the dependencies of this unknown *Lagrange* multiplier associated with the internal variable κ. Turning now to models (I) and (II), the following issues are worth noting:

- The counterparts of class 2 results contain for models (I) and (II) more conditions than given in (28): Both are complemented by the additional condition

$$\frac{\partial}{\partial \dot{\kappa}}\left(\frac{\partial \eta}{\partial \theta}\right) = 0\,, \tag{31}$$

  an expression which has a non-vanishing right-hand side in model (III) and hence belongs to a different class, see the first expression in (32) below. Thus, the total number of results gathered in class 2 amounts for model (II) to twelve equations. Exploiting the class 2 results for model (II) yields however no additional information beyond what has been achieved in (26) so that no result corresponding to (29) can be derived.

- In case of model (I), these 12 conditions are, moreover, complemented by four additional constraints, which read

$$\frac{\partial}{\partial \nabla \kappa}\left(\frac{\partial \eta}{\partial \kappa}\right) = 0, \quad \frac{\partial}{\partial \nabla \kappa}\left(\frac{\partial \eta}{\partial \mathbf{B}}\right) = 0, \quad \frac{\partial}{\partial \nabla \kappa}\left(\frac{\partial \eta}{\partial \mathbf{Z}}\right) = 0, \quad \frac{\partial}{\partial \nabla \kappa}\left(\frac{\partial \eta}{\partial \theta}\right) = 0\,.$$

  As was the case for model (II), a splitting of the constitutive dependencies of η as in (29) cannot be achieved so that we again have to be content with the representation of η as given in (27).

- It is further important to mention that for models (I) and (II) no result corresponding to (30) can be derived. This is seen if the vanishing of the coefficient function preceding $\ddot{\kappa}$ is evaluated: This *Liu*-equations reduces to the statement that η is independent of $\dot{\kappa}$. Models (I) and (II) have thus the disadvantage that the dependencies of λ^{I} are not restricted at all so that $\lambda^{\mathrm{I}} = \hat{\lambda}^{\mathrm{I}}(\mathbb{S})$ holds, whereas $\hat{\lambda}^{\mathrm{II}}$ is given by (20). The advantage that η is in these models independent of $\dot{\kappa}$ is at least in the case of model (I) not considered to be of such influence that drawbacks of this model (which are discussed in detail in [3]) are compensated. As far as model (II) is concerned, the situation requires more detailed investigation, which would however exceed the scope of this article.

7.4 Class 3 results

Within class 3, five subclasses (characterized by an increasing complexity of the equations and labelled "3 i" to "3 v") can be formed for all models. For model (III) the *Poincaré* conditions gathered in classes 3 i – 3 v amount to a total of thirteen equations and are now explicitly given (frequent use is made of (30)).

Class 3 i:

$$\begin{aligned}
&\frac{\partial}{\partial \dot{\kappa}}\left(\frac{\partial \eta}{\partial \theta}\right) = k\frac{\partial \lambda^{\mathrm{III}}}{\partial \theta},\\
&\frac{\partial}{\partial \dot{\theta}}\left(\frac{\partial \eta}{\partial \kappa}\right) = \lambda^{\varepsilon}\frac{\partial^2 \varepsilon}{\partial \dot{\theta}\,\partial \kappa}, \quad \frac{\partial}{\partial \dot{\theta}}\left(\frac{\partial \eta}{\partial \mathbf{Z}}\right) = \lambda^{\varepsilon}\frac{\partial^2 \varepsilon}{\partial \dot{\theta}\,\partial \mathbf{Z}},\\
&\frac{\partial}{\partial \dot{\theta}}\left(\frac{\partial \eta}{\partial \mathbf{B}}\right) = \lambda^{\varepsilon}\frac{\partial^2 \varepsilon}{\partial \dot{\theta}\,\partial \mathbf{B}}.
\end{aligned} \tag{32}$$

Class 3 ii:

$$\begin{aligned}
&\frac{\partial}{\partial \nabla\theta}\left(\frac{\partial \eta}{\partial \kappa}\right) = \lambda^{\varepsilon}\frac{\partial^2 \varepsilon}{\partial \nabla\theta\,\partial \kappa} - \frac{1}{\rho}\frac{\partial \mathbf{q}}{\partial \kappa}\frac{\partial \lambda^{\varepsilon}}{\partial \dot{\theta}},\\
&\frac{\partial}{\partial \nabla\theta}\left(\frac{\partial \eta}{\partial \mathbf{B}}\right) = \lambda^{\varepsilon}\frac{\partial^2 \varepsilon}{\partial \nabla\theta\,\partial \mathbf{B}} - \frac{1}{\rho}\frac{\partial \mathbf{q}}{\partial \mathbf{B}}\frac{\partial \lambda^{\varepsilon}}{\partial \dot{\theta}},\\
&\frac{\partial}{\partial \nabla\theta}\left(\frac{\partial \eta}{\partial \mathbf{Z}}\right) = \lambda^{\varepsilon}\frac{\partial^2 \varepsilon}{\partial \nabla\theta\,\partial \mathbf{Z}} - \frac{1}{\rho}\frac{\partial \mathbf{q}}{\partial \mathbf{Z}}\frac{\partial \lambda^{\varepsilon}}{\partial \dot{\theta}},\\
&\frac{\partial}{\partial \dot{\theta}}\left(\frac{\partial \eta}{\partial \theta}\right) = \lambda^{\varepsilon}\frac{\partial^2 \varepsilon}{\partial \dot{\theta}\,\partial \theta} + \frac{\partial \lambda^{\varepsilon}}{\partial \theta}\frac{\partial \varepsilon}{\partial \dot{\theta}}.
\end{aligned} \tag{33}$$

Class 3 iii:

$$\begin{aligned}
&\frac{\partial}{\partial \kappa}\left(\frac{\partial \eta}{\partial \nabla\kappa}\right) = \lambda^{\varepsilon}\frac{\partial^2 \varepsilon}{\partial \kappa\,\partial \nabla\kappa} + \frac{1}{\rho}\frac{\partial \mathbf{h}}{\partial \kappa}\frac{\partial \lambda^{\mathrm{III}}}{\partial \dot{\kappa}} - \frac{1}{\rho}\lambda^{\varepsilon}\frac{\partial \mathbf{h}}{\partial \kappa},\\
&\frac{\partial}{\partial \mathbf{B}}\left(\frac{\partial \eta}{\partial \nabla\kappa}\right) = \lambda^{\varepsilon}\frac{\partial^2 \varepsilon}{\partial \mathbf{B}\,\partial \nabla\kappa} + \frac{1}{\rho}\frac{\partial \mathbf{h}}{\partial \mathbf{B}}\frac{\partial \lambda^{\mathrm{III}}}{\partial \dot{\kappa}} - \frac{1}{\rho}\lambda^{\varepsilon}\frac{\partial \mathbf{h}}{\partial \mathbf{B}},\\
&\frac{\partial}{\partial \mathbf{Z}}\left(\frac{\partial \eta}{\partial \nabla\kappa}\right) = \lambda^{\varepsilon}\frac{\partial^2 \varepsilon}{\partial \mathbf{Z}\,\partial \nabla\kappa} + \frac{1}{\rho}\frac{\partial \mathbf{h}}{\partial \mathbf{Z}}\frac{\partial \lambda^{\mathrm{III}}}{\partial \dot{\kappa}} - \frac{1}{\rho}\lambda^{\varepsilon}\frac{\partial \mathbf{h}}{\partial \mathbf{Z}}.
\end{aligned} \tag{34}$$

Class 3 iv:

$$\frac{\partial}{\partial \nabla\theta}\left(\frac{\partial \eta}{\partial \theta}\right) = \lambda^{\varepsilon}\frac{\partial^2 \varepsilon}{\partial \nabla\theta\,\partial \theta} + \frac{\partial \lambda^{\varepsilon}}{\partial \theta}\frac{\partial \varepsilon}{\partial \nabla\theta} - \frac{1}{\rho}\frac{\partial \mathbf{q}}{\partial \theta}\frac{\partial \lambda^{\varepsilon}}{\partial \dot{\theta}} - \frac{1}{\rho}\mathbf{q}\frac{\partial^2 \lambda^{\varepsilon}}{\partial \theta\,\partial \dot{\theta}}. \tag{35}$$

Class 3 v:

$$\begin{aligned}
\frac{\partial}{\partial \nabla\kappa}\left(\frac{\partial \eta}{\partial \theta}\right) = {}& \lambda^{\varepsilon}\frac{\partial^2 \varepsilon}{\partial \nabla\kappa\,\partial \theta} + \frac{\partial \lambda^{\varepsilon}}{\partial \theta}\frac{\partial \varepsilon}{\partial \nabla\kappa} +\\
& + \frac{1}{\rho}\frac{\partial \mathbf{h}}{\partial \theta}\frac{\partial \lambda^{\mathrm{III}}}{\partial \dot{\kappa}} - \frac{1}{\rho}\mathbf{h}\frac{\partial \lambda^{\varepsilon}}{\partial \theta} - \frac{1}{\rho}\lambda^{\varepsilon}\frac{\partial \mathbf{h}}{\partial \theta}.
\end{aligned} \tag{36}$$

Let us now compare results (32) – (36) to those obtained for models (I) and (II):

- Generally speaking, the class 3 results capture the model specific influence most obviously, and a first indication for this is the total number of conditions gathered in the respective classes.
  For model (II), class 3 contains only twelve equations (compared to model (III), the first expression of (32) is "missing" – it has become of class 2), and only a total number of eight equations is of class 3 as far as model (II) is concerned.
- Apart from the mentioned change in class 3 i due to the "loss" of one condition, only the results gathered in classes 3 iii and 3 v change for model (II), they are now given by the following expressions:

  Class 3 iii (model (II)):

$$
\begin{aligned}
\frac{\partial}{\partial \kappa}\left(\frac{\partial \eta}{\partial \nabla \kappa}\right) &= \lambda^{\varepsilon}\frac{\partial^2 \varepsilon}{\partial \kappa\, \partial \nabla \kappa} + \frac{1}{\rho}\frac{\partial \mathbf{h}}{\partial \kappa}\frac{\partial \lambda^{\mathrm{II}}}{\partial \dot{\kappa}} + \frac{1}{\rho}\mathbf{h}\frac{\partial^2 \lambda^{\mathrm{II}}}{\partial \kappa\, \partial \dot{\kappa}}, \\
\frac{\partial}{\partial \mathbf{B}}\left(\frac{\partial \eta}{\partial \nabla \kappa}\right) &= \lambda^{\varepsilon}\frac{\partial^2 \varepsilon}{\partial \mathbf{B}\, \partial \nabla \kappa} + \frac{1}{\rho}\frac{\partial \mathbf{h}}{\partial \mathbf{B}}\frac{\partial \lambda^{\mathrm{II}}}{\partial \dot{\kappa}}, \\
\frac{\partial}{\partial \mathbf{Z}}\left(\frac{\partial \eta}{\partial \nabla \kappa}\right) &= \lambda^{\varepsilon}\frac{\partial^2 \varepsilon}{\partial \mathbf{Z}\, \partial \nabla \kappa} + \frac{1}{\rho}\frac{\partial \mathbf{h}}{\partial \mathbf{Z}}\frac{\partial \lambda^{\mathrm{II}}}{\partial \dot{\kappa}}.
\end{aligned}
\tag{37}
$$

  Class 3 v (model (II)):

$$
\begin{aligned}
\frac{\partial}{\partial \nabla \kappa}\left(\frac{\partial \eta}{\partial \theta}\right) = \lambda^{\varepsilon}\frac{\partial^2 \varepsilon}{\partial \nabla \kappa \partial \theta} + \frac{\partial \lambda^{\varepsilon}}{\partial \theta}\frac{\partial \varepsilon}{\partial \nabla \kappa} + \frac{1}{\rho}\frac{\partial \mathbf{h}}{\partial \theta}\frac{\partial \lambda^{\mathrm{III}}}{\partial \dot{\kappa}} - \\
-\frac{\mathbf{h}}{\rho}\frac{\partial \lambda^{\varepsilon}}{\partial \theta} + \frac{\mathbf{h}}{\rho}\frac{\partial^2 \lambda^{\mathrm{II}}}{\partial \theta\, \partial \dot{\kappa}}.
\end{aligned}
\tag{38}
$$

- Bearing in mind that for model (I), a flux $\mathbf{h}$ related to the internal variable does not exist and that, moreover, ε is independent of $\nabla \kappa$, it is clear that for model (I), class 3 contains eight conditions only: (37) and (38) generate the additional conditions entering class 2!

7.5 Class 4 results

Finally, class 4 comprises four equations for models (II) and (III), respectively, and only a single one in case model (I) is applied. For model (III), the conditions read as follows:

Class 4:

$$
\begin{aligned}
\lambda^{\varepsilon}\frac{\partial \mathbf{h}}{\partial \dot{\kappa}} &= \frac{\partial \mathbf{h}}{\partial \dot{\kappa}}\frac{\partial \lambda^{\mathrm{III}}}{\partial \dot{\kappa}} + \mathbf{h}\,\frac{\partial^2 \lambda^{\mathrm{III}}}{\partial \dot{\kappa}^2}\,,\\
\lambda^{\varepsilon}\frac{\partial \mathbf{h}}{\partial \nabla\theta} &= \frac{\partial \mathbf{h}}{\partial \nabla\theta}\frac{\partial \lambda^{\mathrm{III}}}{\partial \dot{\kappa}} + \mathbf{h}\,\frac{\partial \mathbf{q}}{\partial \nabla\kappa}\frac{\partial \lambda^{\varepsilon}}{\partial \dot{\theta}}\,,\\
\lambda^{\varepsilon}\frac{\partial \mathbf{h}}{\partial \dot{\theta}} &= \frac{\partial \mathbf{h}}{\partial \dot{\theta}}\frac{\partial \lambda^{\mathrm{III}}}{\partial \dot{\kappa}} - \mathbf{h}\,\frac{\partial \lambda^{\varepsilon}}{\partial \theta} + \rho\frac{\partial \lambda^{\varepsilon}}{\partial \dot{\theta}}\frac{\partial \varepsilon}{\partial \nabla\kappa}\,,\\
\mathbf{q}\,\frac{\partial^2 \lambda^{\varepsilon}}{\partial \nabla\theta^2} &= \rho\,\frac{\partial \lambda^{\varepsilon}}{\partial \dot{\theta}}\frac{\partial \varepsilon}{\partial \nabla\theta} - \frac{\partial \mathbf{q}}{\partial \theta}\frac{\partial \lambda^{\varepsilon}}{\partial \dot{\theta}}\,.
\end{aligned}
\tag{39}
$$

With the above 36 equations, the *Poincaré* conditions for model (III) are exploited.

Encouraged by the far-reaching reduction of complexity for the *Lagrange* multiplier λ^{III} as presented in (30), we now seek an explicit form of λ^{III} that can in a first approach be used when tackling a given field problem. To this end, we recall that λ^{III} is a function of θ and $\dot{\kappa}$ only, see (30). We wish to replace this general relation by a representation which is more explicit, and in a first step we are guided by the following argumentation: Since it can be shown (see [3]) that $\lambda^{\mathrm{III}}\big|_{\mathrm{E}} := 0$ is required to guarantee that the entropy production π assumes its minimum in thermodynamic equilibrium, λ^{III} has the following representation:

$$
\lambda^{\mathrm{III}} = \breve{\lambda}^{\mathrm{III}}(\theta, \dot{\kappa})\dot{\kappa}\,. \tag{40}
$$

In fact, as $\dot{\kappa} \to 0$ (that is, as $\dot{\kappa}$ approaches its equilibrium value) λ^{III} thus vanishes. To simplify notation we replace $\breve{\lambda}^{\mathrm{III}}$ now by F such that (40) reads

$$
\lambda^{\mathrm{III}} = F(\theta, \dot{\kappa})\dot{\kappa}\,. \tag{41}
$$

It should be noted that this result can again be combined with one of the *Poincaré* conditions, namely the first one belonging to class 3i as given in (32): Taking the partial derivative of λ^{III} with respect to θ in (41) yields

$$
\frac{\partial \lambda^{\mathrm{III}}}{\partial \theta} = \frac{\partial F}{\partial \theta}\dot{\kappa} \overset{(32)}{=} \frac{1}{k}\frac{\partial^2 \eta}{\partial \dot{\kappa}\,\partial \theta}\,,
$$

which provides us with an additional constraint of class 2, whose explicit form depends however still on F. If in a first approach F is treated as a linear function of $\dot{\kappa}$ and θ that is, $F(\dot{\kappa}, \theta) := c_1\,\dot{\kappa} + c_2\,\theta$ with constants c_1 and c_2 we have thus found that

$$
\lambda^{\mathrm{III}} = c_1\,\dot{\kappa}\,\theta + c_2\,\dot{\kappa}^2
$$

holds. We conclude this section with commenting upon the corresponding results obtained for models (I) and (II):

- For model (II), $(39)_4$ remains unchanged but the first three conditions simplify and read

$$
\begin{aligned}
0 &= \frac{\partial \mathbf{h}}{\partial \dot{\kappa}} \frac{\partial \lambda^{\mathrm{II}}}{\partial \dot{\kappa}} + \mathbf{h} \frac{\partial^2 \lambda^{\mathrm{II}}}{\partial \dot{\kappa}^2}, \\
0 &= \frac{\partial \mathbf{h}}{\partial \nabla \theta} \frac{\partial \lambda^{\mathrm{II}}}{\partial \dot{\kappa}} + \mathbf{h} \frac{\partial \mathbf{q}}{\partial \nabla \kappa} \frac{\partial \lambda^{\varepsilon}}{\partial \dot{\theta}}, \\
0 &= \frac{\partial \mathbf{h}}{\partial \dot{\theta}} \frac{\partial \lambda^{\mathrm{II}}}{\partial \dot{\kappa}} - \mathbf{h} \frac{\partial \lambda^{\varepsilon}}{\partial \dot{\theta}} \rho \frac{\partial \lambda^{\varepsilon}}{\partial \dot{\theta}} \frac{\partial \varepsilon}{\partial \nabla \kappa}.
\end{aligned}
\tag{42}
$$

- Clearly then, for model (I), equations (42) reduce to empty statements due to the non-existence of the flux $\mathbf{h}$, (39) remains again unchanged.
- With regard to the *Lagrange* multipliers λ^{II} and λ^{I}, it is worth mentioning that so far, no means has been found to reduce the complexity of λ^{I}, which can thus only be given in its most general form $\lambda^{\mathrm{I}} = \hat{\lambda}^{\mathrm{I}}(\mathbb{S})$. The situation is different for the multiplier λ^{II}, whose complexity reduces due the postulate made in Table 2 to $\lambda^{\mathrm{II}} = \hat{\lambda}^{\mathrm{II}}(\kappa, \dot{\kappa}, \theta)$. Analogous investigations to those performed in the context of (40) show that λ^{II} can be represented as

  $$\lambda^{\mathrm{II}} = G(\theta, \kappa, \dot{\kappa})\dot{\kappa},$$

  with a so far unspecified function G, compare also (41).

8 Conclusions

Dealing with a granular material, the introduction of internal variables is often useful or even necessary to adequately capture the material behaviour. Yet no common "rules" exist according to which equations governing the evolution of these internal variables should be formulated. To shed light on the adequacy of various modelling approaches we have investigated and compared three different approaches describing the evolution of a scalar valued internal variable in a continuum-thermodynamical framework. First hints on the ability and limitation, respectively, of the presented models can be obtained by the exploitation of the entropy principle according to *Müller* and *Liu* (this can in detail be found in [3]), suggesting that model (I) should be abandoned whenever spatial inhomogeneities of the internal variable should be detectable with thermodynamic quantities such as e. g. the *Helmholtz* free energy, see also (10). However, from the exploitation of the entropy principle alone it is difficult to favour one model over another, and this is why we have focused here on the exploitation of the *Poincaré* conditions. The latter provide a powerful tool to restrict the constitutive quantities in such a way that entropy is in fact a well-defined scalar potential, and the results presented in Sections 5 and 7 indicate that model (III) is not only more "robust" than model (II) but also yields the sharpest constraints for the *Lagrange* multiplier

λ^{III} associated with the internal variable κ. "Robust" is to be understood in the sense explained in Section 5, that is model (III) is able to capture different effects caused by κ and $\nabla\kappa$ in the equilibrium response of the granular material even if a few simplifying assumptions are made. The latter however lead for models (I) and (II) to a loss of certain pieces of information which have essentially been related to spatial variations of κ and which may in many problems be of considerable importance. Due to this robustness and the fact that in case of model (III), an explicit expression for the *Lagrange* multiplier λ^{III} – though only in a first approximation – can be deduced, it is believed that model (III) is very apt to account for an additional scalar degree of freedom introduced in a granular material provided – but this applies to all models proposed – a suitable interpretation and justification is given why the evolution of κ should be modelled in exactly the way suggested by [2].

Acknowledgement

This work was supported by the Deutsche Forschungsgemeinschaft (DFG) through the special collaborative research project SFB 298 "Deformation and Failure of Metallic and Granular Media".

References

1. Bowen, R. M., Wang, C.-C.: *Introduction to vectors and tensors.* 1st ed., Plenum Press, New York 1976.
2. Goodman, M. A., Cowin, S. C.: A continuum theory for granular materials. *Arch. Rational Mech. Anal.* **44** (1972), 249–266.
3. Kirchner, N. P.: *Thermodynamics of structured granular materials.* Ph. D. thesis, Institute of Mechanics III, Darmstadt University of Technology, Germany 2001.
4. Kirchner, N. P., Hutter, K.: Thermodynamic modelling of granular continua exhibiting quasi-static frictional behaviour with abrasion. In Capriz, G., Ghionna, V. N., Giovine, G. (eds.): *Modelling and Mechanics of Granular and Porous Materials*, Modelling and Simulation in Science, Engineering and Technology Series, Birkhäuser Verlag, Basel 2002, pp. 65–85.
5. Liu, I-S.: Method of Lagrange multipliers for exploitation of the entropy principle. *Arch. Rational Mech. Anal.* **46** (1972), 131–148.
6. Liu, I-S.: On the entropy supply in a classical and a relativistic fluid. *Arch. Rational Mech. Anal.* **50** (1973), 111–117.
7. Müller, I.: Die Kältefunktion, eine universelle Funktion in der Thermodynamik viskoser wärmeleitender Flüssigkeiten (German) [The coldness function, a universal function in the theory of viscous, heatconducting fluids.] *Arch. Rational Mech. Anal.* **40** (1971), 1–36.
8. Svendsen, B.: On the thermodynamics of thermoelastic materials with additional scalar degrees of freedom. *Continuum Mechanics and Thermodynamics* **4** (1999), 247–262.

9. Svendsen, B., Hutter, K.: On the thermodynamics of a mixture of isotropic materials with constraints. *Int. J. Engng. Sci.* **33** (1995), 2021–2054.
10. Svendsen, B., Hutter, K., Laloui, L.: Constitutive models for granular materials including quasi-static frictional behaviour: Toward a theory of plasticity. *Continuum Mechanics and Thermodynamics* **11** (1999), 263–275.
11. Wang, Y., Hutter, K.: Shearing flows in a Goodman-Cowin type granular material – Theory and numerical results. *J. Particulate Materials* **17** (1999), 97–124.
12. Wilmański, K.: *Mechanics of Continuous Media.* 1 st ed., Springer-Verlag, Berlin 1999.

Mechanical aspect on drying of wet porous media

Stefan J. Kowalski

Poznań University of Technology, Institute of Technology and Chemical Engineering, 60-965 Poznań, Poland

Abstract. An aspect of thermomechanics of fluid saturated capillary-porous media concerning the mechanical phenomena accompanying the heat and mass transfer during drying processes is presented. The wet materials tend to shrink during drying and the shrinkage generates internal stresses which may cause fracturing of the dried body. The constitutive relations and the heat and mass rate equations are developed. The forces that produce shrinkage and the mechanisms responsible for transport of heat and moisture are examined. The considerations are based on the thermodynamics of irreversible processes and the continuum mechanics of porous media. The studies are quite general but the final form of the drying model is simplified by the assumptions admissible in practical applications. The example of convective drying of a ceramic cylinder is presented for illustration of the theory.

1 Introduction

Removal of liquid from wet porous bodies during drying processes results in some mechanical effects. First of all the dried bodies have a tendency to shrink, and the non-uniform shrinkage involves the internal self-stresses. The stresses may warp the dried body or deform it irreversibly and in some cases may cause even their cracking. It is clear that such effects are strongly undesirable in drying processes as they upset the quality of dried products or even make them useless. One of the remedies to avoid these undesirable mechanical effects is to use inconveniently slow drying rates. Slow drying rates, however, elongates the time of drying and get worse the efficiency of the productiveness, what is undesirable from the economic point of view. Therefore, one has to look for some optimal drying processes with respect to the drying time and the strength of dried material. In other words, the drying processes ought to be controlled and optimized, for example, in some periods accelerated if it not violates the material strength and in other periods slowed down to avoid fracture of the body.

This paper presents the thermomechanical model of drying, which may be helpful for construction of optimal drying processes. The model is developed on the basis of the principles of continuum mechanics of porous media and the thermodynamics of irreversible processes with the respective assumptions suitable for dried materials. Some contributions to this topic were developed earlier by Cairncross et al. [5], Coussy et al. [7], Kowalski [14–19], Kowalski et al. [21], Kowalski and Strumiłło [22]. A relevant contribution concerning phase transitions, a problem familiar with drying processes, is presented by de Boer and Kowalski [4]. One can meet in the literature other papers concerning stresses in dried materials, as e. g. Augier et al. [1], Perre and May [24], Scherer [27], but they present rather simplified and intuitively developed models than a systematically constructed thermomechanical theory of drying. The works concerning the heat and mass transfer only, and not the mechanical aspects of drying processes, are not discussed here.

The present considerations concern isotropic materials. The first part of this paper is quite general, and the second part presents a simplified model of drying with reasonable assumptions suitable to the most drying processes used in practical applications. An example of a convectively dried cylinder is solved for illustration of the presented here theory.

2 Preliminaries

The medium under consideration is a porous solid whose pores are filled with liquid (water) and gas (humid air). As the constituents are immiscible the structure of this medium is described by the volume fractions ϕ^a defined as the ratio of the constituent volume to the volume of the porous body, where $\alpha = S$ (solid), L (liquid), G (gas). The volume fractions has to fulfill the constrained condition:

$$\phi^S + \phi^L + \phi^G = 1 . \tag{1}$$

The porosity of the body is determined by the volume fractions in the following way:

$$\phi = \phi^L + \phi^G = 1 - \phi^S . \tag{2}$$

The volumetric saturation of the body S is termed as the fraction of the pore space that is filled with liquid, that is

$$S = \phi^L / \phi . \tag{3}$$

The gas phase is a miscible mixture of water vapour and dry air. The evaporation of water and thus the drying process proceeds if the air is unsaturated. The amounts of vapour and dry air in humid air use to be determined by the mole fractions x^α , defined as the number of vapour moles n^V (or dry air moles n^A) divided by the total number of moles of the humid air n^G, that is

$$x^V = \frac{n^V}{n^G} = \frac{p^V}{p^G} \quad \text{and} \quad x^A = \frac{n^A}{n^G} = \frac{p^A}{p^G} , \tag{4}$$

where p^V and p^A are the partial pressures of vapour and dry air, and p^G is the total pressure of the gas phase, respectively.

The relative humidity of air φ is the mole fraction for vapour x^V divided by the mole fraction of vapour in a saturated state x_s^V, that is

$$\varphi = \left(\frac{x^V}{x_s^V}\right)_T = \left(\frac{p^V}{p_s^V}\right)_T . \tag{5}$$

It is clear that there is no phase transitions of water into vapour in a saturated state ($\varphi = 1$) of the drying medium, that is, no drying occurs in such conditions.

Since each molecule of the individual species has a mass, a true mass density $\rho^{\alpha r}$ as well as a mass concentration ρ^α for each species can be defined. The mass concentrations (partial mass density) is defied as the mass of individual species per unit volume of the body as a whole. The relation between mass concentration and the true mass density is

$$\rho^\alpha = \phi^\alpha \rho^{\alpha r} . \tag{6}$$

A non-dimensional measure of a constituent content in the porous body X^α will be used in further considerations, which is defined as a mass concentration of α constituent referred to the mass concentration of the porous solid, that is

$$X^\alpha = \rho^\alpha / \rho^S . \tag{7}$$

The mass (or molar) flux of a species is a vector quantity denoting the amount of the particular species, in either mass or molar units, that passes per given increment of time through a unit area normal to the vector. In the present considerations, the mass flux $\mathbf{w}^\alpha$ will be used, defined with reference to the coordinates which are moving with the porous solid velocity $\mathbf{v}^S$. The absolute velocities of individual species relative to stationary coordinate axes are defined as

$$\mathbf{v}^\alpha \stackrel{\text{def}}{=} \mathbf{v}^S + \mathbf{w}^\alpha / \rho^\alpha . \tag{8}$$

The accelerations of individual species relative to stationary coordinate axes are denoted by $\mathbf{a}^\alpha$. In usual drying processes these accelerations can be considered for negligibly small, except may be for very special drying processes as, for example, the thermal shocks. Such processes, however, are beyond the scope of the present considerations.

Removal of liquid from a porous body exposes the solid network, what raises the energy of the system. The liquid flows from the interior to prevent exposure of the solid. As the liquid stretches toward exterior, it goes into tension which is balanced by the compressive stress in the network that causes its shrinkage. The shrinkage is generally non-uniform because of moisture gradients. The strongest shrinkage usually occurs at the surface of the dried body. When the body dries, the drier surface attempts to shrink but is restrained by the wet core. The surface is stressed in tension and the core in compression.

The stresses reflect the interaction forces between molecules of the solid network, and can be expressed by the true stress vector t_i^{Sr} or the force concentration t_i^S (partial stress), which is defined as the force acting on the solid network per unit surface of the body as a whole. The true and partial stresses are related to each other through the volume fraction, and the components of the partial stress are linear functions of the normal unit vector n_i, that is

$$t_i^S = \phi^S \, t_i^{Sr} = \sigma_{ji}^S \, n_j \,, \tag{9}$$

where σ_{ji}^S is the *Cauchy* stress tensor in the solid network. In fact, if there exists a meaningful pore pressure one has to used the notion "effective stress" or "drained network stress" as a force acting on the porous solid. The effective stress $\sigma_{ji}^{\mathrm{ef}}$ is a macroscopic averaged stress that controls the stress-strain law for a wet porous solid as well as the volume change and the strength behaviour of a porous medium, independent of the magnitude of the pore pressure. However, in the global balance of momentum appears the total stress tensor, which is the sum of the partial stress tensor in the solid network and the pore pressure, that is

$$\sigma_{ji} = \sigma_{ji}^S + (-p^{\mathrm{por}}) \, \phi \, \delta_{ij} = \sigma_{ji}^{\mathrm{ef}} + (-p^{\mathrm{por}}) \, \delta_{ij} \,, \tag{10}$$

where pore pressure is a combination of liquid pressure p^L, gas pressure p^G, and the saturation S, that is

$$p^{\mathrm{por}} = S \, p^L + (1 - S) \, p^G \,. \tag{11}$$

In this paper the stress in tension is considered as positive and the stress in compression as negative.

3 Global balance equations

The development of an analytical description of moisture flow through a porous body is based upon the fundamental physical laws of conservation of mass, momentum and energy, and the second law of thermodynamics concerning the entropy production in irreversible processes.

To develop a suitable working form of these laws let us separate mentally from the wet porous medium a finite control volume $V^S(t)$ fixed to the porous solid network through which the moisture flows. Note, that such a control volume varies in time following the porous solid volume change. Let the control volume be bounded by a regular surface $A^S(t)$, the position of which is determined by the external normal unit vector $\mathbf{n}$.

With respect to the control volume, the law of balance of mass for an α constituent may be stated as

$$\frac{\mathrm{d}}{\mathrm{d}t}\int\limits_{V^S(t)} \rho^\alpha \,\mathrm{d}V + \int\limits_{A^S(t)} (\mathbf{w}^\alpha \cdot \mathbf{n})\,\mathrm{d}A = \int\limits_{V^S(t)} \overset{*}{\rho}{}^\alpha \,\mathrm{d}V\,, \tag{12}$$

where $\mathbf{w}^\alpha$ is the mass flux of α constituent, defined with reference to the coordinates which are moving with the porous solid velocity $\mathbf{v}^S$, and term $\overset{*}{\rho}{}^\alpha$represents mass production per unit volume associated with phase transitions of liquid into vapour. The law of conservation of mass states that mass may be neither created nor destroyed, and therefore,

$$\sum_\alpha \int\limits_{V^S(t)} \overset{*}{\rho}{}^\alpha \,\mathrm{d}V = 0 \quad \text{with} \quad \overset{*}{\rho}{}^S \equiv 0 \quad \text{and} \quad \overset{*}{\rho}{}^A \equiv 0\,. \tag{13}$$

The time rate of change of α constituent momentum is equal to the force acting on this constituent and takes place in the direction of the net force, that is

$$\begin{aligned}\frac{\mathrm{d}}{\mathrm{d}t}\int\limits_{V^S(t)} \rho^\alpha\, v_i^\alpha \,\mathrm{d}V + \int\limits_{A^S(t)} v_i^\alpha\, (\mathbf{w}^\alpha \cdot \mathbf{n})\,\mathrm{d}A =& \\ = \int\limits_{A^S(t)} t_i^\alpha \,\mathrm{d}A + \int\limits_{V^S(t)} (\rho^\alpha\, g_i + \overset{*}{f}{}_i^\alpha)\,\mathrm{d}V\,,&\end{aligned} \tag{14}$$

where $t_i^S = \sigma_{ji}^S\, n_j$ is the stress in the solid network, and $t_i^\alpha = -p^\alpha\, \phi^\alpha\, n_i$ is the stress in others constituents having only the spherical part of the stress tensor, $\overset{*}{f}{}_i^\alpha$ characterizes the supply of momentum per unit volume to constituent α from the rest of constituents. Since the total momentum for the medium as a whole has to be conserved, it has to be

$$\sum_\alpha \int\limits_{V^S(t)} \overset{*}{f}{}_i^\alpha \,\mathrm{d}V = 0\,. \tag{15}$$

The expression for the balance of energy of α constituent may be written as

$$\begin{aligned}\frac{\mathrm{d}}{\mathrm{d}t}\int\limits_{V^S(t)} \rho^\alpha\, e^\alpha \,\mathrm{d}V + \int\limits_{A^S(t)} e^\alpha (\mathbf{w}^\alpha \cdot \mathbf{n})\,\mathrm{d}A =& \\ = \int\limits_{A^S(t)} (t_i^\alpha\, v_i^\alpha - q_i^\alpha\, n_i)\,\mathrm{d}A + \int\limits_{V^S(t)} \left[\rho^\alpha (g_i\, v_i^\alpha + r^\alpha) + \overset{*}{e}{}^\alpha\right]\mathrm{d}V\,,&\end{aligned} \tag{16}$$

where $e^\alpha = u^\alpha + \mathbf{v}^\alpha \cdot \mathbf{v}^\alpha/2$ is the internal and kinetic energy per unit of the constituent mass, q_i^α the heat flux due to conduction, r^α the heat supply per unit mass (radiation), and $\overset{*}{e}{}^{\alpha}$ the energy per unit volume supplied to constituent α from the rest of constituents. The law of conservation of energy for the medium as a whole requires that

$$\sum_{\alpha} \int_{V^S(t)} \overset{*}{e}{}^{\alpha} \, \mathrm{d}V = 0 \,. \tag{17}$$

The balance of entropy for the α constituent reads

$$\begin{aligned} &\frac{\mathrm{d}}{\mathrm{d}t} \int_{V^S(t)} \rho^{\alpha} s^{\alpha} \, \mathrm{d}V + \int_{A^S(t)} s^{\alpha} \left(\mathbf{w}^{\alpha} \cdot \mathbf{n} \right) \mathrm{d}A = \\ &\quad = - \int_{A^S(t)} \frac{q_i^{\alpha}}{T^{\alpha}} n_i \, \mathrm{d}A + \int_{V^S(t)} \left(\frac{\rho^{\alpha} r^{\alpha}}{T^{\alpha}} + \overset{*}{s}{}^{\alpha} \right) \mathrm{d}V \,, \end{aligned} \tag{18}$$

where s^{α} is the entropy per constituent unit mass, T^{α} the absolute temperature, and $\overset{*}{s}{}^{\alpha}$ the entropy production term per unit volume. According to the second law of thermodynamics the entropy production for the medium as a whole has to be positive, that is

$$\sum_{\alpha} \int_{V^S(t)} \overset{*}{s}{}^{\alpha} \, \mathrm{d}V \geq 0 \,. \tag{19}$$

The first integral in the above balance equations expresses the rate of accumulation of a thermodynamic quantity within the control volume, the second integral is the net outward flow of the quantity across the control surface (or the net efflux from the control volume), the surface integral on the right hand side expresses the external supply of the quantity through the control surface, and the last one expresses both the external volumetric supply of this quantity and the internal exchange between constituents.

4 Local balance equations

Based on the principles of mechanics of continua, the above global balance equations for mass, momentum, energy and entropy can written in the local form as follows:

$$\frac{\mathrm{d}\rho^S}{\mathrm{d}t} + \rho^S v_{i,i}^S = 0 \quad \text{and} \quad \rho^S \frac{\mathrm{d}X^{\alpha}}{\mathrm{d}t} = -w_{i,i}^{\alpha} + \overset{*}{\rho}{}^{\alpha} \,, \tag{20}$$

$$\sigma_{ji,j}^{\alpha} + \rho^{\alpha} \left(g_i - a_i^{\alpha} \right) + \overset{*}{f}{}_i^{\alpha} = \overset{*}{\rho}{}^{\alpha} v_i^{\alpha} \,, \tag{21}$$

$$\rho^S \frac{\mathrm{d}\bar{u}^\alpha}{\mathrm{d}t} = \sigma^\alpha_{ji}\, d_{ij} + \rho^S \mu^\alpha \frac{\mathrm{d}X^\alpha}{\mathrm{d}t} - (q^\alpha_i + s^\alpha\, T^\alpha\, w^\alpha_i)_{,i} + \rho^\alpha\, r^\alpha - \tag{22}$$

$$- w^\alpha_i\, \hat{\mu}^\alpha_{,i} - \overset{*}{\rho}{}^\alpha \mu^\alpha + \overset{*}{e}{}^{\alpha M} + \overset{*}{e}{}^{\alpha T},$$

$$\rho^S \frac{\mathrm{d}\bar{s}^\alpha}{\mathrm{d}t} = - \left(\frac{q^\alpha_i + s^\alpha\, T^\alpha\, w^\alpha_i}{T^\alpha} \right)_{,i} + \frac{\rho^\alpha\, r^\alpha}{T^\alpha} + \overset{*}{s}{}^\alpha, \tag{23}$$

where $\bar{u}^\alpha = X^\alpha u^\alpha$ and $\bar{s}^\alpha = X^\alpha s^\alpha$ denote the internal energy and the entropy of the α constituent per unit mass of the bone dry body, $\mu^\alpha = u^\alpha - s^\alpha\, T^\alpha + p^\alpha / \rho^{\alpha r}$ is the free enthalpy per unit mass of the α constituent, $\overset{*}{e}{}^{\alpha M}$ and $\overset{*}{e}{}^{\alpha T}$ are the mechanical and the thermal interaction terms between constituent α and the other constituents, $\hat{\mu}^\alpha = \mu^\alpha + \mu^{\text{grav}}$ is a generalized constituent potential with $\mu^{\text{grav}}_{,i} = -g_i$. The constituent acceleration and the relative kinetic energy were neglected in the above formulas as a small importance. The expression

$$d_{ij} = \frac{1}{2} \left(v^S_{i,j} + v^S_{j,i} \right) = d^{(r)}_{ij} + d^{(i)}_{ij} \tag{24}$$

is the strain rate of the porous body, generally consisting of the reversible (r) and irreversible (i) parts. The time derivative $\mathrm{d}(\,\cdot\,)/\mathrm{d}t$ is the material derivative with convection velocity $\mathbf{v}^S$.

For reversible and adiabatic processes the first law of thermodynamics can be reduced to an extremely simple relation following the *Caratheodory*'s principle: *there are adiabatically inaccessible states in the neighbourhood of every thermodynamic state*. This means that for a reversible and adiabatic process the deferential (*Pfaff* ian) separated from (22) of the form

$$\frac{\mathrm{d}\bar{Q}^\alpha}{\mathrm{d}t} = \frac{\mathrm{d}\bar{u}^\alpha}{\mathrm{d}t} - \frac{1}{\rho^S}\, \sigma^\alpha_{ji}\, d^{(r)}_{ij} - \mu^\alpha \frac{\mathrm{d}X^\alpha}{\mathrm{d}t} - \overset{*}{e}{}^{\alpha M} \tag{25}$$

has to be equal to zero (Gumiński [10]). However, the differential (25) is integrable for an arbitrary reversible process (see e. g. Naphtali [23], Hutter [11]). It means that there exists an integrating factor $T^\alpha(\vartheta) \geq 0$ which turns the above differential form into a potential which is the constituent entropy, that is

$$\frac{1}{T^\alpha(\vartheta)} \left(\frac{\mathrm{d}\bar{u}^\alpha}{\mathrm{d}t} - \frac{1}{\rho^S}\, \sigma^\alpha_{ji}\, d^{(r)}_{ij} - \mu^\alpha \frac{\mathrm{d}X^\alpha}{\mathrm{d}t} - \overset{*}{e}{}^{\alpha M} \right) = \frac{\mathrm{d}\bar{s}^\alpha}{\mathrm{d}t}\,. \tag{26}$$

One identifies (26) as the *Gibbs* identity. It may be written as a function of free energy, taking the following form for the porous network and for the other constituents:

$$\frac{\mathrm{d}f^S}{\mathrm{d}t} = -s^S \frac{\mathrm{d}T^S}{\mathrm{d}t} + \frac{1}{\rho^S}\, \sigma^S_{ji}\, d^{(r)}_{ij} + \frac{\overset{*}{e}{}^{SM}}{\rho^S}, \tag{27}$$

$$\frac{\mathrm{d}\bar{f}^{\alpha}}{\mathrm{d}t} = -\bar{s}^{\alpha}\,\frac{\mathrm{d}T^{\alpha}}{\mathrm{d}t} - \frac{1}{\rho^{S}}\,p^{\alpha}\,\phi^{\alpha}\,d_{ii}^{(r)} + \mu^{\alpha}\,\frac{\mathrm{d}X^{\alpha}}{\mathrm{d}t} + \frac{\overset{*}{e}{}^{\alpha M}}{\rho^{S}}\,, \tag{28}$$

where $\bar{f}^{\alpha} = \bar{u}^{\alpha} - \bar{s}^{\alpha}\,T^{\alpha}$ is the free energy of α constituent referred to the unit mass of dry material. Having in mind the definition of the free enthalpy per unit mass of α constituent one can rearrange (28) and find the following expression for the mechanical interaction term:

$$\overset{*}{e}{}^{\alpha M} = -p^{\alpha}\,\frac{\mathrm{d}\phi^{\alpha}}{\mathrm{d}t}\,. \tag{29}$$

Using the definitions of porosity and saturation given in Section 2, we may write the interaction term for individual constituents as follows:

$$\overset{*}{e}{}^{LM} = -p^{L}\,\frac{\mathrm{d}\phi^{L}}{\mathrm{d}t} = -p^{L}\left(S\,\frac{\mathrm{d}\phi}{\mathrm{d}t} + \frac{\mathrm{d}S}{\mathrm{d}t}\,\phi\right),$$

$$\overset{*}{e}{}^{GM} = -p^{G}\,\frac{\mathrm{d}\phi^{G}}{\mathrm{d}t} = -p^{G}\left[(1-S)\,\frac{\mathrm{d}\phi}{\mathrm{d}t} - \frac{\mathrm{d}S}{\mathrm{d}t}\,\phi\right],$$

$$\overset{*}{e}{}^{SM} = -\left(\overset{*}{e}{}^{LM} + \overset{*}{e}{}^{GM}\right) = p^{\mathrm{por}}\,\frac{\mathrm{d}\phi}{\mathrm{d}t} + p^{\mathrm{cap}}\,\phi\,\frac{\mathrm{d}S}{\mathrm{d}t}\,,$$

where p^{por} denotes the pore pressure expressed by (11), and $p^{\mathrm{cap}} = p^{L} - p^{G}$ defines the capillary pressure.

Gibbs' expression (27) may now be written as

$$\frac{\mathrm{d}f^{S}}{\mathrm{d}t} = -s^{S}\,\frac{\mathrm{d}T^{S}}{\mathrm{d}t} + \frac{1}{\rho^{S}}\,\sigma_{ji}^{\mathrm{ef}}\,d_{ij}^{(r)} - p^{\mathrm{por}}\,\frac{\mathrm{d}}{\mathrm{d}t}\left(\frac{1}{\rho^{Sr}}\right) + \frac{1}{\rho^{S}}\,P^{\mathrm{cap}}\,\frac{\mathrm{d}S}{\mathrm{d}t}\,, \tag{30}$$

where $\sigma_{ji}^{\mathrm{ef}} = \sigma_{ji}^{S} + p^{\mathrm{por}}(1-\phi)\delta_{ij}$ is the effective stress acting on the porous network, and $P^{\mathrm{cap}} = p^{\mathrm{cap}}\phi$ is the bulk capillary pressure. It is visible that the time alteration of free energy in the porous solid depends on time alterations of the temperature, the reversible strains of the porous solid, the true density of the solid matrix, and the saturation. If the irreversible deformations occur in the porous solid, the work of the effective stress on these deformations will dissipate, causing an increase of the body temperature. Thus, the free energy in (30) will be influenced from the dissipated work through an increase of the temperature. The term expressing the work of stress on irreversible deformations will occur as the heat source in the differential equation of heat conduction.

Gibbs' identity (28) for the other constituents reads

$$\frac{\mathrm{d}\bar{f}^{\alpha}}{\mathrm{d}t} = -\bar{s}^{\alpha}\,\frac{\mathrm{d}T^{\alpha}}{\mathrm{d}t} - p^{\alpha}\,\frac{\mathrm{d}}{\mathrm{d}t}\left(\frac{\phi^{\alpha}}{\rho^{S}}\right) + \mu^{\alpha}\,\frac{\mathrm{d}}{\mathrm{d}t}\left(\frac{\rho^{\alpha}}{\rho^{S}}\right). \tag{31}$$

Thus, the free energy of α constituent referred to the unit mass of dry body depends on the constituent temperature, the constituent volume fraction and the constituent mass concentration, both per unit mass of the dry body.

(30) and (31) constitute the basic relations for the analysis of constitutive equations describing the physical behaviour of wet materials during drying.

The balance of energy (22) after substituting *Gibbs*' identity (26) takes the following form for the solid network and for the other constituents:

$$\rho^S\, T^S\, \frac{\mathrm{d}s^S}{\mathrm{d}t} = \sigma_{ji}^{\mathrm{ef}}\, d_{ij}^{(i)} - q_{i,i}^S + \rho^S\, r^S + \overset{*}{e}{}^{ST}\,,$$

$$\rho^S\, T^\alpha\, \frac{\mathrm{d}\bar{s}^{\,\alpha}}{\mathrm{d}t} = -\left(q_i^\alpha + s^\alpha\, T^\alpha\, w_i^\alpha\right)_{,i} + \rho^\alpha\, r^\alpha - w_i^\alpha\left(\mu_{,i}^\alpha - g_i\right) - \overset{*}{\rho}{}^\alpha \mu^\alpha + \overset{*}{e}{}^{\alpha T},$$

$$\sum_\alpha \overset{*}{e}{}^{\alpha T} = 0\,. \tag{32}$$

The second law of thermodynamics (19), with applications of the first law (32), will yield the following final form:

$$\begin{aligned}&\sum_\alpha \frac{1}{T^\alpha}\left[-\frac{q_i^\alpha + s^\alpha\, T^\alpha\, w_i^\alpha}{T^\alpha}\; T_{,i}^\alpha - w_i^\alpha\left(\mu_{,i}^\alpha - g_i\right) - \overset{*}{\rho}{}^\alpha \mu^\alpha + \overset{*}{e}{}^{\alpha T}\right] + \\ &\quad + \frac{1}{T^S}\,\sigma_{ji}^{ef}\, d_{ij}^{(i)} \geq 0\,.\end{aligned} \tag{33}$$

The above inequality consists of terms having different tensor representation. According to *Courier*'s principle (see e. g. Gumiński [10]): *it is not possible for the cause of greater tensorial symmetry to produce an effect of smaller tensorial symmetry*. For example, a cause of scalar type cannot involve an effect of vector type. This principle allows us to write three independent inequalities instead one, namely

$$\sigma_{ji}^{ef}\, d_{ij}^{(i)} \geq 0\,, \tag{34}$$

$$\sum_\alpha \left[\left(q_i^\alpha + h^\alpha w_i^\alpha\right)\left(\frac{1}{T^\alpha}\right)_{,i} - w_i^\alpha\left(\left(\frac{\mu^\alpha}{T^\alpha}\right)_{,i} - \frac{g_i}{T^\alpha}\right)\right] \geq 0\,, \tag{35}$$

$$\sum_\alpha \sum_\beta \frac{1}{2}\left[\rho^{\alpha\beta}\left(\frac{\mu^\beta}{T^\beta} - \frac{\mu^\alpha}{T^\alpha}\right) + e^{\alpha\beta}\left(\frac{1}{T^\beta} - \frac{1}{T^\alpha}\right)\right] \geq 0\,, \tag{36}$$

where $h^\alpha = \mu^\alpha - s^\alpha\, T^\alpha$ is the enthalpy of the α constituent per unit mass, and μ^α / T^α is the isothermal *Planck* potential (see e. g. Prigogine and Defay [25], Berry et al. [3]). In the last inequality, use was made of some mathematical expressions stated that the heat and mass exchange between the constituents have the form

$$\overset{*}{\rho}{}^{\alpha} = \sum_{\beta} \rho^{\alpha\beta}, \qquad \overset{*}{e}{}^{\alpha T} = \sum_{\beta} e^{\alpha\beta},$$

where $\rho^{\alpha\beta} = -\rho^{\beta\alpha}$ expresses the supply of mass and $e^{\alpha\beta} = -e^{\beta\alpha}$ the supply of energy (heat) from constituent β to constituent α.

(34) to (36) constitute the basic relations for the analysis of irreversible deformations and the heat and mass flow via the porous solid (dried body). The rate equations for heat and mass flow may be constructed with the use of the above constrains followed from the second law of thermodynamics.

5 Physical relations for a wet porous body

We shall develop from *Gibbs'* identity (30) the physical relations for wet porous solid, suitable for dried materials. *Gibbs'* equation suggests that the free energy ought to be dependent on: The *Green* strain tensor $E_{KL}^{(r)}$ ($\dot{E}_{KL}^{(r)} = d_{ij}^{(r)}\, x_{i,K}\, x_{j,L}$), where the term $x_{i,K}$ is the deformation gradient, the temperature T^S, the specific volume of the porous matrix $V^{Sr} = 1/\rho^{Sr}$, and the saturation S, that is

$$f^S = f^S\left(E_{KL}^{(r)}, T^S, V^{Sr}, S\right). \tag{37}$$

Calculating the time derivative of free energy f^S by making use of the chain-rule expansion and combining this with the one expressed by (30), will yield the equations of state

$$\sigma_{ij}^{\text{ef}} = \rho^S\, x_{i,K}\, x_{j,L} \left(\frac{\partial f^S}{\partial E_{KL}^{(r)}}\right)_{T^S, V^{Sr}, S} = \sigma_{ij}^{\text{ef}}\left(E_{KL}^{(r)}, T^S, V^{Sr}, S\right), \tag{38}$$

$$s^S = -\left(\frac{\partial f^S}{\partial T^S}\right)_{E_{KL}^{(r)}, V^{Sr}, S} = s^S\left(E_{KL}^{(r)}, T^S, V^{Sr}, S\right), \tag{39}$$

$$p^{\text{por}} = -\left(\frac{\partial f^S}{\partial V^{Sr}}\right)_{E_{KL}^{(r)}, T^S, S} = p^{\text{por}}\left(E_{KL}^{(r)}, T^S, V^{Sr}, S\right), \tag{40}$$

$$P^{\text{cap}} = \rho^S \left(\frac{\partial f^S}{\partial S}\right)_{E_{KL}^{(r)}, T^S, V^{Sr}} = P^{\text{cap}}\left(E_{KL}^{(r)}, T^S, V^{Sr}, S\right). \tag{41}$$

The above equations of state are helpful by the construction of physical relations as they indicate whose parameters of state ought to be present in these relations. The interpretation of two first equations is well known in thermomechanics of continua. The third equation indicates that pore pressure p^{por} is the main reason for the compression of the solid matrix. The compression of solids is usually assumed to be insignificant and neglected. In such a case the free energy for solid (30) is not changed due to the action of the pore pressure and the equation of state (40) falls out.

The forth equation point out that the capillary pressure depends on the ratio of saturation. In the literature dealing with the physics of fluid flow through porous media a lot of attention was devoted searching of a relation between the capillary pressure and the saturation ratio (see e. g. Scheidegger [26], Kirkham and Powers [13], Czolbe et al. [8], Cairncross et al. [5]). Figure 1 presents a typical plot of capillary pressure versus saturation ratio, drawn on the base of relation given in Kowalski et al. [20].

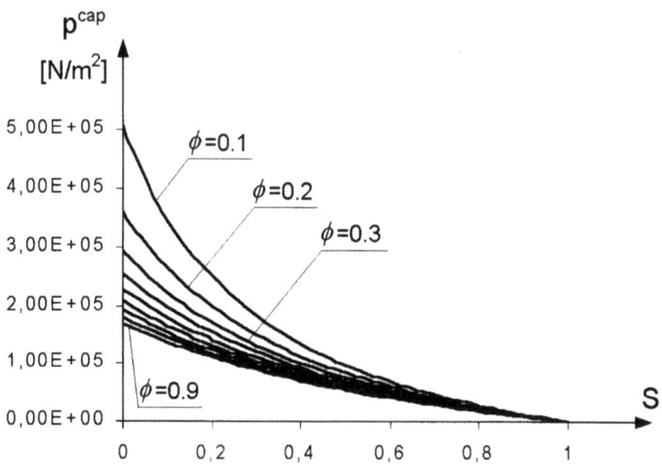

Fig. 1. Capillary pressure versus saturation for different porosities.

A great number of initial-boundary-value problems met in industrial drying concerns materials which suffer rather small deformations, for example, ceramics, wood, etc. The solid matrix for such materials is rather incompressible. Having in mind these simplifications, one can write the physical relation between effective stresses and the reversible small strains $\varepsilon_{ij}^{(r)}$, similarly as in the theory of elasticity, that is

$$\sigma_{ij}^{\mathrm{ef}} = \sigma_{ij} + p^{\mathrm{por}}\,\delta_{ij} = 2\,M\,\varepsilon_{ij}^{(r)} + \left(A\,\varepsilon^{(r)} - \gamma^{(T)}\,\vartheta^{(S)}\right)\delta_{ij}\,, \tag{42}$$

where $M(\vartheta^{(S)}, S)$ and $A(\vartheta^{(S)}, S)$ are the shear and bulk modules of the wet porous body dependent on the temperature $\vartheta^{(S)} = T^S - T_0^S$ and saturation S, and $\gamma^{(T)} = (2\,M + 3\,A)\,\kappa^{(T)}$ is the thermal modulus with $\kappa^{(T)}$ being the coefficient of linear thermal expansion.

Remember, that the effective stress σ_{ji}^{ef} is a macroscopic averaged stress that controls the stress-strain law for a wet porous solid as well as the volume change and the strength behaviour of a porous medium, independent of the magnitude of the pore pressure. Sometimes, it is convenient to use the total instead effective stress and the pore pressure to express by the saturation. To this aim one can use the experimental observation allowing to state that a porous material swells while wetted and shrinks while dried. Thus, the wetting and drying processes involve volume changes of a body similar to cooling and heating.

Let us consider a process of unconstrained wetting during which the total pressure is equal to the atmospheric pressure $\sigma_{ij} = -p^{\mathrm{atm}}\,\delta_{ij}$, and the volumetric strain is a function of saturation,

$$\varepsilon^{(r)} = 3\,\kappa^{(S)}\,S\,, \tag{43}$$

where $\kappa^{(S)}$ may be termed the coefficient of linear swelling expansion (a quantity similar to the coefficient of linear thermal expansion). Assuming the wetting process to be carried out at the temperature T_0^S, one obtains from (42),

$$p^{\mathrm{por}} = p^{\mathrm{atm}} + \gamma^{(S)}\,S\,, \tag{44}$$

where $\gamma^{(T)} = (2M + 3A)\,\kappa^{(T)}$ is the wetting modulus. Note, that in dry state, that is for $S = 0$, the pore pressure $p^{\mathrm{por}} \equiv p^G = p^{\mathrm{atm}}$, and for a fully saturated body ($S = 1$) the pore pressure is $p^{\mathrm{por}} = p^{\mathrm{atm}} + \gamma^{(S)}$. In mechanics, significant is the pressure above the atmospheric one, therefore the physical relation for the total stress may now be written as

$$\sigma_{ij} = 2\,M\,\varepsilon_{ij}^{(r)} + \left[A\,\varepsilon^{(r)} - \gamma^{(T)}\,\vartheta^{(S)} - \gamma^{(S)}\,S\right]\delta_{ij}\,. \tag{45}$$

The relation between stresses and the irreversible strains has to satisfied the inequality (34). A general form of this relation may be written as

$$\sigma_{ji}^{\mathrm{ef}} = C_{ijkl}\,d_{kl}^{(i)} \quad \text{with} \quad C_{ijkl}\,d_{ij}^{(i)}d_{kl}^{(i)} \geq 0\,, \tag{46}$$

where C_{ijkl} is the tensor of material constants. Relation (46) expresses the generalized *Newton-Cauchy-Poisson* law for a viscous or plastic body.

Physical relations for the other constituents (liquid and gas) may be developed from the *Gibbs* identity (31), which suggests that the free energy is to be dependent on: The constituent temperature T^α, the constituent mass concentration $X^\alpha = \rho^\alpha/\rho^S$ and the constituent volume fraction $Y^\alpha = \phi^\alpha/\rho^S$, both per unit mass of the dry body, that is

$$\bar{f}^\alpha = \bar{f}^\alpha\,(T^\alpha, X^\alpha, Y^\alpha)\,. \tag{47}$$

Again, calculating the time derivative of free energy $\bar{f}^\alpha$ by making use of the chain-rule expansion and combining this with the one expressed by (31), will yield further equations of state:

$$\bar{s}^\alpha = -\left(\frac{\partial \bar{f}^\alpha}{\partial T^\alpha}\right)_{Y^\alpha, X^\alpha} = \bar{s}^\alpha\left(T^\alpha, X^\alpha, Y^\alpha\right), \tag{48}$$

$$p^\alpha = -\left(\frac{\partial \bar{f}^\alpha}{\partial Y^\alpha}\right)_{T^\alpha, X^\alpha} = \bar{s}^\alpha\left(T^\alpha, X^\alpha, Y^\alpha\right), \tag{49}$$

$$\mu^\alpha = \left(\frac{\partial \bar{f}^\alpha}{\partial X^\alpha}\right)_{T^\alpha, Y^\alpha} = \mu^\alpha\left(T^\alpha, X^\alpha, Y^\alpha\right), \tag{50}$$

An explicit shape of these functions may be constructed, for example, by their expansion in the *Taylor* series.

6 Phenomenological rate transport equations

The basic tool we shall use in developing the rate transport equations is the second law of thermodynamics represented by inequalities (35) and (36). The rate equations have to be constructed in such a way to preserve positive definition of these inequalities. Here, we assume that each individual term of the polynomial (35) is a positive defined quadratic form, that is

$$w_i^\alpha = -\Lambda_{11}^\alpha \left[\left(\frac{\mu^\alpha}{T^\alpha}\right)_{,i} - \frac{g_i}{T^\alpha}\right] + \Lambda_{12}^\alpha \left(\frac{1}{T^\alpha}\right)_{,i}, \tag{51}$$

$$q_i^\alpha + h^\alpha w_i^\alpha = -\Lambda_{21}^\alpha \left[\left(\frac{\mu^\alpha}{T^\alpha}\right)_{,i} - \frac{g_i}{T^\alpha}\right] + \Lambda_{22}^\alpha \left(\frac{1}{T^\alpha}\right)_{,i}, \tag{52}$$

where the coefficients appeared in these relations have to satisfy the *Silvester* criterion (see e. g. Jefimov and Rozendorn [12])

$$\Lambda_{11}^\alpha \geq 0 \quad \text{and} \quad \Lambda_{11}^\alpha \Lambda_{22}^\alpha \geq \left(\frac{\Lambda_{12}^\alpha + \Lambda_{21}^\alpha}{2}\right)^2. \tag{53}$$

The expression in square brackets and the gradient of inverse temperature in relations (51) and (52) may be termed as the mass and heat transfer potentials. These relations describe the diffusions and thermodiffusion (*Soret*'s effect), heat conduction and the so-called *Dufoure* effect. They have to be accomplished in all possible thermodynamic conditions, for example, also for zero value heat and mass fluxes at non-zero transfer potentials. In such a case, the respective gradients have to be related to each other and the determinant of the coefficients standing by these gradients has to be zero. The in this way obtained additional relation between the coefficients, combined with that of *Silvester*'s criterion, yields the conclusion that $\Lambda_{12}^{\alpha} = \Lambda_{21}^{\alpha}$.

The thermodiffusion is a very relevant effect in drying processes, and in particular in convective drying where the heat and mass fluxes are in opposite directions. The thermodiffusion flux of mass blockades the outflow of moisture from the interior of the dried body and causes the boundary surface to dry very quick. Therefore, we take into account in our considerations both the thermodiffusion effect and the heat transfer by the mass fluxes. However, it is more convenient to use gradient of temperature instead gradient of inverse temperature and gradient of constituent potential instead of gradient of the *Planck* potential in our mathematical description of heat and mass transfer. Taking $\Lambda_{12}^{\alpha} = \mu^{\alpha} \Lambda_{11}^{\alpha}$ one can obtain another form of heat and mass transfer relation, namely,

$$w_i^{\alpha} = -\Lambda_m^{\alpha} \left[(\mu^{\alpha})_{,i} - g_i \right] \quad \text{with} \quad \Lambda_{11}^{\alpha}/T^{\alpha} = \Lambda_m^{\alpha} \geq 0 \,, \tag{54}$$

$$q_i^{\alpha} + s^{\alpha} T^{\alpha} w_i^{\alpha} = -\Lambda_T^{\alpha} (T^{\alpha})_{,i} \quad \text{with} \quad (\Lambda_{22}^{\alpha} - \mu^{\alpha} \Lambda_{12}^{\alpha})/(T^{\alpha})^2 = \Lambda_T^{\alpha} \geq 0 \,. \tag{55}$$

The influence of temperature field on the mass transport is hidden now in the constituent potential μ^{α} and in the transport coefficient (mobility) Λ_m^{α}. Note, that for the constituents movable with respect to the solid network the total heat flux, that is the heat conducted and the heat transported by the mass flux, is proportional to the gradient of temperature. In some place the heat transported by mass may balance the heat due to conduction, and the gradient of temperature may disappear there. Such a situation may happen during the so called constant drying rate period, where the heat supply to a dried body from the hot surroundings is transported back by the vapour flux and no heat is conducted towards the interior of the dried body. The temperature of the dried body is kept constant during this period and is equal to the wet bulb temperature.

Similar analysis as to polynomial (35) can be made with respect to polynomial (36). Assuming again that each individual term of this polynomial is a positive defined quadratic form, we can write

$$\rho^{\alpha\beta} = \lambda_{11}^{\alpha\beta} \left(\frac{\mu^{\beta}}{T^{\beta}} - \frac{\mu^{\alpha}}{T^{\alpha}} \right) + \lambda_{12}^{\alpha\beta} \left(\frac{1}{T^{\beta}} - \frac{1}{T^{\alpha}} \right) , \tag{56}$$

$$e^{\alpha\beta} = \lambda_{21}^{\alpha\beta} \left(\frac{\mu^{\beta}}{T^{\beta}} - \frac{\mu^{\alpha}}{T^{\alpha}} \right) + \lambda_{22}^{\alpha\beta} \left(\frac{1}{T^{\beta}} - \frac{1}{T^{\alpha}} \right) , \tag{57}$$

where the coefficients appeared in these relations have to satisfy the *Silvester* criterion:

$$\lambda_{11}^{\alpha\beta} \geq 0 \quad \text{and} \quad \lambda_{11}^{\alpha\beta} \lambda_{22}^{\alpha\beta} \geq \left(\frac{\lambda_{12}^{\alpha\beta} + \lambda_{21}^{\alpha\beta}}{2} \right)^2 . \tag{58}$$

The same arguments as above can be used to state that $\lambda_{12}^{\alpha\beta} = \lambda_{21}^{\alpha\beta}$.

As it is seen in relations (56) and (57), the differences between *Planck* potentials and inverse temperatures of individual constituents are the thermodynamic forces for heat and mass transfer between constituents. If the coupling effects can be omitted ($\lambda_{12}^{\alpha\beta} = \lambda_{21}^{\alpha\beta} = 0$), then the expression for the heat exchange between constituents takes the form familiar with the *Newton* rate equation for convective heat transfer, namely,

$$e^{\alpha\beta} = \lambda_T^{\alpha\beta}\left(T^\alpha - T^\beta\right) \quad \text{with} \quad \lambda_{22}^{\alpha\beta}/T^\alpha T^\beta = \lambda_T^{\alpha\beta} \geq 0\,. \tag{59}$$

The rate of mass transfer due to phase transitions reads

$$\overset{*}{\rho}{}^L = -\overset{*}{\rho}{}^V = -\lambda_{11}^{LV}\left(\frac{\mu^L}{T^L} - \frac{\mu^V}{T^V}\right) \quad \text{with} \quad \lambda_{11}^{LV} = \lambda_{11}^{VL} \geq 0\,. \tag{60}$$

Relation (60) states that the phase transition of liquid into vapour, and vice versa, takes place if the *Planck* potentials differ from each other (see de Boer and Kowalski [4], Benet and Jouanna [2]). The temperature of the vapour in air may be assumed to be the same as the temperature of gas, $T^V \equiv T^A \equiv T^G$.

7 A simplified model of drying

Many drying processes encountered in practical applications allow to introduce more simplifications by their modelling that it was done up to now. In the previous chapters, we have taken the accelerations and the kinetic energies of the species to be negligibly small. Now, we introduce another realistic assumption, namely, that the temperatures of individual constituents of a dried body are equal to each other. This is a reasonable assumption as the most drying processes proceed very slow (quasi-statically), and thus, the heat exchange between the constituents may be considered for sufficiently fast. Due to this simplification there is no need to calculate several constituent temperatures but only one for the body as a whole, and thus only one energy equation is sufficient to write. For the specified conditions the overall *Gibbs'* identity, the energy equation, and the second law of thermodynamics, become:

- the *Gibbs* identity

$$\dot{f} = -s\,\dot{T} + \frac{1}{\rho^S}\,\sigma_{ij}\,d_{ij}^{(r)} + \sum_\alpha \mu^\alpha\,\dot{X}^\alpha \tag{61}$$

- energy equation (after substituting the *Gibbs* identity)

$$\begin{aligned}\rho^S\,\dot{s}\,T = \sigma_{ij}\,d_{ij}^{(i)} - \left(q_i + T\sum_\alpha s^\alpha\,w_i^\alpha\right)_{,i} + \rho\,r\,- \\ -\sum_\alpha w_i^\alpha\left(\mu_{,i}^\alpha - g_i\right) - \overset{*}{\rho}{}^L\left(\mu^L - \mu^V\right)\end{aligned} \tag{62}$$

- the second law of thermodynamics

$$T \sum_{\alpha} \overset{*}{s}^{\alpha} = \sigma_{ij}\, d_{ij}^{(i)} - \left(q_i + T \sum_{\alpha} s^{\alpha}\, w_i^{\alpha} \right) \frac{T_{,i}}{T} - \sum_{\alpha} w_i^{\alpha} \left(\mu_{,i}^{\alpha} - g_i \right) - \\ - \overset{*}{\rho}^{L} \left(\mu^{L} - \mu^{V} \right) \geq 0\,, \tag{63}$$

where f and s denote the total free energy and entropy referred to the unit mass of dry body, σ_{ij} is the total stress tensor, q_i total heat flux due to conduction, $\rho\, r$ total volumetric heat supply, and dot over the symbol denotes the substantial derivatives with the convection velocity of the porous solid, that is $\dot{(\cdot)} \equiv \mathrm{d}(\cdot)/\mathrm{d}t = \partial(\cdot)/\partial t + (\cdot)_{,i}\, v_i^S$.

Next simplification concerns the amount of gas with respect to the amount of liquid in pores. Note that during the first stage of drying the moisture flows in a form of liquid flux towards the boundary, where it evaporates. Only insignificant amount of vapour is stored in gas bubbles inside the body during this period. The second drying period is characterized by evaporation and condensation inside the body, that is at the boundary between the funicular (continuous liquid) and pendular (isolated pockets of liquid) regions. During this period moisture can migrate by creeping along the capillary walls or by successive evaporation and condensation mechanism between liquid bridges (pendular regions). It means that there is no direct outflow of vapour from the interior to the outside. The outflow proceeds from the meniscus close the boundary layer which is stable for long time due to evaporation-condensation mechanism. In this period, similar as in the previous period, only insignificant amount of vapour is stored inside the body. Finally, the direct removal of vapour from the pore space to outside take place quite at the end of drying process, that is in the third drying period, where the moisture is held in multimolecular layers or monolayer on the pore walls (see e. g. Fortes and Okos [9], Strumiłło [28]). Therefore, we can state that the mass of gas inside the dried body is negligibly small with respect to the mass of liquid almost through the hole period of drying, that is $X^L \gg X^G = X^V + X^A$. To justify this, let us assume that the total volume of dried body is 1 m^3. If the porosity of the body is, say $\phi = 0.2$ then, the pore volume is 0.2 m^3. The mass of water in fully filled pore space at room temperature is then ca. 200 kg. The mass of vapour, on the other hand, fully filling the same pore space is 1000 times less, and equals ca. 0.2 kg, because the volume of vapour condensed to the liquid phase shrink ca. 1000 times (see e. g. Cottrell [6]). This allows us to state that the mass of gas inside the dried body is negligibly small with respect to the mass of liquid, and that the most of energy needed for evaporation of moisture supplied from the surroundings is consumed at the boundary layer. During the evaporation and condensation process inside the body, the internal energy used for evaporation is giving back by the condensation.

Another simplification follows from the fact that drying processes proceed close the equilibrium states. In such a situation, the gas in pore space is almost fully saturated with vapour, and thus the chemical potential of vapour is equal to that of liquid, that is $\mu^V \approx \mu^L$. This equality may hold both during the constant drying rate period, where the vapour in gas bubbles is in equilibrium with the liquid, and during the falling rate period where liquid in pendular state (isolated pockets of liquid) and adsorbed films is in equilibrium with the vapour inside pores, see Figure 2.

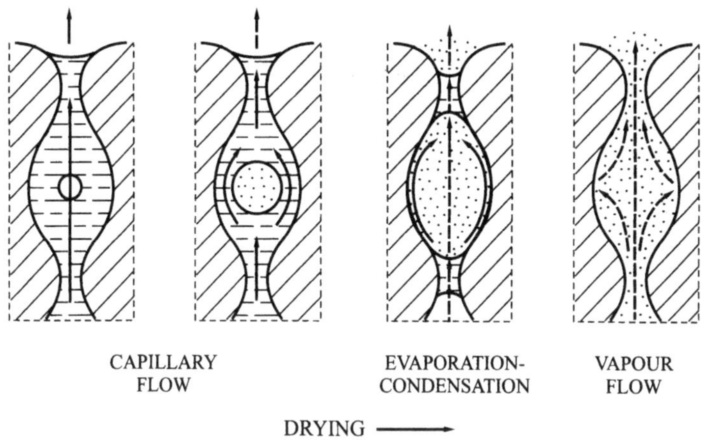

Fig. 2. Moisture state during the constant and the falling drying rate periods.

It is obvious that the pores are full of liquid at the first stage of drying, and the liquid is moving towards the boundary due to capillary forces where it evaporates. This stage of drying is called the constant drying rate period, because the rate of evaporation at stable drying conditions is independent of time. The temperature of the body in this period is constant and equal to the wet bulb temperature. The mass balance equations (20) for the constant rate period may be written as follows:

$$\rho^S \dot{X}^L = -w^L_{i,i} + \overset{*}{\rho}{}^L , \quad \rho^S \dot{X}^V = \overset{*}{\rho}{}^V , \quad \rho^S \dot{X}^A = 0 .$$

The above equations state: First, the liquid inside the body is altered due to liquid flux and the phase transition; second, the amount of vapour in pores increases due to the phase transition; and third, the amount of air in pores is constant.

In the falling rate period, the rate of evaporation decreases and the temperature of the body surface rises above the wet bulb temperature, even at stable drying conditions. Moisture is held in the finest capillarities and can migrate by creeping along the capillary walls or by successive evaporation and condensation between liquid bridges. The mass balance for the falling rate period (20) may be written as

$$\rho^S \dot{X}^L = \overset{*}{\rho}{}^L , \quad \rho^S \dot{X}^V = -w^V_{i,i} + \overset{*}{\rho}{}^V , \quad \rho^S \dot{X}^A = -w^A_{i,i} \approx 0 .$$

We may interpret these equations as follows: First, the phase transition of liquid into vapour proceeds inside the body; the vapour is transported in a form of vapour flux trough the body mainly due to evaporation and condensation mechanism; the amount of air may change insignificantly at the very final stage of drying.

Summarizing the above statements, we can neglect in our considerations the alteration of air inside the pore space ($\dot{X}^A = 0$), and assume the chemical potentials for vapour and liquid inside the pore space to be equal to each other ($\mu^V = \mu^L \equiv \mu$) during the almost whole drying process. For these specified conditions the *Gibbs* identity (61) becomes

$$\dot{f} = -s\dot{T} + \frac{1}{\rho^S}\sigma_{ij}\,d_{ij}^{(r)} + \mu\dot{X}\,, \tag{64}$$

where $\dot{X} = \dot{X}^L + \dot{X}^V$ with $\dot{X}^L \gg \dot{X}^V$. Furthermore, under these conditions we may apply one mass balance equation for the constant and the falling rate periods, namely

$$\rho^S\dot{X} = -w_{i,i}\,, \tag{65}$$

where w_i denotes the total moisture flux expressed by

$$w_i = -\Lambda_m\,(\mu_{,i} - g_i) \quad \text{with} \quad \Lambda_m \geq 0\,. \tag{66}$$

Under these circumstances the final form of the balance energy (62) reads

$$\rho^S\,\dot{s}\,T = \sigma_{ij}\,d_{ij}^{(i)} + (\Lambda_T\,T_{,i})_{,i} + \rho r - w_i\,(\mu_{,i} - g_i)\,, \tag{67}$$

where ρr is the total volumetric heat supply. The rate equation for heat transport was found to be

$$q_i^{\text{total}} = q_i + T\sum_\alpha s^\alpha\,w_i^\alpha = -\Lambda_T\,T_{,i} \quad \text{with} \quad \Lambda_T^\alpha \geq 0\,. \tag{68}$$

As it is seen, the free energy depends now on three parameters: Strain tensor $E_{KL}^{(r)}$ ($\dot{E}_{KL}^{(r)} = d_{ij}^{(r)}x_{i,K}\,x_{j,L}$), temperature T, and moisture content X. The equations of state may now be written as

$$\sigma_{ij} = \rho^S\,x_{i,K}\,x_{j,L}\left(\frac{\partial f}{\partial E_{KL}^{(r)}}\right)_{T,X} = \sigma_{ij}\left(E_{KL}^{(r)}, T, X\right), \tag{69}$$

$$s = -\left(\frac{\partial f}{\partial T}\right)_{E_{KL}^{(r)}, X} = s\,(E_{KL}^{(r)}, T, X)\,, \tag{70}$$

$$\mu = -\left(\frac{\partial f}{\partial X}\right)_{E_{KL}^{(r)}, T} = \mu\,(E_{KL}^{(r)}, T, X)\,. \tag{71}$$

In most drying processes (e. g. ceramics, clay), the drying strains can be consider for small, $E_{KL}^{(r)} \rightarrow \varepsilon_{ij}^{(r)}$. The stress strain relation in this case is similar to that of (45):

$$\sigma_{ij} = 2\,M\,\varepsilon_{ij}^{(r)} + \left[A\,\varepsilon^{(r)} - \gamma^{(T)}\,\vartheta - \gamma^{(X)}X\right]\delta_{ij}\,, \tag{72}$$

where the physical meaning of the material coefficients M, A, $\gamma^{(T)} = (2M + 3A)\,\kappa^{(T)}$, and $\gamma^{(X)} = (2M+3A)\,\kappa^{(X)}$ was explained above.

The total moisture flux (66) may now be written as

$$w_i = -\Lambda_m\left[\left(\frac{\partial\mu}{\partial T}\right)_{\varepsilon,X} T_{,i} + \left(\frac{\partial\mu}{\partial\varepsilon}\right)_{T,X}\varepsilon_{,i} + \left(\frac{\partial\mu}{\partial X}\right)_{\varepsilon,T} X_{,i} - g_i\right]. \tag{73}$$

The transport coefficients appearing in the above equation can be determined experimentally through the measure of moisture flux at different flow conditions. Coefficient Λ_m, for example, can be estimated for the moisture flow involved by gravity force. Coefficient $(\partial\mu/\partial X)_{\varepsilon,T} = C^{(X)} \geq 0$, which is responsible for the capillary uplift, can be measured from the equilibrium between gravity force and the capillary force, that is $C^{(X)} = g/\|\operatorname{grad} X\|$. Coefficient $(\partial\mu/\partial\varepsilon)_{T,X} = -\gamma^{(X)}/\rho^S$ results from the symmetry conditions. The thermodiffusion coefficient $(\partial\mu/\partial T)_{\varepsilon,X} = C^{(T)} \geq 0$ is possible to estimate through the measure of moisture gradient caused by temperature gradient, that is $C^{(T)} = C^{(X)}\|\operatorname{grad} X\|/\|\operatorname{grad} T\|$. All the discussed coefficients depend generally on the parameters of state.

In order to complete the energy equation (67), the time derivative of the entropy is required. Making use of the chain rule expansion with respect to the entropy function (70), one obtains,

$$\dot{s} = \left(\frac{\partial s}{\partial T}\right)_{\varepsilon,X}\dot{T} + \left(\frac{\partial s}{\partial\varepsilon}\right)_{T,X}\dot{\varepsilon} + \left(\frac{\partial s}{\partial X}\right)_{\varepsilon,T}\dot{X}\,. \tag{74}$$

The respective coefficients are: $(\partial s/\partial T)_{\varepsilon,X} = c_v/T$ with c_v being the total specific heat of the body per unit mass of dry body, $(\partial s/\partial\varepsilon)_{T,X} = \gamma^{(T)}/\rho^S$, $(\partial s/\partial X)_{\varepsilon,T} = -C^{(T)}$. The first coefficients results from the well known *Maxwell* equations in classical thermodynamics, and the two others from the symmetry conditions in constitutive relations.

Summing the individual momentums of the species (21) under neglect of accelerations ($a_i^\alpha \approx 0$) and momentum exchange due to phase transitions, $\overset{*}{\rho}{}^\alpha \cdot v_i^\alpha \approx 0$, one arrives at the equation of total equilibrium of internal forces:

$$\sigma_{ji,j} + \rho\,g_i \approx 0\,. \tag{75}$$

where ρ is the total density of the body.

The above set of equations, supplemented by the initial and boundary conditions, constitute the basic relations for the analysis of drying processes. In the next section, we shall present a simple example of convective drying of a cylinder to illustrate the application of the above presented model and the formulation of boundary conditions.

8 Stresses in a cylinder dried convectively

Let us analyze the deformations and the drying induced stresses in a wet cylinder during its convective drying. For simplicity, consider one dimensional problem, assuming symmetry with respect to the cylinder axis (see Figure 3).

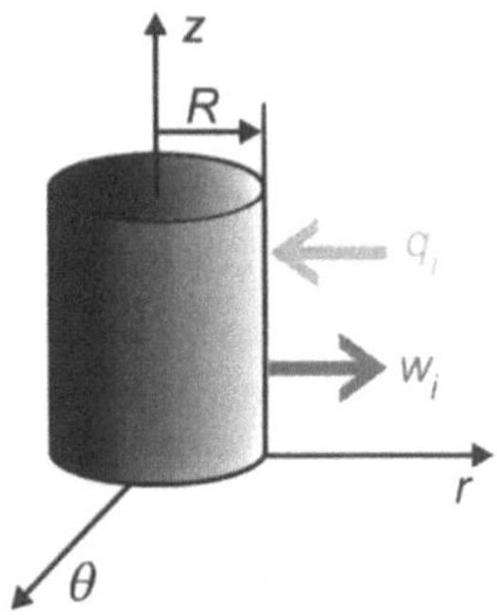

Fig. 3. A sample of cylindrical shape.

In this case all functions depend on the radial coordinate r and time t only. It is easy to state that for these specified conditions the displacements in radial, tangential and longitudinal directions are $u_r \neq 0$, $u_\varphi = 0$, $u_x =$ const., and the overall internal equilibrium equations (75), after neglecting the gravitational forces, become

$$\frac{\partial \sigma_{rr}}{\partial r} + \frac{\sigma_{rr} - \sigma_{\varphi\varphi}}{r} = 0\,.$$

The respective physical relations are

$$\sigma_{rr} = 2\,M\,\frac{\partial u_r}{\partial r} + A\,\varepsilon - \Psi\,,$$

$$\sigma_{\varphi\varphi} = 2\,M\,\frac{u_r}{r} + A\,\varepsilon - \Psi\,,$$

where

$$\varepsilon = \frac{1}{r}\frac{\partial}{\partial r}\,(r\,u_r)\,, \quad \Psi = \gamma^{(T)}\,\vartheta + \gamma^{(X)}X\,.$$

Substituting the physical relations into the equilibrium equation, we obtain the differential equations from which the displacement in radial direction can be calculated. Assuming the boundary conditions of the form

$$u_r = 0 \;\text{ for }\; r = 0 \;\text{ and }\; \sigma_{rr} = 0 \;\text{ for }\; r = R\,,$$

we have

$$u_r = \frac{1}{2M+A}\left(\frac{M}{M+A}\frac{r}{R^2}\int\limits_0^R r\,\Psi\,\mathrm{d}r + \frac{1}{r}\int\limits_0^r r\,\Psi\,\mathrm{d}r\right),$$

$$\sigma_{rr} = \frac{2M}{2M+A}\left(\frac{1}{R^2}\int\limits_0^R r\,\Psi\,\mathrm{d}r - \frac{1}{r^2}\int\limits_0^r r\,\Psi\,\mathrm{d}r\right),$$

$$\sigma_{\varphi\varphi} = \frac{2M}{2M+A}\left(\frac{1}{R^2}\int\limits_0^R r\,\Psi\,\mathrm{d}r + \frac{1}{r^2}\int\limits_0^r r\,\Psi\,\mathrm{d}r - \Psi\right).$$

Note, that

$$\sigma_{rr}\Big|_{r=0} = \sigma_{\varphi\varphi}\Big|_{r=0} = \frac{2M}{2M+A}\left(\frac{A}{M+A}\frac{1}{R^2}\int\limits_0^R r\,\Psi\,\mathrm{d}r - \frac{1}{2}\Psi\Big|_{r=0}\right).$$

It is visible that all the above functions depend on the expression Ψ, that is on the temperature and moisture content.

The distribution of the moisture content and the temperature can be determined from the mass balance equation (65) after substituting moisture flux (66), and from the energy balance equation (67). These equations, after adoption to our boundary problem and neglecting the nonlinear terms, become

$$\rho^S \dot{X} = \Lambda_m\left(\frac{\partial^2\mu}{\partial r^2} + \frac{1}{r}\frac{\partial\mu}{\partial r}\right),$$

$$\rho^S\, T\,\dot{s} = \Lambda_T\left(\frac{\partial^2 T}{\partial r^2} + \frac{1}{r}\frac{\partial T}{\partial r}\right).$$

Combining these equations with the physical relations for the moisture and entropy and rearranging slightly will yield the following form of heat and mass transfer equations:

$$\dot{\mu} - C^{(T)}\dot{T} + \frac{\gamma^{(T)}}{\rho^S}\,\dot{\varepsilon} = \frac{\Lambda_m\, C^{(X)}}{\rho^S}\left(\frac{\partial^2\mu}{\partial r^2} + \frac{1}{r}\frac{\partial\mu}{\partial r}\right),$$

$$-\rho^S\,\frac{C^{(T)}}{C^{(X)}}\,T_0\,\dot{\mu} + \rho^S\left(c_v + \frac{(C^{(T)})^2}{C^{(X)}}\,T_0\right)\dot{T} + \left(\gamma^{(T)} - \frac{C^{(T)}}{C^{(X)}}\right)T_0\,\dot{\varepsilon} =$$
$$= \Lambda_T\left(\frac{\partial^2 T}{\partial r^2} + \frac{1}{r}\frac{\partial T}{\partial r}\right).$$

The boundary conditions for the problem at hand are specified as follows:

- symmetry conditions (no flow of heat and mass across the middle of the cylinder)

$$\left.\frac{\partial \mu}{\partial r}\right|_{r=0} = 0 \quad \text{and} \quad \left.\frac{\partial T}{\partial r}\right|_{r=0} = 0$$

- convective heat transfer at the external surface of the cylinder

$$q_r^{\text{total}} = -\Lambda_T \left.\frac{\partial T}{\partial r}\right|_{r=R} = \alpha_T \left(T_a - \left.T\right|_{r=R}\right) - l\, w_a$$

- variable moisture flux, being maximal at the beginning and zero at the end of the drying process (the *Sommerfeld* radiation" condition)

$$\left.\frac{\partial(\mu - \mu_a)}{\partial r}\right|_{r=R} + \varkappa\,(\mu - \mu_a)\Big|_{r=R} = 0\,.$$

The new quantities in the above conditions are: α_T the coefficient of convective heat transfer, T_a the temperature of the drying medium (air), l the latent heat of evaporation, w_a the flux of vapour from the boundary to the drying medium, $\varkappa = \delta/\Lambda_m$ a proportionality "radiation" coefficient, μ_a the chemical potential of the vapour in drying medium (air). The vapour flux for the constant drying rate period differs from that of the falling rate period. For the first rate period the flux is defined as

$$w_a^{(I)} = \alpha_m^{(I)}\,(\mu_n - \mu_a) = \text{const.}\,,$$

where α_T is the coefficient of convective mass transfer, μ_n and μ_a are the chemical potentials of the vapour in drying medium (air) at the boundary of the dried body and far from the boundary.

During the second rate period the vapour flux reads

$$w_a^{(II)} = \alpha_m^{(II)} \left(\mu\Big|_{r=R} - \mu_a\right) \neq \text{const.}$$

It is known that during the constant rate period the whole heat supply to the dried body from the hot ambient medium is consumed on the boundary for water evaporation. The total heat flux flowing towards the interior of the body equals zero, and thus the temperature of the body equals the wet bulb temperature T_{wb}

$$T_{wb} = T^{(I)}\Big|_{r=R} = T_a - \frac{l}{\alpha_T}\, w_a^{(I)} \quad \text{and} \quad \left.\frac{\partial T^{(I)}}{\partial r}\right|_{r=R} = 0\,.$$

It is worth to mention, that before drying the body has a initial temperature T_0. Each drying process begins with heating of dried material during which the body temperature rises up to the wet bulb temperature. This is usually a short time period and therefore we do not take care of it now. We start our considerations from the constant drying rate period and assume the initial temperature to be equal to the wet bulb temperature. The initial conditions in our problem are

$$T(r,0) = T_{wb} \quad \text{and} \quad \mu(r,0) = \mu_0 .$$

We want to present an analytical solution of the problem in hand, and this will be possible if we neglect the influence of the body volume alteration on the change of temperature and the moisture potential. Some experiences indicate that such an influence is really negligibly small.

Here, we confine our considerations to the constant drying rate period. It is worth to mention that many dried materials as, for example, clay, ceramics, etc., suffer the most shrinkage and the most dangerous stresses just during this period. The body temperature in this period becomes constant for the whole body and equal to the wet bulb temperature. Thus, the differential equation describing the moisture potential distribution in this period becomes

$$\dot{\mu} = K_m \left(\frac{\partial^2 \mu}{\partial r^2} + \frac{1}{r}\frac{\partial \mu}{\partial r} \right), \quad K_m = \frac{\Lambda_m \, C^{(X)}}{\rho^S} .$$

Using the method of variable separation, we find the general solution in the form

$$\mu(r,t) = \mu_a + [A\, J_0(\omega\, r) + B\, Y_0(\omega\, r)] \exp(-\omega^2 K_m\, t) ,$$

where $J_0(\omega\, r)$ and $Y_0(\omega\, r)$ are the *Bessel* functions of order zero of the first and second kind, ω is the constant of separation, and A and B are the integration constants. The requirement that the gradient of moisture potential in the middle of the cylinder has to be zero (the symmetry boundary condition) stipulates that B must be zero, because $Y_0(\omega\, r)$ tends to infinity at $r = 0$. To satisfy the second boundary condition ω must be a root of the characteristic equation

$$\omega J_1(\omega R) = \varkappa J_0(\omega R) .$$

$J_1(\omega R)$ is the first order and the first kind *Bessel* function. This equation has an infinite number of real positive roots $\{\omega_n\}$.

The general solution is obtained by summing the particular solutions for each individual ω_n, giving

$$\mu(r,\, t) = \mu_a + \sum_{n=1}^{\infty} A_n J_0(\omega_n\, r) \exp(-\omega_n^2\, K_m\, t) .$$

The orthogonality condition for the *Bessel* functions reads

$$\int_0^R r\, J_0(\omega_n r)\, J_0(\omega_m r)\, \mathrm{d}r = \begin{bmatrix} 0 & \text{for } n \neq m \\ \int_0^R r\, [J_0(\omega_n r)]^2\, \mathrm{d}r & \text{for } n = m \end{bmatrix} .$$

The orthogonality condition and the initial condition for the moisture potential are used to evaluate A_n. The final solution to the moisture potential in constant rate period is

$$\mu(r,t) = \mu_a + (\mu_0 - \mu_a)\,\frac{2\varkappa}{R}\sum_{n=1}^{\infty}\frac{J_0(\omega_n r)}{(\omega_n^2 + \varkappa^2)\,J_0(\omega_n R)}\,\exp(-\omega_n^2\,K_m\,t)\,.$$

The above expression for moisture potential is valid up to the moment t_n, when the moisture potential at the boundary reach the value of vapour chemical potential, that is $\mu(R, t_n) = \mu_n$. The values of vapour chemical potentials μ_a and μ_n are known a priori for given drying conditions.

The displacement u_r and the stresses σ_{rr} and $\sigma_{\varphi\varphi}$ may be evaluated now after substituting Ψ given by

$$\Psi = (2\,M + 3\,A)\left(\tilde{\kappa}^{(T)}\vartheta_{wb} + \tilde{\kappa}^{(X)}\mu\right),$$

where $\tilde{\kappa}^{(T)} = \kappa^{(T)} - \tilde{\kappa}^{(X)}C^{(T)}$, $\tilde{\kappa}^{(X)} = \kappa^{(X)}/C^{(X)}$, and the function μ is determined above.

Integrating the respective integrals in the expressions u_r, σ_{rr}, and $\sigma_{\varphi\varphi}$ and rearranging slightly will yield the following final forms of these expressions, referred to the constant drying rate period

$$u_r(r,\,t) - u_r(r,\,0) =$$
$$= -\Omega\,\frac{2\varkappa}{R}\sum_{n=1}^{\infty}\left[\frac{M}{M+A}\frac{\varkappa}{\omega_n}\frac{r}{R} + \frac{J_1(\omega_n r)}{J_0(\omega_n R)}\right]\frac{1 - \exp(-\omega_n^2\,K_m\,t)}{\omega_n\,(\omega_n^2 + \varkappa^2)}\,,$$

$$\sigma_{rr}(r,\,t) = 2\,M\,\Omega\frac{2\varkappa}{R}\sum_{n=1}^{\infty}\left[\frac{\varkappa}{\omega_n R} - \frac{J_1(\omega_n r)}{r J_0(\omega_n R)}\right]\frac{\exp(-\omega_n^2\,K_m\,t)}{\omega_n\,(\omega_n^2 + \varkappa^2)}\,,$$

$$\sigma_{\varphi\varphi}(r,\,t) =$$
$$= 2M\Omega\frac{2\varkappa}{R}\sum_{n=1}^{\infty}\left[\frac{\varkappa}{\omega_n^2\,R} + \frac{J_1(\omega_n r)}{\omega_n\,r\,J_0(\omega_n R)} - \frac{J_0(\omega_n r)}{J_0(\omega_n R)}\right]\frac{\exp(-\omega_n^2\,K_m\,t)}{(\omega_n^2 + \varkappa^2)}\,,$$

where $\Omega = (2\,M + 3\,A)\,\tilde{\kappa}^{(X)}\,(\mu_0 - \mu_a)\,/\,(2\,M + A)$. It is easy to state that

$$\sigma_{rr}(0,t) = \sigma_{\varphi\varphi}(0,t) = 2M\Omega\,\frac{2\varkappa}{R}\sum_{n=1}^{\infty}\left[\frac{\varkappa}{\omega_n^2 R} - \frac{J_0(0)}{2J_0(\omega_n R)}\right]\frac{\exp(-\omega_n^2 K_m t)}{(\omega_n^2 + \varkappa^2)}\,.$$

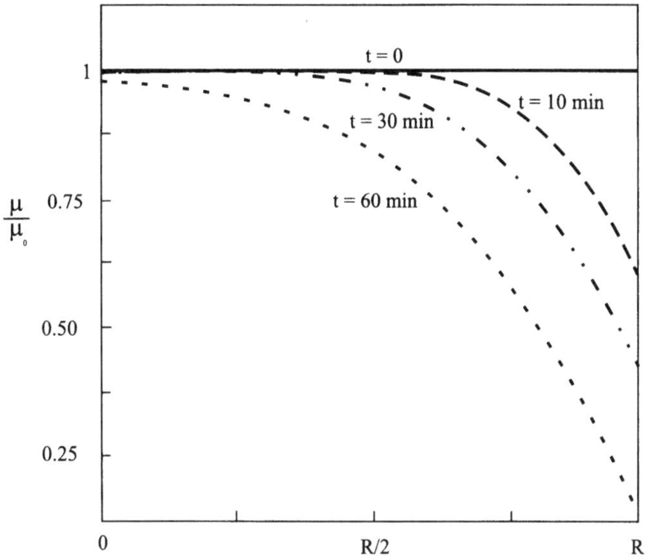

Fig. 4. Distribution of moisture potential μ in several instances of time.

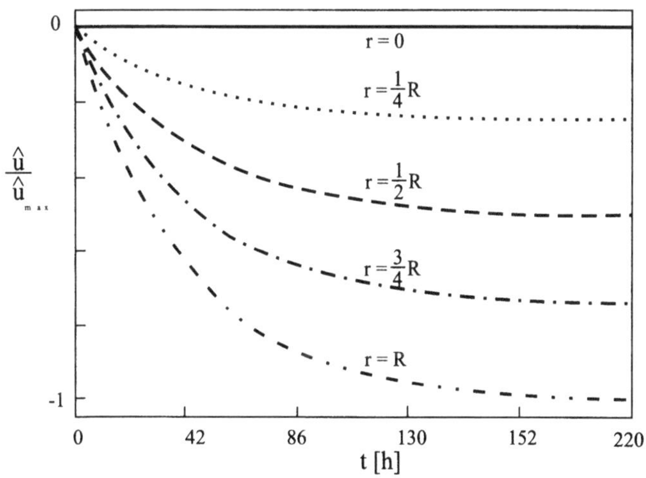

Fig. 5. Evolution of displacements u_r in several points of the cylinder.

The final solutions for the falling rate period can be obtained in a similar way.

The distribution of moisture potential, uniform at the beginning, becomes non-uniform during drying. This means that the shrinkage is also non-uniform and this raises generation of stresses.

Figure 5 illustrate the evolution of cylinder shrinkage. The cylinder at the beginning is in a swelled state, and the drying causes the decrease of

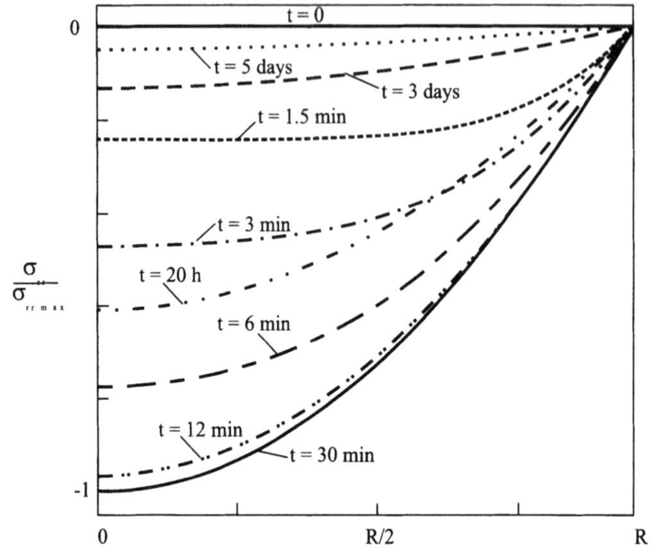

Fig. 6. Alteration of radial stresses distributed along radius.

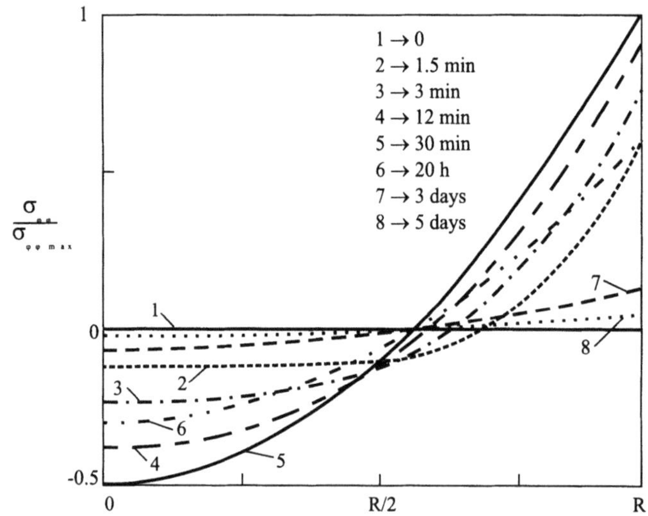

Fig. 7. Alteration of circumferential stresses distributed along radius.

the cylinder radius. The curves in Figure 5 presents the time alteration of cylinder displacements in several points, where $\hat{u}_r = u_r(r, t) - u_r(r, 0)$. It is clear that the most displacement suffer the external surface of the cylinder.

Figure 6 presents the distribution of radial stresses in some instants of time. It is visible that the maximal radial stresses appear in the middle of the cylinder, while they are zero at the surface. They are compressible in the whole area during the whole drying process. The stresses tends to zero in the course of drying, that is when the moisture distribution tends to zero.

Figure 7 presents the distribution of circumferential stresses. They are compressible inside the cylinder and tensional in the layer close the boundary. The circumferential, or the tangential stresses $\sigma_{rr} - \sigma_{\varphi\varphi}$, are the most dangerous in drying, as they are responsible for the destruction of the dried material. Note, that both the radial and the circumferential stresses are zero at the beginning of the drying process, next growth to their maximal values which depend on the drying conditions, and finally they tend to zero when the moisture distribution becomes more uniform.

9 Final remarks

The theory presented in this paper offers a systematic approach to modelling of drying processes of wet deformable porous solids. It is based on the theory of balance equations and the principles of irreversible thermodynamics. Such an approach gives a deeper insight into the mechanical phenomena occurring in materials during their drying. Since dried materials shrink, with consequent changes in size and shape, shrinkage and in particular the shrinkage stresses are of fundamental importance in drying processes. Although the principles of drying have been recognized for decades and some models of drying developed and presented in the literature, the thermomechanical theories of drying enabling calculation of drying induced stresses and strains have been developed relatively recently. Based on what was presented in the present paper, one can say that still more work has to be done to find a more adequate thermomechanical drying model. More effort should be given to develop models in which the heat and mass transfer properties and the material properties are variable with change of moisture content and temperature. Particularly, the irreversible deformations of dried materials ought to be taken into account.

On the other hand, we have to be aware that including more physics to drying models involves serious numerical troubles, and makes difficult their analysis. In many applications, however, a simpler form of drying theory will suffice. Therefore, one has to seek for pertinent drying models with a reasonable choose between physical accuracy and the numerical possibilities. Such an attempt was also presented in this paper. A variety of other strategies ought to be developed to find optimized drying models.

Acknowledgements

This work has been financially supported by the research project DS 32/013/2001 sponsored by the Poznań University of Technology.

References

1. Augier, F., Coumans, W. J., Hugget, A., Kaasschieter, E. F.: On the study of cracking in clay drying. *Proceedings of the 12th International Drying Symposium*, Noordwijkerhout, The Netherlands, 28-31 Aug. 2000.
2. Benet, J. C., Jouanna, P.: Phenomenological relation of phase change of water in a porous medium: experimental verification and measurement of the phenomenological coefficient. *Int. J. Heat Mass Transfer* **25**(11) (1984), 1747–1754.
3. Berry, R. S., Kazakov, V. A., Sieniutycz, S., Szwast, Z., Tsirlin, A. M.: *Thermodynamic optimization of finite-time processes.* John Wiley & Sons, New York 2000.
4. de Boer, R., Kowalski, S. J.: Thermomechanics of fluid saturated porous media with phase change. *Acta Mechanica* **109** (1995), 167–189.
5. Cairncross, R. A., Schunk, P. R., Chen, K. S., Prakash, S. S., Samuel, J., Hurd, A. J., Brinkner, C. J.: Drying in deformable partially-saturated porous media: Sol-Gel Coatings. *Sandia Report SAND96-2149, UC-905* (1996).
6. Cottrell, A. H.: *The mechanical properties of matter.* John Wiley & Sons, New York 1964.
7. Coussy, O., Eymard, R., Lassabatere, T.: Constitutive modeling of unsaturated drying deformable materials. *Journal of Engineering Mechanics* (June 1998), 658–667.
8. Czolbe, P., Jucha, S., Zawisza, L.: Comparison of experimental and theoretical curves for relative permeability. *Archives of Mining* **31**(1) (1986), 55–61 (in Polish).
9. Fortes, M., Okos, M. R.: Drying theories: Their bases and limitations as applied to foods and grains. In Mujumdar, A. S. (ed.): *Advances in Drying*, Vol. 1, Hemisphere Publishing Corp., New York 1980, pp. 119–154.
10. Gumiński, K.: *Thermodynamics of irreversible processes.* PWN, Warszawa 1962 (in Polish).
11. Hutter, K.: The foundations of thermodynamics, its basic postulates and implications. Review of modern thermodynamics. *Acta Mechanica* **27** (1977), 1–54.
12. Jefimov, N. W., Rozendorn, E. R.: *Linear algebra with multidimensional geometry.* PWN, Warszawa 1976 (in Polish).
13. Kirkham, D., Powers, W. L.: *Advanced soil physics.* John Wiley & Sons, New York 1972.
14. Kowalski, S. J.: Thermomechanics of constant drying rate periods. *Archives of Mechanics* **39**(3) (1987), 157–176.
15. Kowalski, S. J.: Thermomechanics of dried materials. *Archives of Mechanics* **42**(2) (1990), 123–149.
16. Kowalski, S. J.: Thermomechanics of the drying processes of fluid-saturated porous media. *Drying Technology* **12**(3) (1994), 453–482.
17. Kowalski, S. J.: Drying processes involving permanent deformations of dried materials. *International Journal of Engineering Science* **34**(13) (1996), 1491–1506.
18. Kowalski, S. J.: Toward thermodynamics and mechanics of drying processes. *Chemical Engineering Science* **55** (2000), 1289–1304.

19. Kowalski, S. J.: Thermomechanical approach to shrinking and cracking phenomena in drying. In Kudra, T., Mujumdar, A. S. (eds.): *Drying Technology - a special issue on "Progress in Drying Technologies"* **19**(4) (2001).
20. Kowalski, S. J., Musielak, G., Kyziol, L.: Non-linear model for wood saturation. *Transport in Porous Media* (2001) (in print).
21. Kowalski, S. J., Rajewska, K., Rybicki, A.: Destruction of wet materials by drying. *Chemical Engineering Science* **55** (2000), 5755–5762.
22. Kowalski, S. J., Strumiłło Cz.: Moisture transport, thermodynamics, and boundary conditions in porous materials in presence of mechanical stresses. *Chemical Engineering Science* **52**(7) (1997), 1141–1150.
23. Naphtali, L. M.: The second law. Caratheodory's principle with simplified mathematics. *A. I. Ch. Journal* **12**(1) (1966), 195–197.
24. Perre, P., May, B. K.: A numerical drying model that accounts for the hape modification of highly deformable products. In *Proceedings of the 12th International Drying Symposium*, Noordwijkerhout, The Netherlands, 28-31 Aug. 2000.
25. Prigogine, I., Defay, R.: *Chemical thermodynamics.* Longmans Green and Co., London 1954.
26. Scheidegger, A. E.: *The physics of flow through porous media.* University of Toronto Press, Toronto 1957.
27. Scherer, G. W.: Theory of drying. *J. Am. Ceram. Soc.* **73**(1) (1990), 3–14.
28. Strumiłło, Cz.: *Fundamentals of drying and its technology.* WNT, Warszawa 1983 (in Polish).

Coupling between the evolution of a deformable porous medium and the motion of fluids in the connected porosity

Renato Lancellotta

Technical University of Torino (Politecnico), Department of Structural and Geotechnical Engineering, 10129 Torino, Italy

Abstract. This contribution is aimed at presenting a consistent mathematical description of porous media. Basic assumptions concerned with the representation of a porous medium as a continuum are discussed in detail, making a clear distinction between two geometric scales: at microscale each constituent occupies a specific domain, while at macroscale soil and fluid particles are superimposed at the same geometric point. At macroscale, a unified *Lagrange*an formulation is given, by assuming the soil skeleton as a material reference volume and by referring the fluid motion to the soil skeleton. Finally, two problems are analysed, both of relevant interest in soil mechanics, i. e. the propagation of body waves in undrained conditions and the consolidation of a soft clay stratum.

1 Introduction

Since the pioneering work of Terzaghi [40, 41], there has been a growing interest in consolidation theory. As outlined in the recent book of de Boer [11], where the reader can find historical developments on this subject, this interest arises from both theoretical requirements, linked to the mechanics of porous media, as well as to requirements linked to engineering applications.

These latter include prediction of settlement rate of structures interacting with the soil, subsidence phenomena, oil production, diffusion of pollutants, transient phenomena occurring in earthquakes and wave propagation.

It is also relevant to observe that similar problems are of interest in biomechanics [24], where the bone structure can be represented by a porous medium with the circulating blood.

Being the solid and fluids motion characterized by different velocities, interaction effects arise, which influence the response of the porous medium, also due to different material properties of the constituents.

Therefore, in its essence, the problem is relevant in all scientific areas where a need arises of modelling the "coupling between the evolution of a deformable porous medium and the motion of the fluids in the connected porosity".

This paper is divided in the following sections: Section 2 presents the general framework of the strategies followed when modelling porous materials, i. e. the "averaging approaches", emphasizing the related basic assumptions;

Section 3 derives a consistent *Lagrangean* model for both the soil skeleton and the fluid; Section 4 discusses the "effective stress" concept; Section 5 deals with the problem of body waves propagation in undrained conditions; Section 6 formulates the initial boundary value problem related to the consolidation of a stratum between draining and impervious boundaries subjected to stress or velocity conditions.

2 Modelling porous media

2.1 Preliminaries

At present, the mechanics of a porous medium is described by two different approaches.

The "averaging approach" takes into account all the phases constituting the porous medium and each of them is treated in its own domain as a single body. Then the obtained results can be related to the macroscopic domain by using an averaging strategy.

The "macroscopic approach" starts from an "a priori" assumption of homogenized phases and is based on axioms of the theory of mixtures supplemented by the concept of volume fraction.

It is noted that the final goal of both these approaches is the description of some macroscopic features of the evolution of the porous medium that are predictable and repeatable. In fact, even if it would be desirable to achieve a detailed knowledge of the flow of the fluid in the pores and the displacement of the solid particles at each point, this appears as an impossible task, because we do not know the detailed geometry at the microscopic scale and, more important, this geometry will be different from specimen to specimen, i. e. there is a lack of repeatability.

For these reasons, it is relevant to focus the attention on some macroscopic features that are presumed to be independent of the exact configuration of the pores and can be verified by repeatable macroscopic experiments.

The approximation represented by the continuum mechanics approach reflects these basic aspects, and any experimental determination of the constitutive response of a given medium yields relationships between "gross properties".

Therefore, in their essence, the procedures are aimed at "the replacement of a micro-heterogeneous medium with a homogeneous one, which macroscopically behaves in the same manner".

Such a process poses a fundamental problem: How to describe the macroscopic behaviour of a medium which exhibits microscopic heterogeneity on the base of microstructural information.

In the present paragraph, we first discuss the conceptual passage from a microscopic to a macroscopic viewpoint through an averaging procedure, with the aim to clarify the meaning and the role of several terms appearing into field equations.

2.2 Volume and ensemble averaging approach

Methodologically, we consider the real non-homogeneous structure of the porous medium, the scale of heterogeneity being of the same order of magnitude of the pore or grain dimension. The attention is then focused at fields describing the status of a single phase, defined only at the points occupied by that phase. Because, as already mentioned, this level of detail is not needed, we use then an average technique in order to get a macroscopic description (see [3, 28, 39]).

Central to this process is the concept of a *Representative Elementary Volume* (*REV*).

This concept was already introduced and discussed in 1909 by Lorentz, in his book on the theory of electrons [30]. Successively, Bear (1972) [2] discussed the same concept in relation to porous media and further rigorous contributions were given by Nemat-Nasser and Hori (1993) [35] and Drugan and Willis (1996) [21]. Because a comprehensive discussion on this subject can be found in the paper of Marcov (2000) [31], in the following we introduce the *REV* definition in a rather heuristic manner:

"It is a volume that, at a macroscopic scale, is small enough to be treated as a point of the heterogeneous medium, and, on the reverse, at a microscale, is large enough to contain a significant number of single heterogeneities".

From the above definition it follows that the position of a *REV* is identified within the domain which mathematically defines the medium of interest by the vector position $\mathbf{x}$.

In addition, it must be observed that any field q of interest at the micro-level, i. e. within the *REV*, also depends on the local coordinate $\mathbf{r}$, spanning the *REV* so that

$$q := q\left(\mathbf{x},\, \mathbf{r},\, t\right) . \tag{1}$$

The dependence expressed by (1) of any internal field from both the macro-coordinate $\mathbf{x}$ and the micro-coordinate $\mathbf{r}$ implies that the internal field can behave in different ways within different *REV*s of the medium.

The relation between macro and micro quantities is obtained through the *volume averaging with respect to the micro-coordinate* $\mathbf{r}$, which allows to define the so-called *phase average* and *intrinsic phase average* of any field q:

$$\overline{q_\alpha\left(\mathbf{x},\, t\right)} := \frac{1}{V}\int_v H_\alpha\left(\mathbf{r},\, t\right) q\left(\mathbf{r},\, t\right)\, \mathrm{d}\mathbf{r}\,, \tag{2}$$

$$\overline{q_\alpha\left(\mathbf{x},\, t\right)}^{\alpha} := \frac{1}{V^\alpha}\int_v H_\alpha\left(\mathbf{r},\, t\right) q\left(\mathbf{r},\, t\right)\, \mathrm{d}\mathbf{r}\,, \tag{3}$$

where v is the *REV* of volume V centered on $\mathbf{x}$, and H_α is an *indicator function* for the α-th constituent, its value being one if at time t the constituent is in $\mathbf{x}$, and zero otherwise.

According to the definition of this function, the volume of the α constituent within a REV is given by

$$V^{\alpha}(\mathbf{x}, t) := \int_{v} H_{\alpha}(\mathbf{r}, t)\, \mathrm{d}\mathbf{r} \tag{4}$$

and the *volume fraction* n^{α} of the constituent is expressed by

$$n^{\alpha}(\mathbf{x}, t) := \frac{V^{\alpha}}{V} = \frac{1}{V}\int_{v} H_{\alpha}(\mathbf{r}, t)\, \mathrm{d}\mathbf{r}\,. \tag{5}$$

Once the REV average property has been obtained, in order to define the property of the whole porous medium, we should in principle work (for example to perform experiments) with N different REVs and obtain the so-called *ensemble average*:

$$\overline{q_i}^{*} := \frac{1}{N}\left(\overline{q_{i,v_1}} + \overline{q_{i,v_2}} + \ldots + \overline{q_{i,v_N}}\right)\,. \tag{6}$$

Provided that the microscale length l of a typical inhomogeneity is small enough as compared with the volume V of the medium, it can be expected that the *ensemble* and the *volume averages* coincide. This implies the introduction of the *hypothesis of ergodicity*, i. e. a *macroscopic* or *statistical homogeneity* of the porous medium.

Roughly speaking, this hypothesis means that the macroscopic properties of all the REVs are the same and the REV has the same properties of the porous medium as a whole. For this reason, in the following, we make use of the volume averaging approach.

Further, in some cases we also introduce the hypothesis that the medium is *statistically isotropic*, i. e. the macroscopic properties are independent of direction.

2.3 Averaging rules

Suppose that q is a continuous vectorial or scalar function over the domain occupied by the phase α. Then it can be proved that the derivatives with respect to time and to the position vector $\mathbf{x}$ of an average value can be related to the average value of the derivative through the following formulas:

$$\frac{\partial \overline{q}}{\partial t} := \overline{\frac{\partial q}{\partial t}} + \frac{1}{V}\int_{S_{\alpha\beta}} q(\mathbf{r}, t)\, \mathbf{v}(\mathbf{r}, t) \cdot \mathbf{n}\, \mathrm{d}S\,, \tag{7}$$

$$\mathrm{grad}_{\mathbf{x}}\overline{q} := \overline{\mathrm{grad}_{\mathbf{r}} q} - \int_{S_{\alpha\beta}} q(\mathbf{r}, t) \otimes \mathbf{n}\, \mathrm{d}S\,, \tag{8}$$

where the surface $S_{\alpha\beta}$, moving with velocity $\mathbf{v}(\mathbf{r}, t)$, separates the phase α from all other phases β and $\mathbf{n}(\mathbf{r}, t)$ is the outwards unit normal.

2.4 Kinematics

When considering the porous medium as a continuum, it is relevant to make a clear distinction between two geometric scales: At *microscale* each constituent occupies a specific domain, while at *macroscale* soil and fluid particles are superimposed at the same geometric point. Accordingly, the basic assumption of the theory of mixture is that the individual components of the porous medium are all statistically distributed over the control space, i. e. *each spatial point* $\mathbf{x}$ *of the control space is simultaneously occupied by particles of all components.*

As a consequence, mathematical functions describing both geometrical and physical properties of each constituent are *field functions defined all over the control space.*

Within the aim of this note, reference is made to a two phases medium, i. e. a solid skeleton saturated by water, the case of major interest in soil mechanics, and the fluid is assumed without viscosity. The methodological approach can be anyway extended to more than two phases media.

The structure composition of the porous medium is described throughout the introduction of *volume fractions*:

$$n^f := \frac{\text{pore volume}}{\text{total volume}} = n = \frac{e}{1+e}, \qquad (9)$$
$$n^s := \frac{\text{solid volume}}{\text{total volume}} = (1-n) = \frac{1}{1+e},$$

corresponding to the fluid phase and to the solid matrix respectively, and e being the void index, i. e. the volume occupied by fluid referred to the one occupied by solids.

The *porosity* n in (9) corresponds to the so-called effective or *connected porosity* represented by the interconnected voids, strictly related to the motion of the fluid relative to the solid skeleton.

In addition, the porous medium is assumed to be statistically isotropic, i. e. for any cross-section, we observe the same ratio of the fluid area referred to the solid area so that the porosity appearing in (9) has also the meaning of ratio of fluid area to cross-section of the porous medium.

Remark 1. The total porosity is referred to the voids connected or not to the external ambient and the difference between the total and the connected porosity is indicated as disconnected porosity. In sedimentary rocks and in soils, the disconnected porosity is negligible, while in artificially made porous media can reach values up to 10 %.

(9) leads to the *saturation condition*

$$n^f + n^s = 1\,, \qquad (10)$$

which has to be considered as an internal constraint for any thermodynamical process.

The introduction of volume fraction allows to define two different density functions for each constituent, i. e. the *material density*

$$\rho^{\alpha R} := \frac{dm_\alpha}{dv_\alpha} \tag{11}$$

and the *partial density*

$$\rho^{\alpha} := \frac{dm_\alpha}{dv} \, . \tag{12}$$

From the above definitions, it follows that the *material incompressibility* is not equivalent to the bulk incompressibility of each component since the partial density can change due to changes in the volume fraction, being

$$\rho^{\alpha} := n^{\alpha} \, \rho^{\alpha R} \, . \tag{13}$$

Starting from the general assumption that each spatial point $\mathbf{x}$ of the current configuration at any time t is simultaneously occupied by material particles of all constituents, for each phase the *motion function* is defined in a *Lagrange*an description by

$$\mathbf{x} := \chi^{\alpha} \left(\mathbf{X}^{\alpha}, t \right) . \tag{14}$$

In order to have a continuous and bijective mapping, the *Jacobi*an of the transformation (14) must be non-zero, and, because it is equal to the determinant of the deformation gradient tensor, it must also be strictly positive:

$$J^{\alpha} = \det \left(\frac{\partial \chi^{\alpha}}{\partial \mathbf{X}^{\alpha}} \right) \geq 0 \, . \tag{15}$$

The *Lagrange*an description (14) is non-singular so that it is possible to use its inverse and to work in terms of an *Euler*ian description:

$$\mathbf{X}^{\alpha} = \chi^{\alpha^{-1}} \left(\mathbf{x}, t \right) . \tag{16}$$

From (14) it follows that each phase has its own velocity and acceleration field, given in the *Lagrange*an description by

$$\mathbf{V}^{\alpha} := \frac{\partial \chi^{\alpha} \left(\mathbf{X}^{\alpha}, t \right)}{\partial t} \, , \tag{17}$$

$$\mathbf{A}^{\alpha} := \frac{\partial^2 \chi^{\alpha} \left(\mathbf{X}^{\alpha}, t \right)}{\partial t^2} \, . \tag{18}$$

By introducing (16) into (17) and (18) the alternative *Euler*ian description is obtained in terms of velocity $\mathbf{v}^{\alpha} \left(\mathbf{x}, t \right)$ and acceleration $\mathbf{a}^{\alpha} \left(\mathbf{x}, t \right)$ fields.

As in classical continuum mechanics, the *deformation rate tensor*, referred to the current configuration, is defined as

$$\mathbf{d}^{\alpha} := \frac{1}{2} \left(\nabla \mathbf{v}^{\alpha} + \mathbf{v}^{\alpha} \nabla \right) , \tag{19}$$

and it is useful to recall that

$$\operatorname{tr}\mathbf{d}^\alpha = \frac{1}{V}\frac{\mathrm{d}^\alpha V}{dt} = \nabla \cdot \mathbf{v}^\alpha \,. \tag{20}$$

In a coupled problem, it is convenient to assume as a basic kinematic variable the solid displacement $\mathbf{u}^s$ and to express the motion of the fluid relative to the solid. The relative velocity is represented by the *Euler*ian vector

$$\mathbf{w}\,(\mathbf{x},\,t) := \mathbf{v}^f\,(\mathbf{x},\,t) - \mathbf{v}^s\,(\mathbf{x},\,t)\,. \tag{21}$$

Finally, the material time derivative of any differentiable function g related to the fluid, given in its spatial description, can be expressed in terms of the material derivative referred to the soil skeleton:

$$\frac{\mathrm{d}^f g}{\mathrm{d}t} = \frac{\mathrm{d}^s g}{dt} + \mathbf{w} \cdot \nabla g \,. \tag{22}$$

2.5 Mass balance equations

The mass balance equation for the phase at the *microscale* is given by

$$\frac{\partial \rho^{\alpha R}}{\partial t} + \nabla \cdot \left(\rho^{\alpha R}\,\mathbf{v}^\alpha\,(\mathbf{r},\,t)\right) = 0\,. \tag{23}$$

By using the averaging rules (7) and (8), the following mass balance equations are obtained for the soil skeleton and the fluid:

$$\frac{\partial \rho^s}{\partial t} + \nabla \cdot \left(\rho^s\,\mathbf{v}^s\,(\mathbf{x},\,t)\,\right) = 0\,, \tag{24}$$

$$\frac{\partial \rho^f}{\partial t} + \nabla \cdot \left(\rho^f\,\mathbf{v}^f\,(\mathbf{x},\,t)\,\right) = 0\,, \tag{25}$$

where the *mass averaged velocity* is defined by

$$\rho^\alpha \mathbf{v}^\alpha\,(\mathbf{x},\,t) := \overline{\rho^{\alpha R}\,\mathbf{v}^\alpha\,(\mathbf{r},\,t)}\,. \tag{26}$$

In (23), the divergence operates at the microscale, i. e. $\nabla_\mathbf{r} \cdot (\,\cdot\,)$, while in (24) and (25) it operates at the macroscale, i. e. $\nabla_\mathbf{x} \cdot (\,\cdot\,)$.

The mass averaged velocity defined by (26) merges into the intrinsic average for material incompressibility.

Since the surface $S_{\alpha\beta}$ in (7) and (8) is a material surface, there is no mass flux across it.

2.6 Momentum balance equation of a phase

By using similar arguments, if the body force per unit mass is assumed to be uniform and identical for the two constituents and is denoted by $\mathbf{b}$, the *macroscopic momentum balance* for each phase will be

$$n^\alpha \, \overline{\rho}^\alpha \, \frac{\mathrm{d}\overline{\mathbf{v}}^\alpha}{\mathrm{d}t} = \nabla \cdot \mathbf{T}^\alpha + \frac{1}{V} \int_{S_{\alpha\beta}} \mathbf{T} \cdot \mathbf{n} \, \mathrm{d}S + n^\alpha \, \overline{\rho \mathbf{b}}^\alpha . \tag{27}$$

The surface integral term in the right-hand side represents the *interaction force*, related to the local interaction between the phases across the microscopic interface that separates them (note that in the sequel it will be denoted by $\mathbf{f}^{\mathrm{int}}$).

In general, such interaction occurs between wetting and non-wetting fluid phases, as well as between solid and fluid. Usually, the assumption of continuity of traction at the interface between the solid and the fluids is made so that no jump exists across this microscopic interface, i. e.

$$[\![\mathbf{T}]\!]_{s,w} = [\![\mathbf{T}]\!]_{s,n} = 0 . \tag{28}$$

On the contrary, such a jump exists at the interface between two immiscible fluids, giving rise to a *surface tension.*

The partial stress tensor $\mathbf{T}^\alpha$ in (27) differs from the mean value of the microscale tensor:

$$\mathbf{T}^\alpha = n^\alpha \, \overline{\mathbf{T}}^\alpha - \overline{\rho \left(\mathbf{v}\left(\mathbf{r},\, t\right) - \overline{\mathbf{v}}^\alpha \right) \otimes \left(\mathbf{v}\left(\mathbf{r},\, t\right) - \overline{\mathbf{v}}^\alpha \right)} , \tag{29}$$

being the second term in the right-hand side the *Reynold* stress.

Remark 2. The *Reynold* stress in (29) comes from averaging the inertial term $\rho \frac{\mathrm{d}\mathbf{v}}{\mathrm{d}t}$, also noting that, by combining the material time derivative operator $\left(\frac{\mathrm{d}}{\mathrm{d}t} = \frac{\partial}{\partial t} + \mathbf{v} \cdot \nabla\right)$ with the mass balance equation (23) at microscale, one can write

$$\rho \, \frac{\mathrm{d}\mathbf{v}\left(\mathbf{r},\, t\right)}{\mathrm{d}t} = \frac{\partial \rho \, \mathbf{v}\left(\mathbf{r},\, t\right)}{\partial t} + \nabla_{\mathbf{r}} \cdot \left(\rho \, \mathbf{v}\left(\mathbf{r},\, t\right) \otimes \mathbf{v}\left(\mathbf{r},\, t\right)\right) .$$

3 *Lagrange*an description of porous media

3.1 Kinematics

Usually, the kinematics of the fluid phase is described by using an *Euler*ian formulation, whereas for the kinematics of the solid structure reference is made to a *Lagrange*an one. In order to overcome shortcomings deriving from this mixed description, Coussy [18, 19], Bourgeois and Dormieux [12] and Wilmański [46] have suggested to introduce a unified *Lagrange*an description, by assuming the soil skeleton as a material reference volume, and by referring the fluid motion to the soil skeleton.

There are some basic differences, with respect to the usual approach followed for a single phase continuum that need to be outlined. First, a porous medium exchanges mass fluid with the surroundings so that it is a *local thermodynamic open system.*

For such a reason, conservative laws have to be formulate considering the convective transports.

Starting from the general assumptions, stated in Section 2, for each phase is defined the motion function (14) and each phase has a velocity and acceleration field (17) and (18).

The *deformation gradient*, defined by

$$\mathbf{F} := \frac{\partial x_i}{\partial X_j}\, \mathbf{e}_i \otimes \mathbf{e}_j \,, \tag{30}$$

maps the material vector $\mathrm{d}\mathbf{X}$ onto its image $\mathrm{d}\mathbf{x}$ in the current configuration:

$$\mathrm{d}\mathbf{x} = \mathbf{F} \cdot \mathrm{d}\mathbf{X} \,. \tag{31}$$

Similarly, the infinitesimal initial volume $\mathrm{d}V_0$ is transformed in the current configuration in $\mathrm{d}V$ throughout the relation

$$\mathrm{d}V = J\,\mathrm{d}V_0 \,, \tag{32}$$

with the domain of the *Jacobi*an J restricted to positive values, since the motion of each phase is assumed to be unique and also invertible.

Further, a material area $\mathrm{d}\mathbf{A}$, oriented by its normal $\mathbf{N}$ (i. e. $\mathrm{d}\mathbf{A} = \mathbf{N}\,\mathrm{d}A$), is transformed in the current configuration according to the rule

$$\mathbf{n}\,da = J\,(\mathbf{F}^{-1})^T \cdot \mathbf{N}\,\mathrm{d}A \,. \tag{33}$$

The deformation is described by the symmetric *Green-Lagrange*an tensor $\mathbf{E}$, given by

$$2\,\mathbf{E} := \mathbf{F}^T \cdot \mathbf{F} - 1 \,, \tag{34}$$

and the material time derivative of $\mathbf{E}$ is linked to the *Euler*ian strain rate tensor by

$$\mathbf{d}^s = (\mathbf{F}^{-1})^T \cdot \dot{\mathbf{E}} \cdot \mathbf{F}^{-1} \,. \tag{35}$$

If we now denote by $\mathbf{w}^m$ the mass flow relative to the skeleton movement, i. e.

$$\mathbf{w}^m := \rho^f \left(\mathbf{v}^f - \mathbf{v}^s\right) = \rho^f\,\mathbf{w} \tag{36}$$

according to Biot [9], it is possible to introduce the *Lagrange*an mass flow vector $\mathbf{M}$ such that

$$\mathbf{w}^m \cdot \mathbf{n}\,da = \mathbf{M} \cdot \mathbf{N}\,\mathrm{d}A \,. \tag{37}$$

Remark 3. Wilmański [46] observes that the kinematics of the fluid is defined within the domain of the skeleton, because we are not interested in the motion beyond this domain, except for the phenomena appearing on the boundary of the skeleton. In order to describe the motion of the fluid in the reference configuration so that all fields are defined in the same domain, the author introduces the *Lagrange*an velocity of the fluid, given by

$$\mathbf{V}^f(\mathbf{X}, t) = \mathbf{F}^{-1} \cdot \left(\mathbf{v}^f - \mathbf{v}^s\right) . \tag{38}$$

The above expression gives the velocity of the image of the material point of the fluid in the reference configuration of the skeleton. By using (33), it can be proved that (37) is equivalent to (38).

The *Lagrange*an mass flow vector $\mathbf{M}$, defined by (37), has not a real physical meaning in itself, but, unlike $\mathbf{w}^m$, depends only on $\mathbf{X}$ and t, i. e. $\mathbf{M} = \mathbf{M}(\mathbf{X}, t)$.

3.2 Mass balance

In the *Lagrange*an formulation, the conservation of mass of the solid component is identically satisfied, being

$$\frac{\mathrm{d}}{\mathrm{d}t}\left[J\rho^{sR}(1-n)\right] = 0 \tag{39}$$

with

$$J := \det \mathbf{F} = \frac{1+e}{1+e_0} . \tag{40}$$

The **balance of mass for the fluid phase** over a material control volume V_m fixed on the solid matrix and bounded by the surface A_m writes in integral form:

$$\frac{\mathrm{d}}{\mathrm{d}t}\int_{V_m}\left(\rho^{fR}\frac{e}{1+e}\right)\mathrm{d}V + \int_{A_m}\left(\rho^{fR}\frac{e}{1+e}\right)\left(\mathbf{v}^f - \mathbf{v}^s\right)\cdot\mathbf{n}\,\mathrm{d}a = 0 . \tag{41}$$

By transforming the volume integral in the left-hand side of (41) in an integral over the reference volume and by using the *Gauß* theorem in order to transform the surface integral in an integral over the reference volume, one gets

$$\frac{\mathrm{d}}{\mathrm{d}t}\left(\frac{e}{1+e_0}\right) + \mathrm{Div}\left[\frac{e}{1+e_0}\mathbf{F}^{-1}\cdot\left(\mathbf{v}^f - \mathbf{v}^s\right)\right] = 0 , \tag{42}$$

where the operator 'Div' outlines that the operation is performed in terms of *Lagrange*an coordinates.

At this point, it is convenient to introduce, see [19], the change of fluid mass referred to the initial volume, denoted in the sequel by the symbol m.

Accordingly, the quantity $m\,\mathrm{d}V_0$ indicates the difference of the fluid mass passing from the initial to the current configuration so that

$$m := \frac{1}{\mathrm{d}V_0}\left(\rho^{fR}\, n\,\mathrm{d}V - \rho_0^{fR}\, n_0\,\mathrm{d}V_0\right) = J\,\rho^{fR}\, n - \rho_0^{fR}\, n_0\,. \tag{43}$$

By using the *Euler*ian vector (36) and (37), the mass balance of the fluid phase can also be expressed as

$$\frac{\mathrm{d}^s m}{\mathrm{d}t} + \mathrm{Div}\,\mathbf{M} = 0\,. \tag{44}$$

The **mass balance of the porous medium** as a whole is obtained by summing up the two mass conservation expressions

$$\frac{\mathrm{d}^s \rho^s}{\mathrm{d}t} + \rho^s\,\nabla\cdot\mathbf{v}^s = 0 \tag{45}$$

and

$$\frac{\mathrm{d}^s \rho^f}{\mathrm{d}t} + \rho^f\,\nabla\cdot\mathbf{v}^s + \nabla\cdot\mathbf{w}^m = 0\,. \tag{46}$$

With the above introduced definitions, the result can be expressed in the *Lagrange*an formulation (see also [20]):

$$\frac{\mathrm{d}^s R}{\mathrm{d}t} + \mathrm{Div}\,\mathbf{M} = 0\,, \tag{47}$$

where

$$R := \rho_0^f + \rho_0^s + m\,. \tag{48}$$

We note that (48) outlines that the variation of mass of the volume $\mathrm{d}V$ is only due to the mass that the fluid exchanges with the surroundings.

3.3 Momentum balance equation

By denoting with $\rho = \rho^f + \rho^s$ the density of the porous medium, the first *Cauchy* law of motion of the porous medium as a whole writes

$$\int_V \left[\rho\,\mathbf{a} + n\,\rho^{fR}\left(\mathbf{a}^f - \mathbf{a}^s\right)\right]\mathrm{d}V = \int_V \rho\,\mathbf{b}\,\mathrm{d}V + \int_S \mathbf{T}\cdot\mathbf{n}\,\mathrm{d}S\,. \tag{49}$$

Being (49) valid for any control volume, under the assumption of regular fields, the local form is obtained by using *Gauß*' theorem:

$$\nabla\cdot\mathbf{T} + \rho\,(\mathbf{b} - \mathbf{a}^s) - n\,\rho^{fR}\left(\mathbf{a}^f - \mathbf{a}^s\right) = \mathbf{0}\,. \tag{50}$$

In order to obtain the corresponding *Lagrange*an formulation, we introduce the second *Piola-Kirchhoff* stress tensor

$$\widetilde{\mathbf{T}} := J\left(\mathbf{F}^{-1}\cdot\mathbf{T}\cdot(\mathbf{F}^{-1})^T\right) \tag{51}$$

and the transport formula

$$\mathbf{F} \cdot \widetilde{\mathbf{T}} \cdot \mathbf{N}\,\mathrm{d}A = \mathbf{T} \cdot \mathbf{n}\,\mathrm{d}a \tag{52}$$

so that (50) writes

$$\begin{aligned} &\int_{A_0} \mathbf{F} \cdot \widetilde{\mathbf{T}} \cdot \mathbf{N}\,\mathrm{d}A + \\ &\quad + \int_{V_0} \left[(\rho_0 + m)\,(\mathbf{b} - \mathbf{a}) - (\rho_{w_0}\, n_0 + m)\,(\mathbf{a}^w - \mathbf{a})\right] \mathrm{d}V = \mathbf{0}\,, \end{aligned} \tag{53}$$

with the local form

$$\mathrm{Div}\left(\mathbf{F} \cdot \widetilde{\mathbf{T}}\right) + (\rho_0 + m)\,(\mathbf{b} - \mathbf{a}^s) - \left(\rho_0^f + m\right)\left(\mathbf{a}^f - \mathbf{a}^s\right) = \mathbf{0}\,. \tag{54}$$

At this stage, we do not introduce the momentum balance equation of a phase, i. e. (27), because we are not in the position to specify the expression for the interaction force $\mathbf{f}^{\mathrm{int}}$. For this reason, reference is first made to energy considerations.

3.4 Energy balance

As already mentioned in the introduction to this section, there are two main features characterizing a porous medium: The first one is the presence of a connected porosity, a parameter which accounts of voids independently of their shape, intended to be a global and not a local quantity.

Secondly, the connected porosity is partially or completely filled by fluids.

Because such fluids can be exchanged with the surroundings, the porous medium has to be considered as a local thermodynamic open system.

With such a premise, the first law in its *rate form* can be expressed as follows:

The rate of increase of internal energy per unit volume is equal to the input power not converted into kinetic energy (i. e. the stress power) plus the internal supply of heat per unit volume and the inflow per unit volume of heat through the boundaries of the element.

In particular, for a thermomechanic continuum, the rate at which thermal energy is added is given by

$$Q = \int_V \rho\, r\,\mathrm{d}V - \int_S \mathbf{q} \cdot \mathbf{n}\,\mathrm{d}S\,, \tag{55}$$

where the scalar field r specifies the rate of internal heat production per unit mass, also known as *heat supply*, and the vector $\mathbf{q}$, *heat flux*, is the measure of the rate at which heat is conducted into the system, per unit area per unit time (the negative sign is introduced because $\int_S \mathbf{q} \cdot \mathbf{n}\,\mathrm{d}S$ is the outward flux).

The stress power assumes the *Euler*ian expression $\mathbf{T} : \mathbf{D}$ for a *non-polar medium*, being $\mathbf{D}$ the rate of the deformation tensor, i. e. the symmetric part of the velocity gradient tensor $\mathbf{L} := \nabla \mathbf{v}$.

By denoting e^f and e^s as the specific internal energy, associated with the fluid and the solid particles, the *Lagrange*an density of the internal energy E and of the entropy S (volume density) can be defined, according to Coussy et al. [20] as

$$E \, \mathrm{d}V_0 := \left(\rho^f \, e^f + \rho^s \, e^s\right) \mathrm{d}V \,, \tag{56}$$

$$S \, \mathrm{d}V_0 := \left(\rho^f \, s^f + \rho^s \, s^s\right) \mathrm{d}V \,. \tag{57}$$

Recalling that the specific enthalpy of the fluid is given by

$$h := e^f + \frac{p}{\rho^{fR}} \,, \tag{58}$$

the *Lagrange*an local form of the energy balance assumes the form

$$\begin{aligned} \frac{\mathrm{d}^s E}{\mathrm{d}t} = \mathrm{tr}\big(\widetilde{\mathbf{T}} \cdot \dot{\mathbf{E}}\big) - \\ - \mathrm{Div}\left[\left(e^f + \frac{p}{\rho^{fR}}\right)\mathbf{M}\right] - \mathrm{Div}\,\mathbf{Q} + \mathbf{M} \cdot \mathbf{F}^T \cdot \left(\mathbf{b} - \mathbf{a}^f\right) . \end{aligned} \tag{59}$$

The term $\mathrm{Div}\left[\left(e^f + p/\rho^{fR}\right)\mathbf{M}\right]$ in (59) arises from the fact that, being the porous medium a local *thermodynamic open system*, we have to account first that mass flow serves as an additional mechanism to change energy so that the variation of the internal energy of the system will be $\mathrm{d}^s/\mathrm{d}t \int_{V_0} E \, \mathrm{d}V_0 + \int_{A_0} e^f \, \mathbf{M} \cdot \mathbf{N} \, \mathrm{d}A_0$; secondly, we have to account the power of the forces in the relative motion of the fluid with respect to the skeleton, i. e. $\dot{W} = \int_{A_0} -p/\rho^{fR} \, \mathbf{M} \cdot \mathbf{N} \, \mathrm{d}A_0$.

For any macroscopic system, the state function *entropy* can be introduced as the sum of two terms

$$\mathrm{d}S = \mathrm{d}S^e + \mathrm{d}S^i \,, \tag{60}$$

being $\mathrm{d}S^e$ the entropy supplied to the system and $\mathrm{d}S^i$ the entropy produced inside the system.

The second law of thermodynamics states that *the entropy produced inside the system must be zero for reversible processes and positive for irreversible transformations*, i. e.

$$\mathrm{d}S^i \geq 0 \,. \tag{61}$$

For a closed system, which can only exchange heat with its surroundings, by accounting for the theorem of *Carnot-Clausius*, i. e. $\mathrm{d}S^e = \frac{\delta Q}{T}$, the second law can also be expressed in the form

$$\mathrm{d}S \geq \frac{\delta Q}{T} \,. \tag{62}$$

For an open system, which can exchange both heat and matter with the surroundings, $\mathrm{d}S^e$ must also contain a term related to the transfer of matter so that the second law assumes the form

$$\frac{\mathrm{d}^s S}{\mathrm{d}t} + \mathrm{Div}\left(\mathbf{M}\, s^f\right) = -\mathrm{Div}\,\frac{\mathbf{Q}}{T} + \frac{\Phi}{T}, \tag{63}$$

where Φ is the total dissipation, Φ/T is the internal production rate of entropy and $\mathbf{Q}$ is defined such that $\mathbf{Q}\cdot\mathbf{N}\,\mathrm{d}A_0 = \mathbf{q}\cdot\mathbf{n}\,\mathrm{d}a$.

(63) shows that, for open systems, the entropy flow consists of two parts: The first one is connected with the heat flow and the second one with the diffusion flow of matter.

In order to relate the entropy production to the various irreversible processes occurring in the system, the assumption is introduced that, even if the total system is not in equilibrium, there exists a *state of local equilibrium* for which the local entropy is a function of the same state variables needed to define the macroscopic state of the system.

With such an assumption, by extracting $-\mathrm{Div}\,\mathbf{Q}$ from (59) and introducing into (63), gives the identification of the dissipation terms

$$\Phi_f := \mathbf{M}\cdot\mathbf{F}^T\cdot\left(-\nabla p + \rho^{fR}\left(\mathbf{b}-\mathbf{a}^f\right)\right), \tag{64}$$

$$\Phi_{th} := -\frac{\mathbf{Q}}{T}\cdot\mathrm{Grad}\,T, \tag{65}$$

$$\Phi_l := \mathrm{tr}\left(\widetilde{\mathbf{T}}\cdot\dot{\mathbf{E}}\right) + g^f\,\dot{m} - S\,\dot{T} - \dot{\Psi}, \tag{66}$$

being Φ_f the dissipation associated with the fluid mass transport, Φ_{th} the thermal dissipation associated with the heat conduction and Φ_l the intrinsic dissipation associated to the movement of the skeleton.

In (66), the *free enthalpy per unit mass of the fluid* is defined by

$$g^f = e^f + \frac{p}{\rho^{fR}} - T\,s^f, \tag{67}$$

and the *free energy per unit volume of the porous medium* is given by

$$\Psi = E - T\,S. \tag{68}$$

Because of their different nature, the three dissipation terms are assumed to be positive, each independently of the other, and in particular, if we write them in *Euler*ian form, we have

$$\varphi_f := n^f\left(\mathbf{v}^f - \mathbf{v}^s\right)\cdot\left(-\nabla p + \rho^{fR}\left(\mathbf{b}-\mathbf{a}^f\right)\right) \geq 0, \tag{69}$$

$$\varphi_{th} := -\frac{\mathbf{q}}{T}\cdot\nabla T \geq 0. \tag{70}$$

The bilinear structure of (69) and (70) is the product of two factors: The first one is a flow quantity, already introduced in balance laws; the second factor is related to the gradient of an intensive quantity and may contain an external force, and is called *thermodynamic force* or *affinity*.

As outlined by [20], (69) and (70) allow to introduce *a posteriori* the simplest conduction laws, i. e. the *Darcy* law and the *Fourier* law:

$$n^f \left(\mathbf{v}^f - \mathbf{v}^s\right) = \frac{\mathbf{K}}{\mu} \left(-\nabla p + \rho^{fR} \left(\mathbf{b} - \mathbf{a}^f\right)\right) , \tag{71}$$

$$\mathbf{q} = -\mathbf{k} \cdot \nabla T . \tag{72}$$

Furthermore, a comparison of (71) with (29) for the fluid phase,

$$\rho^f \, \mathbf{a}^f = \nabla \cdot \mathbf{T}^f + \rho^f \, \mathbf{b} + \mathbf{f}^{\text{int}} , \tag{73}$$

with

$$\mathbf{T}^f = -n^f \, p \, \mathbf{I} , \tag{74}$$

allows to identify the interaction force

$$\mathbf{f}^{\text{int}} = -n^2 \, \mu \, \mathbf{K}^{-1} \left(\mathbf{v}^f - \mathbf{v}^s\right) + p \, \nabla n . \tag{75}$$

Remark 4. As observed by [19] and [20], the balance equations presented in this section, supplemented by the *Darcy* and *Fourier* laws expressed in terms of *Lagrange*an vectors **M** and **Q**, constitute a closed sets of equations, to be solved with the added boundary conditions.

A central point which deserves further investigation is related to the entropy inequality.

The entropy principle used in this section is the *Clausius-Duhem inequality*, with the procedure of exploitation due to Coleman and Noll [17].

Hutter [25] and Wang and Hutter [44] have discussed in great detail the differences between this formulation and the one suggested by Müller [33] and Liu [29].

In particular, whereas the *Clausius-Duhem* approach makes *a priori postulates* about entropy flux and entropy supply, *Müller*'s and *Liu*'s approach postulate the entropy flux to be a general constitutive variable and all field equations are considered to be constraints for the exploitation of the entropy principle.

According to Wang and Hutter [44], the *Clausius-Duhem* approach with the exploitation due to Coleman and Noll [17] should be abandoned for models concerned with polar continua, coupled field theories, and structured continua.

4 The effective stress

The interaction law between the soil skeleton and the pore water was introduced by Terzaghi [40] as the *principle of effective stress*, stating that the total stress can be decomposed in two parts: One part acts in the water and in the solid in every direction and is called *pore water pressure*; the balance, total stress minus neutral stress, represents an excess over the neutral one that has its seat exclusively in the solid phase and is called *effective stress.*

In a more formal presentation of this principle, *Terzaghi* also stated that "porous materials (such as sand, clay and concrete) react to a change of pore pressure as if they were incompressible and as if their internal friction were equal to zero. All the measurable effects of a change of stress, as a compression, distortion and a change of shearing resistance are exclusively due to changes in the effective stress".

Remark 5. Note that the *Terzaghi* definition of effective stress does not derive from any theoretical investigation of basic mechanics of porous media, but it is introduced in a rather heuristic manner in terms of cause and measurable effects.

In order to clarify the meaning of this definition within the present framework, we consider the case of a single fluid phase that saturates the void space.

In this case, the macroscopic total stress tensor is given by

$$\overline{\mathbf{T}} = (1 - n)\,\overline{\mathbf{T}}^{s} - n\,\overline{p}^{f}\,\mathbf{I}\,. \tag{76}$$

According to the definition of *effective stress*, this equation is usually written in soil mechanics in the form

$$\overline{\mathbf{T}} = \overline{\mathbf{T}}' - \overline{p}^{f}\,\mathbf{I}\,, \tag{77}$$

and it follows that for the effective stress the expression holds:

$$\overline{\mathbf{T}}' = (1 - n)\,\overline{\mathbf{T}}^{s} + (1 - n)\,\overline{p}^{f}\,\mathbf{I}\,. \tag{78}$$

At this stage, it is relevant to note that there is a basic difference between mixtures and porous media: The interaction between the constituents in a saturated porous medium is due not only to friction effects but also to space displacement, as described by the saturation condition.

In particular, if the two constituent are considered incompressible, by summing up the two mass balance equation the following condition arises

$$\nabla \cdot \left(n\,\mathbf{v}^{F} + (1 - n)\,\mathbf{v}^{s}\right) = 0\,, \tag{79}$$

which is a constraint on the velocity field, which reveals an indeterminacy and brings a *Lagrange*an multiplier in.

It can be shown [14] that this multiplier is related to the isotropic part of the tensor $\mathbf{T}$ and can therefore be interpreted as a hydrostatic pressure acting on the porous medium.

As a consequence, the partial stress tensor assumes in the porous media theory the more general expression

$$\mathbf{T}^S = -(1-n)\,p\,\mathbf{I} + \mathbf{T}_E^S\,, \tag{80}$$

$$\mathbf{T}^F = -n\,p\,\mathbf{I} + \mathbf{T}_E^F\,,$$

where the partial stress for each phase is divided into two parts: The first one is the reaction to the constraint (79) and the second one (indicated in (80) with the index $(\cdot)_E$ as an extra term) is linked to the deformations of the soil skeleton, corresponding to the effective stress or to the fluid dissipation.

5 Body waves propagation in a porous medium

With the assumption of linearized theory, the effective stress tensor of the soil skeleton is given by the generalized *Hooke*'s law

$$\mathbf{T}_E^s = \lambda\,(\nabla\cdot\mathbf{u})\,\mathbf{I} + \mu\,(\nabla\mathbf{u} + \mathbf{u}\nabla)\;. \tag{81}$$

If the assumption is introduced that during the wave propagation no relative motion occurs between the pore water and the solid skeleton ($\mathbf{w} = \mathbf{0}$), i. e. the time scale of any consolidation process is higher if compared with the propagation velocity, it can be shown that the motion of the porous medium as a whole assumes the form

$$\left(\rho^s + \rho^f\right)\ddot{\mathbf{u}} = \left(\lambda + 2\mu + \frac{1}{n\,\alpha}\right)\nabla\left(\nabla\cdot\mathbf{u}\right) - \mu\nabla\times\left(\nabla\times\mathbf{u}\right)\,, \tag{82}$$

where the identity

$$\nabla^2\mathbf{u} = \nabla\left(\nabla\cdot\mathbf{u}\right) - \nabla\times\left(\nabla\times\mathbf{u}\right) \tag{83}$$

has been used and the compressibility α of the fluid has been introduced.

The displacement field can be expressed, as usual, by using the *Helmholtz* decomposition as the sum of the gradient of a scalar potential and the curl of a vector potential:

$$\mathbf{u} = \nabla\varphi + \nabla\times\psi\,. \tag{84}$$

So that, by substituting this expression into (82), we derive the following wave equations:

$$\left(\rho^s + \rho^f\right)\frac{\partial^2\varphi}{\partial t^2} = \left(\lambda + 2\mu + \frac{1}{n\,\alpha}\right)\nabla^2\varphi\,, \tag{85}$$

$$\left(\rho^s + \rho^f\right)\frac{\partial^2\psi}{\partial t^2} = \mu\nabla^2\psi\,.$$

(85) represents a dilational wave and a pure rotational wave.

The stiffness parameter affecting the pure rotational wave depends only on the solid skeleton, according with *Kelvin*'s theorem, stating that in a frictionless fluid without body forces (both assumption have been introduced here) non-circulation can be generated.

As far as the dilational wave propagating in the porous medium as a whole is concerned (i. e. in undrained condition), due to the constraint of no relative motion, the coupling between the two components is reflected in an increase of the skeleton stiffness, and the velocity of propagation is given by

$$v_p = \sqrt{\frac{1}{\rho}\left(\lambda + 2\mu + \frac{1}{n\,\alpha}\right)}\,. \tag{86}$$

A formal proof of (82) can be given by the following arguments.

According to Biot [5], let define by the increase in fluid content, i. e. the amount of fluid which has flowed in a given element of the solid skeleton

$$\varsigma := \frac{m}{\rho_0^{fR}} - \nabla \cdot \left[n\left(\mathbf{v}^f - \mathbf{v}^s\right)\right]\,. \tag{87}$$

If the small perturbation produced by wave propagation is *adiabatic*, in virtue of the above mentioned time scales, according to Biot [6] a reduced potential can be introduced, which depends on the small deformation tensor ε and on ς:

$$\psi = \psi\left(\varepsilon, \varsigma\right)\,. \tag{88}$$

In virtue of the existence of this potential, the macroscopic stress tensor and the thermodynamic pore pressure are given by

$$\boldsymbol{\sigma} = \frac{\partial \psi}{\partial \varepsilon}\,, \qquad p = \frac{\partial \psi}{\partial \varsigma}\,. \tag{89}$$

Within the framework of a linearized theory, the potential depends on the first two scalar invariants of the deformation tensor and on ς so that, by using the notation introduced by *Biot*, we can write

$$\begin{aligned} 2\,\psi = {} & (\lambda_U + 2\,\mu)\,(\mathrm{tr}\,\varepsilon)^2 + \mu\left[2\,(\mathrm{tr}\,\varepsilon\cdot\varepsilon) - (\mathrm{tr}\,\varepsilon)^2\right] - \\ & -2\,\beta\,M\,\varsigma\,\mathrm{tr}\,\varepsilon + M\,\varsigma^2. \end{aligned} \tag{90}$$

From (89), it follows:

$$\boldsymbol{\sigma} = \lambda_U\,\mathrm{tr}\,\varepsilon\,\mathbf{I} + 2\,\mu\,\varepsilon - \beta\,M\,\varsigma\,\mathbf{I}\,, \tag{91}$$

$$p = M\left(\varsigma - \beta\,\mathrm{tr}\,\varepsilon\right)\,. \tag{92}$$

By inserting ς from (92) into (91), we also obtain:

$$\sigma = \lambda_{dr}\,\mathrm{tr}\,\varepsilon\,\mathbf{I} + 2\,\mu\,\varepsilon - \beta\,p\,\mathbf{I}\,, \tag{93}$$

$$\varsigma = \frac{p}{M} + \beta \operatorname{tr} \varepsilon . \tag{94}$$

The experimental determination of the coefficients in the above equations has been discussed by Biot and Willis [7]. In particular, λ_{dr} is the *Lamé* coefficient in drained conditions, whereas $\lambda_U = \lambda_{dr} + \beta^2 M$ is the same coefficient in undrained conditions. The coefficient M is the pressure on the fluid needed to increase the fluid content of a unit value, when the macroscopic volume is preserved ($\varepsilon_{kk} = 0$).

If we introduce the bulk modulus K_s of the solid matrix and the bulk modulus $K_f = \frac{1}{\alpha}$ of the fluid, it can be proved, by referring to the definitions, that the following relation holds:

$$\frac{1}{M} = \frac{\beta - n}{K_s} + \frac{n}{K_f} . \tag{95}$$

Finally, from (94) it can be deduced that the coefficient β represents, in drained conditions ($p = 0$), the proportion of the macroscopic volumetric strain due to change in fluid content. In this case, the following relation can also be proved:

$$\beta = 1 - \frac{K_{sk}}{K_s} , \tag{96}$$

being K_{sk} the bulk modulus of the soil skeleton as a whole.

Remark 6. With the above specified meaning of the coefficients, (93) gives the more general expression for the effective stress tensor in poroelasticity:

$$\boldsymbol{\sigma}' = \boldsymbol{\sigma} + \left(1 - \frac{K_{sk}}{K_s}\right) p \mathbf{I} . \tag{97}$$

This expression merges in the one suggested by *Terzaghi* if the ratio K_{sk}/K_s can be neglected, which is the case for soils. In addition, it also proves that the effective stress concept is a constitutive one.

Under the assumption of undrained conditions, $\varsigma = 0$, and, setting for soils $\beta = 1$, (91) reduces to

$$\boldsymbol{\sigma} = \left(\lambda_{dr} + \frac{1}{n\,\alpha}\right) \varepsilon_{kk} \mathbf{I} + 2\,\mu\,\boldsymbol{\varepsilon} , \tag{98}$$

which proves (82) and (86).

In order to validate (86), an examination of the results of field tests is here presented.

The tests under consideration refer to the measurements in Figure 1 of shear and dilational wave velocity by means of cross-hole technique (see Lancellotta [26] for a description of these tests).

(a) By considering the silty clay stratum at depth $12 \leq z \leq 16$ m, the shear wave velocity v_s is equal to 150 m/s. With the assumption of a *Poisson* ratio

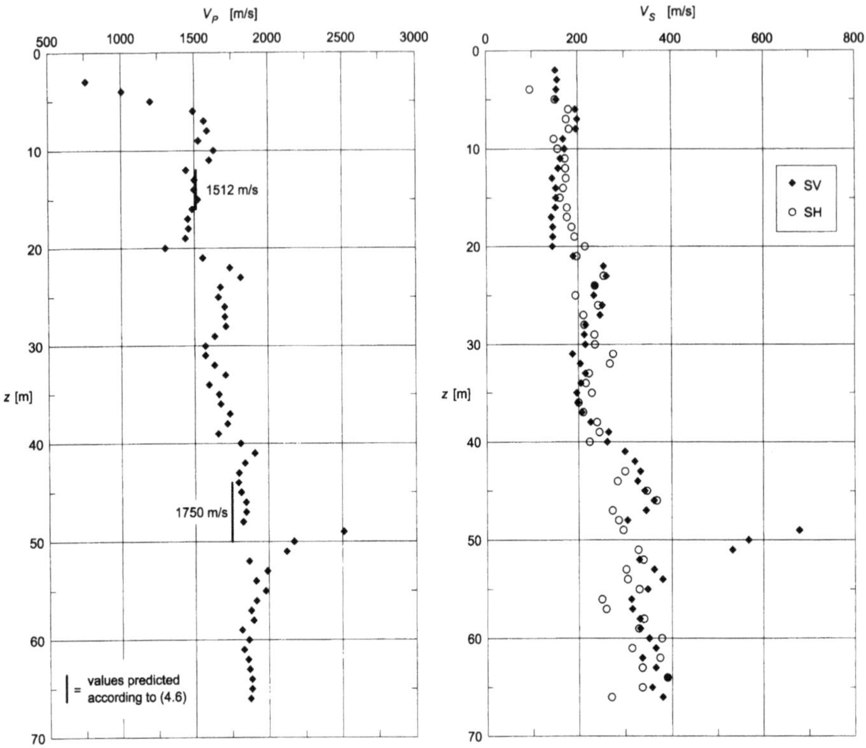

Fig. 1. Dilational and shear wave velocity at *Pisa* site, MURST [34].

of 0.15 and considering the porosity value of 0.6, the density of the porous medium is equal to 1 680 kg/m^3 and the following values are deduced for *Lamé*'s constants:

$\mu = 37\,800$ kN/m^2; $\lambda = 16\,170$ kN/m^2 (recall that $\frac{\lambda + 2\,\mu}{\mu} = \frac{2 - 2\,\upsilon}{1 - 2\,\upsilon}$).

By using (86), a dilational wave velocity is obtained of 1 512 m/s.

(b) At depth $44 \leq z \leq 50$ m, a fine silty sand stratum is present, with a shear wave velocity of about 340 m/s and a porosity of 0.4. By using the same procedure the following values are obtained:

$\rho = 2\,020$ kg/m^3; $\mu = 233\,512$ kN/m^2; $\lambda = 99\,896$ kN/m^2;
$1/(n\,\alpha) = 5\,625\,000$ kN/m^2; $\upsilon_p = 1\,750$ m/s.

In both cases, the computed values are in agreement with the measured ones, and, in addition, it can also be observed that (86) predicts the behaviour of the porous medium for limiting values of the porosity: If n tends to unity, both *Lamé* constants are very small and the density of the porous medium is equal to the fluid density so that the velocity of the dilational wave assume the value in the fluid. On the contrary, if n tends to zero, the dilational wave velocity tends to infinity, because of the assumption of incompressibility of the solid matrix.

6 Consolidation theories

The first attempt to describe the consolidation of a deformable porous medium with pores completely filled by water is due to Terzaghi [40, 41].

The suggested theory introduced the concept of effective stress and was limited to the one-dimensional case.

Later, Biot [5] generalized to the three-dimensional case in a framework consistent with the basic principles of continuum mechanics. The presented theory deals with small strains and elastic behaviour of the soil skeleton, and it is briefly recalled in the following in order to outline fields of further developments.

6.1 *Biot*'s theory

Within the general framework presented in the previous sections, the *Biot* formulation has to be considered already as macroscopic one so that the porous medium is equivalent to two superimposed continua. In particular, the behaviour of the soil skeleton is described by global deformation characteristics, which include all local deformations (i. e. rolling and sliding of particles).

The momentum balance equation of the fluid phase is expressed by the *Darcy* law, and by combining with the mass balance equation one gets

$$\frac{\partial\left(\nabla\cdot\mathbf{u}^{s}\right)}{\partial t}-\left(\frac{k}{\rho_{w}g}\nabla^{2}p\right)=0\,,\tag{99}$$

where $\mathbf{u}^s$ is the displacement vector of the soil skeleton, p is the pore pressure in excess to the initial equilibrium value and k is the hydraulic conductivity of the porous medium.

With the convention adopted in soil mechanics that compressive stresses are considered positive, the momentum balance equation of the porous medium as a whole in quasi-static motion writes

$$\nabla\cdot\mathbf{T}-\rho\,\mathbf{b}=\mathbf{0}\,,\tag{100}$$

being the total stress tensor decomposed into the effective stress tensor and the pore pressure according to the *Terzaghi* equation

$$\mathbf{T}=\mathbf{T}'+p\,\mathbf{I}\,.\tag{101}$$

Assuming an elastic behaviour for the soil skeleton, the equilibrium equations can be expressed in terms of displacement components (*Navier* equations) so that, by introducing the effective stress definition, one obtains

$$(\lambda+\mu)\,\nabla\left(\nabla\cdot\mathbf{u}^{s}\right)+\mu\,\nabla^{2}\,\mathbf{u}^{s}-\nabla p+\rho\,\mathbf{b}=\mathbf{0}\,.\tag{102}$$

(99) and (102) give a system of four differential equations for the unknown fields $\mathbf{u}^s$ and p.

With the assumption made on the pore pressure, the *initial conditions* can simply state that all quantity $\mathbf{u}^s$ and p are zero. The *boundary conditions* can be specified as:

- prescribed pore pressure: $p = h$,
- prescribed flux: $k/(\rho_w\, g)\, \nabla p \cdot \mathbf{n} = f$,
- prescribed traction: $\mathbf{T} \cdot \mathbf{n} = \mathbf{t}$,
- prescribed displacement: $\mathbf{u}^s = \mathbf{a}$.

Remark 7. Exact solutions of the consolidation problem are limited to simple boundary conditions under the assumption of elastic behaviour of soil skeleton. There are few examples of general treatment of the problem, involving a more realistic soil behaviour, coupled with the diffusion process of the fluid phase (one of these examples is provided by Carter et al. [16]). A formal proof should be also given about the uniqueness of solution for the initial boundary problem given by (99) and (102).

6.2 1-d finite deformation theory

There are two major areas of developments, where departures from results predicted by *Terzaghi* and *Biot*'s theories need to be explored: Finite deformations and non-linear soil response. The first aspect is dealt with in the following with reference to the one-dimensional case.

Consider a soil stratum of infinite extent in the 0XY-plane and transversely isotropic about the 0Z-axis.

The upper surface is supposed to be a free draining boundary so that

$$p\,(t,\, Z = H) = 0\,. \tag{103}$$

The lower boundary is supposed to be impermeable and the related boundary condition is given by

$$\frac{\partial p}{\partial Z}\,(t,\, Z = 0) = 0\,. \tag{104}$$

At the instant $t = 0$, the upper boundary is being loaded with a uniform load

$$\mathbf{t}\,(t \geq 0,\, Z = H) = -\Delta q\, \mathbf{e}_z\,. \tag{105}$$

By accounting for the one-dimensional nature of the problem, the deformation gradient reduces to

$$\mathbf{F} = \mathbf{I} + \frac{\partial u}{\partial Z}\, \mathbf{e}_z \otimes \mathbf{e}_z\,, \tag{106}$$

and the vertical component of the *Piola-Kirchhoff* tensor is linked to the *Cauchy* tensor according to

$$\widetilde{T}_{zz} = \frac{T_{ZZ}}{1 + \partial u / \partial Z} \,. \tag{107}$$

The *Lagrange*an mass balance equation for the solid phase reduces to

$$\frac{\partial z}{\partial Z} = \frac{1+e}{1+e_0} \,, \tag{108}$$

being e_0 the void ratio in the reference configuration.

The *Lagrange*an mass balance formulation for the fluid phase reduces to

$$\frac{\partial}{\partial t}\left(\rho^{fR}\frac{e}{1+e}\frac{\partial z}{\partial Z}\right) + \frac{\partial}{\partial Z}\left[\rho^{fR}\frac{e}{1+e}\left(\mathbf{v}^f - \mathbf{v}^s\right)\right] = 0 \,, \tag{109}$$

and by using (108) one gets

$$\frac{\partial e}{\partial t} + (1+e_0)\frac{\partial}{\partial Z}\left[\frac{e}{1+e}\left(\mathbf{v}^f - \mathbf{v}^s\right)\right] = 0 \,. \tag{110}$$

The balance of linear momentum of the porous medium, neglecting inertial terms, gives

$$\mathrm{Div}\left(\mathbf{F}\cdot\widetilde{\mathbf{T}}\right) + (\rho_0 + m)\,\mathbf{b} = \mathbf{0} \,, \tag{111}$$

and by recalling the definition of m one obtains

$$\frac{\partial T_{zz}}{\partial Z} + g\frac{e\,\rho^{fR} + \rho^{sR}}{1+e_0} = 0 \,. \tag{112}$$

The *Lagrange*an formulation of the *Darcy* law is given by

$$\frac{e}{1+e}\left(\mathbf{v}^f - \mathbf{v}^s\right) = -\frac{K}{g\rho^{fR}}\left[\left(1+\frac{\partial u}{\partial Z}\right)^{-1}\frac{\partial u}{\partial Z} + g\rho^{fR}\right] . \tag{113}$$

Thus, by inserting (113) into (110) and by using the definition of effective stress, one obtains the *Lagrange*an equation of one-dimensional consolidation (Lancellotta and Preziosi [27]):

$$\frac{\partial e}{\partial t} + (1+e_0)\frac{\partial}{\partial Z}\left[\frac{K}{g\,\rho^{fR}}\left(\frac{\rho^{sR} - \rho^{fR}}{1+e}g + \frac{\partial T'_{zz}}{\partial Z}\frac{1+e_0}{1+e}\right)\right] = 0 \tag{114}$$

The above equation represents the most general formulation, see [23, 27, 37], and it can be proved that all other 1-d models suggested in the literature can be obtained as a special case of it.

Remark 8. As shown by Carter et al. [16], the solution of (114) under the assumption of an elastic soil behaviour characterized by *Young*'s modulus E' and *Poisson*'s ratio v' depends on the following dimensionless parameters (see Figure 2):

$$\frac{\Delta q}{E'} \,, \quad \frac{g\rho^{fR}H}{E'} \,, \quad v' \,, \quad e_0 \,, \quad \frac{\rho^{sR}}{\rho^{fR}} \,.$$

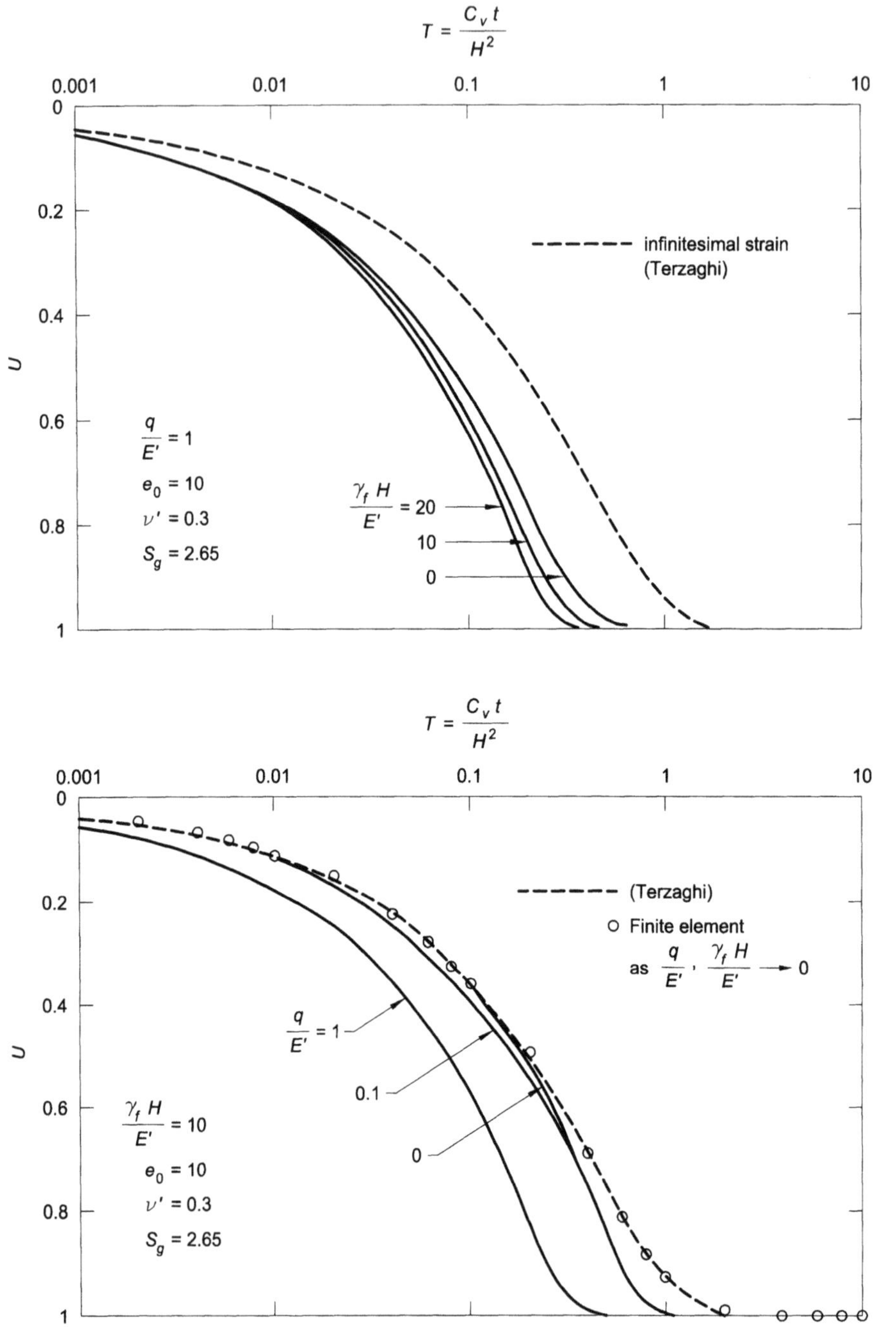

Fig. 2. Large strain 1-d consolidation model (Carter et al. [16]).

In general, to obtain the *Terzaghi* solution, both the parameters $\Delta q/E'$ and $g\,\rho^{fR}\,H/E'$ have to approach zero, indicating that the behaviour of deep soft layers can be significantly different from the one predicted by *Terzaghi*'s theory.

6.3 Finite non-linear consolidation

An example of theoretical formulation and numerical solution method for the consolidation of an elasto-plastic soil with finite deformation has been given by Carter et al. [16], with constitutive rate equations satisfying the objectivity principle. Here, again the general conclusion is resumed that the need to account for large deformation as well as non-linear soil behaviour arises for soft materials and when the imposed load is large if compared with the soil stiffness. Both these cases are of interest in soil mechanics, when dealing with soft clay strata.

References

1. Atkin, R. J., Craine, R. E.: Continuum theories of mixtures: basic theory and historical development. *Q. J. Mech. Appl. Math.* **29** (1976), 209–244.
2. Bear, J.: *Dynamics of fluids in porous media.* Elsevier, 1972.
3. Bear, J., Bachmat, Y.: *Introduction to modeling of transport phenomena in porous media.* Kluwer, 1991.
4. Bedford, A., Drumheller, D. S.: Theory of immiscible and structured mixtures. *Int. J. Eng. Sci.* **21** (1983), 863–960.
5. Biot, M. A.: General theory of three-dimensional consolidation. *J. Appl. Phy.* **12** (1941), 155–165.
6. Biot, M. A.: Theory of propagation of elastic waves in a fluid-saturated porous solid. *Journal of the Acoustica Society of America* **28**(2) (1956), 168–191.
7. Biot, M. A., Willisi, D. G.: The elastic coefficients of theory of consolidation. *J. Appl. Mech.* (1957), 594–601.
8. Biot, M. A.: Mechanics of deformation and acoustic propagation in porous media. *J. Appl. Phys.* **23** (1962), 1482–1498.
9. Biot, M. A.: Variational Lagrangean thermodynamics of nonisothermal finite strain mechanics of porous solids and thermomolecular diffusion. *Int. J. Solids Structures* **13** (1977), 579–597.
10. de Boer, R., Ehlers, W.: *Theorie der Mehrkomponentenkontinua mit Anwendung auf bodenmechanische Probleme.* Forschungsberichte aus dem Fachbereich Bauwesen, Heft 40, Universität GH Essen 1986.
11. de Boer, R.: *Theory of Porous Media.* Springer-Verlag, Berlin 2000.
12. Bourgeois, E., Dormieux, L.: Consolidation of a nonlinear poroelastic layer in finite deformations. *Eur. J. Mech. A/Solids* **15**(4) (1996), 575–598.
13. Bowen, R. M.: Theory of mixtures. In Eringen, A. C. (ed.): *Continuum physics* III/1, Academic Press, 1976, p. 317.
14. Bowen, R. M.: Incompressible porous media models by the use of the theory of mixtures. *Int. J. Eng. Sc.* **18** (1980), 1129–1148.

15. Bowen, R. M.: Compressible porous media models by the use of the theory of mixtures. *Int. J. Eng. Sc.* **20** (1982), 697–735.
16. Carter, J. P., Booker, J. R., Small, J. C.: The analysis of finite elasto-plastic consolidation. *Int. Journal for Num. An. Methods in Geomechanics* **3** (1979), 107–129.
17. Coleman, B. D., Noll, W.: The thermodynamics of elastic materials with heat conduction and viscosity. *Arch. Rat. Mech. and Anal.* **13** (1963), 167–178.
18. Coussy O.: Thermodynamics of saturated porous solids in finite deformations. *Eur. J. Mech. A/Solids* **8** (1989), 1–14.
19. Coussy O.: *Mechanics of porous continua*. John Wiley & Sons, 1995.
20. Coussy, O., Dormieux, L., Detournay, E.: From mixture theory to Biot's approach for porous media. *Int. J. Solids Structures* **35**(34-35) (1998), 4619–4635.
21. Drugan, W. J., Willis, J. R.: A micromechanics-based nonlocal constitutive equation and estimates of representative element size for elastic composites. *J. Mech. Phys. Solids* **44** (1996), 497–524.
22. Ehlers, W.: Constitutive equations for granular materials in geomechanical context. In Hutter, K. (ed.): *Continuum mechanics in environmental sciences and geophysics*, CISM Courses and Lectures No. 337, Springer-Verlag, 1993, pp. 313–402.
23. Gibson, R. E., England, G. L., Hussey, M. J. L.: The theory of one dimensional consolidation of saturated clays. *Géotechnique* **17** (1967), 261–273.
24. Gu, W. Y., Lai, W. M., Mow, V. C.: Transport of multi-electrolytes in charged hydrated biological soft tissues. In de Boer, R. (ed.): *Porous media: Theory and Experiments*, Kluwer Academic Press, 1999, pp. 143–157.
25. Hutter, K.: The foundations of thermodynamics, its basic postulates and implications. A review of modern thermodynamics. *Acta Mechanica* **27** (1977), 1–54.
26. Lancellotta, R.: *Geotechnical Engineering*. Balkema, 1993.
27. Lancellotta, R., Preziosi, L.: A general nonlinear mathematical model for soil consolidation problems. *Journal of Engineering Sciences* **35** (1997), 1045–1063.
28. Lewis, R. W., Schrefler, B. A.: *The Finite Element Method in the Static and Dynamic Deformation and Consolidation of Porous Media*. 2nd ed., John Wiley & Sons, Chichester 1998.
29. Liu, I-S.: Method of Lagrange multipliers for exploitation of entropy principle. *Arch. Rat. Mech. and Anal.* **46** (1972), 131–148.
30. Lorentz, H. A.: *The theory of electrons*. Teubner, Leipzig 1909, reprint Dover 1952.
31. Markov, K. Z.: Elementary micromechanics of heterogeneous media. In Markov, Preziosi (eds.): *Heterogeneous Media*, Birkhäuser Verlag, 2000, pp. 1–162.
32. Müller, I.: A thermodynamic theory of mixture of fluids. *Arch. Rat. Mech. Anal.* **28** (1968), 1–39.
33. Müller, I.: Die Kältefunktion, eine universelle Funktion in der Thermodynamik viskoser warmeleitender Flüssigkeiten. *Arch. Rat. Mech. and Anal.* **40** (1971), 1–36.
34. MURST: Analisi geotecnica della vulnerabilità sismica dei monumenti storici. Unità operativa di Torino: Caratterizzazione in sito mediante misure della propagazione di onde sismiche superficiali, 2000.
35. Nemat-Nasser, S., Hori, M.: *Micromechanics: Overall properties of heterogeneous solids*. Elsevier, 1993.

36. Neuman, S. P.: Theoretical derivation of Darcy's law. *Acta Mechanica* **25** (1977), 153–170.
37. Pane, V.: *Sedimentation and consolidation of clays.* Ph. D. Thesis, University of Colorado, Boulder 1985.
38. Rajagopal, K. R., Tao, L.: *Mechanics of Mixtures.* World Scientific, 1995.
39. Slattery, J. C.: Flow of viscoelastic fluids through porous media. *Am. Inst. Chem. Eng. J.* **13** (1967), 1066.
40. von Terzaghi, K.: Die Berechnung der Durchlässigkeitsziffer des Tones aus dem Verlauf der hydrodynamischen Spannungserscheinungen. *Sitz. Akad. Wissen. Wien*, Math.-Naturw. Kl. Abt. IIa **132** (1923), 125–138.
41. von Terzaghi, K.: *Erdbaumechanik auf bodenphysikalischer Grundlage.* Deuticke, Leipzig 1960. See also: From Theory to Practice, John Wiley & Sons, 1925, 146–148.
42. Truesdell, C.: Thermodynamics of diffusion. In Truesdell, C. (ed.): *Rational Thermodynamics*, Springer-Verlag, 1984, pp. 219–236.
43. Truesdell, C., Toupin, R. A.: The classical field theories. In Flügge, S. (ed.): *Handbuch der Physik*, III/1, Springer-Verlag, 1960, pp. 226–902.
44. Wang, Y., Hutter, K.: Comparison of two entropy principles and their applications in granular flows with/without fluid. *Arch. Mech.* (1999), 1–18.
45. Whitaker, S.: The equation of motion in porous media. *Chem. Eng. Sc.* **21** (1966), 291.
46. Wilmański, K.: Porous media at finite strains: The new model with balance equation for porosity. *Arch. Mech.* **48**(4) (1996), 591–628.
47. Wilmański, K.: *Thermomechanics of Continua.* Springer-Verlag, Berlin 1998.
48. Wilmański, K.: *Mathematical theory of porous media – Lectures notes.* WIAS preprint 602, Berlin 2000.
49. Wilmański, K.: *Note on the notion of incompressibility in theories of porous and granular materials.* WIAS preprint 465, Berlin 2000.

Fixed negative charges modulate mechanical behaviours and electrical signals in articular cartilage under unconfined compression – a triphasic paradigm

Van C. Mow, Daniel D. Sun, X. Edward Guo,
Morakot Likhitpanichkul, and W. Michael Lai

Columbia University, Departments of Biomedical Engineering,
Orthopaedic Surgery and Mechanical Engineering, 10032 New York, USA

Abstract. The unconfined compression test has been frequently used to study the mechanical behaviour of articular cartilage. Recently, it has also been used in explant and gel-cell-complex studies in tissue engineering. Mechanical responses in these experiments have been analyzed using the biphasic theory as well as fibril reinforced poroelastic theory (Armstrong et al. [1], Brown and Singerman [5], Spilker et al. [45, 46], Soulhat et al. [44], Li et al. [30, 31], Fortin et al. [12], DiSilvestro et al. [10, 11]). Using an optical technique and testing cartilage samples in unconfined compression, the apparent *Poisson*'s ratio of articular cartilage has also been determined (Jurvelin et al. [23]). In the biphasic and poroelastic theory, the effect of fixed charges is embodied in the *apparent* compressive *Young*'s modulus and *Poisson*'s ratio of the tissue, and the fluid pressure is considered to be that which is over and above the osmotic pressure. In order to understand the effects of fixed charges on the mechanical behaviours of articular cartilage, and in order to predict the osmotic pressure and electric fields inside the tissue in this experimental configuration, it is necessary to use a model that explicitly takes into account the charged nature of the tissue and ion flow. In this paper, the triphasic theory is used to study how the fixed charges within a porous-permeable soft tissue modulate its mechanical and electrochemical responses under a step displacement load. The results showed that: 1) A charged tissue always supports a larger load than an uncharged tissue of the same intrinsic elastic moduli. 2) The apparent *Young*'s modulus (ratio of equilibrium axial stress to axial strain) is always more than the intrinsic *Young*'s modulus of an uncharged tissue. 3) The apparent *Poisson*'s ratio (negative ratio of lateral strain to axial strain) is always larger than the intrinsic *Poisson*'s ratio of an uncharged tissue. 4) Load support derives from three sources: intrinsic matrix stiffness, hydraulic pressure and osmotic pressure. Under unconfined compression, the *Donnan* osmotic pressure can constitute between 13–22 % of the total load support at equilibrium. 5) During the stress-relaxation and recoiling processes following the initial instant of loading, diffusion potential (due to the gradient of the fixed charge density or FCD, and the gradient of ion concentrations) and streaming potential (due to fluid pressure gradient) compete against each other. Within physiological range of material parameters, the polarity of the electric potential depends on both the mechanical properties and FCD of the tissue. For softer tissue, the diffusion potential dominates while the streaming potential dominates in a stiffer tissue. 6) Fixed charges do not affect the instantaneous strain field relative to the initial equilibrium state. However, there is a sudden increase in the fluid pressure above the initial equilibrium state. These new findings are relevant and necessary for the understanding of cartilage mechanics, cartilage biosynthesis, electromechanical signal transduction by chondrocytes, and tissue engineering.

1 Introduction

The unconfined compression experimental test (Figure 1) has been frequently used to investigate the mechanical behaviours of articular cartilage, and to determine its material property (Armstrong et al. [1], Brown and Singerman [5], Jurvelin et al. [23], Soulhat et al. [44], Li et al. [30, 31], Fortin et al. [12], Soltz and Ateshian [43], Huang et al. [21, 22], DiSilvestro et al. [10, 11]). From this testing configuration, mechanical properties of cartilage such as the apparent *Young*'s modulus and apparent *Poisson*'s ratio have been determined. This test configuration is also widely used in live explants for tissue engineering and gel-cell-complex studies (i. e. in chondrocyte mechano-signal transduction experiments) for the reason that the lateral edge of explant that is exposed to the surrounding bathing solution provides a convenient pathway for easier nutrient transport into, and metabolic waste product out of, the tissue (Sah et al. [39], Kim et al. [26], Buschmann et al. [6]).

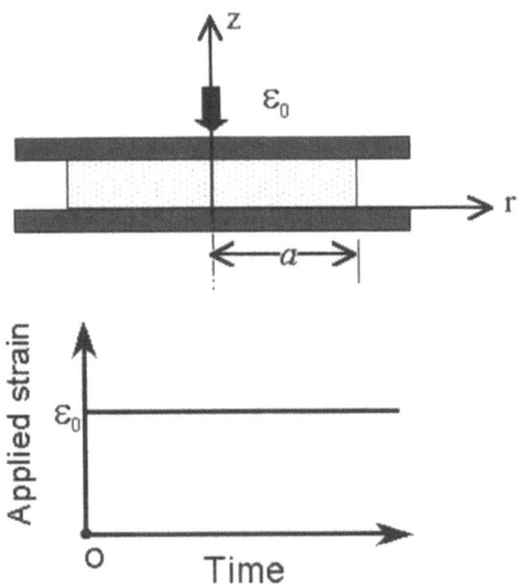

Fig. 1. The schematic representation of the unconfined compression test configuration for stress relaxation; a sudden step displacement is imposed.

Mechanical responses (such as deformation, stress, fluid pressurization and fluid flow within the tissue) in the unconfined compression configuration were first studied by Armstrong and co-workers [1] using the biphasic theory (Mow et al. [36]). These investigators provided mathematical creep and stress-relaxation solutions for a thin cylindrical cartilage specimen compressed between two smooth-frictionless and impermeable platens. Different unconfined compression test conditions, including platen permeability and friction, and different tissue properties that include transverse isotropy or strain-dependent permeability have also been studied (e. g. Lai et al. [27], Spilker et al. [45, 46], Cohen et al. [8]). Using a conewise-linear elastic matrix symmetry (Curnier et al. [9]), Soltz and Ateshian [43], Huang and coworkers [21, 22] have studied theoretically, and experimentally, the tension-compression nonlinear effects due to the large differences between the collagen tensile modulus and proteoglycan compressive modulus under unconfined compression. An important finding from these theoretical and experimental studies is that although this popular unconfined compression experimental configuration seems intuitively simple, it produces unanticipated complex field distributions, such as stresses, strains, pressure, flow velocities and fluid pressures, within the extracellular matrix (ECM). Moreover, in the biphasic theory, the effect of fixed charges is implicitly included in the *apparent* elastic moduli and *apparent Poisson*'s ratio of the tissue, and the fluid pressure is considered to be that which is over and above the *Donnan* osmotic pressure.

In order to better understand the effects of proteoglycan fixed charges on the mechanical and electrochemical behaviours of the tissue and to be able to predict the osmotic pressure and electric fields inside such tissues, it is necessary to use a constitutive model that explicitly takes into account the charged nature of the tissue, and the movement of electrolytes (ions) through the tissue during loading. In recent times, two equivalent sets of mixture theories have been developed by 1) Lai et al. [29], Gu et al. [19] (the triphasic theory) and 2) Huyghe and Janssen [24], and Frijns [15] (the quadriphasic theory), to account for the charged nature of biologic tissues, and movement of ions through these tissues. In this paper, using the triphasic theory (also known as the mechano-electrochemical theory—MEC) developed by Lai et al. [29], we investigate the effects of fixed charge density (FCD) and ion movement on both the mechanical and electrochemical behaviours of articular cartilage under an unconfined compression experimental configuration. Since the environments surrounding the chondrocytes inside the tissue include mechanical (stress, strain, hydraulic pressure etc.) as well as electrochemical fields (ion concentrations, osmotic pressure and electric potential/current), studies such as this are important steps toward understanding the "mechano-signal transduction" mechanism inside the tissue where cells reside (Schneiderman et al. [42], Grodzinsky et al. [16]).

The source of all the electrochemical events derives from the fixed negative charge groups (SO_3^- and COO^-) distributed along the chondroitin sulfate, keratan sulfate and hyaluronan molecules comprising the proteoglycan aggrecans inside the articular cartilage (Frank and Grodzinsky [13, 14], Lai et al. [29], Gu et al. [17, 19], Maroudas [33, 34]). Normally, for the time scale of the electrical events and mechanical deformations considered, these proteoglycans may be assumed to be "immobilized and trapped" inside the ECM, and therefore they are assumed to be firmly attached to the ECM (Maroudas [33, 34], Hascall and Hascall [20], Muir [38], Lai et al. [28]). The density of these fixed charges is commonly known in the literature as the fixed charge density measured in mEq/ml units; in normal articular cartilage, ranges from 0.05 to 0.35 mEq/ml for wet and dry tissue (Maroudas [34]). Together with the surrounding collagen network, these proteoglycan macromolecules form a cohesive, strong, porous-permeable, charged, collagen-proteoglycan solid matrix that normally contains 65–85 % water (Bollet and Nance [4], Mankin and Thrasher [32], Maroudas [34], Torzilli et al. [50]).

By virtue of the electro-neutrality law, there is always a cloud of counter-ions (e. g. Ca^{++}, Na^{+}) and co-ions (e. g. Cl^{-}) dissolved in the interstitial water surrounding the fixed charges in the porous-permeable ECM. Unlike the fixed charges, these ions are free to move with the interstitial fluid by convection and through the interstitial fluid by diffusion. These fixed charges produce profound effects not only on tissue hydration but also control of fluid and ion transport through the interstitium (Lai et al. [29], Gu et al. [17–19], Maroudas [34], Muir [38]), but also on a broad spectrum of other observed MEC responses such as streaming potentials.

The recent development of the constitutive modelling of the charged-hydrated articular cartilage enabled a comprehensive investigation of the MEC fields within the tissue. The objective of this paper is to predict the MEC responses of a thin circular wafer of tissue sample under a Heaviside step displacement loading in the unconfined compression configuration using the triphasic theory (Lai et al. [29]), and its finite element formulation (Sun et al. [47]). Since the mechanical and electrochemical events are nonlinearly and inextricably coupled inside the tissue, it is of interest to determine exactly how the charged property of the tissue influences the tissue responses and apparent tissue properties. Specifically, we seek answers to the following fundamental questions: 1) What are the mechanical, chemical and electrical responses of a charged articular cartilage, and how are they different from those of an uncharged tissue? 2) How does the electrical potential inside the tissue depend on the FCD and intrinsic elastic properties of the ECM? 3) How does the FCD affect the apparent, measurable mechanical and electrical properties of articular cartilage such as apparent *Young*'s modulus and apparent *Poisson*'s ratio?

2 Triphasic mixture theory

To simulate a step displacement application onto the tissue in the unconfined compression configuration (Figure 1), we formulate the problem as follows: a thin circular-cylindrical cartilage sample with thickness $= h$ and radius $= a$ is bathed in a uni-univalent salt (e. g. NaCl) solution with a concentration c^* and placed between two rigid, frictionless and impermeable loading platens.

The tissue is squeezed via a *Heaviside* compression with strain ε_o given by $\varepsilon_o \, \mathrm{H}(t)$ that is applied along the negative axis z from the top platen while the bottom platen is fixed.

2.1 Governing equations

For the triphasic mixture modelling of articular cartilage, four basic governing equations for the four basic unknowns (solid displacement $\mathbf{u}^s$, modified chemical potential for water ε^w, modified electrochemical potentials for cation and anions ε^+ and ε^-) are given by (Sun et al. [47])

$$\nabla \cdot \mathbf{v}^s + \nabla \cdot \mathbf{J}^w = 0\,, \tag{1}$$

$$\nabla \cdot \boldsymbol{\sigma} = 0\,, \tag{2}$$

$$\frac{\partial(\phi^w c^+)}{\partial t} + \nabla \cdot \mathbf{J}^+ + \nabla \cdot (\phi^w c^+ \mathbf{v}^s) = 0\,, \tag{3}$$

$$\frac{\partial(\phi^w c^-)}{\partial t} + \nabla \cdot \mathbf{J}^- + \nabla \cdot (\phi^w c^- \mathbf{v}^s) = 0\,. \tag{4}$$

These four governing equations are mixture continuity equation, mixture momentum equation, cation continuity equation and anion continuity equation, respectively, where $\mathbf{v}^s$ is the velocity of the solid matrix (for infinitesimal deformation case, $\mathbf{v}^s = \partial \mathbf{u}^s / \partial t$), c^+ and c^- are cation and anion concentrations respectively, ϕ^w is the porosity of the tissue, and $\boldsymbol{\sigma}$ is the mixture stress tensor. $\mathbf{J}^w$, $\mathbf{J}^+$, and $\mathbf{J}^-$ are the water flux, cation flux, and anion flux, respectively (see also the Appendix).

In this formulation, the constitutive equations for the triphasic mixture ($\boldsymbol{\sigma}$, ε^w, ε^+, and ε^-) are given by

$$\boldsymbol{\sigma} = -p\mathbf{I} + \lambda_s \, e\, \mathbf{I} + 2\,\mu_s \, \mathbf{E}\,, \tag{5}$$

$$\varepsilon^w = \frac{p}{RT} - \phi\,(c^+ + c^-) + \frac{B_w}{RT}\, e\,, \tag{6}$$

$$\varepsilon^+ = \gamma_+ \, c^+ \, \exp\left(\frac{F_c \psi}{RT}\right)\,, \tag{7}$$

$$\varepsilon^- = \gamma_- \, c^- \, \exp\left(-\frac{F_c \psi}{RT}\right)\,, \tag{8}$$

where $\mathbf{E}$ is the solid matrix strain tensor, p is the fluid pressure, e is the solid matrix dilatation (trace($\mathbf{E}$)), λ_s and μ_s are *Lamé* constants of the solid matrix. The symbol R is the universal gas constant, T is the absolute temperature, F_c is the *Faraday* constant, γ_+ and γ_- are the activity coefficients of cation and anion respectively, ϕ is the osmotic coefficient, B_w is a coupling coefficient. Due to the conservation of the fixed charges on the solid matrix, a relationship exists between the FCD and the dilatation e:

$$c^F = c_0^F/(1 + e/\phi_0^w)\,, \tag{9}$$

where c^F is the fixed charge density (the subscript 0 indicates the reference state), and ϕ_0^w is the porosity of the tissue at the reference state. At any point of the tissue, the electroneutrality condition is satisfied:

$$c^+ = c^- + c^F\,. \tag{10}$$

The water flux $\mathbf{J}^w$, cation flux $\mathbf{J}^+$ and anion flux $\mathbf{J}^-$ can be expressed from their corresponding momentum equations respectively as (Sun et al. [47])

$$\mathbf{J}^w = -\frac{RT}{\alpha}\phi^w\left(\nabla\varepsilon^w + \frac{c^+}{\varepsilon^+}\nabla\varepsilon^+ + \frac{c^-}{\varepsilon^-}\nabla\varepsilon^-\right), \tag{11}$$

$$\begin{aligned}\mathbf{J}^+ = &-\frac{RT\,\phi^w c^+}{\alpha}\nabla\varepsilon^w - \left[\frac{\phi^w c^+ D^+}{\varepsilon^+} + \frac{RT\,\phi^w (c^+)^2}{\alpha\,\varepsilon^+}\right]\nabla\varepsilon^+ - \\ &-\frac{RT\,\phi^w c^+ c^-}{\alpha\,\varepsilon^-}\nabla\varepsilon^-\,,\end{aligned} \tag{12}$$

$$\begin{aligned}\mathbf{J}^- = &-\frac{RT\,\phi^w c^-}{\alpha}\nabla\varepsilon^w - \frac{RT\,\phi^w c^+ c^-}{\alpha\,\varepsilon^+}\nabla\varepsilon^+ - \\ &-\left[\frac{\phi^w c^- D^-}{\varepsilon^-} + \frac{RT\,\phi^w (c^-)^2}{\alpha\,\varepsilon^-}\right]\nabla\varepsilon^-\,,\end{aligned} \tag{13}$$

where α is the drag coefficient between the solid and water phases.

2.2 Initial conditions

Initially, the tissue is to be in a free (i. e. no externally applied load) swollen state at equilibrium in the external NaCl bathing solution with concentration c^*. The modified chemical potential for water (ε^w) and the electrochemical potential for cation and anions (ε^+ and ε^-) are all homogeneous within the tissue, which are identical to their counterparts in the external solution respectively. Choosing this free swollen state as the reference configuration, the initial solid displacement is zero, and the tissue has an initial isotropic tensile pre-stress p_0 from the osmotic pressure (Lai et al. [29], Gu et al. [17]).

2.3 Boundary conditions

$$At\ r = 0\,,\ u_r = 0\,,\ J_r^w = 0\,,\ J_r^+ = 0\,,\ J_r^- = 0\,, \tag{14}$$

$$At\ r = a\,,\ \sigma_{rr} = 0\,,\ \varepsilon^w = \varepsilon^{w*}\,,\ \varepsilon^+ = \varepsilon^{+*}\,,\ \varepsilon^- = \varepsilon^{-*}\,, \tag{15}$$

$$At\ z = 0\,,\ u_z = 0\,,\ J_z^w = 0\,,\ J_z^+ = 0\,,\ J_z^- = 0\,, \tag{16}$$

$$At\ z = h\,,\ u_z = \varepsilon_0 H(t)\,,\ J_z^w = 0\,,\ J_z^+ = 0\,,\ J_z^- = 0\,, \tag{17}$$

where the subscripts r and z indicate the r and z components, respectively (Figure 1). During the whole process, ε^{w*}, ε^{+*} and ε^{-*} in the external solution are kept constant (e. g. no additional salt is added).

3 Solution method

The finite element method was used to obtain the tissue response with time. A mesh of 30×2 ($r \times z$) uniform quadrilateral elements was used in the analysis. The following numerical parameters were used in the calculations: the tissue radius $a = 1.5$ mm and the height $h = 1.0$ mm, cation diffusivity $D^+ = 0.5 \times 10^{-9}$ m^2/s, anion diffusivity $D^- = 0.8 \times 10^{-9}$ m^2/s, the external concentration $c^* = 0.15$ M, the shear modulus of the solid matrix $\mu_s = 0.15$ MPa, the porosity $= 0.75$ and the drag coefficient $\alpha = 0.7 \times 10^{15}$ Ns/m^4. Two values of initial FCD (i. e. prior to compression) were used in the calculation: zero and 0.2 mEq/ml. Finally, the applied strain was $\varepsilon_o = -0.1$. In the calculation, the time step was set as 1.0 s. The details of the triphasic finite element formulation have been given by Sun and co-workers [47].

4 Finite element numerical results

Under the assumption of the frictionless-impermeable loading platens, the lateral movement of the tissue is independent of the depth. The time history of the normalized lateral edge displacement is shown in Figure 2. The tissue's instantaneous lateral expansion is described by an apparent *Poisson*'s ratio of 0.5. Under axial compression, a radial tensile stress is generated in the solid matrix instantaneously, after that the solid matrix recoils laterally until an equilibrium is reached. The lateral strain for a charged tissue recoils at a slower rate (flatter slope) than that for an uncharged tissue for all intrinsic *Poisson*'s ratio ν_s of the solid matrix (Figure 2). For an uncharged tissue, the tissue's equilibrium lateral expansion is controlled entirely by the intrinsic *Poisson*'s ratio of the solid matrix (Armstrong et al. [1]), i. e. the ratio of the equilibrium lateral expansion to the applied axial strain defines the intrinsic *Poisson*'s ratio ν_s. In a charged tissue, the larger equilibrium lateral expansion (i. e. less lateral recoil) is due to the increase in osmotic pressure in the swollen state, therefore, resulting in a larger apparent equilibrium *Poisson*'s ratio (e. g. compare the two $\nu_s = 0$ cases in Figure 2 for the charged and uncharged tissue). Only when the intrinsic *Poisson*'s ratio is equal to 0.5, both the charged and the uncharged tissues behave like incompressible elastic materials without the lateral recoil.

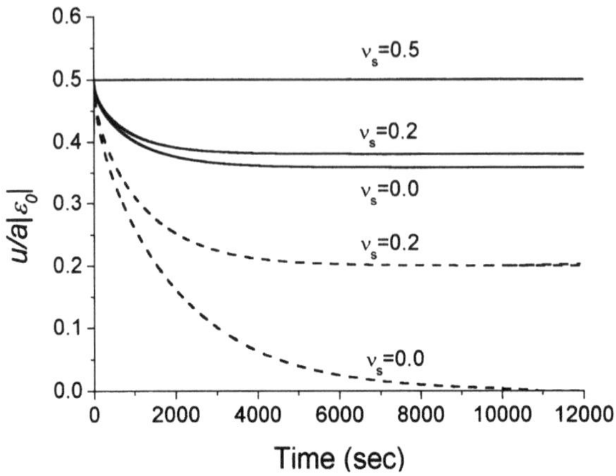

Fig. 2. Comparison of the histories of the normalized radial displacement at the lateral edge of the cylindrical explant after an imposed step-vertical displacement. The solid line is the case with initial FCD of 0.2 mEq/ml (i. e. the triphasic case); the dash line is the case with FCD = 0 (i. e. the biphasic case); the magnitude of the suddenly applied axial strain is $\varepsilon_0 = -0.1$. The following numerical parameters were used in the calculations: the tissue radius $a = 1.5$ mm and the height $h = 1.0$ mm, cation diffusivity $D^+ = 0.5 \times 10^{-9}$ m^2/s, anion diffusivity $D^- = 0.8 \times 1^{-9}$ m^2/s, the external concentration $c^* = 0.15$ M, the shear modulus of the solid matrix $\mu_s = 0.15$ MPa, the porosity = 0.75 and the drag coefficient $\alpha = 0.7 \times 10^{15}$ Ns/m^4. Due to the osmotic (swelling) pressure, the lateral strain for a charged tissue recoils at a slower rate (flatter slope) than that for an uncharged tissue for all intrinsic *Poisson*'s ratio ν_s of the solid matrix.

The histories of the normalized axial compressive load (Figure 3) for the charged and uncharged tissues are compared with the same intrinsic *Young*'s modulus E_s but three different values of the intrinsic *Poisson*'s ratio (0, 0.2, and 0.5). The compressive load is supported by three sources: the intrinsic stiffness of the ECM, the hydraulic pressure and the osmotic pressure. At the initial instant, i. e. at $t = 0^+$, both the charged and the uncharged tissues behave as an incompressible elastic medium with an instantaneous apparent *Young*'s modulus being equal to $3\,\mu_s = E_s$, and the load on the platen is $F(0^+) = 3\,\mu_s\,\pi\,a^2\,\varepsilon(0^+) = \frac{3\,E_s}{2\,(1+v_s)}\,\pi\,a^2\,\varepsilon(0^+)$. Subsequent to that, i. e. for $t > 0^+$, the tissue recoils radially inward with concomitant exudation of fluid across the lateral surface at $\bar{r} = 1$, and depressurization of the hydraulic pressure inside the tissue $0 \leq \bar{r} < 1$. The load decreases with time for all values of the *Poisson*'s ratio $\nu_s < 0.5$, until equilibrium is reached, at which time, the load is determined by the intrinsic *Young*'s modulus of the ECM, and the increase of osmotic pressure due to a decrease of tissue volume (i. e. an increase in tissue FCD—see (9)) from its initial swollen state. For the uncharged tissue, as expected, the equilibrium load is only determined by the intrinsic *Young*'s modulus E_s and independent of the intrinsic *Poisson*'s ratio. This result indicates that the apparent *Young*'s modulus of a charged tissue is larger than that of the uncharged tissues by virtue of the increase in osmotic pressure due to an increase in the FCD. Furthermore, this apparent *Young*'s modulus also depends on the intrinsic *Poisson*'s ratio, as manifested by the different equilibrium value of F(∞)—see Figure 3. For a tissue with $\nu_s = 0.5$, no recoiling process takes place and the load remains a constant at all time since no fluid flow nor decrease of tissue volume take place.

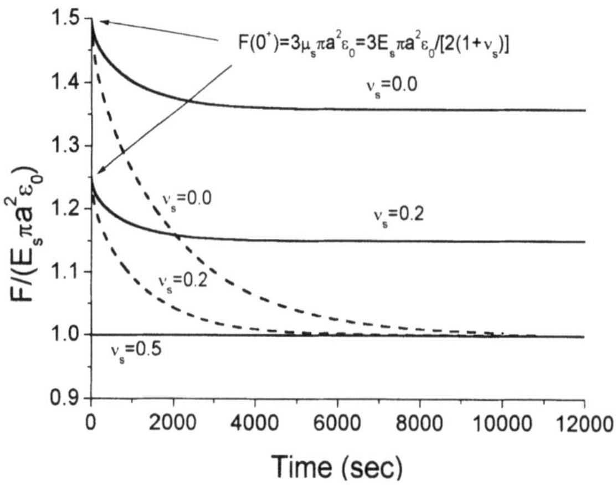

Fig. 3. Comparison of the histories of the total compressive load on the loading platens of charged and uncharged tissues of same intrinsic *Young*'s modulus but with three *Poisson*'s ratios (0.0, 0.2, and 0.5) following the sudden step application of compression. The solid line is the case with initial FCD of 0.2 mEq/ml (i. e. the triphasic case); the dash line is the case with FCD = 0 (i. e. the biphasic case). Other parameter values are the same as in Figure 2.

The distributions of fluid pressure (relative to the free swollen state) in the radial direction at four different times (0 s, 100 s, 1,000 s, and 10,000 s) for a charged tissue and an uncharged tissue with the intrinsic solid matrix *Young*'s modulus $E_s = 0.36$ MPa and *Poisson*'s ratio $\nu_s = 0.2$ are shown in Figure 4. At $t = 0^+$, both the charged and the uncharged tissues develop an instantaneous pressure (above the free-swollen state, $p - p_0 = \mu_s\,\varepsilon(0^+) = \frac{E_s}{2\,(1+\nu_s)}\,\varepsilon(0^+)$). For $t > 0^+$, as the tensile stress in the solid matrix pulls the tissue back toward $\bar{r} = 0$, the interstitial water must be exuded from the tissue through its lateral boundary at $\bar{r} = 1$. A negative pressure gradient is generated inside the tissue along the radial direction. Initially, a compression boundary layer is established near the lateral edge, which with time, spreads throughout the entire tissue until the boundary layer dissipates. This is similar to the uncharged case (Armstrong et al. [1]). In the uncharged tissue, the pressure eventually vanishes, while in the charged tissue, a uniform pres-

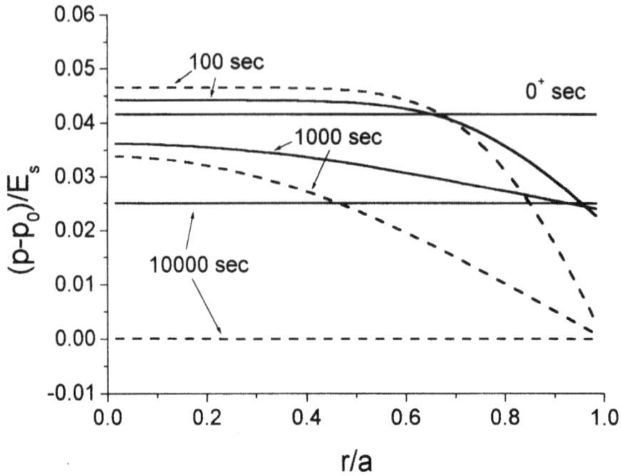

Fig. 4. The radial distribution of the fluid pressure inside the explant after the sudden application of loading. The solid line is for the triphasic case with initial FCD of 0.2 mEq/ml; the dash line is for the biphasic case (initial FCD = 0); the *Poisson*'s ratio $\nu_s = 0.2$. Here p_0 is the pre-stress at the free swollen state due to the FCD. Other parameter values are the same as in Figure 2.

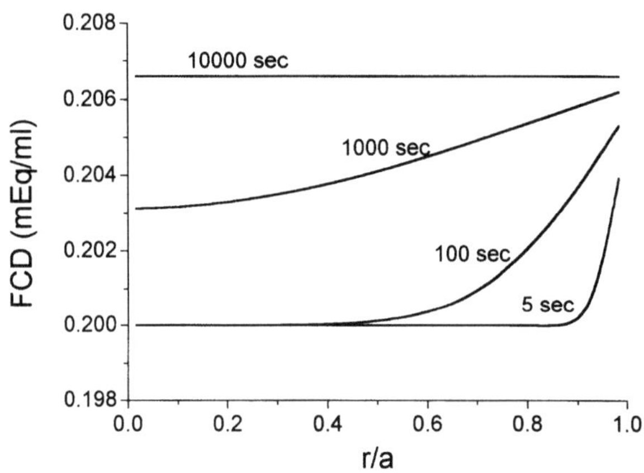

Fig. 5. The radial distribution of FCD inside the tissue after the application of a suddenly applied vertical load. The initial FCD is 0.2 mEq/ml, and the *Poisson*'s ratio $\nu_s = 0.2$. Other parameters are the same as in Figure 2. Vertical compression would eventually increase the FCD inside the tissue explant in an uniform manner (as predicted in this homogeneous model).

sure persists at equilibrium (i. e. the osmotic pressure). In the charged tissue, the equilibrium increase in the osmotic pressure (t = 10,000 s) results from a compression-induced increase of FCD; the transient local variation of the FCD under unconfined compression is shown in Figure 5.

For a charged tissue, because of the existence of the gradients of cation and anion concentrations as well as the convective ion transport induced by the interstitial fluid flow, electrical potential generated within the cartilage tissue can be quite complicated, and is shown in Figure 6. With the electrical potential in the external bathing solution chosen as the zero reference, Figure 6a shows the history of the electrical potential at the lateral edge inside the tissue. Across the lateral interface boundary, the electrical potential

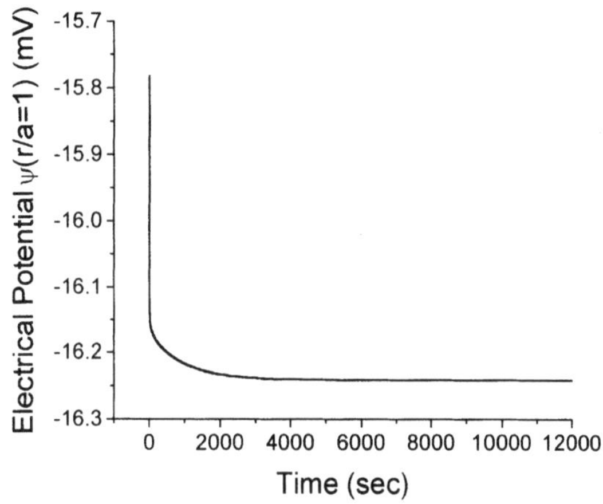

(a)

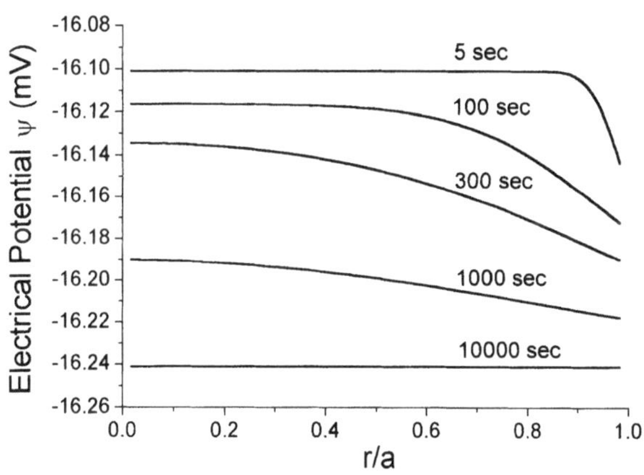

(b)

Fig. 6. (a) The electrical potential history at the lateral edge inside the tissue. The external solution is taken as the zero (reference) electrical potential. At any given time instant, the electrical potential is not continuous across the lateral interface boundary between the external solution and the tissue. (b) The radial distribution of the electrical potential inside the explant (relative to the internal lateral edge of the tissue) after the application of a step displacement. The initial FCD is 0.2 mEq/ml and the *Poisson*'s ratio $\nu_s = 0.2$. Other parameters are the same as in Figure 2.

is not continuous, and this jump value across the interface boundary varies with time under loading until the equilibrium is reached. Within the tissue, a radial distribution of the electrical potential *difference* is generated initially but eventually vanishes (Figure 6 b). For example, at 5 s, the potential *difference* between the center of the specimen and the internal lateral edge ($\bar{r} = 1^-$) is 0.05 mV; this *difference* becomes essentially zero at 10,000 s.

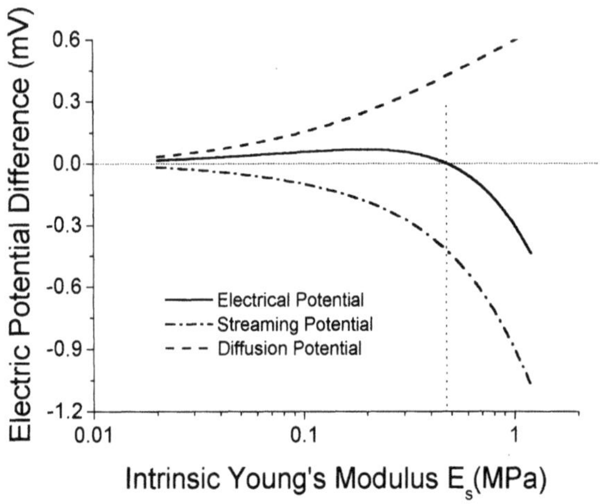

Fig. 7. The electrical potential difference inside the tissue and its streaming potential and diffusion potential components calculated at 10 s vs. the *Young*'s modulus of the solid matrix with the initial FCD of 0.2 mEq/ml and a *Poisson*'s ratio of 0.2. Other parameter values are the same as in Figure 2. The electrical potential difference is the electrical potential at the center of the tissue minus the one at the lateral edge inside the tissue. This electrical potential difference is zero when $E_s = 0.46$ MPa.

The electrical potential results from the competition between the diffusion potential due to ion concentration gradients and a streaming potential due to ion transport (Lai et al. [28]; also refer to Appendix). The electrical potential difference between the center of the tissue and the internal lateral edge ($\bar{r} = 1^-$), and its streaming and diffusion potential components, calculated at 10s vs. the intrinsic *Young*'s modulus of the charged solid matrix (with FCD = 0.2 mEq/ml and an intrinsic *Poisson*'s ratio of $\nu_s = 0.2$) are shown in Figure 7. From this figure, we see that the electrical potential can change polarity (sign), i. e. for a softer tissue with a *Young*'s modulus less than 0.46 MPa, the resultant electrical potential has the same polarity as that of the diffusion potential, whereas for a stiffer tissue, it has the same polarity as that of the streaming potential.

From an unconfined compression test of articular cartilage, the equilibrium apparent *Young*'s modulus and the equilibrium apparent *Poisson*'s ratio as well as the transient electric potential can be used to characterize the mechano-electrochemical properties of the tissue. As shown previously, they depend on the FCD of the tissue, which was implicitly included in a biphasic or poroelastic theory. From these theories, one cannot explicitly determine the effects of the FCD during tissue deformation, nor use these FCD effects explicitly as a measure of the tissue proteoglycan content from a mechanical experiment. The *apparent Poisson*'s ratio, which has been measured optically in the unconfined compression experiment (Jurvelin et al. [23]) is always larger than the *intrinsic Poisson*'s ratio (ratio of lateral strain to axial strain bathed in a hypertonic solution); this apparent *Poission*'s ratio also depends on the FCD of the tissue, see Figure 8. The measurable *apparent Young*'s modulus is also greater than the *intrinsic Young*'s modulus and varies with the FCD, Figure 9.

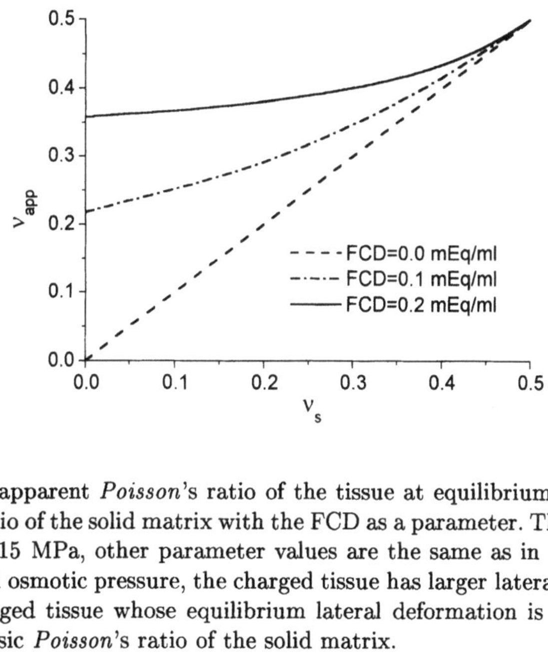

Fig. 8. The apparent *Poisson*'s ratio of the tissue at equilibrium vs. the intrinsic *Poisson*'s ratio of the solid matrix with the FCD as a parameter. The shear modulus is fixed as 0.15 MPa, other parameter values are the same as in Figure 2. Due to the increased osmotic pressure, the charged tissue has larger lateral expansion than the non-charged tissue whose equilibrium lateral deformation is determined only by the intrinsic *Poisson*'s ratio of the solid matrix.

The contribution of fixed charges on the apparent *Young*'s modulus indicates the significance of osmotic pressure in load support by articular cartilage. Table 1 shows the contribution of osmotic pressure to the equilibrium load support provided by a charged tissue with various initial FCD (c_o^F), and with intrinsic *Poisson*'s ratio = 0.2, porosity = 0.75 and three values for the intrinsic *Young* modulus E_s; the bathing solution is 0.15 M NaCl. The applied axial compressive strain is 10 % measured from the initial swollen state in a 0.15 M NaCl bathing solution . We see that for a tissue with a given initial FCD of 0.1 mEq/ml and intrinsic *Young*'s modulus of 360 kPa, the osmotic

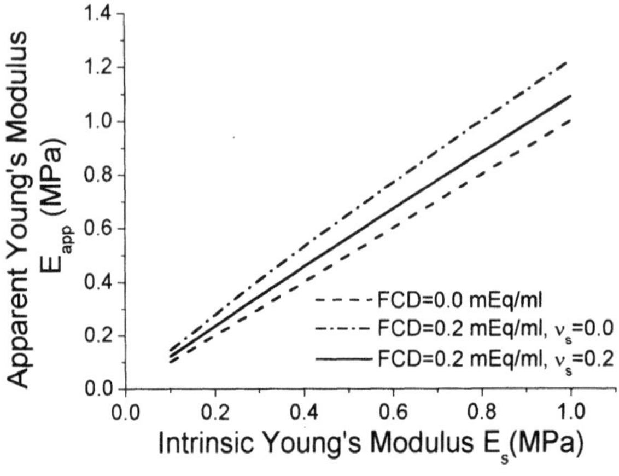

Fig. 9. The apparent *Young*'s modulus of the charged and uncharged tissues at equilibrium vs. the intrinsic *Young*'s modulus of the solid matrix with the same intrinsic *Poisson*'s ratio. For the non-charged case, the apparent *Young*'s modulus is the same as the intrinsic *Young*'s modulus of the solid matrix and is independent of the intrinsic *Poisson*'s ratio of the solid matrix. Due to the increased osmotic pressure, the charged tissue has larger apparent *Young*'s modulus than the uncharged tissue, and the enhancement of the apparent *Young*'s modulus is more remarkable at the low value of the intrinsic *Poisson*'s ratio of the solid matrix.

pressure at the equilibrium is 4.63 kPa, and a total load support of 38.8 kPa. That is, about 12 % of the support at equilibrium is derived from the osmotic pressure. This may be compared with an 18 % support from osmotic pressure in the confined compression case obtained by Lai and co-workers [28]. Under a 10 % surface-to-surface compression, the equilibrium dilatation for the confined compression case is −10 % whereas for the unconfined compression case, it is −4.1 % (Table 1). Therefore there is a smaller osmotic pressure in the unconfined compression case.

		Intrinsic *Young*'s modulus of the solid matrix								
		360 [kPa]			720 [kPa]			1080 [kPa]		
		$-\sigma_{zz}$ [kPa]	$p-p_0$ [kPa]	e [%]	$-\sigma_{zz}$ [kPa]	$p-p_0$ [kPa]	e [%]	$-\sigma_{zz}$ [kPa]	$p-p_0$ [kPa]	e [%]
FCD	0.05	37.0	1.60	−5.4	73.0	1.70	−5.7	109.1	1.74	−5.8
	0.10	38.8	4.63	−4.1	75.3	5.54	−4.9	111.6	5.94	−5.2
	0.15	39.5	7.21	−3.1	77.8	9.61	−4.1	114.5	10.84	−4.6
	0.20	41.4	9.01	−2.4	79.8	13.02	−3.4	117.2	15.33	−4.0

Table 1. The equilibrium mixture stress σ_{zz}, fluid pressure $p-p_0$, and dilatation e under a 10 % unconfined compression. The parameter values are $c^* = 0.15$ M, porosity $= 0.75$, and $\nu_s = 0.2$. The unit for the FCD is mEq/ml.

5 Discussion

We have shown that the fixed charges associated with the proteoglycans in articular cartilage contribute significantly to the deformational and electrical responses of the tissue under unconfined compression. In addition, our results have shown that these fixed charges give rise to significantly new and unanticipated electrochemical behaviours that undoubtedly will have important implications in delineating physiological differences between normal and diseased cartilages. In this paper using the triphasic theory, we have studied the effects of fixed charges on the mechanical and electrochemical behaviours of articular cartilage under unconfined compression. This is a commonly employed explant loading condition for mechano- or electro-mechanical transduction studies (e. g. Sah et al. [39], Kim et al. [26], Grodzinsky et al. [16]). This simple testing configuration has also been frequently used by investigators to experimentally determine the *Young*'s modulus and *Poisson*'s ratio of articular cartilage (Brown and Singerman [5], Jurvelin et al. [23], DiSilvestro et al. [10]). Therefore, it is important to know just exactly how the mechanical and electrical components contribute to the measured properties or how the mechanical and electrical fields interact with each other within the tissue. Several new findings need to be emphasized:

First, the apparent mechanical properties of such tissues that can be experimentally measured in an unconfined compression experiment are: the *Young*'s modulus, *Poisson*'s ratio and permeability. Our focus in this study is on how the apparent *Young*'s modulus and apparent *Poisson*'s ratio depend on the fixed charges in articular cartilage. The interested reader should consult the publications by Gu et al. [17–19] for a detailed triphasic description of cartilage permeability. From Figure 8, we see that the *apparent Poisson*'s ratio (e. g. as measured by an indentation method—Mow et al. [37] or by optical method—Jurvelin et al. [23]) is always higher than the *intrinsic Poisson*'s ratio of the ECM. This is due to the swelling effect above the hypertonic reference state that is induced by the fixed charges. Also, from Figure 9, we see that the *apparent Young*'s modulus of a charged tissue is always higher than the *intrinsic Young*'s modulus of the ECM. The measured apparent *Young*'s modulus is a dependent variable, that depends not only on the *intrinsic Young*'s modulus of the solid matrix but also the FCD and the *intrinsic Poisson*'s ratio. Similarly, this dependence holds true for the *apparent Poisson*'s ratio. These new findings are essential in the proper interpretation of all experimental measurements on charged-hydrated tissues such as articular cartilage.

Second, the dependence of these apparent mechanical properties on the FCD indicates the importance of the osmotic pressure in the mechanical load support provided by articular cartilage. More important, we can now quantitatively determine the osmotic pressure effects on cartilage's apparent compressive properties. Indeed, Mow et al. [35] provided the first theoretical result demonstrating that under *confined* compression, *Donnan* osmotic pressure provides approximately 50 % compressive aggregate modulus of articular cartilage. In our present analysis of *unconfined* compression, this osmotic pressure can constitute between 13–22 % of the total load support at the equilibrium, depending on the intrinsic *Young*'s modulus and the FCD of the solid matrix. Thus after the introduction of physicochemical concepts to describe articular cartilage more than 30 years ago (Maroudas [33, 34], Katchalsky and Curran [25]), we now have a definite idea on how much and exactly how the FCD in articular cartilage will function in providing the load support, and in modulating fluid and ion transport for the tissue (Gu et al. [17–19], Mow et al. [35], Lai et al. [28]).

Third, the FCD generates a variety of electrical phenomena that might be important in the understanding the biology of the tissue for the mechano-signal transduction process in the regulation of cartilage biosynthesis (Kim et al. [26], Buschmann et al. [6], Grodzinsky et al. [16], Chen et al. [7]). The electrical potential inside the tissue comes from two competing sources (Lai et al. [28]): diffusion potential (due to non-uniform distributions of FCD—by deformation or natural) and streaming potential (due to ion convection by interstitial fluid flow). Both potentials depend not only on the values of the FCD and its gradients, but also on the intrinsic mechanical properties such as the intrinsic *Young*'s modulus and intrinsic *Poisson*'s ratio. Within physiological range of material parameters, the polarity of the potential may be different depending on the values of the intrinsic material properties. For softer tissue (such as in osteoarthritis), the diffusion potential dominates, while the streaming potential dominates for stiffer tissue (normal cartilage). This phenomenon was first pointed out by Lai and co-workers [28] in two other commonly used experimental configurations, i. e. confined compression and direct permeation. Indeed, our preliminary calculations (not shown) including strain-dependent intrinsic *Young*'s modulus (Zhu et al. [52]) demonstrate that the electrical potential can even change its polarity during mechanical loading, given certain material parameters within the physiological range. The dependency of the electrical potential on the intrinsic *Young*'s modulus and *Poisson*'s ratio, and the FCD are also reflected in the possible experimental measures of the electrochemical potential using Ag/AgCl electrodes. This suggests that the measures of electrochemical potential taken across cartilage specimens and the calculation of electrical potentials within the tissue may provide an additional assay for characterizing the intrinsic properties of normal and pathological articular cartilage.

In comparison with the previous biphasic analyses of unconfined compression studies (Armstrong et al. [1], Spilker et al. [45, 46], Cohen et al. [8]), it is of interest to note that the fixed charges do not affect the instantaneous strain field inside the tissue relative to the initial equilibrium state such that the instantaneous load support is determined entirely by the intrinsic shear modulus of the solid matrix. However, there is a sudden increase in the water chemical potential inside the tissue due to pressurization of fluid over and above the initial swollen state, which increases the hydraulic fluid pressure.

Certain theoretical limitations should be discussed regarding these predictions. First, the triphasic theory used here only considers an isotropic, linearly elastic solid matrix undergoing infinitesimal strain; the effects of the well-known strain-dependent permeability have not been examined either. Future studies that include tension-compression nonlinearity (Soltz and Ateshian [43], Huang et al. [21, 22]) as well as finite deformation with strain-dependent permeability (e. g. Ateshian et al. [2]) will be helpful in providing more and generally valid insightful knowledge of cartilage response in unconfined compression. Second, cartilage tissue has been assumed to be homogeneous, ignoring the depth-dependent distribution of elastic properties as well as the fixed charges (Maroudas [34], Schinagl et al. [40, 41], Wang et al. [51]). With our current finite element capabilities (Sun et al. [47]), the influence of these spatial distributions of material properties can be easily incorporated in our future studies. However, the simplifications made in this study should not affect the major conclusions or the general trends reported in this paper. Indeed, ongoing experimental and computational studies are currently underway along this direction to experimentally validate cartilage mechano-electrochemical phenomena (Sun et al. [48, 49]).

In summary, the current triphasic analyses of articular cartilage under unconfined compression provide new and insightful information regarding the mechano-eletrochemical behaviours of cartilage under unconfined compression loading. The findings of dependence of the apparent *Young*'s modulus and *Poisson*'s ratio on the FCD in the tissue will have significant implications in interpreting experimental measurements from unconfined compression tests of articular cartilage. The demonstrated contribution of osmotic pressure in cartilage load support in a quantitative manner, and the nature of electrical potential within the charged-hydrated cartilage will have a significant impact in the basic science of cartilage mechanics and mechano-signal transduction in articular cartilage biosynthesis.

Acknowledgements

This study was supported by NIH grants AR-41913, AR-45832 and a Biomedical Engineering Research Grant from the Whitaker Foundation (97-0086).

Appendix

The electrical current density and electrical potential gradient inside the tissue

Define the water volume flux $\mathbf{J}^w$, cation molar flux $\mathbf{J}^+$ and anion molar flux $\mathbf{J}^-$ as,

$$\mathbf{J}^w = \phi^w \left(\mathbf{v}^w - \mathbf{v}^s\right), \tag{18}$$

$$\mathbf{J}^+ = \phi^w c^+ (\mathbf{v}^+ - \mathbf{v}^s)\,, \tag{19}$$

and

$$\mathbf{J}^- = \phi^w c^- (\mathbf{v}^- - \mathbf{v}^s)\,. \tag{20}$$

The electrical current density inside the tissue is

$$\mathbf{I}_e = \mathbf{J}^+ - \mathbf{J}^- = \phi^w \left[c^+(\mathbf{v}^+ - \mathbf{v}^s) - c^-(\mathbf{v}^- - \mathbf{v}^s)\right]. \tag{21}$$

Using the triphasic momentum equations (Lai et al. [29], Sun et al. [47]), the electrical current density can be expressed with the modified electrochemical potentials as,

$$\mathbf{I}_e = \left(-\frac{\phi^w c^+ D^+}{\varepsilon^+} \nabla \varepsilon^+ + \frac{\phi^w c^- D^-}{\varepsilon^-} \nabla \varepsilon^- \right) + c^F \mathbf{J}^w . \tag{22}$$

By considering the triphasic constitutive equations, the electrical current density can be expressed as

$$\mathbf{I}_e = \phi^w \left(D^- \nabla c^- - D^+ \nabla c^+\right) - \frac{F_c}{RT} \phi^w (c^+ D^+ + c^- D^-) \nabla \psi + c^F \mathbf{J}^w . \tag{23}$$

In the above equation, the three terms are the diffusion current due to the ion concentration gradients, conductive current due to the electrical potential gradient and the streaming current due to the water convection, respectively. Under no current condition, $\mathbf{I}_e = \mathbf{0}$, the electrical potential gradient can be expressed as,

$$\nabla \psi = \frac{RT}{F_c} \frac{D^- \nabla c^- - D^+ \nabla c^+}{c^+ D^+ + c^- D^-} + \frac{RT}{F_c} \frac{c^F}{\phi^w (c^+ D^+ + c^- D^-)} \mathbf{J}^w , \tag{24}$$

that is, the electrical potential difference inside the tissue consists of two components, the diffusion potential due to the ion concentration gradients and streaming potential due to the water convection. This is consistent with the literature (Lai et al. [28]).

References

1. Armstrong, C. G., Lai, W. M., Mow, V. C.: An analysis of the unconfined compression of articular cartilage. *J. Biomech. Eng.* **106** (1984), 165–173.
2. Ateshian, G. A., Warden, W. H., Kim, J. J., Grelsamer, R. P., Mow, V. C.: Finite deformation biphasic material properties of bovine articular cartilage from confined compression experiments. *J. Biomechanics* **30** (1997), 1157–1164.
3. Bachrach, N. M., Mow, V. C., Guilak, F.: Incompressibility of the solid matrix of articular cartilage under high hydrostatic pressures. *J. Biomechanics* **31** (1998), 445–451.

4. Bollet, A. J., Nance, J. L.: Biochemical findings in normal and osteoarthritic articular cartilage: II-Chondroitin sulfate concentration and chain length, water and ash contents. *J. Clin. Invest.* **45** (1966), 1170–1177.
5. Brown, T. D., Singerman, R. J.: Experimental determination of the linear biphasic constitutive coefficients of human fetal proximal femoral chondroepiphysis. *J. Biomech.* **19** (1986), 597–605.
6. Buschmann, M. D., Gluzband, Y. A., Grodzinsky, A. J., Hunziker, E. B.: Mechanical compression modulates matrix biosynthesis in chondrocyte/agarose culture. *J. Cell. Sci.* **108** (1995), 1497–1508.
7. Chen, A. C., Nguyen, T. T., Sah, R. L.: Streaming potential during the confined compression creep test of normal and proteoglycan-depleted cartilage. *Ann. Biomed. Eng.* **25** (1999), 269–277.
8. Cohen, B., Lai, W. M., Mow, V. C.: A transversely isotropic biphasic model for unconfined compression of growth plate and chondroepiphysis. *J. Biomech. Eng.* **120** (1998), 491–496.
9. Curnier, A., He, Q-C., Zysset, P.: Conewise linear elastic materials. *J. Elasticity* **37** (1995), 1–38.
10. DiSilvestro, M. R., Zhu, Q., Wong, M., Jurvelin, J. S., Suh, J. K.: Biphasic poroviscoelastic simulation of the unconfined compression of articular cartilage: I–Simultaneous prediction of reaction force and lateral displacement. *J. Biomech. Eng.* **123** (2001), 191–197.
11. DiSilvestro, M. R., Zhu, Q., Suh, J. K.: Biphasic poroviscoelastic simulation of the unconfined compression of articular cartilage: II–Effect of variable strain rates. *J. Biomech. Eng.* **123** (2001), 198–200.
12. Fortin, M., Soulhat, J., Shirazi-Adl, A., Hunziker, E. B., Buschmann, M. D.: Unconfined compression of articular cartilage: nonlinear behavior and comparison with a fibril-reinforced biphasic model. *J. Biomech. Eng.* **122** (2000), 189–195.
13. Frank, E. H., Grodzinsky, A. J.: Cartilage electromechanics – I. Electrokinetic transduction and the effects of pH and ionic strength. *J. Biomechanics* **20** (1987), 615–627.
14. Frank, E. H., Grodzinsky, A. J.: Cartilage electromechanics – II. A continuum model of cartilage electrokinetics and correlations with experiments. *J. Biomechanics* **20** (1987), 629–639.
15. Frijns, A. J. H.: *A four-component mixture theory applied to cartilaginous tissues: Numerical modelling and experiments*. Ph. D. Thesis, Eindhoven University of Technology 2000.
16. Grodzinsky, A. J., Frank, E. H., Kim, Y. J., Buschmann, M. D.: The role of specific macromolecules in cell-matrix interactions and in matrix function: Physicochemical and mechanical mediators of chondrocyte biosynthesis. In: *Extracellular Matrix*, Harwood Academic Publishers, Melbourne 1996, pp. 310–334.
17. Gu, W. Y., Lai, W. M., Mow, V. C.: Transport of fluid and ions through a porous-permeable charged-hydrated tissue, and streaming potential data on normal bovine articular cartilage. *J. Biomech.* **26** (1993), 709–723.
18. Gu, W. Y., Lai, W. M., Mow, V. C.: A triphasic analysis of negative osmotic flows through charged hydrated soft tissues. *J. Biomech.* **30** (1997), 71–78.
19. Gu, W. Y., Lai, W. M., Mow, V. C.: A mixture theory for charged-hydrated soft tissues containing multi-electrolytes: Passive transport and swelling behaviors. *J. Biomech. Engng.* **120** (1998), 169–180.

20. Hascall, V., Hascall, G.: Proteglycans. In: *Cell Biology of Extracellular Matrix*, Plenum Press, 1983, pp. 39–63.
21. Huang, C. Y., Mow, V. C., Ateshian, G. A.: Role of flow-independent viscoelasticity in the tensile and compressive responses of biphasic articular cartilage. *J. Biomech. Engng.* **123** (2001a), 410–417.
22. Huang, C. Y., Soltz, M. A., Kopacz, M., Mow, V. C., Ateshian, G. A.: Experimental verification of the role of intrinsic matrix viscoelasticity and tension-compression nonlinearity in the biphasic response of cartilage in unconfined compression. *J. Biomech. Engng.* (2001b), in review.
23. Jurvelin, J. S., Buschmann, M. D., Hunziker, E. B.: Optical and mechanical determination of Poisson's ratio of adult bovine humeral articular cartilage. *J. Biomech.* **30** (1997), 235–241.
24. Huyghe, J. M., Janssen, J. D.: Quadriphasic mechanics of swelling in compressible porous media. *Int. J. Eng. Sci.* **35** (1997), 793–802.
25. Katchalsky, A., Curran, P.: *Nonequilibrium Thermodynamics in Biophysics.* 4th ed., Harvard University Press, Cambridge 1975.
26. Kim, Y. J., Sah, R. L., Grodzinsky, A. J., Plaas, A. H., Sandy, J. D.: Mechanical regulation of cartilage biosynthetic behavior: Physical stimulation. *Arch. Biochem. Biophys.* **311** (1994), 1–12.
27. Lai, W. M., Mow, V. C., Roth, V.: Effects of a nonlinear strain dependent permeability and rate of compression on the stress behavior of articular cartilage. *J. Biomech. Engng.* **103** (1981), 61–66.
28. Lai, W. M., Mow, V. C., Sun, D. D., Ateshian, G. A.: On the electric potentials inside a charged soft hydrated biological tissue: Streaming potential vs. diffusion potential. *J. Biomech. Engng* **122** (2000), 336–346.
29. Lai, W. M., Hou, J. S., Mow, V. C.: A triphasic theory for the swelling and deformation behaviors of articular cartilage. *J. Biomech. Eng.* **113** (1991), 245–258.
30. Li, L. P., Buschmann, M. D., Shirazi-Adl, A.: A fibril reinforced nonhomogeneous poroelastic model for articular cartilage: inhomogeneous response in unconfined compression. *J. Biomech.* **33** (2000), 1533–1541.
31. Li, L. P., Soulhat, J., Buschmann, M. D., Shirazi-Adl, A.: Nonlinear analysis of cartilage in unconfined ramp compression using a fibril reinforced poroelastic model. *Clin. Biomech. (Bristol, Avon)* **14** (1999), 673–682.
32. Mankin, H. J., Thrasher, A. Z.: Water content and binding in normal and osteoarthritic human carilage. *J. Bone Jt. Surg* **57A** (1975), 76–79.
33. Maroudas, A.: Physicochemical properties of cartilage in the light of ion-exchange theory. *Biophysical J.* **8** (1968), 575–595.
34. Maroudas A.: Physicochemical properties of articular cartilage. In: *Adult Articular Cartilage*, Pitman Medical, Kent 1979, pp. 215–290.
35. Mow, V. C., Ateshian, G. A., Lai, W. M., Gu, W. Y.: Effects of fixed charges on the stress-relaxation behavior of hydrated soft tissues in a confined compression problem. *Int. J. Solids Structure* **35** (1998), 4945–4962.
36. Mow, V. C., Kuei, S. C., Lai, W. M., Armstrong, C. G.: Biphasic creep and stress relaxation of articular cartilage in compression: theory and experiments. *J. Biomech. Eng.* **102** (1980), 73–84.
37. Mow, V. C., Gibbs, M. C., Lai, W. M., Zhu, W., Athanasiou, K. A.: Biphasic indentation of articular cartilage - Part II. A numerical algorithm and an experimental study. *J. Biomechanics* **22** (1989), 853–861.

38. Muir, H.: Proteoglycans as organizers of the intercellular matrix. *Biochem. Soc. Trans.* **9** (1983), 613–622.
39. Sah, R. L., Doong, J. Y., Grodzinsky, A. J., Plaas, A. H., Sandy, J. D.: Effects of compression on the loss of newly synthesized proteoglycans and proteins from cartilage explants. *Arch. Biochem. Biophys.* **286** (1991), 20–29.
40. Schinagl, R. M., Gurskis, D., Chen, C. C., Sah, R. L-Y.: Depth-dependent confined compression modulus of full-thickness bovine articular cartilage. *J. Orthop. Res.* **15** (1997), 499–506.
41. Schinagl, R. M., Ting, M. K., Price, J. H., Sah, R. L-Y.: Video microscopy to quantitate the inhomogeneous equilibrium strain within articular cartilage during confined compression. *Ann. Biomed. Engng.* **24** (1996), 500–512.
42. Schneiderman, R., Keret, D., Maroudas, A.: Effects of mechanical and osmotic pressure on the rate of glycosaminoglycan synthesis in the human adult femoral head cartilage: An *in vitro* study. *J. Orthop. Res.* **4** (1986), 393–408.
43. Soltz, M. A., Ateshian, G. A.: A Conewise Linear Elasticity mixture model for the analysis of tension-compression nonlinearity in articular cartilage. *J. Biomech. Eng.* **122** (2000), 576–586.
44. Soulhat, J., Buschmann, M. D., Shirazi-Adl, A.: A fibril-network-reinforced biphasic model of cartilage in unconfined compression. *J. Biomech. Eng.* **121** (1999), 340–347.
45. Spilker, R. L., Suh, J. K., Mow, V. C.: A linear biphasic finite element analysis of the unconfined compression of articular cartilage. In: *Advances in Bioengineering*, ASME Winter Annual Meeting 1987, pp. 49–50.
46. Spilker, R. L., Suh, J. K., Mow, V. C.: Effects of friction on the unconfined compressive response of articular cartilage: a finite element analysis. *J. Biomech. Eng.* **112** (1990), 138–146.
47. Sun, D. N., Gu, W. Y., Guo, X. E., Lai, W. M., Mow, V. C.: A mixed finite element formulation of triphasic mechano-electrochemical theory for charged, hydrated biological soft tissues. *Int. J. Num. Meth. Engng.* **45** (1999), 1375–1402.
48. Sun, D. N., Guo, X. E., Lai, W. M., Mow, V. C.: The fixed charge inhomogeneity modulates the mechano-electrochemical behaviors of articular cartilage under compression. (2001), in preparation.
49. Sun, D. N., Guo, X. E., Lai, W. M., Mow, V. C.: The fixed charge inhomogeneity influences the mechano-electrochemical behaviors of articular cartilage under osmotic loading. (2001), in preparation.
50. Torzilli, P. A., Rose, D. E., Dethmers, D. A.: Equilibrium water partition in articular cartilage. *Biorheology* **19** (1982), 519–537.
51. Wang, C. B., Hung, C. T., Mow, V. C.: An analysis of the effects of depth-dependent aggregate modulus on articular cartilage stress-relaxation behavior in compression. *J. Biomech.* **34** (2001), 245–258.
52. Zhu, W., Mow, V. C., Koob, T. J., Eyre, D. R.: Viscoelastic Shear Properties of Articular Cartilage and the Effects of Glycosidase Treatments. *J. Orthop. Res.* **11** (1993), 771–781.

III. Experiments and Numerical Applications

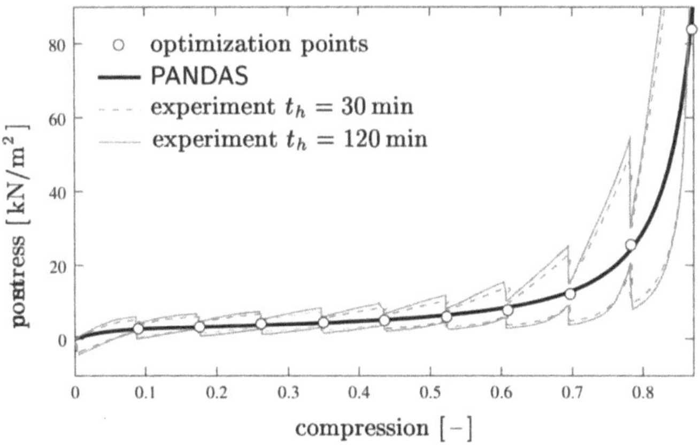

Time adaptive analysis of saturated soil by a discontinuous-Galerkin method

Harald Cramer[1], Rudolf Findeiß[2], and Walter Wunderlich[2]

[1] Universität Rostock, Fachgebiet Baustatik und Baudynamik, 18059 Rostock, Germany
[2] Technische Universität München, Lehrstuhl für Statik, 80333 München, Germany

Abstract. The topic of this presentation is the numerical analysis of saturated soil by the finite element method. As the solution procedure should be extended to describe the flow of the pore fluid through the deforming solid skeleton, a time dependency is introduced into the problem. Therefore, the time coordinate has to be discretized and treated by an appropriate integration scheme. In contrast to adaptive mesh refinement strategies in the spatial domain which are well founded for elasticity and have also been successfully applied to elastic-plastic problems only few papers deal with time adaptive procedures for the quasi-static analysis of consolidation.

Therefore, in this presentation a time discretization dependent on the specific problem is emphasized. For this purpose the time-discontinuous-*Galerkin* method is applied to the differential equations of first order in time. It is based on a variational form permitting jumps in the temporal evolution of the field variables, where the continuity is satisfied in a weak sense. It can be shown that these jumps may then be used to define a natural error indicator for the temporal discretization error. On the other hand, attention is drawn to another error which arises from the numerical integration of the rate equations of plasticity. In this context, an indicator is derived from the residual of the *Kuhn-Tucker* conditions within the time interval.

The numerical examples of a one dimensional consolidation problem and a strip footing on a half space demonstrate the applicability of the method to problems in geomechanics. Both indicators are combined to improve the efficiency of the time stepping scheme.

1 Introduction

The modelling of water saturated soil may be performed in the well founded framework of the Mixture Theory and the Theory of Porous Media. Following these approaches, saturated soil is considered as a two phase material consisting of an elastic-plastic porous solid skeleton and a pore fluid which fills out the voids completely. First theoretical investigations for this binary medium were performed by Biot [2, 3] who extended *Terzaghi*'s consolidation theory to the three-dimensional case. A more general concept based on continuum theories was given by Truesdell and Toupin [20] and Bowen [7] who developed a thermodynamical consistent theory of mixtures taking into account interaction forces between the single constituents. Later, de Boer and Ehlers applied these strategies to the analysis of porous materials like soils using the concept of volume fractions to achieve a homogenization [4, 5, 11].

As the deformation of the solid skeleton leads to the time dependent process of fluid flow the coupled finite element analysis of water saturated soils requires the solution of discretized equations in space and time. Most approaches are based on a semi-discretization by finite elements in space leading to a system of first order differential equations in time. Very often, the *Crank-Nicolson* scheme or a generalized trapezoidal rule is applied to perform the time integration.

In this contribution, the initial boundary-value problem is treated by a combined space-time finite element method. The displacement field as well as the pore water pressures are approximated by an appropriate product form including spatial and temporal shape functions. The space discretization is based on standard isoparametric finite elements. In the time domain a linear or quadratic evolution is assumed for both fields. Following the concept of the time-discontinuous-*Galerkin* procedure, discontinuities are permitted at the discrete time levels. The continuity between the intervals is then satisfied in a weak sense by the applied variational formulations. This solution technique for initial boundary-value problems was used for example by Wiberg and Li in structural dynamics [21] and has been successfully applied to the quasi-static consolidation problem by Cramer et al. [10].

Basic requirements for an accurate and reliable computation are the application of elastic-plastic constitutive laws and non-associated flow rules to describe the inelastic deformation characteristics of soil. This presentation uses the concept of classical rate independent plasticity. As localization phenomena play an important role in the analysis of saturated soils, especially, when the load carrying capacity is reached, the efficiency of the computation is closely related to a spatial and temporal discretization dependent on the given problem [9]. On the one hand, the spatial mesh size in regions of strain localization has to be very fine to capture the evolution of shearbands properly. On the other hand, time domains characterized by a rapid growth of plastic zones or by an initiation of shearband formations have to be treated by smaller time steps to ensure a certain accuracy.

Strategies of spatial mesh refinement on the basis of a posteriori error estimation are well founded for elasticity and have also been applied successfully in elastic-plastic problems. In this context, the work of Babŭska and Miller [1], Zienkiewicz and Zhu [24] and Johnson and Hansbo [16] have to be mentioned. In the framework of a *Cosserat* regularization Perić et al. [17] and Cramer et al. [8] proposed extended indicators including the micropolar variables into the error measures. However, only few works deal with the determination of measures for the temporal discretization error in the case of the quasi-static consolidation problem. Special approaches have been proposed by Ehlers et al. in the framework of implicit *Runge-Kutta* methods and embedded error estimators [14] and by Sloan and Abbo applying a *Thomas-Gladwell* scheme [19]. The present paper focuses on an adaptive control of the step sizes using temporal error indicators derived from the time-discontinuous-*Galerkin* procedure.

In detail, one indicator captures the temporal discretization error by limiting the value of the L_2-norm of the jumps in the displacement and pore water pressure fields, respectively. A second indicator provides a measure for the error in the numerical integration of the constitutive equations and is based on the evaluation of the residual in the *Kuhn-Tucker* conditions over the time interval. As for these heuristical indicators a rigorous mathematical foundation is lacking an equal error distribution over all time steps cannot be ensured in the case of elastic-plastic constitutive laws. However, the numerical examples show that reasonable time discretizations may be achieved which improve the global accuracy with a reduced computational effort. Thus, the time adaptive strategy results in a higher efficiency of the space-time finite element method. Moreover, the computational cost which is necessary for an evaluation of the indicators is very low.

2 Initial boundary-value problem

In the context of the Theory of Porous Media, soil is considered as a two phase medium consisting of an elastic-plastic solid skeleton (solid phase φ^s) and a viscous pore fluid (fluid phase φ^f) that fills out the pores between the single grains as illustrated in Figure 1. This binary material is modelled from a macroscopic point of view applying the concept of volume fractions. Then, the condition of full saturation is given by

$$\mathrm{d}V = \mathrm{d}V^s + \mathrm{d}V^f \,. \tag{1}$$

This allows a formulation in the frame of continuum mechanics without explicitely modelling the complex micro structure of soils. Both constituents are assumed to be incompressible. Using a *Lagrange*-setting for the solid skeleton and an *Euler*-setting for the transport of the fluid flow, the governing equations of the coupled problem are defined by

$$\begin{aligned} \operatorname{div}(\boldsymbol{\sigma} + p\,\mathbf{I}) + \varrho\,\bar{\mathbf{b}} &= 0\,, \\ \operatorname{div}\left(\frac{\partial \mathbf{u}}{\partial t} + \frac{\mathbf{k}}{\mu}\left(\operatorname{grad} p + \varrho^f\,\bar{\mathbf{b}}\right)\right) &= 0\,, \end{aligned} \tag{2}$$

which define the local equilibrium conditions and the balance of mass. In (2) $\boldsymbol{\sigma}$ represents the effective stress tensor acting on the solid skeleton, p is the pore water pressure, $\mathbf{k}$ is the permeability tensor of the solid phase, and μ the dynamic viscosity of the pore fluid. In addition, appropriate boundary

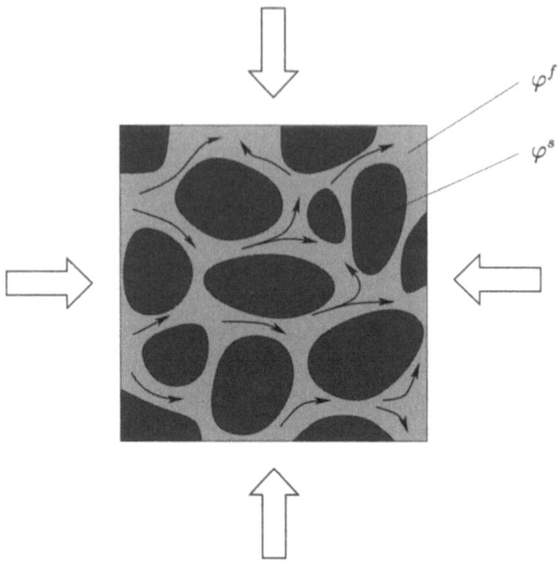

Fig. 1. Fluid flow in a deforming porous media.

and initial conditions have to be prescribed:

$$\begin{aligned} \mathbf{u}(t = t_0) &= \mathbf{u}_0 \text{ in } \Omega\,, \\ p(t = t_0) &= p_0 \text{ in } \Omega\,, \end{aligned} \tag{3}$$

$$\begin{aligned} \mathbf{u} - \bar{\mathbf{u}} &= 0 \text{ on } \Gamma_u\,, \\ (\boldsymbol{\sigma} + p\mathbf{I}) \cdot \mathbf{n} - \bar{\mathbf{t}} &= 0 \text{ on } \Gamma_t\,, \qquad \Gamma_u \cup \Gamma_t = \Gamma_p \cup \Gamma_Q = \Gamma\,, \\ p - \bar{p} &= 0 \text{ on } \Gamma_p\,, \qquad \Gamma_u \cap \Gamma_t = \Gamma_p \cap \Gamma_Q = \emptyset\,, \\ \mathbf{Q} \cdot \mathbf{n} - \bar{Q} &= 0 \text{ on } \Gamma_Q\,, \end{aligned} \tag{4}$$

where $\bar{\mathbf{b}}$ is the volume force density, $\bar{\mathbf{t}}$ are imposed surface tractions and $\bar{Q}$ represents the fluid flow over the boundary Γ_Q.

Additionally, kinematic and constitutive relations have to be formulated to describe the behaviour of the elastic-plastic solid skeleton. In the framework of a geometric linear theory the relation between strain rates $\dot{\varepsilon}$ and displacements $\mathbf{u}$ is given by:

$$\dot{\varepsilon}(t) = \frac{\partial}{\partial t} \nabla^S \mathbf{u} = \frac{\partial}{\partial t} \frac{1}{2} \left(\nabla \mathbf{u} + (\nabla \mathbf{u})^T \right). \tag{5}$$

3 Non-linear material behaviour

In order to describe the non-linear material behaviour, the classical rate-independent plasticity concept is applied. This includes the following aspects. The elastic stress strain relation is defined in rate form and is based on an additive decomposition of strain rates in elastic and plastic parts:

$$\dot{\boldsymbol{\sigma}} = \mathbf{C} : \left(\dot{\varepsilon} - \dot{\varepsilon}^{pl}\right). \tag{6}$$

A non-associated flow rule is applied to determine the plastic strain rates from a plastic potential function g, where the absolute value is defined by the plastic multiplier $\dot{\gamma}$. Moreover, an evolution law has to be prescribed for the internal parameters $\mathbf{q}$. The hardening and softening moduli, respectively, are defined by the matrix $\mathbf{D}$:

$$\dot{\varepsilon}^{pl} = \dot{\gamma}\,\frac{\partial g}{\partial \boldsymbol{\sigma}}\,, \qquad \dot{\mathbf{q}} = -\dot{\gamma}\,\mathbf{D}\,\frac{\partial g}{\partial \mathbf{q}}\,. \tag{7}$$

A yield function is used to bound the space of admissible stresses. In most plasticity models, the yield surface is formulated in terms of the internal parameters and the invariants of the stress tensor and the deviator tensor, respectively:

$$f(\boldsymbol{\sigma}, \mathbf{q}) = f(I_1, J_2, J_3, \mathbf{q}) = 0\,. \tag{8}$$

A single surface formulation is applied which originates from a hyperbolic

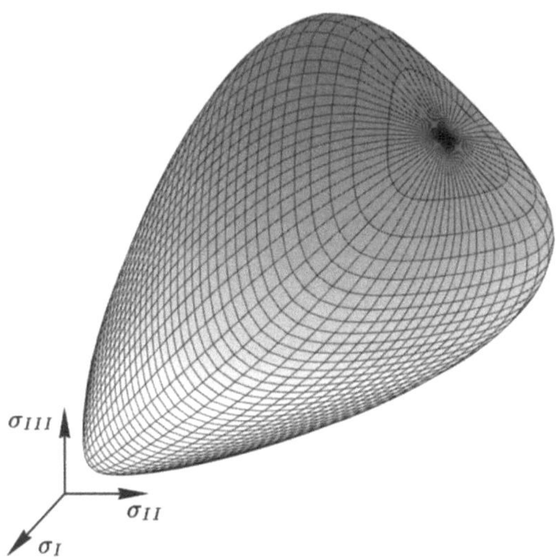

Fig. 2. Single surface yield function in principal stress space.

approximation to the *Drucker-Prager* criterion combined with a cap in hydrostatic compression. This formulation takes the form

$$f = \left(\left(m(\gamma,\theta)^2 J_2\right)^n + \left(\beta^2 I_1^4\right)^n + \sigma_{ms}^{2n}\right)^{1/(2n)} + \alpha I_1 - k = 0\,, \tag{9}$$

in which

$$m(\gamma,\theta) = \left(\frac{1+\gamma^4 + (1-\gamma^4)\,\sin(3\,\theta)}{2\,\gamma^4}\right)^{1/4} \tag{10}$$

controls the shape in the deviatoric plane depending on the *Lode* angle θ, which may be given by

$$\theta = \frac{1}{3}\arcsin\left(\frac{\sqrt{27}}{2}\frac{J_3}{J_2^{3/2}}\right) \qquad \text{with} \qquad -\frac{\pi}{6} \leq \theta \leq \frac{\pi}{6}\,. \tag{11}$$

Figure 3 shows the effect of different values $0.6109 \leq \gamma \leq 1.0$. The parameters α and k may be derived from the *Drucker-Prager* parameters while β, σ_{ms}, and the exponent n define the size and the shape of the caps in hydrostatic extension and compression, respectively. A representation of the yield surface in principal stress space is shown in Figure 2. A similar material model introducing an additional parameter for the deviatoric shape has been proposed by Ehlers [12]. The yield function given by (9) is quite robust with respect to a formulation of convex yield surfaces.

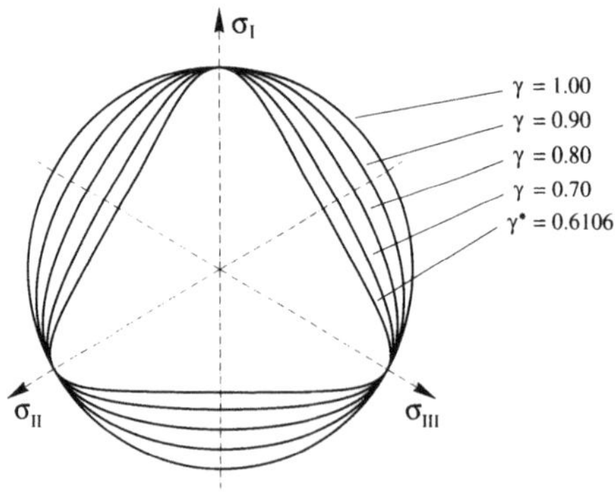

Fig. 3. Parameter γ in the deviatoric plane.

The direction of plastic flow is derived from a non-associated flow rule which is given by

$$g = \left(J_2^{n_g} + \left(\beta_g^2 I_1^4\right)^{n_g} + \sigma_{ms,g}^{2\,n_g}\right)^{1/(2\,n_g)} + \alpha_g I_1\,, \tag{12}$$

where a separate set of parameters α_g, β_g, $\sigma_{ms,g}$, and n_g is used.

In the case of plastic response, the consistency condition ensures that the stress state remains on the boundary of the elastic domain. For a distinction between plastic loading and unloading the *Kuhn-Tucker* conditions are used:

$$\dot{\gamma} \geq 0\,, \qquad f(\boldsymbol{\sigma}, \mathbf{q}) \leq 0\,, \qquad \dot{\gamma}\, f(\boldsymbol{\sigma}, \mathbf{q}) = 0\,. \tag{13}$$

A more detailed discussion of the equations of plasticity may be found in Simo and Hughes [18]. Different approaches in rock- and soilmechanics have been mentioned by Wunderlich et al. [23]. The most important requirement for a correct determination of limit-load states and the post-critical load-displacement path is that the integration schemes are unconditionally stable. Therefore, in this paper, *Newton* schemes are applied based on a consistent tangent operator. In extension to the standard case the properties of the discontinuous-*Galerkin* formulation have to be taken into account. In this context, it becomes necessary to include the temporal jumps in the field-variables into a variational form of the kinematic and constitutive relations.

4 Space-time finite element formulation

4.1 Application of the time-discontinuous-*Galerkin* method

In this Section, a space-time finite element method is proposed, where the time stepping scheme is based on a time-discontinuous-*Galerkin* procedure. In the spatial domain a standard finite element discretization is applied as illustrated by Figure 4. The displacement field as well as the pore water pressure field are approximated by piecewise linear shape functions within each interval $I_n = [t_n, t_{n+1}]$. In contrast to the semi-discretization method, a discontinuous approximation is permitted at the discrete time levels. It is

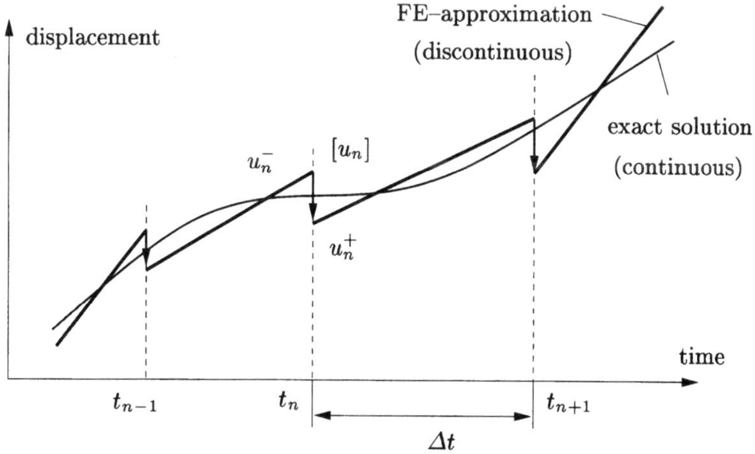

Fig. 4. Illustration of the time-discontinuous-*Galerkin* method.

necessary to introduce the following notation in order to distinguish between the two values at a discrete time t_n:

$$
\begin{aligned}
u_n^+ &= \lim_{\epsilon \to 0^+} u(t_n + \epsilon) = u(t_n^+) \,, \\
u_n^- &= \lim_{\epsilon \to 0^-} u(t_n + \epsilon) = u(t_n^-) \,, \\
[u_n] &= u_n^+ - u_n^- \,.
\end{aligned}
\tag{14}
$$

As a result, the whole space-time formulation may be considered as a combination of isoparametric elements in space and rectangular elements in time. The local form of the basic equations (2) has to be transformed into an equivalent variational form in which the discontinuities have to be included in a weak sense. By the use of virtual velocities $\delta\dot{\mathbf{u}}$ and virtual pore water pressures δp as *Galerkin* type weighting functions the following expression is obtained

$$
\begin{aligned}
&\int_{I_n} \Bigg\{ \int_\Omega \delta\dot{\mathbf{u}}^T \mathbf{D}_u^T \boldsymbol{\sigma} \, \mathrm{d}\Omega + \int_\Omega \delta\dot{\mathbf{u}}^T \mathbf{D}_u^T \mathbf{I} p \, \mathrm{d}\Omega - \int_\Omega \delta\dot{\mathbf{u}}^T \rho \bar{\mathbf{b}} \, \mathrm{d}\Omega - \int_{\Gamma_\sigma} \delta\dot{\mathbf{u}}^T \bar{\mathbf{t}} \, \mathrm{d}\Gamma + \\
&\quad + \int_\Omega \delta p \, \mathbf{I}^T \mathbf{D}_u \, \dot{\mathbf{u}} \, \mathrm{d}\Omega - \int_\Omega \delta p \, \mathbf{D}_p^T \, \mathbf{K}^{\mathrm{Darcy}} \, \mathbf{D}_p \, p \, \mathrm{d}\Omega + \int_{\Gamma_Q} \delta p \, \bar{Q} \, \mathrm{d}\Gamma \Bigg\} \, dt + \\
&\quad + \int_\Omega \delta p_n^+ \, \mathbf{I}^T \mathbf{D}_u \, [\mathbf{u}_n] \, \mathrm{d}\Omega = 0 \,,
\end{aligned}
\tag{15}
$$

which summarizes all internal and external virtual work expressions. In (15), $\mathbf{K}^{\mathrm{Darcy}}$ is the *Darcy* permeability as a combination of the permeability tensor and the fluid viscosity, $\mathbf{I} = [1,\, 1,\, 1,\, 0]^T$, $\mathbf{D}_u$, and $\mathbf{D}_p$ are the standard differential operators. In this variational formulation, the spatial coordinates $\mathbf{x}$ as well as the time coordinate t are independent variables. For a discretization in space and time shape functions are used in an appropriate product form:

$$
\begin{aligned}
\mathbf{u}(\mathbf{x}, t) &= \mathbf{N}_u(\mathbf{x}) \, \mathbf{T}_u(t) \, \hat{\mathbf{u}} \,, \\
p(\mathbf{x}, t) &= \mathbf{N}_p(\mathbf{x}) \, \mathbf{T}_p(t) \, \hat{\mathbf{p}} \,.
\end{aligned}
\tag{16}
$$

The discrete values $\hat{\mathbf{u}}$ and $\hat{\mathbf{p}}$ represent the generalized vector at the element nodes for the times t_{n+1} and t_n. In the context of a *Galerkin* type weighting concept, the same approximations are applied to their corresponding weighting functions. $\mathbf{N}_u$ and $\mathbf{N}_p$ are standard finite element shape function for the spatial coordinate. Due to the *Babuška-Brezzi* condition, the shape functions for the displacements should be chosen one order higher than those for the pore water pressures. In the calculations, quadratic shape functions have been applied for the displacements and linear ones for the pore water pressures in order to avoid locking effects. $\mathbf{T}_u$ and $\mathbf{T}_p$ are the temporal shape functions which are chosen linear for both fields. Figure 5 shows the 18-node element resulting from this approximation.

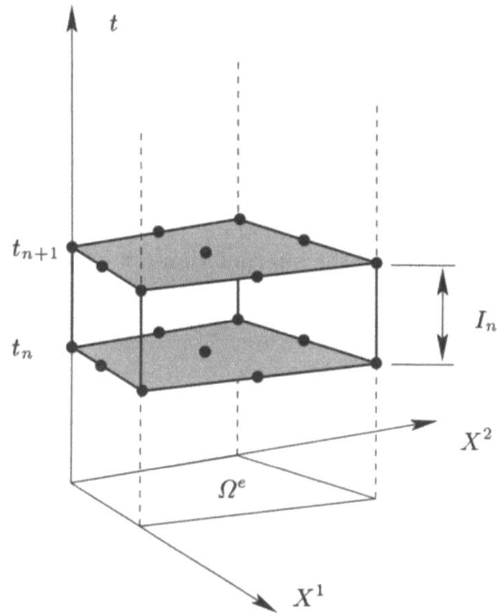

Fig. 5. 18-node space-time finite element.

4.2 Integration of the path-dependent variables

In the context of a discontinuous approximation of the variables in time, the standard backward *Euler* integration scheme is no longer applicable to determine the path-dependent variables. Therefore, an extended integration algorithm is based on variational formulations with respect to the actual time interval I_n. In this way, it is possible to include the jumps at the discrete time levels in a consistent manner. While a standard 3×3 integration is used in the space coordinate, a two-point *Gauß* scheme is applied in the time domain, where the *Kuhn-Tucker* conditions are enforced at the two integration points $\tau = \tau_1$ and $\tau = \tau_2$, simultaneously. All values within the interval are obtained by a linear interpolation of the integration point values demonstrated by Figure 6. The kinematic relations (5) have to be rewritten in the form

$$\int_{t_n^+}^{t_{n+1}^-} \delta\boldsymbol{\sigma} : (\nabla\dot{\mathbf{u}} - \dot{\boldsymbol{\varepsilon}})\, \mathrm{d}t + \delta\boldsymbol{\sigma}(t_n) : \nabla\,[\mathbf{u}_n] = 0\,, \tag{17}$$

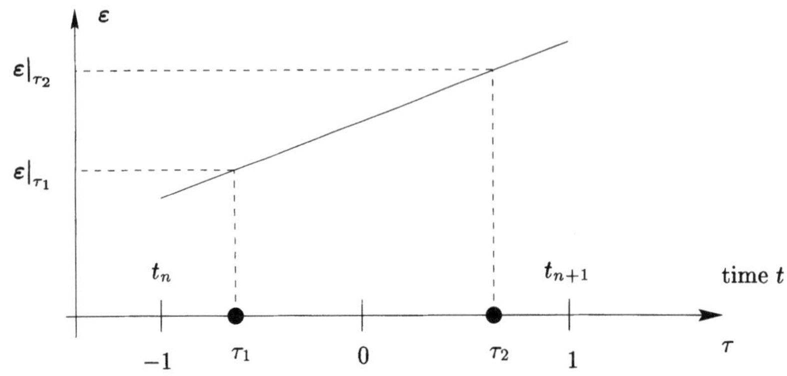

Fig. 6. Local coordinate τ and linear interpolation (two-point scheme).

where $\delta\boldsymbol{\sigma}$ is used as a weighting function. The vector of strain rates $\dot{\boldsymbol{\varepsilon}}$ at the *Gauß* points may then be calculated by

$$\dot{\tilde{\boldsymbol{\varepsilon}}} = \begin{bmatrix} \dot{\boldsymbol{\varepsilon}}|_{\tau_1} \\ \dot{\boldsymbol{\varepsilon}}|_{\tau_2} \end{bmatrix} = \tilde{\mathbf{B}}\,\hat{\mathbf{u}}\,, \tag{18}$$

where the matrix $\tilde{\mathbf{B}}$ is given by

$$\tilde{\mathbf{B}} = \left[\int_{I_n} \mathbf{T}_\sigma^T\,\mathbf{T}_\varepsilon\,\mathrm{d}t\right]^{-1} \left[\int_{I_n} \mathbf{T}_\sigma^T\,\mathbf{D}_u\,\mathbf{N}_u\dot{\mathbf{T}}_u\,\mathrm{d}t + \mathbf{T}_\sigma^T(t_n)\,\mathbf{D}_u\,\mathbf{N}_u\,\mathbf{T}_u(t_n)\right], \tag{19}$$

in which $\mathbf{T}_\sigma$ and $\mathbf{T}_\varepsilon$ represent shape functions in time on the basis of the *Gauß* point values. In the special case of a linear approximation, the strain rates are calculated by

$$\dot{\tilde{\boldsymbol{\varepsilon}}} = \frac{1}{\Delta t}\begin{bmatrix} \sqrt{3} & 1 \\ -\sqrt{3} & 1 \end{bmatrix} \begin{bmatrix} \mathbf{D}_u\mathbf{N}_u & 0 \\ 0 & \mathbf{D}_u\mathbf{N}_u \end{bmatrix} \begin{bmatrix} [\mathbf{u}_n] \\ \Delta\mathbf{u}_n \end{bmatrix}. \tag{20}$$

Additionally, the rate equations of plasticity (6) and (7) have to be formulated in a variational expression leading to

$$\begin{aligned} &\int_{t_n^+}^{t_{n+1}^-} \delta\boldsymbol{\varepsilon} : \left(\dot{\boldsymbol{\sigma}} - \mathbf{C} : \dot{\boldsymbol{\varepsilon}} + \mathbf{C} : \frac{\partial g}{\partial \boldsymbol{\sigma}}\,\dot{\gamma}\right) \mathrm{d}t + \delta\boldsymbol{\varepsilon}(t_n) : [\boldsymbol{\sigma}_n] + \\ &\quad + \int_{t_n^+}^{t_{n+1}^-} \delta\boldsymbol{\alpha} \cdot \left(\dot{\mathbf{q}} + \mathbf{D}\,\frac{\partial g}{\partial \mathbf{q}}\,\dot{\gamma}\right) \mathrm{d}t + \delta\boldsymbol{\alpha}(t_n) \cdot [\mathbf{q}_n] = 0\,, \end{aligned} \tag{21}$$

in which $\delta\boldsymbol{\varepsilon}$ and $\delta\boldsymbol{\alpha}$ serve as corresponding weighting functions. In consequence, this procedure leads to a strong interdependence of the non-linear response at the temporal integration points.

4.3 Consistent linearization

A consistent linearization of the whole algorithm is based on residual equations for the stresses, the internal parameters and the yield conditions at both temporal *Gauß* points. For simplicity, the formulation is presented here for ideal plasticity and the following vectors of generalized *Gauß* point values are defined:

$$\tilde{\mathbf{C}} = \Delta t \begin{bmatrix} \frac{1}{3}\mathbf{C} & \frac{1-\sqrt{3}}{6}\mathbf{C} \\ \frac{1+\sqrt{3}}{6}\mathbf{C} & \frac{1}{3}\mathbf{C} \end{bmatrix}, \tag{22}$$

$$\mathbf{G} = \begin{bmatrix} \left.\frac{\partial g}{\partial \boldsymbol{\sigma}}\right|_{\tau_1} & 0 \\ 0 & \left.\frac{\partial g}{\partial \boldsymbol{\sigma}}\right|_{\tau_2} \end{bmatrix}, \qquad \mathbf{F} = \begin{bmatrix} \left.\frac{\partial f}{\partial \boldsymbol{\sigma}}\right|_{\tau_1} & 0 \\ 0 & \left.\frac{\partial f}{\partial \boldsymbol{\sigma}}\right|_{\tau_2} \end{bmatrix}, \tag{23}$$

$$\dot{\tilde{\gamma}} = \begin{bmatrix} \dot{\gamma}|_{\tau_1} \\ \dot{\gamma}|_{\tau_2} \end{bmatrix}, \qquad \tilde{\mathbf{f}} = \begin{bmatrix} f(\boldsymbol{\sigma})|_{\tau_1} \\ f(\boldsymbol{\sigma})|_{\tau_2} \end{bmatrix} = 0\,. \tag{24}$$

Performing the numerical integration over the time interval, the stresses $\tilde{\boldsymbol{\sigma}}$ at the integration points are given by

$$\tilde{\boldsymbol{\sigma}} = \begin{bmatrix} \boldsymbol{\sigma}|_{\tau=\tau_1} \\ \boldsymbol{\sigma}|_{\tau=\tau_2} \end{bmatrix} = \begin{bmatrix} \boldsymbol{\sigma}_n^- \\ \boldsymbol{\sigma}_n^- \end{bmatrix} + \tilde{\mathbf{C}}\,\dot{\tilde{\boldsymbol{\varepsilon}}} - \tilde{\mathbf{C}}\,\mathbf{G}\,\dot{\tilde{\gamma}}\,. \tag{25}$$

In this context, it has to be checked which of the integration points are active and show plastic behaviour. The set of active points is represented by

$$\mathbb{J}_{\text{act}}^{(i)} = \left\{\alpha \in \{1,2\} \,|\, f\left(\boldsymbol{\sigma}^{(i)}, \mathbf{q}^{(i)}\right)\Big|_{\tau_\alpha} > 0\right\} \tag{26}$$

and may change within the iteration process. As a result, an algorithm has to be applied which is usually used in the context of non-smooth multi-surface plasticity. The residual values in terms of the stresses are given by

$$\mathbf{R}^{(i)} = \tilde{\boldsymbol{\sigma}}^{\text{trial}} - \tilde{\boldsymbol{\sigma}}^{(i)} - \tilde{\mathbf{C}}\,\mathbf{G}^{(i)}\,\dot{\tilde{\gamma}}^{(i)}\,. \tag{27}$$

Starting the iteration by assuming an elastic behaviour and setting $\dot{\tilde{\gamma}}^{(0)} = 0$ and $\tilde{\boldsymbol{\sigma}}^{(0)} = \tilde{\boldsymbol{\sigma}}^{\text{trial}} = \tilde{\mathbf{C}}\,\dot{\tilde{\boldsymbol{\varepsilon}}}$, the iterative corrections in case of a plastic response are obtained by

$$\mathrm{d}\dot{\tilde{\gamma}}^{(i)} = \left(\mathbf{F}^{(i)}\,\mathbf{A}^{(i)\,-1}\,\tilde{\mathbf{C}}\,\mathbf{G}^{(i)}\right)^{-1}\left(\tilde{\mathbf{f}}^{(i)} + \mathbf{F}^{(i)}\,\mathbf{A}^{(i)\,-1}\,\mathbf{R}^{(i)}\right), \tag{28}$$

$$\mathrm{d}\tilde{\boldsymbol{\sigma}}^{(i)} = \mathbf{A}^{(i)\,-1}\left(\mathbf{R}^{(i)} - \tilde{\mathbf{C}}\,\mathbf{G}^{(i)}\,\mathrm{d}\dot{\tilde{\gamma}}^{(i)}\right) \tag{29}$$

with the consistent tangent modulus

$$\mathbf{A}^{(i)} = \mathbf{I} + \tilde{\mathbf{C}} \begin{bmatrix} \left.\dfrac{\partial^2 g^{(i)}}{\partial \boldsymbol{\sigma}^2}\right|_{\tau_1} \left.\dot{\gamma}\right|_{\tau_1} & 0 \\ 0 & \left.\dfrac{\partial^2 g^{(i)}}{\partial \boldsymbol{\sigma}^2}\right|_{\tau_2} \left.\dot{\gamma}\right|_{\tau_2} \end{bmatrix} . \tag{30}$$

The iteration is performed until the convergence criteria in terms of the yield condition and the residual are satisfied. In a next step, the vector of internal forces is assembled assuming a linear approximation of the stresses within the interval.

4.4 Consistent tangent operator

In order to achieve a quadratic rate of convergence in the global iteration procedure, the stiffness matrices have to be assembled using a tangent operator which is consistent with the iteration procedure for the integration of the stresses. This consistent tangent operator $\tilde{\mathbf{C}}^{ct}$ is obtained from the linearized equations of plasticity and may be written in the form:

$$\mathbf{C}^{ct} = \frac{\partial \tilde{\sigma}}{\partial \dot{\tilde{\varepsilon}}} = \mathbf{A}^{-1} \left(\tilde{\mathbf{C}} - \tilde{\mathbf{C}}\, \mathbf{G}\, \mathbf{B}^{-1}\, \mathbf{F}\, \mathbf{A}^{-1}\, \tilde{\mathbf{C}} \right) \tag{31}$$

with

$$\mathbf{B} = \mathbf{F}\, \mathbf{A}^{-1}\, \tilde{\mathbf{C}}\, \mathbf{G} . \tag{32}$$

After a transformation from the integration points to the nodal points of the space-time element using

$$\hat{\mathbf{C}}^{ct} = \begin{bmatrix} \frac{1}{2}(1+\sqrt{3}) & \frac{1}{2}(1-\sqrt{3}) \\ \frac{1}{2}(1-\sqrt{3}) & \frac{1}{2}(1+\sqrt{3}) \end{bmatrix} \mathbf{C}^{ct} \begin{bmatrix} \sqrt{3} & 1 \\ -\sqrt{3} & 1 \end{bmatrix} = \begin{bmatrix} \hat{\mathbf{C}}^{ct}_{11} & \hat{\mathbf{C}}^{ct}_{12} \\ \hat{\mathbf{C}}^{ct}_{21} & \hat{\mathbf{C}}^{ct}_{22} \end{bmatrix} , \tag{33}$$

the stiffness matrices are calculated by

$$\mathbf{K}_{\alpha\beta} = \sum_{e=1}^{N_{el}} \int_{\Omega^e} (\mathbf{D}_u \mathbf{N}_u)^T\, \hat{\mathbf{C}}^{ct}_{\alpha\beta}\, \mathbf{D}_u \mathbf{N}_u \, \mathrm{d}\Omega^e . \tag{34}$$

It has to be mentioned that in the special case of a purely elastic behaviour at both *Gauß* points, the consistent tangent operator reduces to the block-diagonal form

$$\hat{\mathbf{C}}^{ct} = \begin{bmatrix} \mathbf{C} & 0 \\ 0 & \mathbf{C} \end{bmatrix} . \tag{35}$$

4.5 System of equations

Applying these discretizations to (15) and considering the kinematic and constitutive relations (5) and (6), the following system of non-linear equations is obtained

$$\begin{bmatrix} \mathbf{K}_{11} & \mathbf{K}_{12} & \mathbf{L} & \mathbf{0} \\ \mathbf{K}_{21} & \mathbf{K}_{22} & \mathbf{0} & \mathbf{L} \\ \mathbf{L}^T & \mathbf{0} & -\frac{1}{6}\Delta t\mathbf{H} & \frac{1}{6}\Delta t\mathbf{H} \\ \mathbf{0} & \mathbf{L}^T & -\frac{1}{2}\Delta t\mathbf{H} & -\frac{1}{2}\Delta t\mathbf{H} \end{bmatrix} \mathrm{d} \begin{bmatrix} [\mathbf{u}_n] \\ \Delta\mathbf{u}_n \\ [\mathbf{p}_n] \\ \Delta\mathbf{p}_n \end{bmatrix} = \\ = \begin{bmatrix} \mathbf{F}^{\text{ext}}(t_n^+) - \mathbf{F}^{\text{int}}(t_n^+) \\ \mathbf{F}^{\text{ext}}(t_{n+1}^-) - \mathbf{F}^{\text{int}}(t_{n+1}^-) \\ \mathbf{0} \\ \mathbf{F}^p \end{bmatrix} \tag{36}$$

$$\text{with} \quad \mathbf{F}^p = \frac{1}{2}\mathbf{H}\left(p(t_n^+) + p(t_{n+1}^-)\right) - \mathbf{L}^T\left(u(t_{n+1}^-) - u(t_n^-)\right), \tag{37}$$

which is solved by a *Newton-Raphson* iteration. In the special case of elastic behaviour, the submatrices $\mathbf{K}_{12}$ and $\mathbf{K}_{21}$ reduce to zero. In general, however, these matrices include the interdependence of the integration points. Thus, the method presented may be regarded as a straightforward extension to the time-discontinuous-*Galerkin* procedure taking into account the elastic-plastic material laws.

The advantage of the space-time formulation proposed in this Section and, especially, the time-discontinuous-*Galerkin* method relies on improved stability properties. Temporal oscillations which are often encountered when a semi-discretization is applied are consequently avoided. Due to the stiffness of the differential equations that arise especially in the case of an adaptive space discretization due to very fine mesh sizes, explicit integration techniques are not applicable for the quasi-static consolidation problem. The *Crank-Nicolson* scheme is of second order accuracy and shows disadvantages with respect to reduced stability properties (A-stability). The time-discontinuous-*Galerkin* methods, however, reveal the higher L-stability when applied to the problem under study and are therefore much more suitable for time integration. This is the main reason to apply the *Galerkin*-type weighting also in the time domain. Using linear discontinuous approximations, the accuracy is of third order. However, the computational effort is much higher compared to continuous approximations since a second set of unknowns representing the unknowns at the beginning of the time interval is introduced. On the other side, larger time steps may be used due to the higher accuracy of the method. In addition, with respect to adaptive procedures, the TDG-method provides a natural basis for an indicator for the temporal discretization error using the jumps at the discrete time levels.

The model presented in this paper is based on linear discontinuous shape functions in the time domain. An extension to general higher order approximations has been proposed by Findeiß [15]. However, the benefit of quadratic discontinuous shape functions is only minor with respect to the higher computational effort.

5 Time adaptivity

In time dependent problems, not only the spatial coordinate but also the time domain has to be discretized. In most cases, the initial value problem is treated by a time stepping scheme which extrapolates the approximate solution at the end of a subinterval from the data at the beginning of the interval. Subsequently, the state variables are updated and the algorithm proceeds to the next increment. This method avoids an expensive a priori discretization of the whole time domain and directly permits an adaptive adjustment of the time step sizes.

In the analysis of elastic-plastic problems, the efficiency of the computation may be increased when an adaptive time discretization is applied. Obviously, domains characterized by a rapid change in the field variables have to be treated by much smaller increment sizes. Optimal conditions are reached, when an equal error distribution over all intervals is achieved. Therefore, the actual finite element solution must be analysed in order to decide whether the last step can be accepted or has to be rejected and resolved with a reduced step size. As it is often difficult or impossible to evaluate the real error, special measures are used as indicators on a heuristical basis. In this contribution, two different error indicators are presented which try to capture the temporal discretization error and the error in the numerical integration of the constitutive equations.

5.1 Temporal discretization error

The temporal discretization error originates from the approximative character of the time stepping scheme. It is closely related to the step size Δt and the order of the integration scheme. Thus, larger time steps may be applied when the shape functions are enriched from a linear to a quadratic approximation, for example.

Considering the quasi-static consolidation problem, the TDG-procedure offers a natural method to derive an error indicator. The variational equation (15) shows that temporal discontinuities of the unknowns are included in the formulation in a weak sense. Therefore, it is possible to use the L_2-norm of these jumps in the displacement and pore water pressure fields as a measure for the temporal discretization error:

$$\eta_{\Delta t,u} = ||\ [\mathbf{u}]\ ||_{L_2(\Omega)}\,, \qquad \eta_{\Delta t,p} = ||\ [p]\ ||_{L_2(\Omega)}\,. \tag{38}$$

The observation that these values reduce to zero when the step size is reduced directly supports the formulation of a relative error. Thus, it is possible to relate the norm of the jumps to the norm of the whole increments $\Delta\mathbf{u}$ and δp. This results in a reasonable error indicator because the convergence property of the jumps is of a higher order than that of the increment size. To a great extent this temporal error measure is independent of the spatial mesh size for the given problem. The relevant value is then obtained from the maximum of both fields:

$$\eta_{\Delta t} = \max\left(\frac{\eta_{\Delta t,u}}{||\Delta\mathbf{u}||_{L_2}}, \frac{\eta_{\Delta t,p}}{||\Delta p||_{L_2}}\right)\,. \tag{39}$$

As the jumps $[\mathbf{u}]$ and $[p]$ are directly associated with the unknowns of the global system of equations, the evaluation of these indicators is very cheap with respect to the total computational cost.

5.2 Error in the integration of the constitutive equations

When plasticity models are applied, the stress strain relation becomes non-linear and an analytical determination of the path dependent variables is not possible. Thus, numerical integration schemes have to be used as for example the backward *Euler* rule which represents a closest point projection of the trial stresses onto the yield surface. These iterative methods introduce an additional error into the computation which is also related to the time step size. In order to bound this error, the increment sizes have to be controlled. The error in the integration of the constitutive equations in terms of the plastic strains may be given by

$$||\mathbf{e}||_{L_2(\Omega)} = \left|\left| \int_{t_n}^{t_{n+1}} \dot{\varepsilon}^{pl}\,\mathrm{d}t - \Delta\varepsilon^{pl} \right|\right|_{L_2(\Omega)}\,, \tag{40}$$

where $\Delta\varepsilon^{pl}$ is the discrete solution. However, the time integral in (40) cannot be calculated analytically. Therefore, another indicator is proposed in the following to provide a measure for this error.

Due to the discontinuous time approximation, the standard backward *Euler* rule becomes inconvenient in the space-time finite element method. The type of integration scheme which is used in this context may be characterized by a two-point *Gauß* scheme which iteratively satisfies the yield condition and the *Kuhn-Tucker* condition at two integration points in time. This means that not the conditions at the end of the time step but other representative points within the interval are considered for the determination of the path dependent variables. A more detailed description of this technique may be found in Cramer et al. [10].

The error indicator proposed in this Section originates from the residual in the *Kuhn-Tucker* condition $\dot{\gamma} f = 0$, which is strictly enforced at the two temporal integration points. However, at all other points in the interval $\dot{\gamma} f \neq 0$ may be encountered at times $\tau \notin \{\tau_1, \tau_2\}$, where τ_1 and τ_2 indicate the local time coordinates of the integration points. Figure 7 illustrates the typical case. On the basis of a linear interpolation of stresses, internal parameters and plastic multipliers and due to the convexity of the yield surface, the yield condition ($f \leq 0$) is satisfied only between the *Gauß* points. The indicator is defined separately for each spatial sampling point by the integral of the residuum of $\dot{\gamma}_h f_h(t)$:

$$\eta^i_{\sigma\varepsilon} = \int_{I_n} |\underbrace{\dot{\gamma} f(\boldsymbol{\sigma}, \mathbf{q})}_{\to 0} - \dot{\gamma}_h f(\boldsymbol{\sigma}_h, \mathbf{q}_h)| \, \mathrm{d}t = \int_{I_n} |\dot{\gamma}_h f(\boldsymbol{\sigma}_h, \mathbf{q}_h)| \, \mathrm{d}t \,. \tag{41}$$

This integral is evaluated by a numerical integration scheme of a higher order. The finite element solution $\dot{\gamma}_h f_h$ is therefore computed by a three point *Gauß* integration scheme.

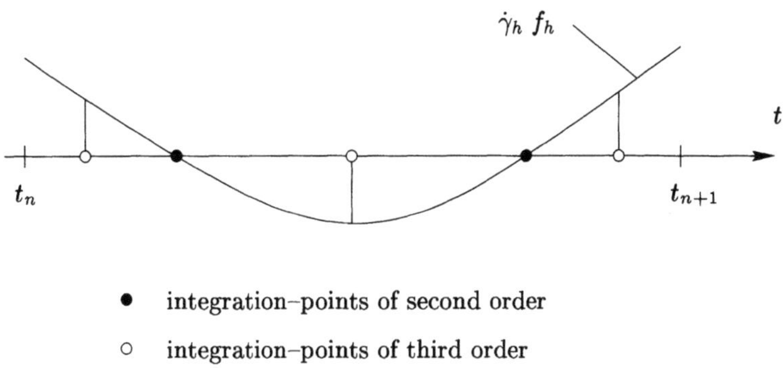

Fig. 7. *Gauß*-point-integration of higher order over the interval I_n.

It is obvious that all elastic points of the domain do not contribute to this error measure as the determination of stresses may be performed exactly. It was found that the indicator responds with slightly higher values at initial plastification. As a result, a rapid change in the number of plastic integration points is avoided, which is a crucial point when a sudden shearband formation is initiated. In general, this process is connected to an unloading in great parts of the system and accompanied by a decrease in the number of plastic points. A global error indicator is given by the average value of all plastic points in relation to the time step size

$$\eta_{\sigma\varepsilon} = \frac{1}{\Delta t\, N_{pl}} \sum_{i=1}^{N_{pl}} \eta^{i}_{\sigma\varepsilon}\,, \tag{42}$$

in which N_{pl} represents the number of spatial integration points showing a plastic response. This ensures that single points with a very high plastic dissipation do not lead to an excessive reduction of the time step sizes. Choosing the maximum value of all single points would be too inefficient for the global analysis.

5.3 Time step adjustment

The two error indicators proposed in the last Sections are now combined to control the time step size automatically. This means that higher error measures will lead to reduced increment sizes. For this purpose, a relative tolerance has to be prescribed for the temporal discretization error $\mathrm{tol}_{\Delta t}$ and an absolute tolerance for the constitutive error $\mathrm{tol}_{\sigma\varepsilon}$. If at least one of the tolerances is exceeded, the actual time step has to be rejected. On the other side, the increments may be increased when the errors fall below the tolerances. The adjustment is controlled by the following relation:

$$\Delta t_{\text{new}} = \min\left(\Delta t_{\text{old}}\,\beta\left(\frac{\mathrm{tol}_{\Delta t}}{\eta_{\Delta t}}\right)^{1/p},\ \Delta t_{\text{old}}\,\beta\left(\frac{\mathrm{tol}_{\sigma\varepsilon}}{\eta_{\sigma\varepsilon}}\right)^{1/q}\right). \tag{43}$$

When the time step is reduced, β which is used as a safety factor prevents that the next step has to be recalculated once more due to a marginal exceeding of

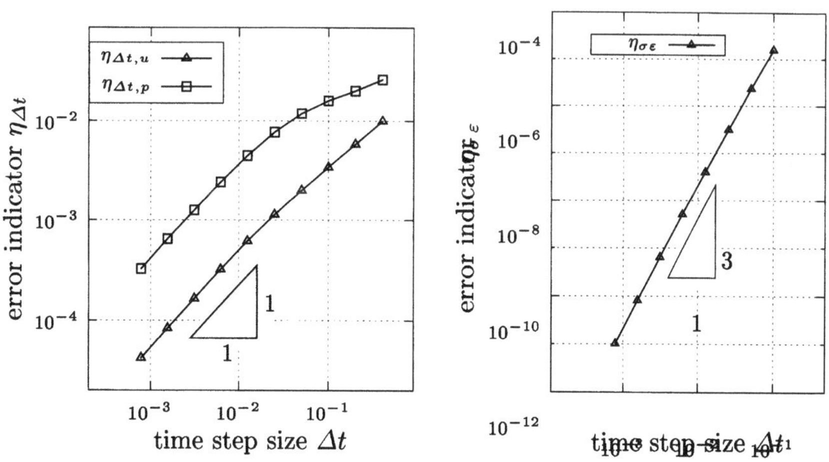

Fig. 8. Rates of convergence p and q.

the tolerances. Usually, it is set to $\beta \approx 0.95$. When the step size is increased, the same equation may be applied but no safety factor is necessary and $\beta = 1.0$ is used. Another important role play the rates of convergence p and q defining the relation between the error indicators and the step size. The values of p and q have been determined by test computations using a sequence of different time step sizes Δt. Figure 8 shows the relation in a logarithmic scale, where the relevant values may be obtained from the gradient of the curve.

This automatic time step control leads to a discretization of the time domain which is not specified a priori. Instead, the increment sizes may be adjusted in a problem dependent way. In the case of a fully elastic response of the system, the constitutive error vanishes and the step size control is based on the temporal discretization error alone. However, when limit load conditions are reached, the constitutive error becomes the restricting factor. The numerical examples will show the behaviour of this time adaptive procedure.

6 Numerical examples

6.1 One dimensional consolidation

In a first example, the one dimensional consolidation problem shown in Figure 9 is analysed. A layer of saturated porous soil is compressed by a ramp-type loading function. The computation is performed until steady state conditions are reached. This example will show that the error indicator $\eta_{\Delta t}$ is well suited to control the step size and that the L_2-norm of the jumps may be used to achieve a constant error distribution over the time domain, approximately. Assuming elastic material behaviour $\eta_{\sigma\varepsilon}$ is neglected.

For the correct determination of the relative temporal discretization error, a reference solution (u_{Ref}) was computed on the basis of very small time steps (overkill solution). Figure 10 shows the evolution of this error over the whole time of consolidation when an adaptive adjustment of the step sizes is applied.

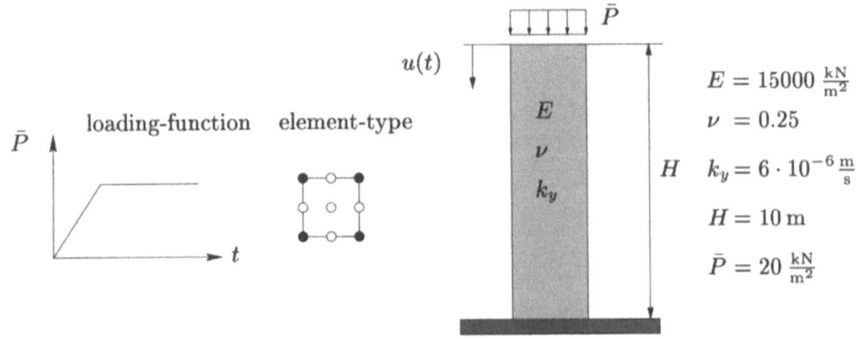

Fig. 9. System for one-dimensional consolidation analysis.

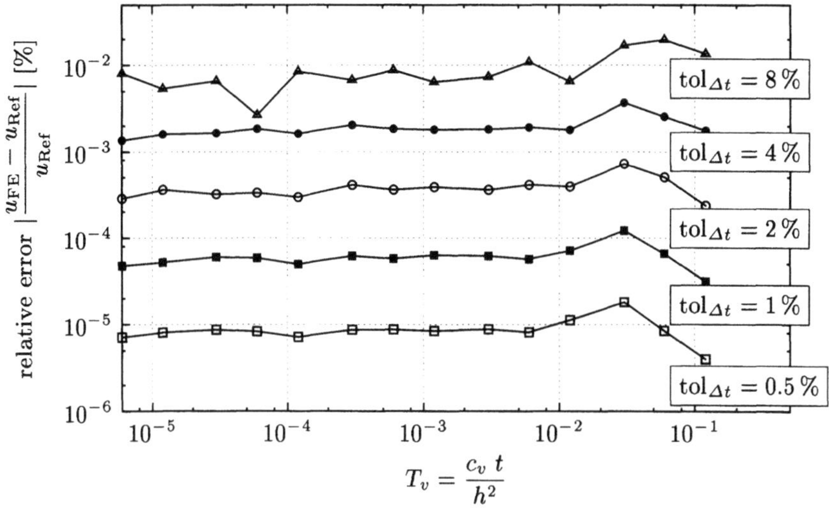

Fig. 10. Relative temporal discretization error.

It is demonstrated on one hand that the error is nearly constant in time which leads to a minimum of the overall computational effort. On the other side, the effect of different tolerances $tol_{\Delta t}$ becomes obvious. A more restrictive choice leads to a decrease in the discretization error. Thus, the accuracy level is defined in terms of the tolerance and may be increased systematically with a reasonable effort.

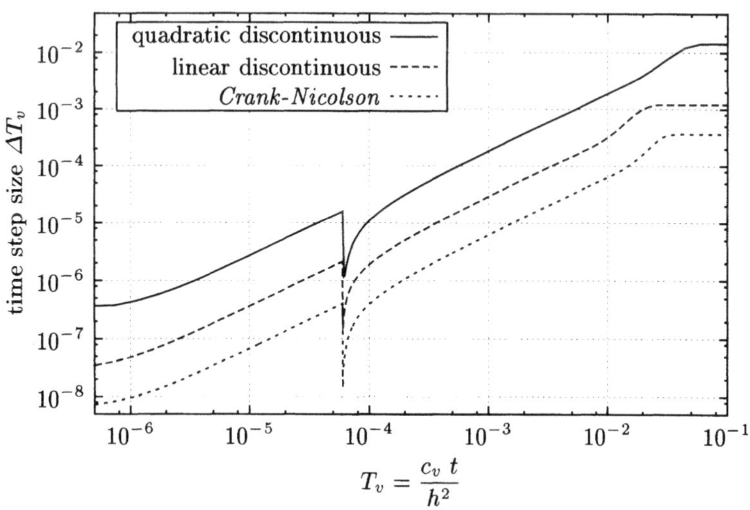

Fig. 11. Successive adjustment to larger time steps.

The evolution of the time step sizes for different approximations and identical accuracy is shown in Figure 11. A comparison between the continuous *Crank-Nicolson* scheme and linear as well as quadratic discontinuous-*Galerkin* methods shows that in all three cases the step sizes are increased over the time domain by a factor of 10^4. Clearly, a computation with constant time steps would be much too inefficient. Another conclusion is that the time steps may be chosen larger by a factor of $5-8$ when the linear TDG-method is applied. About the same factor was obtained when the shape functions are extended to a quadratic approximation. However, the computational effort for one increment raises in the same relation. For that reasons, a linear discontinuous time approximation seems most suitable and combines the advantages of improved stability properties and the applicability of the error indicators proposed in Section 5.

6.2 Strip-footing on a half space

In this Section, a practical geomechanical problem is investigated. A strip-footing on a half space is continuously loaded until failure. The geometric properties and the material parameters of the system are shown in Figure 12. For limit state problems, other advanced solution techniques should additionally be applied in the framework of the finite element method. In this context, a strategy for adaptive mesh refinement and the *Cosserat*-theory as a strong regularization tool were implemented. With these improvements, the final zone of failure characterized by the localized plastic strain increments in the expected shape may be refined properly as shown in Figure 13. However, these techniques are not the main topic of this contribution. We refer to the work of Wunderlich et al. [22], Ehlers and Volk [13] or de Borst [6] for a detailed discussion. In the following, emphasis is laid on the results of time adaptivity.

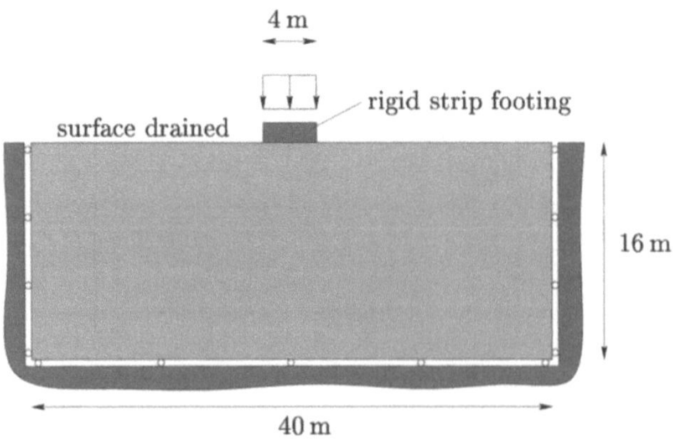

Material parameters:	
E	$5000\ \frac{\mathrm{kN}}{\mathrm{m}^2}$
ν	0.20
k_x, k_y	$10^{-4}\ \frac{\mathrm{m}}{\mathrm{s}}$
ϕ	25°
c	$5\ \frac{\mathrm{kN}}{\mathrm{m}^2}$
ψ	5°

Fig. 12. Geometrical and material properties of the strip-footing on half space.

This example is characterized by a complex deformation process including strain localization. The constitutive law for the solid skeleton was modelled by a non-linear stress-strain relation including a cone-shaped yield surface with a rounded triangular shape in the deviatoric plane. To predict a dilatant material behaviour and a realistic plastic volume change a non-associated flow rule was applied. In comparison to the last example, it is much more essential to capture the state of stress at the time of shearband formation and unloading effects. Hence, the error in the integration of the constitutive equations is considered in more detail now.

The time scale of the computation has to be divided into several Sections. Firstly, for a small period the system response is purely elastic. Then the plastic zone increases and covers a wide area below the strip-footing. After a time of only small changes in the stress state, the shearband formation is initiated leading to failure. This is a quite sudden process and marks the limit load of the system. Obviously, constant step sizes are inefficient in this problem, too, and the very complex deformation process needs an appropriate time discretization. However, without an adaptive control the user has to perform several trial and error computations until he is able to prescribe a reasonable selection of step sizes. It is shown now that the adaptive control proposed in this paper leads to a reasonable time discretization and, therefore, to an acceptable relation between accuracy and computational cost.

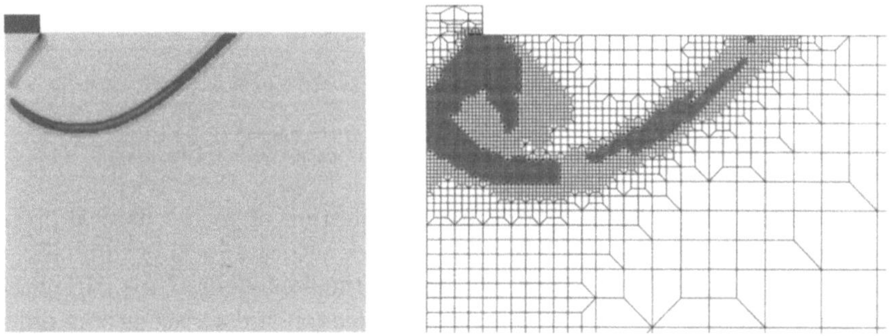

Fig. 13. Shearband formation and adaptively refined mesh.

Figure 14 represents the evolution of the time step sizes until the limit load is reached. As the error in the integration of the constitutive relations becomes the restricting factor, the step size control was based mainly on an evolution of the indicator $\eta_{\sigma\varepsilon}$. The step sizes are consequently reduced when the change in the number of plastic integration points is high. This is especially the fact in the beginning of the loading process and when the shearband formation is initiated connected to an unloading of large domains. A comparison of different tolerance levels $\text{tol}_{\sigma\varepsilon} = 10^{-5}$, 10^{-6} and 10^{-7} shows that a reduction leads to a systematical adjustment to smaller time steps and, thus, to an increase in accuracy. This example shows that critical times are detected and passed by smaller increment sizes to ensure a specific accuracy level.

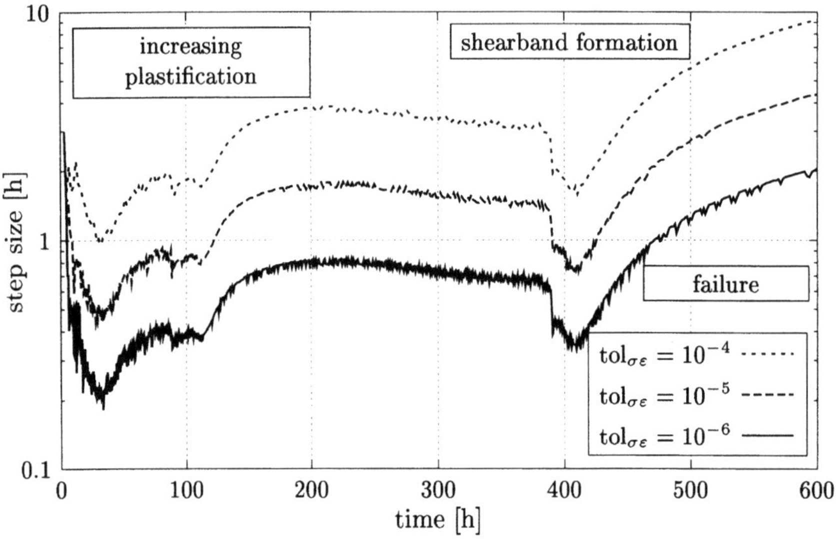

Fig. 14. Evolution of the time step sizes.

7 Conclusions

In this contribution, the time adaptive computation of geomechanical problems has been presented. The analysis of water saturated soil is emphasized using the Theory of Porous Media to capture the behaviour of the two phase material consisting of an elastic-plastic solid skeleton and a viscous pore fluid. As a consequence of the fluid transport, the computation has to be extended to a discretization of the time domain. To obtain the discrete equations in space and time a consistent space-time formulation in the context of the time-discontinuous-*Galerkin* method was presented. Due to the *Galerkin*-type weighting functions and the numerical dissipation of higher modes, the resulting time stepping scheme provides a good combination of accuracy and stability properties and introduces a natural possibility to derive a temporal error indicator. For a good accuracy of the time discretization of the consolidation problem, the approximation of the displacement and pore water pressure fields as well as the numerical integration of the rate-equations of plasticity have to be taken into account simultaneously. Therefore, an automatic time step control has been proposed based on the evaluation of two different error indicators. A measure for the temporal discretization error was formulated using the L_2-norm of the jumps at the discrete time levels. The second error was captured by the residuum of the *Kuhn-Tucker*-conditions. The computational cost for the calculation of both indicators is very low. Therefore, the efficiency of the total analysis may be improved, especially, when space adaptive strategies are included, additionally.

The numerical examples of the one dimensional consolidation problem and a strip-footing on half space demonstrate that both indicators are applicable to practical problems in geomechanics and lead to reasonable time discretizations. While the first indicator is mainly responsible for the time steps in the elastic regime, the second indicator controls the step sizes when large plastic zones become evident. The sudden process of shearband formation may then be detected and passed by smaller steps. Future work is necessary to define the interdependence of temporal and spatial error indicators in a space and time adaptive computation. Thus, the relation between the local temporal error measures and the mesh size has to be specified. Another problem relies on the high number of recalculations connected to a step size reduction leading to a lower efficiency. Therefore, the time step control should be improved to minimize the number of step size changes.

References

1. Babuška, I., Miller, A.: *A-posteriori error estimates and adaptive techniques for the finite element method.* Technical report bn-968, University of Maryland, Institute for Physical Science and Technology, Maryland 1981.
2. Biot, M. A.: General theory of three-dimensional consolidation. *Journal for Applied Physics* **12** (1941), 155–164.
3. Biot, M. A.: General solutions of the equations of elasticity and consolidation for a porous material. *Journal for Applied Mechanics* **23** (1956), 91–96.
4. de Boer, R.: *Theorie poröser Medien – Historische Entwicklung und gegenwärtiger Stand.* Forschungsbericht aus dem Fachbereich Bauwesen 53, Universität-GH-Essen 1986.
5. de Boer, R., Ehlers, W.: *Theorie der Mehrkomponentenkontinua mit Anwendung auf bodenmechanische Probleme.* Forschungsbericht aus dem Fachbereich Bauwesen 40, Universität-GH-Essen 1986.
6. de Borst, R.: Simulation of strain localization: A reappraisal of the Cosserat continuum. *Engineering Computations* **8** (1991), 317–332.
7. Bowen, R. M. (1976): Theory of mixtures. In Eringen, A. C. (ed.): *Continuum Physics* III, Academic Press, New York 1976, pp. 1–127.
8. Cramer, H., Findeiß, R., Steinl, G., Wunderlich, W.: An approach to the analysis of localization in granular materials by extended continuum formulations. In Idelsohn, S. R., Oñate, E., Dvorkin, E. N. (eds.): *World Congress on Computational Mechanics* (CD-ROM), Buenos Aires 1998.
9. Cramer, H., Findeiß, R., Steinl, G., Wunderlich, W.: An approach to the adaptive finite element analysis in associated and non-associated plasticity considering localization phenomena. *Computer Methods in Applied Mechanics and Engineering* **176**(1-4) (1999), 187–202.

10. Cramer, H., Findeiß, R., Wunderlich, W.: Elastic-plastic consolidation analysis of saturated soil by a time discontinuous method. In Wunderlich, W. (ed.): *European Congress on Computational Mechanics* (CD-ROM), München 1999.
11. Ehlers, W.: Poröse Medien – ein kontinuumsmechanisches Modell auf der Basis der Mischungstheorie. Forschungsbericht aus dem Fachbereich Bauwesen 47, Universität-GH-Essen 1989.
12. Ehlers, W.: A single surface yield function for geomaterials. *Archive of Applied Mechanics* **65** (1995), 246–259.
13. Ehlers, W., Volk, W.: On theoretical and numerical methods in the theory of porous media based on polar and non-polar elasto-plastic solid material. *International Journal of Solids and Structures* **35**(34-35) (1998), 4597–4617.
14. Ehlers, W., Ellsiepen, P., Ammann, M.: Localization phenomena in saturated and empty frictional porous materials computed by time- and space-adaptive methods. In Wunderlich, W. (ed.): *European Congress on Computational Mechanics* (CD-ROM), München 1999.
15. Findeiß, R.: *Ein orts- und zeitadaptives Finite-Element-Verfahren zur Traglastanalyse wassergesättigter Böden.* Dissertation, TU München 2001.
16. Johnson, C., Hansbo, P.: Adaptive finite element methods in computational mechanics. *Computational Methods in Applied Mechanics and Engineering* **101** (1992), 143–181.
17. Perić, D., Yu, J., Owen, D. R. J.: On error estimates and adaptivity in elastoplastic solids: Applications to the numerical simulation of strain localization in classical and Cosserat continua. *International Journal for Numerical Methods in Engineering* **37** (1994), 1351–1379.
18. Simo, J. C., Hughes, T. J. R.: *Computational Inelasticity.* Springer-Verlag, New York 1998.
19. Sloan, S. W., Abbo, A. J.: Biot consolidation analysis with automatic time stepping and error control. Part 1: Theory and implementation. *International Journal for Numerical and Analytical Methods in Geomechanics* **23** (1999), 467–492.
20. Truesdell, C., Toupin, R. A.: The Classical Field Theories. In Flügge, S. (ed.): *Handbuch der Physik*, III/1, Springer-Verlag, Berlin 1960, pp. 226–902.
21. Wiberg, N.-E., Li, X. D.: Adaptive finite element procedures for linear and nonlinear dynamics. In Idelsohn, S. R., Oñate, E., Dvorkin, E. N. (eds.): *World Congress on Computational Mechanics* (CD-ROM), Buenos Aires 1998.
22. Wunderlich, W., Cramer, H., Steinl, G. (1998): An adaptive finite element approach in associated and non-associated plasticity considering localization phenomena. In Ladevèze, P., Oden, J. T. (eds.): *Advances in Adaptive Computational Methods in Mechanics*, Elsevier Science, Amsterdam 1998, pp. 293–308.
23. Wunderlich, W., Kutter, H. K., Cramer, H., Rahn, W.: *Finite-Element-Modelle für die Beschreibung des Materialverhaltens von Fels.* Techn.-Wiss. Mitteilungen Nr. 81-10, Institut für Konstruktiven Ingenieurbau, Ruhr-Universität Bochum 1981.
24. Zienkiewicz, O. C., Zhu, J. Z.: Adaptivity and mesh generation. *International Journal for Numerical Methods in Engineering* **32** (1991), 783–810.

Biphasic description of viscoelastic foams by use of an extended Ogden-type formulation

Wolfgang Ehlers, Bernd Markert, and Oliver Klar

University of Stuttgart, Institute of Applied Mechanics (CE),
70550 Stuttgart, Germany

Abstract. Soft polymeric foams exhibit distinct relaxation and creep phenomena which in combination with the cellular micro structure result in the outstanding mechanical characteristics of this type of porous materials. It is the goal of this contribution to present an appropriate biphasic continuum mechanical model based on the Theory of Porous Media (TPM) which allows the description of viscoelastic foams at a suitable means of computational costs. To reproduce the complex behaviour of the cellular polymer skeleton an extended *Ogden*-type viscoelasticity formulation is embedded into the porous media concept. Thus, the macroscopic model accounts for all relevant physical properties, i. e. the porous cell structure, the moving and interacting pore-fluid, and the intrinsic viscoelasticity of the polymeric matrix material.

1 Introduction

Foamed polymers (soft foams) are subject to large viscoelastic deformations, mostly compression, during their practical application, e. g. as bumpers, packaging or seat cushions. They consist of a multiphasic micro structure built by a combination of open and/or closed fluid-filled cells. In particular, each cell of the micro structure undergoes complex deformation mechanisms under external loads resulting in the characteristic high-grade nonlinear stress-strain behaviour of the macroscopic foam, cf. Figure 1.

With new methods, e. g. the computer tomography (CT), it is possible to obtain detailed information of the inner cellular structure. However, even by increased computer power, the numerical simulation of real applications on the micro scale using discrete cell models is not reasonable until now. Furthermore, the description of the moving pore-fluid and, as a consequence, the frictional drag forces on the cell faces is problematic. A further possibility to analyse the response of foams to stress is to combine the beam theory with scaling laws, i. e. to approximate an average cell by a simple beam model and determine the effective material parameters via homogenization, cf. Gibson and Ashby [22]. This leads to a good description concerning honeycombs and open-cell foams but the model reaches its limits in describing closed-cell foams and, in general, in describing the pore-fluid motion properly. Thus, statistical, multiphasic continuum mechanical models are necessary to successfully represent the behaviour of foam materials under external loads.

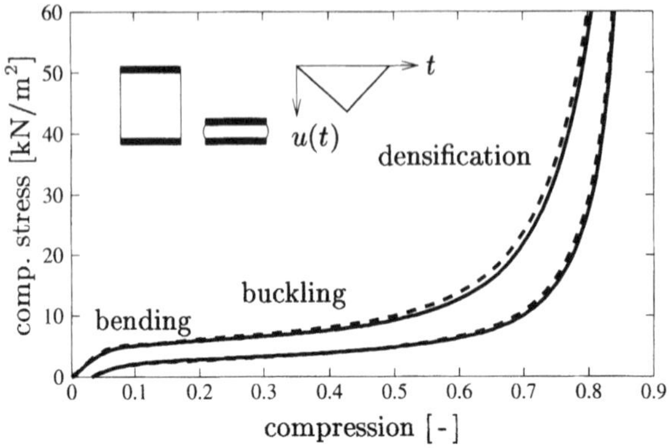

Fig. 1. Characteristic stress-strain behaviour of a cellular polymer foam divided into three regions which are governed by the deformation mechanisms on the micro scale, i. e. bending, buckling and densification of the cell faces. Here: Results of displacement driven uniaxial hysteresis experiments on cubic open-cell polyurethane (PU) foam specimens (size $70 \times 70 \times 70\,\mathrm{mm}^3$, bulk density $48\,\mathrm{kg/m}^3$, deformation rate 7 mm/s, max. compression 86.7 %).

In the present contribution, the macroscopic description of fluid-saturated polymer foams is considered in the framework of the well-founded Theory of Porous Media (TPM), see e. g. de Boer [4], Bowen [5], Ehlers [12–14]. Following this, we proceed from a biphasic description of polymeric foams consisting of a materially incompressible, cellular polymer matrix saturated by an incompressible or a compressible viscous pore-fluid, i. e. a pore-liquid or a pore-gas. The assumption of a materially incompressible solid skeleton is used based on the fact that the material compressibility of the polymer material itself is much smaller than the bulk compressibility of the entire foam. Therefore, there exists a so-called point of compaction, which is defined as that deformation state, where all pores are closed, i. e. where all microscopic cell faces are lying on each other building a single-phase solid, and no further volume reduction is possible. This concept of the compaction point implies the characteristic property of structural densification of solid foams. The energy absorbing behaviour of the cellular polymer matrix is described by an appropriate *Ogden*-type finite viscoelasticity law which can be descriptively deduced from rheological and thermodynamical considerations. Concerning the description of finite deformation viscoelasticity of non-porous solids, a lot of work can be found in the literature, see e. g. Lion [26], Reese and Govindjee [32], and le Tallec et al. [24] and citations therein. The solid-fluid interaction is taken into account by the viscous drag force, where a deformation-dependent permeability is considered, since experiments have shown that it plays an important role on the transient compressive response of open-cell foams at higher deformation rates. For simplicity, thermal effects as well as mass exchanges between the two constituents are excluded. The arising pure mechanical, finite viscoelastic two-phase model is capable to describe the complex material response of open- and closed-cell polymeric foams under mechanical loading, where the biphasic formulation is essential for the consideration of the basic physical properties.

The numerical treatment of the presented model is carried out by use of the finite element method (FEM), where the semi-discretization of the governing field equations in the spatial domain leads to a system of differential-algebraic equations (DAE) in the time domain. The numerical integration of this semi-discrete system can be efficiently performed by suitable time integration schemes, e. g. *Runge-Kutta* methods (Diebels et al. [11], Ellsiepen [20]), where an embedded error estimator and a time step control can be added. For details concerning the general treatment of DAE systems, cf. e. g. Brenan et al. [7] or Hairer and Wanner [23]. For the practical application, the mathematical model has to be adjusted to real material behaviour by use of a numerical optimization technique. In this contribution, the sequential quadratic programming (SQP) method (Spellucci [35, 36]) is used to minimize the deviation between experiment and numerical simulation. In particular, the SQP procedure is applied to adapt the biphasic model to a highly porous, open-cell polyurethane (PU) foam. Finally, numerical simulations of really finite compression experiments show the applicability of the presented macroscopic model and reveal the necessity of the biphasic formulation with an independent fluid-phase.

2 Theoretical framework

In the framework of the Theory of Porous Media (TPM), a fluid-saturated cellular foam can be treated as an immiscible mixture of constituents φ^α ($\alpha = S$: cellular solid skeleton; $\alpha = F$: pore-fluid) which are assumed to be in a state of ideal disarrangement. Following this, the prescription of a

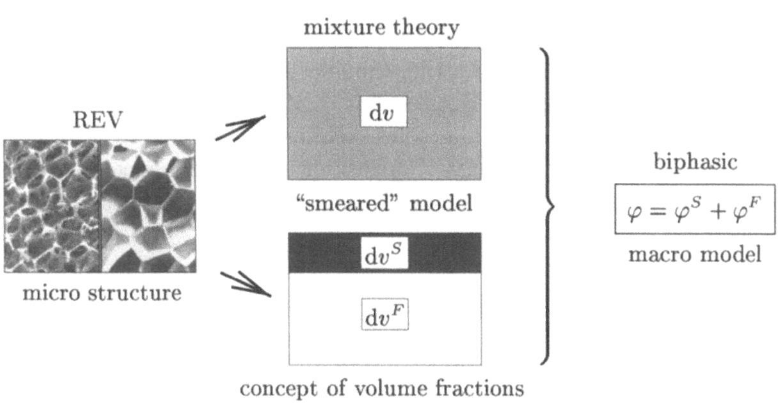

Fig. 2. REV of a foam with real micro structure and biphasic TPM macro model.

real or a virtual averaging process over a representative elementary volume (REV) leads to a model φ of superimposed and interacting continua, i. e. the homogenized or "smeared" model. In the arising biphasic macro model (Figure 2), the local structure of the mixture is represented by scalar variables, the volume fractions

$$n^\alpha = \frac{dv^\alpha}{dv}\,, \tag{1}$$

which are defined as the local ratios of the constituent volumes v^α with respect to the bulk volume v. Thus, assuming fully saturated conditions, i. e. avoiding any vacant space, the saturation condition yields

$$n^S + n^F = 1\,. \tag{2}$$

The model under consideration incorporates three independent fields, the solid displacement $\mathbf{u}_S$, the seepage velocity $\mathbf{w}_F$, and the effective fluid pressure p. The corresponding three equations for quasi-static considerations can be obtained from the kinematics, the balance relations, and the constitutive equations discussed in detail by Ehlers [12–14]:

- Mixture balance of momentum:

$$\operatorname{div}(\mathbf{T}_E^S - p\,\mathbf{I}) + (n^S\rho^{SR} + n^F\rho^{FR})\,\mathbf{b} = \mathbf{0}\,, \tag{3}$$

- Fluid balance of mass:

$$n^F\,(\rho^{FR})'_S + \rho^{FR}\operatorname{div}(\mathbf{u}_S)'_S + \operatorname{div}(n^F\rho^{FR}\,\mathbf{w}_F) = 0\,, \tag{4}$$

- Generalized *Darcy*'s filter law:

$$\frac{1}{\gamma^{FR}}\,\mathbf{K}^F\,(\operatorname{grad} p - \rho^{FR}\,\mathbf{b}) + n^F\,\mathbf{w}_F = \mathbf{0}\,. \tag{5}$$

Herein, $\mathbf{T}_E^S$ is the solid Cauchy extra (effective) stress, $\mathbf{I}$ is the identity tensor, $\mathbf{b}$ is the body force density, and $\gamma^{FR} = \rho^{FR}g$ is the effective fluid weight, where g is the gravitation constant. Furthermore, $\mathbf{K}^F$ is the positive definite permeability tensor and $(\,\cdot\,)'_S$ denotes the material time derivative of $(\,\cdot\,)$ corresponding to the individual motion of φ^S. In this representation, the pore-fluid can either be compressible, i. e. the effective (material) fluid density is a function of the pore-fluid pressure, $\rho^{FR} = \rho^{FR}(p)$, or incompressible, i. e. $\rho^{FR} = \text{const.}$, whereas the cellular solid skeleton is supposed to be materially incompressible, i. e. $\rho^{SR} = \text{const.}$ Concerning gas-filled foams, the pore-gas is assumed to be governed by the ideal gas law (*Boyle-Mariotte*'s law)

$$\rho^{FR}(p) = \frac{M_m^F}{R\,\theta}\,p \qquad \text{or} \qquad p(\rho^{FR}) = \frac{R\,\theta}{M_m^F}\,\rho^{FR}\,, \tag{6}$$

where R is the universal gas constant, θ the absolute *Kelvin*'s temperature, and M_m^F the molecular mass of φ^F. Note that in case of a materially incompressible pore-liquid ($\rho^{FR} = \text{const.}$), the pore pressure p acts as a *Lagrange*an multiplier and is simply determined from the boundary conditions under study.

Restricting to isotropic permeability conditions, the permeability tensor in *Darcy*'s law[1] (5) is given by

$$\mathbf{K}^F = \frac{\gamma^{FR}}{\mu^{FR}} K^S \mathbf{I}, \tag{7}$$

where μ^{FR} is the effective pore-fluid viscosity and K^S the intrinsic permeability. In general, the intrinsic permeability depends on the deformation state, i. e. the value of K^S decreases when the pore volume decreases. Thus, it is assumed that the permeability depends on the porosity n^F of the solid skeleton. The deformation dependency of the permeability parameter can be described by a power function (cf. Eipper [19]):

$$K^S(n^F) = K_0^S \left(\frac{n^F}{n_{0S}^F} \right)^{\kappa}, \tag{8}$$

where K_0^S is the initial permeability and the exponent κ governs the deformation dependency. Therefore, K^S governs the property of an open-cell, i. e. permeable ($K^S > 0$), or a closed-cell, i. e. impermeable ($K^S \to 0$), foam.

3 Finite viscoelasticity law

To describe the intrinsic dissipative phenomena of the cellular polymer skeleton, an adequate viscoelastic material formulation is required including the property of structural densification towards the point of compaction. The fundamental approach is based on the one-dimensional rheological structure of the generalized *Maxwell* model (Figure 3), an elementary rheological model to describe the complex behaviour of a viscoelastic solid. The governing one-dimensional equations describing this parallel assembly of one *Hooke* element (elastic spring) and N *Maxwell* elements (spring and dashpot in series) can easily be obtained from the rheological structure and the constitutive laws representing the individual elements, see e. g. Tschoegl [37]. Restricting to the geometrically linear approach, the three-dimensional formulation results from a formal extension of these governing equations and can be included straight forward into porous media theories (Ehlers and Markert [18]).

In the framework of a finite deformation theory of porous materials with elastic and inelastic (here: viscous) behaviour, it is convenient to proceed from a multiplicative split of the solid deformation gradient

$$\mathbf{F}_S = (\mathbf{F}_{Se})_n \, (\mathbf{F}_{Si})_n \,, \quad n = 1, ..., N \tag{9}$$

into elastic parts $(\mathbf{F}_{Se})_n$ and inelastic parts $(\mathbf{F}_{Si})_n$ (Sidoroff [34]). From elasto-plasticity, it is commonly known that the concept of the multiplicative decomposition of deformation gradients (Lee [25]) is connected with the

[1] Note in passing that in case of high flow rates, the linear *Darcy* ansatz (5) has to be extended towards a *Forchheimer*-type formulation [21] since the increasing inertial forces strongly affect the permeability properties (cf. Bear [3]).

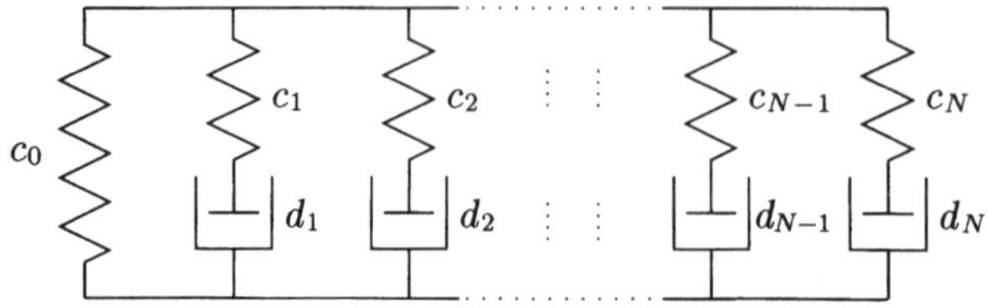

Fig. 3. Generalized *Maxwell* model. c_0, c_n, and d_n $(n = 1, \dots, N)$ denote the elasticity and viscosity (damping) constants of the respective elements.

suggestion of a stress-free intermediate configuration (Figure 4), where the purely inelastic state of deformation is frozen into the memory of the material. Furthermore, from thermodynamics with internal state variables (Coleman

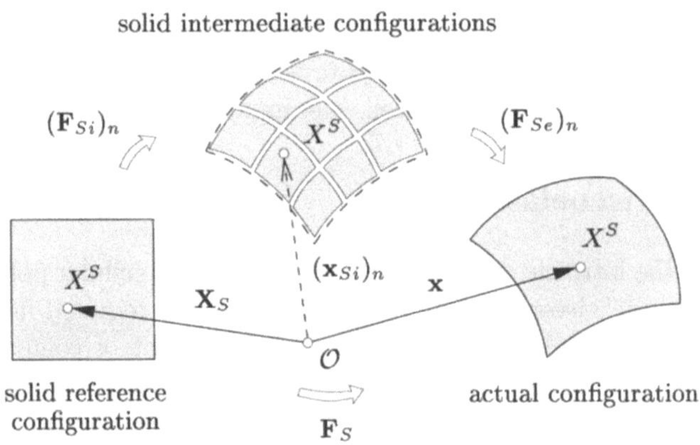

Fig. 4. Solid configurations.

and Gurtin [9]) and rheological considerations (cf. e. g. Reese and Govindjee [32]), one obtains the ansatz of a decomposed solid *Helmholtz* free energy density and a decomposed solid extra stress

$$
\begin{aligned}
\psi^S[\mathbf{F}_S, (\mathbf{F}_{Se})_n] &= \psi^S_{EQ}[\mathbf{F}_S] + \psi^S_{NEQ}[(\mathbf{F}_{Se})_n]\,, \\
\boldsymbol{\tau}^S_E[\mathbf{F}_S, (\mathbf{F}_{Se})_n] &= \boldsymbol{\tau}^S_{EQ}[\mathbf{F}_S] + \boldsymbol{\tau}^S_{NEQ}[(\mathbf{F}_{Se})_n]\,, \quad n = 1, ..., N
\end{aligned}
\tag{10}
$$

into equilibrium parts (Index *EQ*) describing the basic elasticity and non-equilibrium parts (Index *NEQ*) vanishing in the thermodynamic equilibrium. Note that the *Kirchhoff* extra stress is a weighted *Cauchy* stress $\boldsymbol{\tau}^S_E = J_S\,\mathbf{T}^S_E$, where $J_S = \det \mathbf{F}_S$ is the solid *Jacobian*.

In general, the stress-strain behaviour of the cellular polymer matrix is very complex and high-grade non-linear concerning both the basic elasticity and the viscoelastic overstress. Therefore, assuming isotropic material behaviour, we proceed from an *Ogden*-type material formulation (cf. Ogden [28, 29]) which is extended towards an inelastic porous media application. In particular, volumetric extension terms are developed based on the work by Eipper [19] which describe the finite volume change including the concept of the compaction point. Thus, the equilibrium part of the solid extra stress can be obtained by

$$\boldsymbol{\tau}^S_{EQ} = \mu^S_0 \sum_{j=1}^{3} \sum_{k=1}^{K_0} \mu^*_k \left(\lambda_j^{\alpha_k/2} - 1\right) \mathbf{N}_j + \\ + \frac{\Lambda^S_0}{\gamma_0 - 1 + \dfrac{\gamma_0 + 1}{(1 - n^S_{0S})^2}} \left(J_S^{\gamma_0} - \frac{J_S\,(1 - n^S_{0S})^{\gamma_0}}{(J_S - n^S_{0S})^{\gamma_0 + 1}} + \frac{J_S\, n^S_{0S}}{1 - n^S_{0S}} \right) \mathbf{I}, \tag{11}$$

whereas the non-equilibrium part is computed from

$$\boldsymbol{\tau}^S_{NEQ} = \sum_{n=1}^{N} \Bigg(\mu^S_n \sum_{j=1}^{3} \sum_{k=1}^{K_n} (\mu^*_k)_n \left[(\lambda_{ej})_n^{(\alpha_k)_n/2} - 1 \right] (\mathbf{N}_{ej})_n + \\ + \frac{\Lambda^S_n}{\gamma_n - 1 + \dfrac{\gamma_n + 1}{\left[1 - (n^S_i)_n\right]^2}} \\ \left[(J_{Se})_n^{\gamma_n} - \frac{(J_{Se})_n \left[1 - (n^S_i)_n\right]^{\gamma_n}}{\left[(J_{Se})_n - (n^S_i)_n\right]^{\gamma_n + 1}} - \frac{(J_{Se})_n\,(n^S_i)_n}{1 - (n^S_i)_n} \right] \mathbf{I} \Bigg). \tag{12}$$

In (11) and (12), λ_j $(j = 1, \ldots, 3)$ are the eigenvalues of the *Cauchy-Green* deformation tensors $\mathbf{C}_S = \mathbf{F}^T_S \mathbf{F}_S$ or $\mathbf{B}_S = \mathbf{F}_S \mathbf{F}^T_S$, respectively, $(\lambda_{ej})_n$ $(j = 1, \ldots, 3)$ are the eigenvalues of the elastic *Cauchy-Green* deformation tensors $(\mathbf{C}_{Se})_n = (\mathbf{F}^T_{Se})_n (\mathbf{F}_{Se})_n$ or $(\mathbf{B}_{Se})_n = (\mathbf{F}_{Se})_n (\mathbf{F}^T_{Se})_n$, respectively, and $\mathbf{N}_j = \partial\lambda_j / \partial\mathbf{B}_S$ and $(\mathbf{N}_{ej})_n = \partial(\lambda_{ej})_n / \partial(\mathbf{B}_{Se})_n$ denote the eigentensors corresponding to the eigenvalues λ_j and $(\lambda_{ej})_n$. Furthermore, μ^S_0 and μ^S_n are the first macroscopic *Lamé* constants and μ^*_k, α_k $(k = 1, \ldots, K_0)$ and $(\mu^*_k)_n$, $(\alpha_k)_n$ $(k = 1, \ldots, K_n)$ are the dimensionless *Ogden* parameters, where the number of required *Ogden* terms K_0 and K_n depends on the complexity of the material behaviour under study. Moreover, in the volumetric extension terms[2] Λ^S_0 and Λ^S_n are the second *Lamé* constants, γ_0 and γ_n are parameters that influence the volumetric non-linearity, $(J_{Se})_n = \det(\mathbf{F}_{Se})_n$ are the determinants of the elastic parts of the deformation gradient, n^S_{0S} is the initial solidity, and $(n^S_i)_n = n^S_{0S} \det(\mathbf{F}_{Si})^{-1}_n$ are the inelastic solid volume fractions with respect to the intermediate configuration. Note that all *Lamé* constants and *Ogden* parameters are macroscopic material parameters of the cellular solid skeleton structure and not of the microscopic polymer material itself.

[2] Note that the volumetric extension terms are independent from the type of the finite material formulation. In general, they can be added to any reasonable material law, e. g. to a simpler *Neo-Hooke*an ansatz (Ehlers and Markert [17]).

In order to guarantee the downward compatibility to *Hooke*an laws and to maintain stability (cf. Ball [1], Ogden [28, 30], and Reese [31]), the *Ogden* parameters have to fulfil the following conditions:

$$\begin{aligned}
&\sum_{k=1}^{K_0} \mu_k^* \, \alpha_k = 2 \wedge \{\mu_k^* > 0, \alpha_k > 1 \vee \mu_k^* < 0, \alpha_k < -1\}\,, \\
&\sum_{k=1}^{K_n} (\mu_k^*)_n \, (\alpha_k)_n = 2 \wedge \{(\mu_k^*)_n > 0, (\alpha_k)_n > 1 \vee (\mu_k^*)_n < 0, (\alpha_k)_n < -1\}\,.
\end{aligned} \tag{13}$$

Note that the given conditions guarantee the polyconvexity of the *Ogden* terms in the underlying strain energy function without taking into account the volumetric extensions which are itself polyconvex if

$$\gamma_0 \geq 1 \quad \text{and} \quad \gamma_n \geq 1\,, \tag{14}$$

respectively.

The inelastic strains as internal state variables are obtained from linear evolution equations formulated with respect to the intermediate configuration, since this configuration acts as the actual configuration of the viscous deformation:

$$\begin{aligned}
(\hat{\mathbf{D}}_{Si})_n &= \overset{4}{\hat{\mathbf{D}}}{}^{-1}_{n} \, (\hat{\boldsymbol{\tau}}^S_{\scriptscriptstyle NEQ})_n \,, \\
\overset{4}{\hat{\mathbf{D}}}{}^{-1}_{n} &= \frac{1}{2\eta_n^S} \, (\mathbf{I} \otimes \mathbf{I})^{\overset{23}{T}} - \frac{\zeta_n^S}{2\eta_n^S \, (2\eta_n^S + 3\zeta_n^S)} \, (\mathbf{I} \otimes \mathbf{I})\,.
\end{aligned} \tag{15}$$

Therein, $(\hat{\mathbf{D}}_{Si})_n$ are the inelastic solid deformation rates, $\overset{4}{\hat{\mathbf{D}}}{}^{-1}_{n}$ are the positive definite, isotropic, fourth order, viscous compliances, where η_n^S and ζ_n^S are the macroscopic viscosity parameters, and $(\hat{\boldsymbol{\tau}}^S_{\scriptscriptstyle NEQ})_n = (\mathbf{F}_{Se})_n^{-1} \, (\boldsymbol{\tau}^S_{\scriptscriptstyle NEQ})_n \, (\mathbf{F}_{Se})_n^{T-1}$ are the respective non-equilibrium stress tensors. Furthermore, the transposition $(\,\cdot\,)^{\overset{ik}{T}}$ indicates an exchange of the i-th and k-th basis systems included into the tensor basis of higher order tensors and the superscript $(\hat{\cdot})$ indicates the belonging to the intermediate configuration.

4 Numerical treatment

For the numerical treatment of quasi-static coupled solid-fluid problems in the sense of the TPM approach within the finite element method (FEM), weak forms of the governing field equations (3) and (4) are required. Therefore, after eliminating the seepage velocity by use of the *Darcy* filter law (5), the balance relations weighted by independent test functions and integrated over the spatial domain Ω with surface $\partial\Omega$ result in the respective weak formulations $\mathcal{G}_{\mathrm{MM}}$ and $\mathcal{G}_{\mathrm{MF}}$:

$$\begin{aligned} \mathcal{G}_{\mathrm{MM}}(\mathbf{u}_S, \delta\mathbf{u}_S, p) \equiv & \int_\Omega \operatorname{grad}\delta\mathbf{u}_S \cdot (\mathbf{T}_E^S - p\,\mathbf{I})\,\mathrm{d}v - \\ & - \int_\Omega \delta\mathbf{u}_S \cdot (n^S \rho^{SR} + n^F \rho^{FR})\,\mathbf{b}\,\mathrm{d}v - \int_{\partial\Omega} \delta\mathbf{u}_S \cdot \bar{\mathbf{t}}\,\mathrm{d}a = 0\,, \end{aligned} \tag{16}$$

$$\begin{aligned} \mathcal{G}_{\mathrm{MF}}(\mathbf{u}_S, p, \delta p) \equiv & \int_\Omega \delta p \left[n^F (\rho^{FR})'_S + \rho^{FR} \operatorname{div} (\mathbf{u}_S)'_S \right] \mathrm{d}v + \\ & + \int_\Omega \operatorname{grad}\delta p \cdot \left[\frac{K^S}{\mu^{FR}} \rho^{FR} \left(\operatorname{grad} p - \rho^{FR}\,\mathbf{b} \right) \right] \mathrm{d}v + \int_{\partial\Omega} \delta p\, \bar{q}\,\mathrm{d}a = 0\,. \end{aligned} \tag{17}$$

Herein, $\delta\mathbf{u}_S$ and δp are the test functions corresponding to the solid displacement $\mathbf{u}_S$ and the pore pressure p, $\bar{\mathbf{t}} = (\mathbf{T}_E^S - p\,\mathbf{I})\,\mathbf{n}$ is the external load vector, $\bar{q} = n^F \rho^{FR}\,\mathbf{w}_F \cdot \mathbf{n}$ denotes the filter mass flow of the fluid draining through the surface $\partial\Omega$, and $\mathbf{n}$ is the outward oriented unit surface normal. Note that the surface traction $\bar{\mathbf{t}}$ acts on both the solid and the fluid phase. This is essential for the formulation of boundary-value problems as no separation of the boundary conditions into actions on the different phases is needed and, therefore, physically meaningful boundary conditions can be applied. For a more detailed discussion of the weak formulation, see Diebels and Ehlers [10], Ehlers and Ellsiepen [16], and Ellsiepen [20].

Starting from the weak formulations (16) and (17), the spatial semi-discretization is carried out by the finite element method (FEM). Applying the *Bubnov-Galerkin* procedure, the unknown fields $\mathbf{u}_S$ and p and the corresponding test functions $\delta\mathbf{u}_S$ and δp are approximated by the same shape functions. Concerning the treatment of coupled solid-fluid problems within the finite element method, mixed element formulations are used for the spatial discretization. In mixed formulations, the inf-sup condition (*Ladyshenskaya-Babuška-Brezzi* (LBB) condition) is a crucial criterion for the stability of the numerical solution, see e. g. Brezzi and Fortin [8]. One possible choice of a stable discretization is to use quadratic shape functions for the solid displacement $\mathbf{u}_S$ and linear shape functions for the pressure p which is known as the *Taylor-Hood* element. In the three-dimensional case, the quadratic-linear approximation leads to an enormous number of mid nodes in finer meshes which causes a large number of degrees of freedom (DOF) increasing the computational costs rapidly. In order to overcome this problem, one has to reduce the order of the displacement interpolation. This leads to an equal interpolation element with a linear approximation for both the solid displacement $\mathbf{u}_S$ and the pore pressure p. In general, this mixed element can be used for the computation of coupled problems but users have to be aware that their results are strongly mesh dependent. In particular, the pore pressure exhibits a very strange instability due to so-called spurious pressure modes (Brezzi and Fortin [8]). This unpredictable behaviour of equal interpolation methods led to the introduction of the more sophisticated *Taylor-Hood* elements as mentioned above.

However, to get a stable pressure continuous element while keeping the linear approximation for the displacement, one has to add bubble functions to the displacement field. This results in the so-called MINI element, where the additional bubbles enlarge the trial space for the displacement which guarantees stability (Braess [6]). Since the bubble functions are restricted to a single finite element, the barycentric bubble nodes can be removed via static condensation. Therefore, for the numerical treatment of large coupled solid-fluid problems in the sense of the TPM approach, the MINI element is the best choice concerning quality and computational costs in the framework of the finite element method. For illustration (Figure 5), the mixed element formulations are shown in a triangle and can directly be transferred to tetrahedrons.

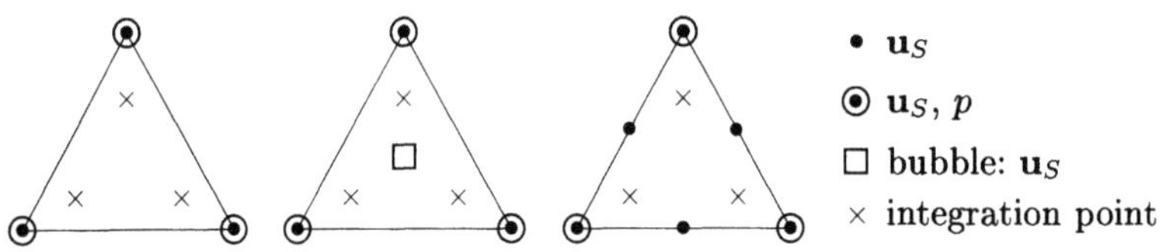

Fig. 5. Mixed element formulations in a triangle: linear-linear approximation T1P1 (left), MINI element (middle), *Taylor-Hood* element T2P1 (right).

For the time discretization, the degrees of freedom $\mathbf{u}_S$ and p at all FEM nodes are collected in a vector $\boldsymbol{u}$ of external variables, whereas the internal (history) variables $(\mathbf{C}_{Si})_n$, i. e. the inelastic deformation tensors, at all integration points are collected in a vector $\boldsymbol{q}$. Then, with the vector of unknowns $\boldsymbol{y} := (\boldsymbol{u}, \boldsymbol{q})^T$, the semi-discrete initial-value problem in time with $t \geq t_0$ and $\boldsymbol{y}(t_0) = \boldsymbol{y}_0$ can be formulated as follows (Ehlers and Ellsiepen [15]):

$$\begin{aligned} \boldsymbol{F}\big[\,\boldsymbol{y}, (\boldsymbol{y})'_S, t\big] &\equiv \begin{bmatrix} \boldsymbol{F}_1\big[t, \boldsymbol{u}, (\boldsymbol{u})'_S, \boldsymbol{q}\big] \\ \boldsymbol{F}_2\big[t, \boldsymbol{q}, (\boldsymbol{q})'_S, \boldsymbol{u}\big] \end{bmatrix} \\ &\equiv \begin{bmatrix} \boldsymbol{M}\,(\boldsymbol{u})'_S + \boldsymbol{k}\big[\boldsymbol{u}, \boldsymbol{q}\big] - \boldsymbol{f} \\ (\boldsymbol{q})'_S - \boldsymbol{g}\big[\boldsymbol{q}, \boldsymbol{u}\big] \end{bmatrix} \stackrel{!}{=} \mathbf{0}\,. \end{aligned} \tag{18}$$

Therein, the first equation $\boldsymbol{F}_1$ represents the field equations, where $\boldsymbol{M}$ is the generalized mass matrix, $\boldsymbol{k}$ is the generalized stiffness vector, and $\boldsymbol{f}$ is the generalized external force vector. The second equation $\boldsymbol{F}_2$ represents the evolution equations of the viscoelastic model which are evaluated at the integration points of the finite elements. In case of a quasi-static or a materially incompressible description, the generalized mass matrix does not possess the full rank. In particular, the quasi-static formulation of the mixture balance of momentum (16) delivers no contribution to the generalized mass matrix which is related to $(\mathbf{u}_S)'_S$. Furhermore, concerning the materially incompressible model, no evolution equation for p exists and, as a consequence, $\boldsymbol{M}$ has no entry corresponding to $(p)'_S$. Thus, (18) becomes an index one system of differential-algebraic equations (DAE) of first order in time which must be handled by a suitable time integration method. In order to integrate large FEM systems with internal variables, one-step methods with embedded error estimators, e. g. singly diagonally implicit *Runge-Kutta* methods (SDIRK), provide a suitable means at moderate storage and computational costs (Diebels et al. [11], Ellsiepen [20]).

Within the finite element discretization, the evolution equations (15) for the inelastic deformation tensors $(\mathbf{C}_{Si})_n$ are computed only at the integration points of the numerical quadrature. Therefore, the evolution equations are rewritten with respect to the reference configuration leading to

$$\begin{aligned}
&[(\mathbf{C}_{Si})_n]'_S - \frac{1}{\eta^S_n}\,(\mathbf{C}_{Si})_n\,(\mathbf{S}^S_{NEQ})_n\,(\mathbf{C}_{Si})_n + \\
&+ \frac{\zeta^S_n}{\eta^S_n\,(2\eta^S_n + 3\zeta^S_n)}\,[(\mathbf{S}^S_{NEQ})_n \cdot (\mathbf{C}_{Si})_n]\,(\mathbf{C}_{Si})_n = \mathbf{0}\,,
\end{aligned} \tag{19}$$

where $\mathbf{S}^S_{NEQ} = \mathbf{F}_S^{-1}\,\boldsymbol{\tau}^S_{NEQ}\,\mathbf{F}_S^{T-1}$ is the second *Piola-Kirchhoff* solid stress tensor. The ordinary differential equations (ODE) (19) represent the second equation $\boldsymbol{F}_2$ of (18) and are treated with the same time integration method as is used for the global FEM system. The resulting non-linear set of equations is then solved by a local *Newton* iteration at the *Gauß* point level.

5 Model adaptation

For the application of the presented biphasic viscoelastic model, it is necessary to adapt the model to real material behaviour. Therefore, the material specific and unknown parameters that are theoretically introduced in the constitutive equations (8), (11), (12), and (15) have to be identified. This determination is carried out by comparing the results of well defined experiments that show all relevant physical properties of the considered material with their numerical simulation using the chosen material model. A common procedure for this purpose is to vary strategically the searched material constants until the consistency of the numerical simulation and the experimental data reaches a satisfactory result. Usually, this technique leads to a non-linear optimization problem, i. e. the minimization of a least squares functional that represents the deviation between simulation and experiment in consideration of equality and inequality constraints which built the restrictions for the unknown parameters, e. g. the stability and polyconvexity conditions (13) and (14). Thus, it is obvious that the number of unknowns for the optimization process equals the number of material parameters in the constitutive equations.

In order to get interpretable and manageable solutions, the possibility to reduce the number of unknowns is decisive for the stability and convergence of the overall optimization process. Therefore, the constitutive split of the solid extra stress into an equilibirium and a non-equilibrium part (10) is used to separate the identification into the unknown parameters of the basic elasticity and the one of the intrinsic viscoelasticity as well. But to realize this separation, experiments are needed, where the response of the test piece to an external load can be associated either to the equilibrium or to the non-equilibrium part. Hence, in the case of viscoelastic material behaviour, first experiments are necessary which exhibit the purely elastic response or several points of it, respectively. Subsequently, the viscoelastic stresses (overstresses) can be obtained from standard hysteresis experiments since the time independent part is already known. In particular, for the identification of the basic elasticity one takes advantage of the properties of viscoelastic materials which show distinct hystersis effects depending on the deformation rate but behave purely elastic if the load is applied infinitely slow. Since such tests are technically difficult to realize, experiments with incorporated holding times which allow the complete relaxation of the overstresses at several deformation states are used to determine discrete optimization points belonging to the equilibrium part. For the identification of the intrinsic viscoelasticity, it has to be ensured that the deformation velocity of the identification tests does not cause any pore pressure, i. e. that the pore-gas can escape without developing any pore pressure. Otherwise, if the deformation velocity is too fast, the pore pressure is part of the time-dependent overstress.

In order to improve the consistency of the numerical simulation and the experimental data, a criterion and a strategy for this problem has to be found. Therefore, in general, the deviation, defined as the squared difference between several points ($h = 1, \dots, H$) of the experimental stresses $\sigma_{h,exp}$ from the identification tests and the simulated stresses $\sigma_{h,sim}$, is used and has to be minimized. This procedure leads to the minimization of the least squares functional in consideration of N_{ec} equality $h_j(\boldsymbol{\kappa})$ and N_{ic} inequality $g_k(\boldsymbol{\kappa})$ constraints for the set of unknown parameters $\boldsymbol{\kappa}$,

$$
\begin{aligned}
& f(\boldsymbol{\kappa}) = \sum_{h=1}^{H} w_h \left[\sigma_{h,sim}(\boldsymbol{\kappa}) - \sigma_{h,exp}\right]^2 \quad \rightarrow \quad \min \\
& \text{with} \quad \begin{cases} h_j(\boldsymbol{\kappa}) = 0\,; & j = 1, \dots, N_{ec}, \\ g_k(\boldsymbol{\kappa}) \geq 0\,; & k = 1, \dots, N_{ic}, \end{cases}
\end{aligned}
\tag{20}
$$

wherein H is the number of optimization points and w_h are weighting factors. The weighting factors can be used to emphasize some segments of the stress-strain relation and to differentiate between certain and uncertain experimental results. Note in passing that a weighting is already included by the choice of discrete optimization points.

An efficient procedure for the determination of local minima is the sequential quadratic programming (SQP) method. Although the SQP method as any other gradient based procedure does not ensure the determination of the global minimum, the algorithm is used due to the desired convergence behaviour. The disadvantage to probably end up in a local minimum can be reduced by investigating the intial parameter dependency. Starting from the optimization problem (20), a dual *Lagrange*an equation can be set up which summarizes the function $f(\boldsymbol{\kappa})$ and the equality and inequality constraints $h_j(\boldsymbol{\kappa})$ and $g_k(\boldsymbol{\kappa})$ with the help of the *Lagrange*an multipliers μ_j and λ_k, where the minimization problem has changed into a saddle point problem:

$$\mathcal{L}(\boldsymbol{\kappa}, \boldsymbol{\mu}, \boldsymbol{\lambda}) = f(\boldsymbol{\kappa}) - \sum_{j=1}^{N_{ec}} \mu_j \, h_j(\boldsymbol{\kappa}) - \sum_{k=1}^{N_{ic}} \lambda_k \, g_k(\boldsymbol{\kappa}) \quad \rightarrow \quad \text{stat.} \tag{21}$$

In order to determine the optimal parameter set, the *Karush-Kuhn-Tucker* (KKT) conditions must be fulfilled in the point of solution. Here the KKT conditions are represented by the partial derivatives of first order of (21) with respect to the parameters and the *Lagarange*an multipliers:

$$\nabla_{\boldsymbol{\kappa}} \mathcal{L} = \mathbf{0}\,, \quad \nabla_{\mu} \mathcal{L} = \mathbf{0}, \quad \lambda_k \, \nabla_{\lambda_k} \mathcal{L} = 0\,, \quad \nabla_{\lambda} \mathcal{L} \geq \mathbf{0}\,, \quad \boldsymbol{\lambda} \geq \mathbf{0}\,. \tag{22}$$

For the further calculation, the resulting non-linear set of equations

$$\boldsymbol{F}(\boldsymbol{x}) = \boldsymbol{F}(\boldsymbol{\kappa}, \boldsymbol{\mu}, \boldsymbol{\lambda}) = \begin{bmatrix} \nabla_{\boldsymbol{\kappa}} \mathcal{L} \\ \nabla_{\mu} \mathcal{L} \\ \lambda_k \, \nabla_{\lambda_k} \mathcal{L} \end{bmatrix} = \mathbf{0} \quad \text{with} \quad \boldsymbol{x} := \begin{pmatrix} \boldsymbol{\kappa} \\ \boldsymbol{\mu} \\ \boldsymbol{\lambda} \end{pmatrix} \tag{23}$$

can be solved with *Newton*'s method of approximation. Due to the cost-intensive computation of the *Jacobi* matrix in the standard *Newton* method, a quasi-*Newton* procedure is used, where the *Jacobi*an is approximated by a cheaper computed matrix. A common method is to use the so-called *Broyden-Fletcher-Goldfarb-Shanno* (BFGS) update (Luenberger [27]) which has the advantage that the approximation of the *Hesse*an matrix of the original *Lagrange* function is always positive definite. The positive definiteness of the *Hesse*an matrix represents the sufficient condition for a local minimum in the point of solution, i. e. by use of the *BFGS* method in each iteration step, a decent towards a minimum is ensured. Moreover, an improved parameter set can be computed with an incremental operator by minimizing a merit function together with the *Lagrange*an function leading to a better convergence of the SQP method. For more details on the SQP method, the reader is referred to the works by Bazaraa et al. [2] and Spellucci [35, 36].

6 Application and Examples

In the present contribution, the model is applied to a highly porous, viscoelastic, open-cell polyurethane (PU) foam (bulk density $48\,\text{kg/m}^3 \rightarrow$ porosity

96 %) which finds its application, e. g. as seat cushions, in the automotive industry. Following the ideas of the preceding section, uniaxial compression experiments were performed in our laboratory on cubic PU foam specimens of size $70 \times 70 \times 70\,\text{mm}^3$ (Figure 6). The foam specimens were cemented on aluminium platens at the top side and at the bottom side and were displacement driven loaded and unloaded with a constant deformation rate of $v(t) = 7\,\text{mm/s}$ and a maximal compression of 86.7 %, whereas the lateral walls were unconfined. During the loading and unloading cycles, several holding

Fig. 6. Uniaxial compression experiment performed on cubic PU foam samples.

times of 30 min and 120 min were incorporated while recording the responding force. The experimental results are shown in the stress-strain diagram in Figure 7. It can easily be seen that even holding times of 120 min are too short, since the overstresses are not completely relaxed. Therefore, assuming a purely viscoelastic behaviour, i. e. no plasticity or damage, the mid points of the relaxed stresses between the loading and unloading cycles were taken as input for the numerical adjustment of the *Ogden* law (11) describing the equilibrium stress, where only one *Ogden* term ($K_0 = 1$) is used. In particular, the optimization algorithm DONLP2 (© Spellucci) is applied which is based on a variation of the SQP method and the gradient projection technique (Spellucci [35, 36]). Running a numerical simulation of the uniaxial experiment with PANDAS[3], Figure 8 shows the perfect fit of the basic elasticity to the experimental data, see Figure 7. Note that all finite element simulations are carried out fully three-dimensional, since, in general, the developed pore pressure leads to an inhomogeneous stress state.

After the adjustment of the basic elasticity model, the parameters of the viscous material law are identified from hysteresis experiments without holding times. Since the viscous behaviour is time-dependent and, in case of the PU foam, high-grade non-linear, the numerical optimization of the respective material parameters is unavoidable. For simplicity, only one *Maxwell* element

[3] PANDAS (**P**orous media **A**daptive **N**onlinear finite element solver based on **D**ifferential **A**lgebraic **S**ystems) is an adaptive finite element tool designed for multi-phase problems.

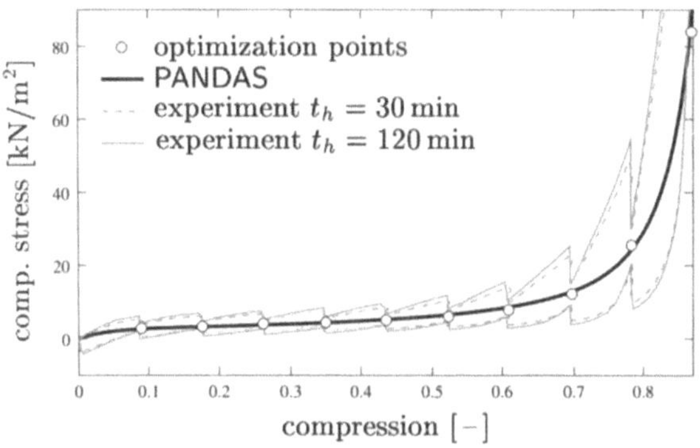

Fig. 7. Experimental results of the uniaxial compression experiment with holding times and numerical simulation of the equilibrium stress with PANDAS after parameter optimization.

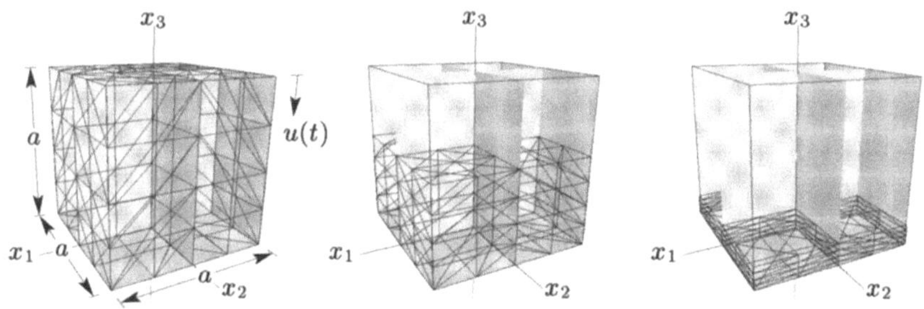

Fig. 8. Initial boundary-value problem of the uniaxial compression test and deformed configurations with $a = 70\,\mathrm{mm}$, $v(t) = 7\,\mathrm{mm/s}$ and $u_{\max} = 60.66\,\mathrm{mm}$ (86.7 % compression). Tetrahedral mesh is generated with NETGEN [33].

($N = 1$) and one *Ogden* term ($K_n = 1$) is used for the description. Again, the respective parameters are optimized with the optimization tool DONLP2 as was also used for the basic elasticity. The result of the parameter adjustment is shown in Figure 9, where a fixed amount of equidistant optimization points (every 0.02 s) and a time step size which is adaptively controlled according to the convergence behaviour of the underlying *Newton* method for the evaluation of the viscous evolution equation is used. As can be seen, the *Ogden* model is able to describe the complex time-dependent behaviour of the considered PU foam. Note that the optimization can be improved by increasing the number of *Ogden* terms ($K_0 > 1$, $K_n > 1$) and in case of the intrinsic viscoelasticity by using more than one *Maxwell* element ($N > 1$) depending on the frequency range that should be described.

Furthermore, to illustrate the deformation-dependent influence of the pore-fluid (here: pore-gas air), the impact of a PU foam cushion of size $40 \times 40 \times 10\,\mathrm{cm}^3$ is simulated by a displacement driven computation, where the prescribed displacement velocity equals an impact velocity of 5 m/s. Fig-

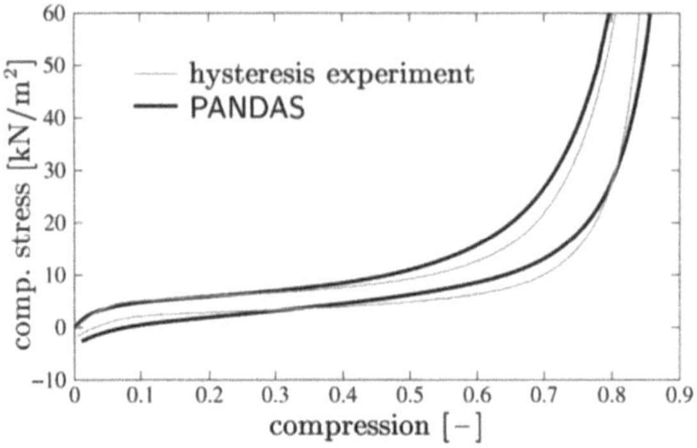

Fig. 9. Experimental result of the uniaxial hysteresis experiment and numerical simulation with PANDAS after parameter optimization.

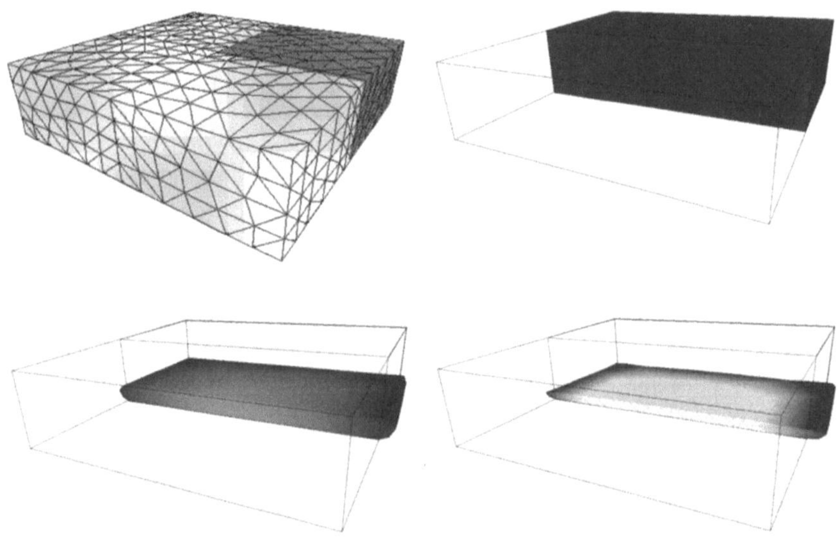

Fig. 10. Finite element mesh and pore pressure distribution within the PU foam cushion at different deformation states (0 %, 65 % and 87 % compression) of the simulated impact experiment (dark shading: $p = p_0 = 1 \cdot 10^5\,\mathrm{N/m^2}$, white: $p_{\max} = 7.9 \cdot 10^5\,\mathrm{N/m^2}$). Tetrahedral mesh is generated with NETGEN [33].

ure 10 shows the computed pore pressure distribution within the foam, where only the lateral surface is permeable for the pore-gas. It can be seen that the pressure only develops significantly beyond 60 % compression due to the assumed deformation-dependent permeability. This property gives the considered open-cell polymer foam its outstanding mechanical characteristics under dynamical loading inclusive the experimental observable size effect which can only be explained by the outstreaming pore-gas. For the sake of completeness, all required model parameters which are used for the computations are listed in the Appendix.

7 Conclusions

Proceeding from the Theory of Porous Media (TPM), the macroscopic model equations for the description of cellular polymer foams have been derived. Therefore, an extended *Ogden*-type viscoelasticity formulation which implies the property of structural densification of the cellular polymer skeleton has been developed. Furthermore, the model accounts for an independent incompressible or compressible pore-fluid constituent and a deformation-dependent permeability, since this is an important property of open-cell polymer foams at large deformations.

In the framework of the finite element method (FEM), a weak form of the governing model equations has been presented. Proceeding from quasi-static situations, the application of the *Bubnov-Galerkin* procedure leads to a system of differential-algebraic equations in the time domain which can efficiently be solved by one-step time integration methods using embedded error estimators for the time-adaptive scheme. Moreover, different mixed element formulations have been presented and discussed.

For the practical application, it is necessary to identify the theoretically introduced unknown material parameters to simulate real material behaviour. Therefore, the consistency of the numerical simulation and the experimental data was improved by minimizing the resulting least squares functional under consideration of equality and inequality constraints. For this optimization problem the sequential quadratic programming (SQP) method has been applied.

To show its suitability, the presented biphasic viscoelastic model has been used for the description of a highly porous, open-cell polyurethane (PU) foam which finds its application, e. g. as seat cushions, in the automotive industry. Therefore, the model has firstly been adapted to experiments with holding times to identify the parameters of the basic elasticity and secondly to hysteresis experiments to adapt the viscous overstresses. Finally, the numerical simulation of large deformation uniaxial compression experiments shows the applicability of the presented model and underlines the necessity of the biphasic formulation with an independent fluid-phase for the description of cellular polymers such as the considered PU foam.

Appendix

Complete list of used model parameters for the considered open-cell polyurethane (PU) foam (bulk density 48 kg/m^3):

General parameters (porous characteristics)

$n_{0S}^S = 0.04$ [−], $\rho^{SR} = 1200$ [kg/m^3], $K_0^S = 1.5 \cdot 10^{-10}$ [m^2],
$\kappa = 14$ [−], $g = 9.81$ [m/s^2], $p_0 = 1 \cdot 10^5$ [N/m^2],
$\bar{R}\Theta = 80000$ [J/kg], $\mu^{FR} = 18 \cdot 10^{-6}$ [Ns/m^2]

Basic elasticity (equilibrium part)

$\mu_0^S = 153253.55$ [N/m^2], $\Lambda_0^S = 585.15$ [N/m^2],
$\mu_1^* = 0.017$ [−], $\alpha_1 = 120.96$ [−],
$\gamma_0 = 1.50$ [−]

Intrinsic viscoelasticity (non-equilibrium part)

$\mu_1^S = 1000.00$ [N/m^2], $\Lambda_1^S = 1000.00$ [N/m^2],
$\eta_1^S = 1000.00$ [Ns/m^2], $\zeta_1^S = 32826.11$ [Ns/m^2],
$\mu_1^* = 0.298$ [−], $\alpha_1 = 6.705$ [−],
$\gamma_0 = 2.70$ [−]

References

1. Ball, J. M.: Convexity conditions and existence theorems in nonlinear elasticity. *Arch. Rational Mech. Anal.* **63** (1977), 337–403.
2. Bazaraa, M. S., Sherali, H. D., Shetty, C. M.: *Nonlinear Programming, Theory and Algorithms.* John Wiley & Sons, New York 1979.
3. Bear, J.: *Dynamics of Fluids in Porous Media.* Dover Publications, New York 1988.
4. de Boer, R.: *Theory of Porous Media.* Springer-Verlag, Berlin 2000.
5. Bowen, R. M.: Incompressible porous media models by use of the theory of mixtures. *Int. J. Engng. Sci.* **18** (1980), 1129–1148.
6. Braess, D.: *Finite Elemente.* Springer-Verlag, Berlin 1997.
7. Brenan, K. E., Campbell, S. L., Petzold, L. R.: *Numerical Solution of Initial-Value Problems in Differential-Algebraic Equations.* North-Holland, New York 1989.
8. Brezzi, F., Fortin, M.: *Mixed and Hybrid Finite Element Methods.* Springer-Verlag, New York 1991.
9. Coleman, B. D., Gurtin, M. E.: Thermodynamics with internal state variables. *J. Chem. Phys.* **47** (1967), 597–613.

10. Diebels, S., Ehlers, W.: Dynamic analysis of a fully saturated porous medium accounting for geometrical and material non-linearities. *Int. J. Numer. Methods Engng.* **39** (1996), 81–97.
11. Diebels, S., Ellsiepen, P., Ehlers, W.: Error-controlled *Runge-Kutta* time integration of a viscoplastic hybrid two-phase model. *Technische Mechanik* **19** (1999), 19–27.
12. Ehlers, W.: Constitutive equations for granular materials in geomechanical context. In Hutter, K. (ed.): *Continuum Mechanics in Environmental Sciences and Geophysics*, CISM Courses and Lectures No. 337, Springer-Verlag, Wien 1993, pp. 313–402.
13. Ehlers, W.: Grundlegende Konzepte in der Theorie Poröser Medien. *Technische Mechanik* **16** (1996), 63–76.
14. Ehlers, W.: Foundations of multiphasic and porous materials. In Ehlers, W., Bluhm, J. (eds.): *Porous Media: Theory, Experiments and Numerical Applications*, Springer-Verlag, Berlin 2002, pp. 3–86.
15. Ehlers, W., Ellsiepen, P.: Zeitschrittgesteuerte Verfahren bei stark gekoppelten Festkörper-Fluid-Problemen. *ZAMM* **77** (1997), S81–S82.
16. Ehlers, W., Ellsiepen, P.: Theoretical and numerical methods in environmental continuum mechanics based on the theory of porous media. In Schrefler, B. A. (ed.): *Environmental Geomechanics*, CISM Courses and Lectures No. 417, Springer-Verlag, Wien 2001, pp. 1–81.
17. Ehlers, W., Markert, B.: Intrinsic viscoelasticity of porous materials. In Sändig, A. M., Schiehlen, W., Wendland, W. L. (eds.): *Multifield Problems – State of the Art*, Springer-Verlag, Berlin 2000, pp. 143–150.
18. Ehlers, W., Markert, B.: On the viscoelastic behaviour of fluid-saturated porous materials. *Granular Matter* **2** (2000), 153–161.
19. Eipper, G.: *Theorie und Numerik finiter elastischer Deformationen in fluidgesättigten porösen Medien.* Dissertation, Bericht Nr. II-1 aus dem Institut für Mechanik (Bauwesen), Universität Stuttgart 1998.
20. Ellsiepen, P.: *Zeit- und ortsadaptive Verfahren angewandt auf Mehrphasenprobleme poröser Medien.* Dissertation, Bericht Nr. II-3 aus dem Institut für Mechanik (Bauwesen), Universität Stuttgart 1999.
21. Forchheimer P.: Wasserbewegung durch Boden. *Z. Ver. Deutsch. Ing.* **45** (1901), Nr.49, 1736–1741 und Nr.50, 1781–1788.
22. Gibson, L. J., Ashby, F.: *Cellular Solids, Structure and Properties.* 2nd ed., University Press, Cambridge 1997.
23. Hairer, E., Wanner, G.: *Solving Ordinary Differential Equations, Vol. 2: Stiff and Differential-Algebraic Problems.* Springer-Verlag, Berlin 1991.
24. le Tallec, P., Rahier, C., Kaiss, A.: Three-dimensional incompressible viscoelasticity in large strains: Formulation and numerical approximation., *Comp. Methods Appl. Mech. Eng.* **109** (1993), 133–258.
25. Lee, E. H.: Elastic-plastic deformation at finite strains. *J. Appl. Mech.* (1969), pp. 1–6.
26. Lion, A.: A physically based method to represent the thermomechanical behaviour of elastomers. *Acta Mechanica* **123** (1997), 1–25.
27. Luenberger, D. G.: *Linear and Nonlinear Programming - Second Edition.* Addison-Wesley Publishing Company, Reading, Massachusetts 1948.
28. Ogden, R. W.: Large deformation isotropic elasticity – On the correlation of theory and experiment for incompressible rubberlike solids. In: *Proceedings of the Royal Society of London, Series A*, Vol. 326 (1972), pp. 565–584.

29. Ogden, R. W.: Large deformation isotropic elasticity – On the correlation of theory and experiment for compressible rubberlike solids. In: *Proceedings of the Royal Society of London, Series A*, Vol. 328 (1972), pp. 323–338.
30. Ogden, R. W.: Inequalities associated with the inversion of elastic stress-deformation relations and their implications. *Math. Proc. Camb. Phil. Soc.* **81** (1977), 313–324.
31. Reese, S.: *Theorie und Numerik des Stabilitätsverhaltens hyperelastischer Festkörper*. Dissertation, Fachbereich Mechanik, Technische Hochschule Darmstadt 1994.
32. Reese, S., Govindjee, S.: A theory of finite viscoelasticity and numerical aspects. *Int. J. Solids Structures* **35** (1998), 3455–3482.
33. Schöberl, J.: NETGEN - An advancing front 2D/3D-mesh generator based on abstract rules. *Comput. Visual. Sci.* **1** (1997), 41–52.
34. Sidoroff, F.: Un modèle viscoélastique non linéaire avec configuration intermédiaire. *Journal de Mécanique* **13** (1974), 679–713.
35. Spellucci, P.: An SQP method for general linear programs using only equality constrained subproblems. *Math. Prog.* **82** (1998), 413–448.
36. Spellucci, P.: A new technique for inconsistent QP problems in the SQP method. *Math. Meth. OR* **47** (1998), 335–400.
37. Tschoegl, N. W.: *The Phenomenological Theory of Linear Viscoelastic Behaviour, An Introduction*. Springer-Verlag, New York 1989.

Experimental measurement of electrical conductivity and electro-osmotic permeability of ionised porous media

Jacques M. Huyghe[1,2], Charles F. Janssen[1,2], Yoram Lanir[2], Corrinus C. van Donkelaar[1], Alice Maroudas[2], and Dick H. van Campen[1]

[1] Eindhoven University of Technology, Engineering Mechanics Institute, The Netherlands
[2] Technion, Julius Silver Institute for Biomedical Engineering, Haifa, Israel

Abstract. Fine grained porous media typically exhibit non-*Darcy*an behaviour. One of them is electro-osmotic flow. The present paper presents an electro-osmotic flow experiment on hydrogel samples. The electric current density is prescribed. The electric potential drop across the sample and the volume flow through the sample are measured. The results show a linear relationship between potential and current and between flow and current. As diffusional properties are highly dependent on deformation, these linear relationships in a highly deformable medium support the theoretical finding that viscous drag is annihilated by opposite electrical forces in an electro-osmotic flow experiment.

1 Introduction

Mineral, biological and synthetic porous media are electroneutral, i. e. they have no net electric charge. However, many of these materials – e. g. clays, hydrogels, and cartilage – have an electrically charged solid and an oppositely charged liquid [7]. The deformational, swelling and diffusional behaviour of these materials depend on several factors. One is the friction between the different components of the material as they move relative to one another. Other factors are the constitutive equations for the effective stress and of the (electro-)chemical potentials. This paper deals with the experimental determination of frictional coefficients in these materials. The other constitutive properties are discussed elsewhere [4, 5]. As a case study we choose for a hydrogel saturated with a sodium chloride solution. Lanir et al. [5] showed that the diffusional properties of a hydrogel depends on its deformational state, and more particularly on the volumetric strain. Huyghe et al. [4] analysed theoretically how to measure the frictional coefficients of a hydrogel in one well-defined deformational state. The conclusion of that paper is that one procedure that is expected to yield satisfying results, is a combination of an electro-osmotic flow experiment, an electro-osmotic pressure experiment, and a diffusion experiment. The present paper presents an experimental set-up for the electro-osmotic flow experiment. As complement to this measurement, the fixed charge density, dry weight and wet weight of the samples are measured at the end of the experiment.

2 Methods

2.1 General set-up

A hydrogel sample is placed between two saline solutions of equal concentration and pressure. A prescribed electric current is applied across the sample through two electrodes immersed in the saline solutions. Fluid flow through the sample and electric potential across the sample are measured. The volume flow j and the electric current density i typically obey the electrokinetic relationships

$$j = -L^p \nabla p - L^{pe} \nabla \xi \,, \tag{1}$$

$$i = -L^{ep} \nabla p - L^e \nabla \xi \,, \tag{2}$$

in which ∇p and $\nabla \xi$ are the pressure gradient and the electrical potential gradient. Because of *Onsager*'s relation, $L^{ep} = L^{pe}$. L^p is the short circuit hydraulic permeability, L^{pe} the electro-osmotic permeability and L^e the electric conductivity. As under the present conditions, there is no pressure gradient we find

$$j = \frac{L^{pe}}{L^e} i \,, \tag{3}$$

$$i = -L^e \nabla \xi \,. \tag{4}$$

Given the measured values of j, i and $\nabla \xi$, the values of the electro-osmotic permeability L^{pe} and the electric conductivity L^p can be determined. As flow is to be measured by means of a precision capillary at the cathode side of the sample, immersion of the cathode into the saline solution is prohibitive, because of gas evolution. Therefore, the cathode is immersed in a separate compartment filled with a silver nitrate solution and communicating with the saline through a ion selective membrane permeable to positive ions and impermeable to negative (nitrate) ions. In stead of gas evolution, solid silver is deposited on the cathode.

2.2 Material preparation

Hydrogel The material consists of a hydrophilic copolymer gel, which has been synthesized by means of polymerization of acrylic acid (AA) and acrylamide (AAm) monomers. The composition of the hydrogel is listed in Table 1.

Substance	Fraction [mol %]	Weight [g]
acrylic acid	2.52	10.91
acrylamide	2.52	10.76
NaOH	2.52	6.06
demi water	92.34	80.00
MBAA	0.05	0.4631
$K_2S_2O_5$	0.005	0.033
$(NH_4)_2S_2O_8$	0.005	0.034

Table 1. Composition for acrylamide acrylic acid hydrogel.

6.06 g NaOH in the form of beads is put into 80 g demineralized water and stirred until the beads are dissolved. 10.91 g of acrylic acid and 10.76 g of acrylamide are added. 0.4631 g of the cross-linker N,N′-methylenebisacrylamide (MBAA) is added. This solution is stirred until the MBAA is dissolved. The solution is degassed with an ultrasound probe. The solution is transferred to a flask and put in a fridge at +8 °C. Every time the solution is polymerized only part of it is used. The solution is polymerized in a glass tube of approximately 10 cm length and 11.2 mm inner diameter with a rubber plug in one end of the tube. 13.54 g of hydrogel solution is put in a vial. In a second vial 0.0418 g of $K_2S_2O_5$ (part one of the initiator) is added to 6.25 g of demineralized water and stirred until the $K_2S_2O_5$ is dissolved. In a third vial 0.0429 g of $(NH_4)_2S_2O_8$ (part two of the initiator) is added to 6.25 g of demineralized water and stirred until the $(NH_4)_2S_2O_8$ is dissolved. The three solutions are degassed with an ultrasound probe. While stirring the hydrogel solution, 1.25 ml dissolved $K_2S_2O_5$ and 1.25 ml dissolved $(NH_4)_2S_2O_8$ are added with a micropipette. A few seconds pass while mixing the hydrogel solution with the two parts of the initiator. The hydrogel solution is poured into the glass tube. After approximately 10 minutes, the glass tube with the polymerized gel is folded into paper, clamped in a vice and is gently broken. The glass pieces are removed. The gel tube is rinsed for a short time with demineralized water to remove the smaller glass pieces and is then dried with paper. The hydrogel tube is cut into shorter tubes of approximately 15 mm length with a 0.05 mm thick steel sheet. The hydrogel tubes are put into 0.15 M NaCl solution and left to equilibrate for at least three days. The length of the hydrogel tubes after equilibration will be approximately 30 mm. The samples will be cut to 20 mm length when they are inserted in the experimental (see Section 2.3). The amounts of acrylic acid, acrylamide, NaOH and water are weighed with an electronic balance, Mettler PE 400 (±0.01 g). The amounts of MBAA, $K_2S_2O_5$ and $(NH_4)_2S_2O_8$ are determined with a more sensitive electronic balance, Mettler AE 200 (±0.0001 g).

Ion selective membrane A highly charged cation-exchange membrane, Nafion 424, obtained from C. G. Processing Inc., Rockland DE, USA, Dr. W. G. Grot, is used to separate the anions in the 0.15 M NaCl and the 0.15 M $AgNO_3$ solutions (see Section 2.3). The sheet of Nafion 424 with dimensions 130 mm × 130 mm is put in a glass tank with 100 % relative humidity for approximately 2 hours. A tray with demineralized water is heated until the temperature reaches approximately +50 °C. The Nafion 424 sheet is placed in the preheated tray with demineralized water for 30 minutes. It is made sure that the whole surface of the membrane is immersed. As the Nafion 424 membrane is taken out of the tray with water, it is assembled in the set-up.

Ag/AgCl electrodes Two Ag/AgCl electrodes are used to measure the potential difference across the hydrogel sample (see Section 2.3). The two Ag/AgCl electrodes are made of a silver wire of 0.5 mm diameter and respectively 70 and 80 mm long. The two silver wires are glued in perspex cylinders that can be connected to the experimental set-up. The two wires are carefully cleaned using extra fine sandpaper and rinsed with water. The two wires are stripped using an electrolysis process to actively drive off any remaining contaminants from the wire surfaces. The stripping process is opposite of the chloriding step in that the electrical polarization of the electrode is reversed. The two wires are both connected to the negative terminal of a current source and placed, with the full length of the wires immersed, into a bath with demineralized water. A piece of platinum wire is placed into the bath with demineralized water and connected to the positive terminal of the current source. The wires are stripped for a period of approximately 5 seconds at an output voltage of approximately 2 V. The two silver wires and the platinum wire are transferred from the bath with demineralized water to a bath with 0.15 M NaCl solution, prepared with demineralized water. Both the silver wires are connected to the positive terminal of the current source and the platinum wire is connected to the negative terminal. The complete surface of the silver wires is immersed in the 0.15 M NaCl solution. The current source is activated for a period of approximately 30 minutes at a voltage of approximately 1.7 V. After the chloriding procedure, the two Ag/AgCl electrodes are shorted and immersed in a bath with almost saturated NaCl solution and kept in this solution until they have to be connected to the experimental set-up. The above procedure was obtained from Neuromedical Supplies, Neurosoft Inc., Sterling VA, USA.

2.3 Experimental set-up

A hydrogel sample is placed between compartments no. 2 and 3 in a hole of 20 mm diameter and 20 mm length (Figure 1). The diameter of the sample is approximately 21 mm so the sample fits somewhat tightly into the hole.

Compartment no. 1 consists of a perspex tube of 20 mm inner diameter. A nylon porous mesh, pore dimensions 0.01 × 0.01 mm, obtained from Merck

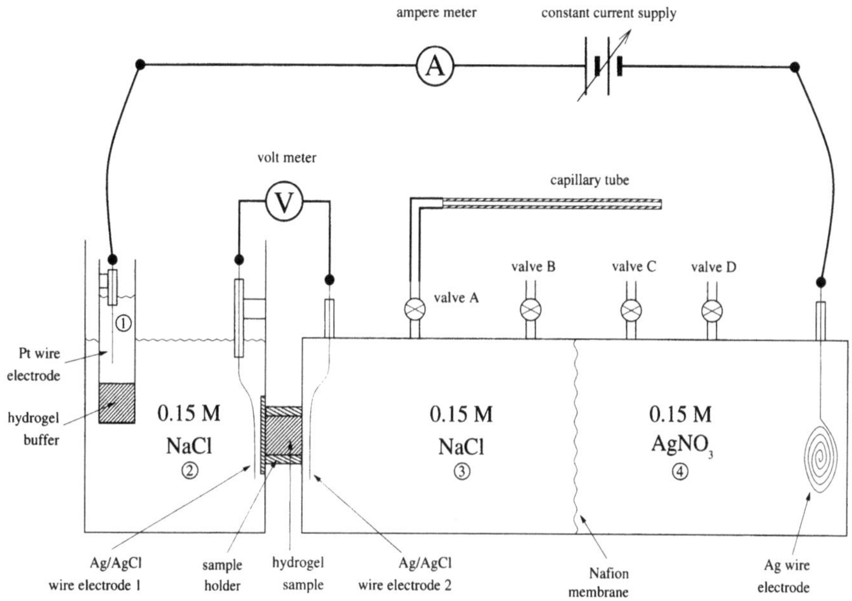

Fig. 1. Schematic representation of the experimental set-up. The hydrogel sample is surrounded by 0.15 M NaCl solution in compartments no. 2 and 3. Constant DC current is applied by the platinum anode in compartment no. 1 and the silver cathode in compartment no. 4. The electrical potential across the sample is measured by means of two Ag/AgCl electrodes at both sides of the sample. Fluid flow through the hydrogel sample is measured with the capillary tube, connected to compartment no. 2. The set-up is placed in a cooling bath and held at a temperature of +4 °C.

Nederland BV, Amsterdam, The Netherlands, is glued to the edge of the bottom of the tube. Another piece of hydrogel, used as pH-buffer, with the same dimensions and the same composition is placed inside the tube against the nylon mesh. The rest of the compartment is filled with 0.15 M NaCl solution, prepared with demineralized water. The solution in compartment no. 1 is circulated to a reservoir (not shown in Figure 1) filled with 5 dm^3 0.15 M NaCl by means of two latex tubes of 5 mm inner diameter and a peristaltic pump, in order to avoid the pH of compartment no. 1 to decrease too quickly under the influence of the electrode reaction. A platinum wire electrode of 70 mm effective length and 1 mm diameter is placed at the top of the tube and is in full contact with the 0.15 M NaCl solution. The platinum wire is glued in a perspex cylinder with a clip that can be attached to the wall of compartment no. 1. Compartment no. 1 and the 5 dm^3 reservoir are in contact with open air.

Compartment no. 2 is made of 10 mm thick perspex plates and has inner dimensions $80 \times 100 \times 160$ mm ($L \times W \times H$). In the right-hand side of the compartment, the sample holder is connected by means of four plastic bolts with washers. A rubber ring is placed between the sample holder and the wall of compartment no. 2. In the left-hand side of the sample holder a glass filter is placed. The pretreated Ag/AgCl electrode no. 1 of 70 mm effective length and 0.5 mm diameter is placed at approximately 1 mm in front of the hole of the sample holder. The Ag/AgCl electrode is glued in a perspex cylinder with a clip that can be attached to the wall of compartment no. 2. A perspex cap with a rubber ring is connected with four plastic bolts with washers in the left hand side of the compartment. Compartment no. 2 is filled with 0.15 M NaCl and in contact with open air.

Compartment no. 3 is a cylindrical perspex tube of 150 mm length, 90 mm inner diameter and 5 mm wall thickness. The pretreated Ag/AgCl electrode no. 2 of 80 mm effective length and 0.5 mm diameter is placed at approximately 1 mm in front of the hole of the sample holder. The Ag/AgCl electrode is glued in a perspex cylinder with a threaded end and a ring, placed at the top of the tube. Two three-way valves are glued at the top of the compartment. The valves are obtained from Elcam Plastic, Kibbutz Bar-Am, Israel. Valve A is connected by a PVC tube to a high-precision capillary tube of 800 mm length and 300 μm inner diameter, obtained from Chase Scientific Glass, Rookwood TN, USA. The capillary tube is placed horizontally. The Nafion membrane is squeezed between compartment no. 3 and 4 with six stainless steel Allen bolts and nuts with washers on both sides of the perspex. Compartment no. 3 is filled with 0.15 M NaCl.

Compartment no. 4 is a cylindrical perspex tube of 150 mm length, 90 mm inner diameter and 5 mm wall thickness. A perspex ring is glued and a rubber gasket is placed at the interface with compartment no. 3. Two three-way valves are glued at the top of the tube. A silver cathode of 20 cm effective length and 0.5 mm diameter is glued in a perspex cylinder with a threaded end and a ring, placed at the top of the tube. The silver cathode is bent in form of a spiral. A perspex ring with 90 mm inner diameter, 120 mm outer diameter and 5 mm thickness is glued to the edge of the tube. A rubber gasket of 90 mm inner diameter, 120 mm outer diameter and 1.4 mm thickness is placed against the perspex ring. A perspex plate of 120 mm diameter and 10 mm thickness is placed against the rubber gasket. The perspex plate and the rubber gasket are clamped to the ring of compartment no. 4 by means of six stainless steel Allen bolts with washers. Compartment no. 4 is filled with 0.15 M $AgNO_3$.

The compartments no. 2, 3, 4, and the 5 dm^3 reservoir are placed in a temperature controlled precision bath Unitronic 100, obtained from Levenson Agencies LTD., Karmiel, Israel. A stirrer is used for circulation of the water. An external cooler is used to control the temperature of the water.

The two Ag/AgCl electrodes are connected to a high impedance preamplifier.

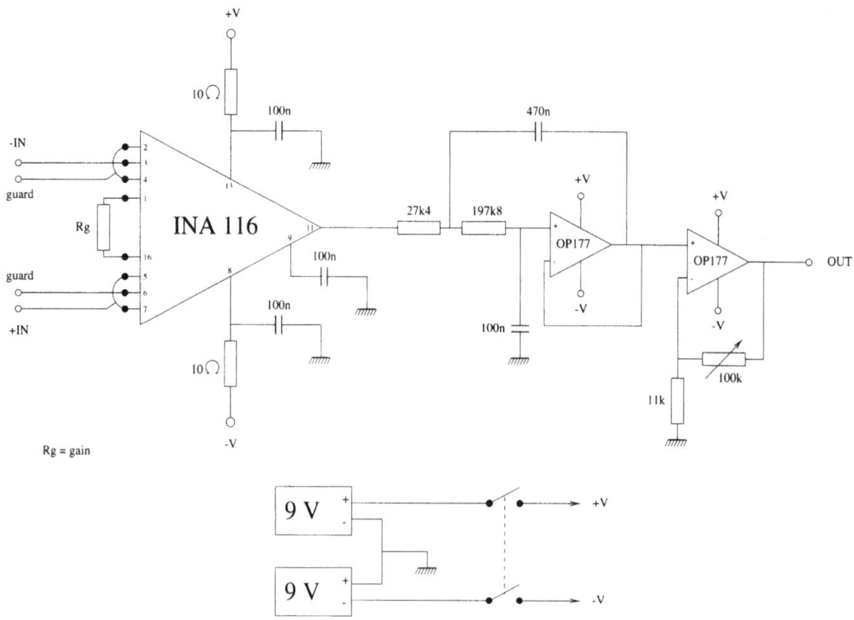

Fig. 2. The high impedance pre-amplifier has an input bias current below 5 fA. Power to the amplifier is supplied by two 9 V batteries to minimize electrical disturbances. The IC has integrated gain that is adjusted by resistor R_g.

The platinum anode and the silver cathode are connected to an adjustable constant current regulator.

Data are measured with a data acquisition board, obtained from National Instruments, connected to a PC. The software package Labview from National Instruments is used to collect the data.

The chemical reactions that occur at the platinum anode and the silver cathode with their reaction potentials are the following:

cathode:	$4\,Ag^+ + 4\,e^-$	$\longleftrightarrow$	$4\,Ag(s)$	+0.80 V
anode:	$O_2(g) + 4\,H^+ + 4\,e^-$	$\longleftrightarrow$	$2\,H_2O$	+1.23 V
overall:	$4\,Ag^+ + 2\,H_2O$	$\longleftrightarrow$	$4\,Ag(s) + O_2(g) + 4\,H^+$	−0.43 V

The overall reaction is achieved by subtracting the reaction at the anode from the reaction at the cathode. At the platinum anode H^+ ions are produced, so the environment in compartment no. 1 becomes more acid when current is applied. The 0.15 M NaCl solution in compartment no. 1 is circulated to a 5 dm^3 reservoir to slow down the rise in H^+ concentration around the platinum anode. The oxygen gas is left to mix with open air. At the silver cathode solid silver is deposited.

3 Experimental protocol

3.1 Initial preparations

8 dm^3 0.15 M NaCl and 1 dm^3 0.15 M $AgNO_3$ solutions are prepared with demineralized water and put in a cooling room with an air temperature of +4 °C. The temperature controlled bath is filled with water. The cooler, the pump of the cooler and the stirrer are switched on. All parts of the set-up are cleaned with water and soap, rinsed with demineralized water and left to dry in open air. A thin layer of silicon gel is put on the threaded end of the perspex plug of Ag/AgCl electrode no. 2. The Ag/AgCl electrode is screwed into compartment no. 3 and the electrode is bent in a way that it is positioned at approximately 1 mm from the hole where the hydrogel sample will be fitted. The silver electrode is screwed into compartment no. 4 and the electrode is bent in the shape of a spiral.

3.2 Placing of Nafion and hydrogel and filling procedures

The end-plate with the rubber gasket is connected to compartment no. 4 with six stainless steel Allen bolts with washers. Compartment no. 4 and compartment no. 3 with two rubber gaskets and the Nafion membrane are assembled with six stainless steel Allen bolts and nuts with washers on both sides of the perspex. Once the Nafion membrane is put in the set-up the NaCl and $AgNO_3$ solutions have to be added rather quickly because the Nafion membrane is drying. The 0.15 M $AgNO_3$ solution is transferred from a vial to compartment no. 4 by a PVC tube that is connected to valve C in compartment no. 4. Valve D in compartment no. 4 is left open. The vial with 0.15 M $AgNO_3$ solution is placed at a higher level than compartment no. 4. This way of slowly filling compartment no. 4 prevents any excess air being left in the compartment. Both valves in compartment no. 4 are closed. A thin layer of silicon gel is put on the sample holder at the places where it is in contact with the wall of compartment no. 2. One hydrogel sample is taken out of 0.15 M NaCl solution and gently put in the sample holder. The hydrogel sample is cut along the edge of the sample holder to the length of 20 mm with a 0.05 mm thick steel sheet. The glass filter with the perspex ring is connected with four plastic bolts and washers. The valves in compartment no. 3 are opened. The sample holder with the hydrogel sample is placed in the hole in compartment no. 2. The sample holder is fixated to the wall of compartment no. 2 with four plastic bolts and washers. The perspex cap is clamped to the left-hand side of compartment no. 2 with four plastic bolts and washers. The 0.15 M NaCl solution is transferred from a vial to compartment no. 3 by a PVC tube that is connected to valve A in compartment no. 3. Valve B in compartment no. 3 is left open. The vial with 0.15 M NaCl solution is placed at a higher level than compartment no. 3. Both valves in compartment no. 3 are closed. compartment no. 2 is also filled with 0.15 M NaCl solution by use of the PVC tube. One 10 ccm syringe filled with 0.15 M NaCl is connected to valve A in compartment no. 3 and the valve is opened. Valve B in compartment no. 3 is opened. The set-up is lifted in an angle so that the remaining air in compartment no. 3 is collected around the opening of valve B. 0.15 M NaCl is injected into compartment no. 3 until all air is removed. Valves A and B are closed. Another 10 ccm syringe filled with 0.15 M $AgNO_3$ is connected to valve C in compartment no. 4 and the valve is opened. Valve D is opened. The set-up is lifted in an angle so that the remaining air in compartment no. 4 is collected around the opening of valve D. 0.15 M $AgNO_3$ is injected into compartment no. 3 until all air is removed. Valves C and D are closed. The 10 ccm syringe connected to valve A is filled again with 0.15 M NaCl, reconnected and valve A is opened. The 10 ccm syringe connected to valve C is emptied in a vial, reconnected and valve C is opened. 0.15 M NaCl is injected into compartment no. 3 at the same rate as 0.15 M $AgNO_3$ is removed from compartment no. 4. Both valves are closed, the NaCl syringe is filled again and the $AgNO_3$ syringe is emptied in a vial. The position of the hydrogel sample is monitored. Any movement of the sample is corrected with addition or removal of 0.15 M NaCl in compartment no. 3. This procedure is repeated until the Nafion 424 membrane is straight again. The 5 dm^3 reservoir is filled with 0.15 M NaCl solution.

3.3 Connection of externals

The set-up with compartments no. 2, 3, 4, and the 5 dm^3 reservoir are immersed in the temperature controlled bath. The set-up is positioned with plastic blocks so that the rings of compartments no. 3 and 4 are just below the water level. The 5 dm^3 reservoir is positioned also with plastic blocks so that the level of the inner solution is equal to the outer cooling water. The position of the hydrogel sample is again monitored and corrected with addition or removal of 0.15 M NaCl in compartment no. 3. Ag/AgCl electrode no. 1 is attached with a clip to the wall of compartment no. 2 and the electrode is bent in a way that it is positioned very close to the hydrogel sample. Another piece of hydrogel, with similar dimensions and composition as the test sample, is taken out of 0.15 M NaCl solution and put gently into compartment no. 1. The piece of hydrogel is moved by blowing air until it is in contact with the nylon mesh and all air is removed between the hydrogel and the mesh. The rest of the tube is filled with 0.15 M NaCl solution. The tube is positioned with clamps so that the nylon mesh is in contact with the 0.15 M NaCl solution in compartment no. 2. The platinum anode is connected to the wall of compartment no. 1 with a perspex clip. The solution in the reservoir is connected to the solution in compartment no. 1 by a latex tube. The latex tube is attached to the top of the reservoir with tape. The solution of the reservoir is sucked into the latex tube with a syringe and the tube is squeezed with a clamp when the tube is filled. The other end of the latex tube is put in the solution of compartment no. 1 and the clamp is released. Another latex tube is put into the solution of the reservoir and taped to the top. The tube is clamped in the peristaltic pump and connected to the solution in compartment no. 1. The peristaltic pump is activated and set to a flow rate of approximately 1.5 cm^3/s. The height of compartment no. 1 is adjusted so that the platinum anode is in full contact with the 0.15 M NaCl solution in compartment no. 1. A three-way valve is connected to valve A in compartment no. 3. A 10 ccm syringe filled with 0.15 M NaCl is connected to the top of the valve. A PVC tube is connected to the other opening of the valve. The other end of the PVC tube is connected to the high precision capillary tube. The capillary tube is placed in two clamps that are positioned at the end of the tube. The height of each clamp can be adjusted, to position the capillary tube exactly horizontally. The Ag/AgCl electrodes are connected to the pre-amplifier, whose output is connected to the data-acquisition board. The platinum anode is connected to the output of the adjustable constant current regulator. The current regulator has an output connected to the data-acquisition board. The input of the current regulator and the silver cathode are connected to a stabilized power supply.

3.4 Capillary tube cleaning and levelling

The capillary tube is cleaned with a detergent solution and rinsed with demineralized water. Air is pumped through the capillary tube using a latex tube and a syringe to remove the remaining water. The valve connected to the syringe in compartment no. 3 is turned in such a way that compartment no. 3 is closed and 0.15 M NaCl can be injected from the syringe into the PVC tube connected to the capillary tube. The PVC tube is filled with 0.15 M NaCl until the solution reaches the entrance of the capillary tube. The capillary tube is positioned vertically and the excess air in front of the entrance of the tube is removed. While part of the capillary tube is filled with 0.15 M NaCl the tube is put back horizontally in the clamps. While compartment no. 3 is closed the valve with the syringe and the PVC tube attached is disconnected. The opening of the valve is put into the solution of compartment no. 2 and the valve is opened so the capillary tube and the solution in compartment no. 2 are directly connected. The height of the clamps is adjusted so that the solution inside the capillary tube is not moving anymore. The valve with the PVC tube and the syringe attached is reconnected to valve A in compartment no. 3. If the solution in the capillary tube is moved back into the PVC tube and air is in front of the entrance of the capillary tube, the capillary tube is positioned vertically and air is removed as described before.

3.5 Measuring electro-osmotic flow

DC current, from 3 to 18 mA with 3 mA steps, is applied to the platinum and the silver electrodes. Fluid movement through the capillary tube is monitored. Time is recorded with a stopwatch when the fluid meniscus passes each marker on the capillary tube. A microscope with lighting is used to make the fluid meniscus more visible. The electric current is measured. The potential difference across the hydrogel sample is recorded during the entire experiment. At the end of the experiment with one electric current level, the fluid is moved back to the beginning of the capillary tube. This procedure is repeated with the other levels of electric current. When the experiment is finished, samples are taken from the solutions in the 5 dm^3 reservoir, compartment no. 2 and no. 3. The conductivity and the pH of these samples are measured. After each experiment the solution of the 5 dm^3 reservoir is adjusted to pH 7 by adding NaOH.

3.6 Fixed Charge Density

As mentioned in Section 2.3, the hydrogel has electrical charges fixed to the matrix of the hydrogel. Fixed Charge Density (FCD) is defined as the concentration of negatively charged fixed groups in the material. It is important to measure FCD, because it causes the electro-osmotic flow [2] and the swelling of the hydrogel. A tracer method , described by Venn and Maroudas [8] is modified to determine the FCD in the hydrogel samples. Instead of determining the FCD in dilute NaCl solution, the FCD for the hydrogel samples is determined in 0.15 M NaCl and with both ^{22}Na and ^{36}Cl tracers. Eight slices of approximately 6 mm diameter and 2 mm thickness are cut from the hydrogel samples used in the electro-osmotic flow experiment. The slices are cut at random locations from different hydrogel tubes. The slices are put into 0.15 M NaCl solution and, while stirring the solutions, left to equilibrate for at least 4 hours at +4 °C. The slices are taken out of the 0.15 M NaCl solution and, after being blotted to remove adherent liquid, weighed to determine the wet slice weight. The slices are transferred to 0.15 M NaCl solution labelled with the radioactive isotopes ^{22}Na and ^{36}Cl. The slices are left to equilibrate for at least 4 hours at +4 °C. After being blotted, the slices are transferred to "cold 0.15 M NaCl solution without the radioactive isotopes (first desorption solution) and left to equilibrate for at least 4 hours at +4 °C. After being blotted, the slices are transferred to a second "cold 0.15 M NaCl desorption solution and left to equilibrate for at least 4 hours at +4 °C. After the slices have been in two "cold desorption solutions almost all radioactive tracer ions that were in the "hot slices are collected in these 2 desorption solutions. After being blotted, the slices are freeze-dried to determine the dry slice weight. The two sets of desorption solutions are analyzed for γ and β radiation by a γ counter (Packard, Cobra-5003) and a Liquid Scintillation Analyzer (Packard 1600 TR) respectively. Because the radioactive isotope ^{22}Na emits γ as well as β radiation a sample of 0.15 M NaCl solution with only ^{22}Na tracer ions is analyzed in the γ counter as well as in the Liquid Scintillation Analyzer. By analysing the 0.15 M NaCl solution with the ^{22}Na tracer ions we know how much γ and β radiation is emitted by the ^{22}Na isotopes. The Na^+ concentration can be directly calculated using the amount of γ radiation in the desorption solutions, the amount of γ radiation in the "hot source solution, the desorption solution weights, the wet slice weight and the outer solution Na^+ concentration. The β radiation measured in the desorption solutions originates from both the ^{22}Na and ^{36}Cl tracer ions. Therefore, the part of β radiation originating from the ^{22}Na tracer ions has to be subtracted from the total β radiation in the desorption solutions. Now the Cl^- concentration can be calculated, using the same data as for the Na^+ concentration. These data are used to calculate partition coefficients K_{Na^+} and K_{Cl^-}:

$$K_{\mathrm{Na}^+} = \frac{\dfrac{\gamma \text{ radioact. in desorpt. sol. [cpm]}}{\text{wet sample weight [g]}}}{\dfrac{\gamma \text{ radioact. in source sol. [cpm]}}{\text{source sol. weight [g]}}}, \tag{5}$$

$$K_{\mathrm{Cl}^-} = \frac{\dfrac{\beta \text{ radioact. in desorpt. sol. [cpm]}}{\text{wet sample weight [g]}} - f_{\beta\gamma} \cdot \dfrac{\gamma \text{ radioact. in desorpt. sol. [cpm]}}{\text{wet sample weight [g]}}}{\dfrac{\beta \text{ radioact. in source sol. [cpm]}}{\text{source sol. weight [g]}} - f_{\beta\gamma} \cdot \dfrac{\gamma \text{ radioact. in source sol. [cpm]}}{\text{source sol. weight [g]}}}. \tag{6}$$

The factor $f_{\beta\gamma}$ is calculated using β and γ readings from the 0.15 M NaCl solution with only ^{22}Na tracer ions by the following formula:

$$f_{\beta\gamma} = \frac{\dfrac{\beta \text{ radioact. in sol. [cpm]}}{\text{sol. weight [g]}}}{\dfrac{\gamma \text{ radioact. in sol. [cpm]}}{\text{sol. weight [g]}}}. \tag{7}$$

FCD is calculated using the following equation:

$$\mathrm{FCD} = (K_{\mathrm{Na}^+} - K_{\mathrm{Cl}^-})\, c_0 , \tag{8}$$

where c_0 is the concentration of Na^+ and Cl^- in the bath solution.

4 Results

4.1 Electro-osmotic flow experiment

Results of fluid flow and electric field strength are presented as function of current density. The fluid pressure on both sides of the hydrogel sample is the same. The electric current density, electric field strength and the fluid velocity are translated from measured data to local data within the hydrogel samples. A careful examination of the data (Figures 3 and 4) shows that the

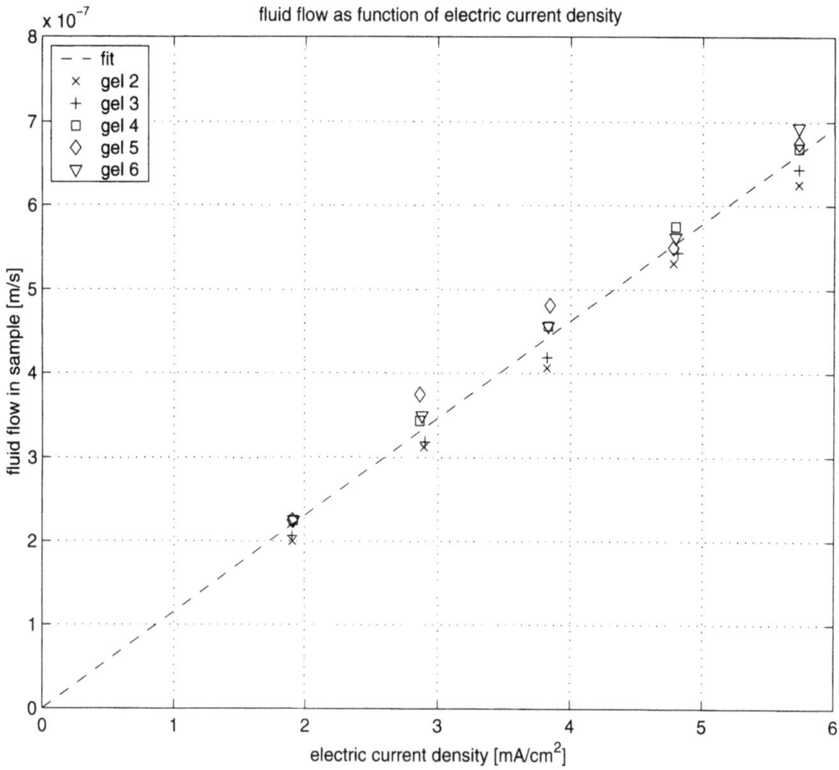

Fig. 3. Electro-osmotic fluid velocity through five hydrogel samples as function of electric current density. The least squares fit gives $(L^{pe}/L^{e}) = 1.2 \cdot 10^{-8}\ \mathrm{m^3\,A^{-1}\,s^{-1}}$.

fluid flow as well as the electric field strength appear to be linear with the applied current density, under these experimental conditions.

Because we measured three variables for each datapoint, the sum of the squared errors between the model predictions and the measured data is normalized. Normalization is performed by calculating the square sum of errors for the fluid velocity (Figure 3) as function of electric current density and dividing this sum by the highest measured fluid velocity. The sum of squared errors for the electric field strength (Figure 4) as function of electric current density is calculated and divided by the electric field strength that belongs to the same datapoint as used with the fluid velocity. The total normalized sum, that is minimized, consists of the addition of these two square sums. In Table 2 the results of these fitting procedures are shown.

5 Fixed Charge Density

Table 3 shows the results from the analysis with ^{22}Na and ^{36}Cl tracer ions to determine the partition coefficients of Na^+ and Cl^- and the FCD of hydrogel

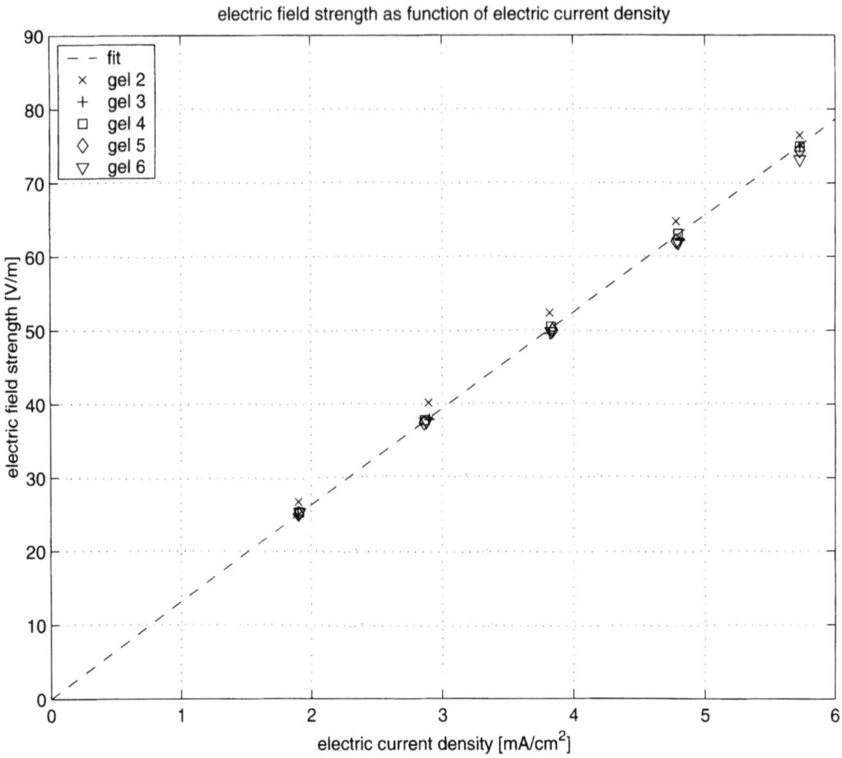

Fig. 4. Electric field strength in five hydrogel samples as function of electric current density. The least squares fit gives $L^e = 0.76\ \mathrm{A\,V^{-1}\,m^{-1}}$.

Parameter	Symbol and unit		Estimated value
Electro-osmotic permeability	L^{pe}	$[\mathrm{m^2\,V^{-1}\,s^{-1}}]$	$8.8 \cdot 10^{-9}$
Electrical conductance	L^e	$[\mathrm{A\,V^{-1}\,m^{-1}}]$	0.76

Table 2. Values of the parameters obtained by fitting of the experimental data.

samples no. 1, 2, 3 and 4. Notice that hydrogel sample no. 1 (gel 1ã and gel 1 b) has a FCD that is significantly lower than the FCD of samples no. 2, 3 and 4. From tabel 3 we get an averaged value for the FCD of 0.156 ± 0.006 mEq/gr water and a hydration of 0.961 ± 0.003 [–]. These averaged values are calculated from the results of hydrogel samples no. 2, 3, and 4, because they are made from the same batch of hydrogel solution.

Sample no.	K_{Na^+} [−]	K_{Cl^-} [−]	FCD [mEq/g water]	Hydration [−]
gel 1 a	1.47	0.86	0.090	0.966
gel 1 b	1.48	0.88	0.091	0.969
gel 2 a	1.81	0.79	0.154	0.958
gel 2 b	1.82	0.78	0.156	0.962
gel 3 a	1.85	0.74	0.166	0.959
gel 3 b	1.82	0.80	0.153	0.966
gel 4 a	1.83	0.76	0.162	0.956
gel 4 b	1.80	0.82	0.148	0.963

Table 3. Results of partition coefficients, fixed charge density and hydration from four hydrogel samples. From each hydrogel sample two slices (a and b) are used for analysis.

Hydration of the hydrogel samples is included in the model and is calculated according to the following equation:

$$\text{Hydration} = \frac{\text{sample water weight [g]}}{\text{total wet sample weight [g]}} . \tag{9}$$

6 Discussion

6.1 Deformation and linearity

The length and displacement of the hydrogel samples during the experiments has not been measured, but visually no change could be detected. The interface of the hydrogel samples is difficult to distinguish from the 0.15 M NaCl solution because of its transparency, but the two Ag/AgCl electrodes were placed so close to the sample that even small displacements of the interface could be detected. When the samples with the sample holder were removed from the set-up after the experiments, their length was measured and no change was observed. The fluid flow as well as the electric field strength appear to be linear with the applied electric current density, indicating that L and L^e are independent of the applied electric current. Considering that diffusional properties of hydrogels depend highly on the deformation of the sample [5], this finding strongly supports the theoretical finding that electro-osmotic flow does not induce deformation of the sample [4]. Viscous drag forces are compensated by equal and opposite electric forces on the ionised solid matrix. The fitted value for $L^{pe}/L^e = 1.2 \cdot 10^{-8}\ \mathrm{m^3\,A^{-1}\,s^{-1}}$ is in the same order of magnitude as found by Grodzinsky [1] for a similar experiment with adult bovine femoropatellar groove cartilage in phosphate buffered saline. This is not surprising, as earlier research showed that acrylamid-acrylic acid polymer gel has a fixed charge density, a porosity and a hydraulic permeability in the same order of magnitude as cartilaginous tissues [7].

6.2 Fixed Charge Density

FCD typically induces osmotic pressure in the sample, and hence a tensile prestressing of the macromolecular network. Therefore, a significant loss in FCD results in shrinking of the hydrogel samples. This effect was indeed observed in the upper part of the piece of hydrogel in compartment no. 1 (Figure 1). The ionization of the hydrogel's fixed carboxyl groups is strongly dependent on the bath pH (Weiss et al. [10]). As a result of H^+ production at the platinum anode, after a few experiments, it could be seen that the upper part of the piece of hydrogel in compartment no. 1 had shrunk. The shrinking of the upper part was only observed when the pH of the 5 dm^3 reservoir was not adjusted to approximately pH 7 after each experiment and when electric current was applied for a long time. This phenomenon is due to counter ion condensation: The hydrogen ions bind ionically onto the fixed charges, hence eliminating the fixed charges. The phenomenon does not occur for the sodium ions because of the hydrophilic nature of sodium. The absence of shrinking in the lower part of the hydrogel buffer indicates that most H^+ ions were trapped by counter-ion condensation in the upper part of the hydrogel. The counter ion condensation frees a sodium ion that diffuses through compartment no. 2, the hydrogel sample and compartment no. 3. Hence, we expect few H^+ ions to pass to compartment no. 2 and further. pH measurements show that the pH in compartment no. 2 after the experiments is almost the same as before the experiments. This indicates that the hydrogel sample between compartments no. 2 and 3 is not affected by counter ion condensation. The shape of the sample after removal from the set-up, was visually inspected and found no abnormalities. The fixed charge density measurements do not indicate significant variations within each sample. Finally, the constancy of the electro-osmotic flow and the electrical potential drop across the sample during each experiment indicates that no changes in fixed charge density occurred during the experiment, either by deformation or by counter ion condensation.

6.3 Silver cathode

The cathode is chosen to be silver, because in the 0.15 M $AgNO_3$ solution the cathode reaction results in silver deposition anyway. The length of the silver wire electrode is chosen so that it is long enough to apply the maximum electric current that we use in the electro-osmotic flow experiment ($I = 18$ mA). At the silver cathode surface Ag^+ ions are depleted from the solution and thus other Ag^+ ions have to diffuse towards the electrode surface. It is obvious that the limiting electric current is dependent on the rate of diffusion, because there is probably hardly any convection. The limiting electric current I_{max} for Ag^+ is calculated according to the following equation [6]:

$$I_{max} = \frac{z\,F\,D\,A\,c_0}{\delta}\,, \tag{10}$$

where z represents the valence of the ion, F is *Faraday*'s constant, D represents the diffusion coefficient, A the exposed surface area of the electrode, c_0 the bulk electrolyte solution concentration and δ the thickness of the boundary layer that forms at the surface of the electrode. We know the geometry of the silver cathode, so we can calculate the exposed surface area A. The diffusion coefficient of Ag^+ at a temperature of 4 °C equals $D_{Ag^+} = 9.37 \cdot 10^{-10}$ m²/s. The thickness of the boundary layer is assumed to be $\delta = 1 \cdot 10^{-4}$ m [6]. The resulting limiting electric current for Ag^+ with this electrode geometry is $I_{max} = 43$ mA. The maximum electric current that we use in the experiment (18 mA) is about half the value of the limiting electric current. If an electric current close to or higher than the limiting current I_{max} is applied, electrolysis of water is initiated, which gives gas evolution at the silver cathode:

cathode:	$4\,H_2O$	$+\,4\,e^-$	$\longleftrightarrow$	$2\,H_2(g) + 4\,OH^-$	-0.83 V
anode:		$O_2(g) + 4\,H^+ + 4\,e^-$	$\longleftrightarrow$	$2\,H_2O$	$+1.23$ V
overall:	$2\,H_2O$		$\longleftrightarrow$	$2\,H_2(g) + O_2(g)$	-2.06 V

The gas evolution (H_2) influences the fluid flow through the capillary tube. In the experiments we did not observe any gas evolution at the silver cathode.

6.4 Ag/AgCl electrodes

The two Ag/AgCl electrodes are chosen for their stability. The observed averaged change in electrical potential is in the order of magnitude of 30 μV/h. In relation to the measured electrical potential difference during experiments, this means that there is an error of less than 0.01 % per hour. In order to verify that the distance between the electrodes and the sample is small enough to measure the potential drop across the sample, the distance between the two Ag/AgCl electrodes and the hydrogel sample was varied between approximately 10 mm and 0 mm from the hydrogel sample. At these distances no change in electrical potential across the sample could be measured. The same hydrogel sample and electric current density were used.

6.5 Mixing and depletion of solutions

The Nafion 424 membrane separates the anions in the 0.15 M NaCl solution from the anions in the 0.15 M $AgNO_3$ solution, because of the high concentration of negative fixed charges inside the membrane. The cations in these solutions can pass the membrane. Because there exists a concentration gradient across the membrane for Na^+ ions as well as for Ag^+ ions, they exchange by diffusion across the membrane. This effect is also observed at the anode side of the Nafion membrane, where in time a layer of solid AgCl is formed. This is evidence that Ag^+ ions travelled from compartment no. 4 to no. 3. Because of electroneutrality Na^+ ions must have moved in the opposite direction. The deposition of solid AgCl on the membrane has an advantage that the Ag^+ ions do not get near the hydrogel sample.

When electric current is applied to the platinum and the silver electrodes, Ag^+ ions are depleted from the solution in compartment no. 4. We can give an estimation of the rate at which this occurs. If we assume that the maximum electric current in the experiment is used, $I = 20$ mA, we can calculate by dividing I by *Faraday*'s constant F that $2.07 \cdot 10^{-7}$ mol/s Ag is deposited at the electrode surface. We know that the starting concentration of Ag^+ is 0.15 mol/l and we know that the volume of compartment no. 4 is approximately 1 l. There is 0.15 mol Ag^+ inside compartment no. 4. The period of time at which the concentration in compartment no. 4 will be 1 % less than the starting concentration is $0.0015/2.07 \cdot 10^{-7} = 7.24 \cdot 10^3$ s ≈ 2 h. As a regular experiment at $I = 20$ mA takes about 10 min, the solutions in the experimental set-up have to be renewed after performing the experiments with one hydrogel sample at different electric current levels.

From conductivity data it appears that the solution concentrations in compartment no. 3 after the experiments are lower than the starting concentrations. The maximum change appeared with hydrogel sample no. 4, where the conductivity dropped by approximately 10 %, so the concentration of NaCl dropped also about 10 % [9]. This is probably because of mixing of cations across the Nafion membrane and deposition of solid AgCl. Cl^- ions are depleted from compartment no. 3 and Na^+ ions are migrating from compartment no. 3 to no. 4. The Nafion membrane has negatively charged fixed groups and therefore produces an electro-osmotic fluid flow in the direction of the cathode. Because compartments no. 3 and 4 are closed, the membrane bulges in the direction of the anode. This was also observed during experiments. As a result of a combination of these effects, the concentration in compartment no. 3 drops.

7 Conclusions

The electro-osmotic flow experiment is an effective method for quantification of the electro-osmotic permeability L^{pe} and the electric conductivity L^e of the gel. The linearity of the flow/current relationship is a confirmation of the no-deformation hypothesis as derived theoretically by [4].

Acknowledgements

The authors thank Amir Landesberg and Chaim Yarnitzky for their valuable advice in this study as well as Chaya Ben-Zaken for her assistance with the lab work. The Technology Foundation STW (the technological branch of the Dutch organisation for Scientific Research NWO and of the Ministry of Economic Affairs, The Netherlands), the Schuurman Schimmel–van Outeren Foundation (The Netherlands), and Technion IIT (Israel). The research of Dr. J. M. Huyghe has been made possible through a fellowship of the Royal Netherlands Academy of Arts and Sciences.

References

1. Grodzinsky A. J.: Mechanical and electrical properties and their relevance to physiological processes: Overview. In Maroudas, A., Kuettner, K. E. (eds.): *Methods for cartilage research*, Academic Press, New York (1990), pp. 275–281.
2. Helfferich F.: *Ion Exchange*. McGraw-Hill, New York 1962.
3. Huyghe J. M., Houben G. B., Drost M. R., Janssen J. D.: An ionised/non-ionised dual porosity model of intervertebral disc tissue: Experimental quantification of model parameters. *Biomechanics and Modelling in Mechanobiology*, submitted.
4. Huyghe J. M., Janssen C. F., van Donkelaar C. C., Lanir Y.: Measuring principles of frictional coefficients in cartilaginous tissues and its substitutes. *Biorheology*, submitted.
5. Lanir Y., Seybold J., Schneiderman R., Huyghe J. M.: Partition and diffusion of sodium and chloride ions in soft charged foam: The effect of external salt concentration and mechanical deformation. *Tissue engineering* **4** (1998), 365–378.
6. Moore W. J.: *Physical chemistry*. 5th ed., Longman, London 1972.
7. Oomens C. W. J., de Heus H. J., Huyghe J. M., Nelissen L. Janssen J. D.: Validation of the triphasic mixture theory for a mimic of intervertebral disk tissue. *Biomimetics* **3** (1995), 171–184.
8. Venn, M., Maroudas, A.: Chemical composition and swelling of normal and osteoarthrotic femoral head cartilage. I. Chemical composition. *Annals of the Rheumatic Diseases* **36** (1977), 121–129.
9. Weast, R. C.: *CRC handbook of chemistry and physics*. 57th ed., CRC Press, Cleveland Ohio 1976.
10. Weiss, A. M., Grodzinsky, A. J., Yarmush, M. L.: Chemically and electrically controlled membranes: Size specific transport of fluorescent solutes through PMAA membranes. *AIChE Symposium Series* **82** (1986), 85–98.

Theory and numerics of localization in a fluid-saturated elasto-plastic porous medium

Ragnar Larsson, Jonas Larsson, and Kenneth Runesson

Chalmers University of Technology, Department of Solid Mechanics,
S-412 96 Göteborg, Sweden

Abstract. In this contribution, we outline a theoretical and numerical approach for describing deformation localization due to hydro-mechanical coupling. In the localization analysis, the concept of "regularized strong discontinuity" is extensively used at the application to the conservation laws of momentum and mass. At the onset of localization, the displacement and pore pressure fields are assumed to contain regularized discontinuities that are superposed on the continuous fields. As a result, we obtain a coupled localization condition, whereby the partly drained situation is discussed and compared to the drained and undrained situations. As to the finite element modelling, it is proposed to capture the development of regularized discontinuities in the displacement and pressure fields it is proposed to use a finite element procedure for the mixture of soil and pore fluid based on the "embedded band approach", where the finite element interpolation allows for discontinuities within the elements. The procedure is based on the enhanced assumed strain concept, and from the pertinent orthogonality condition a coupled set of finite element equations are obtained, where the coupling between continuous and discontinuous response is obtained at the element level. Under certain circumstances, the coupled localization condition may be shown to be preserved by the finite element formulation, and the element response may be characterized like in the continuum situation. It is shown that the algorithm is capable of capturing the onset of localization as well as the post-localized response. In a numerical example, we study the influence of the internal friction angle on the development of a slip surface within a soil slope.

1 Introduction

Porous materials are encountered in a quite broad range of engineering problems within soil mechanics, geophysics, biomechanics, industrial forming processes, etc. The characteristic feature of the continuum theory for porous media is the adoption of a macroscale, whereby the micro-problem, like the detailed flow of pore water in a complex pore structure, is accounted for in an average sense. We note that, parallel with the increasing computer capacity, the porous media theory has been extensively developed and exploited, cf. the historical review by de Boer [6]. As an example, we may mention the development of the continuum theory of porous media undergoing large elastic and inelastic deformations involving compressible or incompressible constituents, as originally formulated by Bowen [4]. Moreover, different constitutive formats can be arrived at depending on the choice of internal variables in the thermodynamic formulation, cf. Larsson and Larsson [15–17] and Diebels [8]. In particular, constitutive models may be formulated based on partial stresses or based on the effective stress principle according to *Terzaghi*. The latter principle seems relatively well established in the engineering community. In the case of incompressible constituents, the effective stress according to *Terzaghi* appears indeed as the natural stress variable in the thermodynamic formulation. However, in the case of compressible constituents the flexibility of the phases influence the effective stress, as discussed by de Boer and Ehlers [7] and more recently by Bluhm and de Boer [2].

An important problem within soil mechanics is the interaction between fluid and saturated soil material. In particular, the soil-fluid interaction influences the failure situation of the material, as reported from experimental tests, e. g. Han and Vardoulakis [10], Vardoulakis [22]. In particular, for a fluid-saturated soil this hydro-mechanical coupling is strong and needs to be addressed in the modelling situation. The soil-fluid interaction also carries over to the possible development of a shear band. In fact, the two-phase interaction introduces a rate-dependence into the problem, although a rate-independent model is used to model the (underlying) effective material. In analogy with viscous one-phase regularization, e. g. Needleman [19], this rate dependence naturally introduces a length scale that may limit the width of the shear band, cf. Loret and Prevost [18]. However, this "interaction induced regularization" depends strongly on the permeability parameter, and it should therefore not be the sole regularization mechanism in an analysis.

To introduce the proper regularizations into the problem, a number of "localization limiters" have been proposed in the literature. Examples of approaches which preserve the continuity of the deformation fields are: rate-dependent visco-plastic, non-local, micro-polar and gradient models. A review of these approaches can be found in de Borst et al. [6]. In contrast to the formulations based on continuous fields, one may augment the conventional continuum with a local regularization that becomes effective only in the post-localized regime. Following the developments in Larsson et al. [11], Simo et al. [20], Larsson and Runesson [12] we introduce a regularized strong discontinuity, which leads to the establishment of cohesive zone type of models.

Numerical simulations on the hydro-mechanically coupled problem with emphasis on localization is presented in the work by Loret and Prevost [18]. For the continuum model they find that, for the occurrence of stationary discontinuities, the localization condition for the porous medium is identical to that of the underlying drained, "effective", material. The discretized problem is regularized using a viscoplastic material model. More recently, Ehlers and Volk [9] presented a formulation where the problem is regularized with *Cosserat* theory. A possible drawback for these models is the re-meshing procedure that is required in order to resolve the size of the localization zone. A one-dimensional formulation based on strong discontinuities is also proposed in Armero and Callari [1]. They follow the ideas of Coussy [5] and assume discontinuities in the skeleton displacement and fluid diffusion (fluid content), whereby the assumed discontinuities are embedded into the finite elements.

In the present contribution dealing with the two-phase continuum, the concept of regularized discontinuity is extensively exploited at the application to the conservation laws of momentum and mass, as discussed in Larsson and Larsson [14]. As a result, we obtain a localization condition that couples the displacement and pressure discontinuities. A finite element procedure for the mixture of soil and pore fluid is proposed on the basis of the "embedded approach", where the finite element interpolation allows for discontinuities within the finite elements. The procedure falls within the range of the enhanced assumed strain methods, where the finite element equations are derived from a three-field variational formulation, of Simo and Rifai [21]. From the pertinent orthogonality condition, a coupled set of finite element equations are obtained, where the coupling between continuous and discontinuous response is obtained at the element level. In the present formulation, the orthogonality condition preserves traction continuity and continuity of mass balance across the embedded band. Under certain circumstances, the coupled localization condition may be shown to be preserved by the finite element formulation, and the element response may be characterized like in the continuum situation.

2 Governing equations for a saturated porous medium

2.1 Balance equations

Let us consider a fluid saturated porous solid skeleton and assume small deformation theory. The constituents are denoted c^α, where the sup-index $\alpha = s$ or $\alpha = f$ denotes the solid and the fluid phase, respectively. We introduce volume fractions $n_\alpha(\mathbf{x}, t)$, defined as the ratio between the local constituent volume v_α and the bulk mixture volume v. The intrinsic density associated with each constituent is denoted ρ_α, whereby the bulk density (per unit bulk volume of mixture) is obtained as $\hat{\rho}_\alpha = n_\alpha \, \rho_\alpha$. We further assume that no voids can develop during deformation, whereby volume fractions must satisfy the saturation constraint

$$n_s + n_f = 1 \,, \tag{1}$$

where $n_f := n$ is the porosity and $n_s = 1 - n$ is the volume fraction of the soil skeleton. Moreover, the balance of momentum for each phase per unit volume of the bulk can be written as

$$\nabla \cdot \boldsymbol{\sigma}_\alpha + \hat{\rho}_\alpha \, \mathbf{g} + \mathbf{h}_\alpha = \mathbf{0} \,, \qquad \alpha = s \,, f \,, \tag{2}$$

where $\boldsymbol{\sigma}_\alpha$ is the partial *Cauchy* stress tensor pertinent to the different constituents, $\mathbf{g}$ is the volume force per unit mass and $\mathbf{h}_\alpha$ is the momentum production force due to local drag interaction between the fluid and the soil. These forces must satisfy the momentum production constraint:

$$\mathbf{h}_s + \mathbf{h}_f = \mathbf{0} \, . \tag{3}$$

Hence, in view of (2) and (3), the momentum balance for the mixture is obtained as

$$\nabla \cdot \bar{\boldsymbol{\sigma}} + \hat{\rho}\,\mathbf{g} = \mathbf{0} \quad \text{with} \quad \bar{\boldsymbol{\sigma}} = \boldsymbol{\sigma}_s + \boldsymbol{\sigma}_f \, , \tag{4}$$

where $\bar{\boldsymbol{\sigma}}$ is the total stress and $\hat{\rho}$ is the saturated density of the soil-fluid mixture.

With the assumption of intrinsically incompressible constituents, mass conservation for the porous medium may be expressed as

$$\nabla \cdot \dot{\mathbf{u}} + \nabla \cdot \mathbf{d}_d = 0 \quad \text{with} \quad \mathbf{v}_d = n\,\mathbf{v}_r \quad \text{and} \quad \mathbf{v}_r = \mathbf{v}_r - \mathbf{v}_s \, , \tag{5}$$

where $\mathbf{v}_d$ is the *Darcy*an, velocity which is the relative volumetric flow of fluid per unit area of the deforming soil mass, and $\dot{\mathbf{u}} \equiv \mathbf{v}_s$.

Following the developments in e. g. Larsson and Larsson [15–17], we establish the entropy inequality for the two-phase material at isothermal conditions in terms of the assumption of incompressible constituents as

$$\boldsymbol{\sigma} : \dot{\boldsymbol{\epsilon}} - \hat{\rho}_s \dot{\Psi}_s - \mathbf{h}^f_e \cdot \mathbf{v}_d \geq 0 \, , \tag{6}$$

where $\boldsymbol{\sigma}$ is the effective stress and $\mathbf{h}^f_e$ is the effective drag force. These are defined as

$$\begin{aligned} &\boldsymbol{\sigma} = \boldsymbol{\sigma}_s + p n_s \mathbf{1}, \\ &\mathbf{h}^f_e = \nabla \bar{p} - \rho_f\, g = \nabla p^e \quad \text{with} \quad p^e = \bar{p} - p^h \quad \text{and} \quad \nabla p^h = \rho_f\, g \, , \end{aligned} \tag{7}$$

where we introduced the total fluid pressure $\bar{p}$ in terms of the excess pore pressure p^e and the hydrostatic pore pressure p^h. Henceforth, we shall drop the sup-index 'e' for the excess pore pressure, i. e. we set $p = p^e$. Moreover, from the definition of the fluid stress, $\boldsymbol{\sigma}_f = -pn\,\mathbf{1}$, and of the total stress $\bar{\boldsymbol{\sigma}}$ in $(4)_2$, we now find the classical identity $\bar{\boldsymbol{\sigma}} = \boldsymbol{\sigma} - p\,\mathbf{1}$ and $\boldsymbol{\sigma}_f = -pn\,\mathbf{1}$.

2.2 Constitutive equations

We shall establish constitutive equations for the effective drag force $\mathbf{h}^e_f$ and the effective stress $\boldsymbol{\sigma}$ on the basis of the entropy inequality (6).

In order to ensure that $-\mathbf{h}^e_f \cdot \mathbf{v}_r > 0$ in (6), we make the constitutive assumption

$$\mathbf{h}^e_f = -\frac{1}{k} v_d \, , \tag{8}$$

where $k > 0$ is the *Darcy*an permeability coefficient. Upon combining (8) and $(7)_2$, we obtain *Darcy*'s law in the form

$$\mathbf{v}_d = -k\,\nabla p\,. \tag{9}$$

As to the effective stress, we establish the appropriate constitutive relation within the plasticity framework, where the *Helmholtz* free energy density (per unit volume) is defined as

$$\Psi_s(\boldsymbol{\epsilon}, \boldsymbol{\epsilon}^p, \kappa) = \frac{1}{2}(\boldsymbol{\epsilon} - \boldsymbol{\epsilon}^p) : \mathbf{E}^e : (\boldsymbol{\epsilon} - \boldsymbol{\epsilon}^p) + \bar{\Psi}(\kappa)\,, \tag{10}$$

where $\mathbf{E}^e$ is the elastic stiffness modulus tensor, $\boldsymbol{\epsilon}$ is the total strain, $\boldsymbol{\epsilon}^p$ is the total plastic strain and κ is a hardening variable. In view of the entropy inequality, we obtain the constitutive equations

$$\boldsymbol{\sigma} = \frac{\partial \Psi_s}{\partial \boldsymbol{\epsilon}} = \mathbf{E}^e : (\boldsymbol{\epsilon} - \boldsymbol{\epsilon}^p)\,, \quad \boldsymbol{\sigma}^p = -\frac{\partial \Psi_s}{\partial \kappa} \equiv \boldsymbol{\sigma}\,, \quad K = -\frac{\partial \bar{\Psi}_s}{\partial \kappa}\,, \tag{11}$$

where $\boldsymbol{\sigma}^p$ and K are dissipative stresses that are energy conjugated to $\boldsymbol{\epsilon}^p$ and κ, respectively. Furthermore, we introduce the convex (but not necessarily smooth) set $\mathcal{B}$ of admissible states:

$$\mathcal{B} = \{\boldsymbol{\sigma}, K : \phi(\boldsymbol{\sigma}, K) \leq 0\}\,, \tag{12}$$

where $\phi(\boldsymbol{\sigma}, K) = 0$ is the state boundary surface. Constitutive rate equations of $\boldsymbol{\epsilon}^p$ and κ may now be defined via the evolution rules

$$\dot{\boldsymbol{\epsilon}}^p = \dot{\lambda}\,\frac{\partial \phi^*}{\partial \boldsymbol{\sigma}}\,, \quad \dot{\kappa} = \dot{\lambda}\,\frac{\partial \phi}{\partial K}\,, \tag{13}$$

where λ is the plastic multiplier which is determined from the loading conditions

$$\dot{\lambda} \geq 0\,, \quad \phi \leq 0\,, \quad \dot{\lambda}\phi = 0\,. \tag{14}$$

In (13), we include the possibility of non-associated plastic flow by choosing the plastic potential $\Phi^* = \Phi$. It may be shown in a standard fashion that the tangent response for the effective stress can be written in terms of the continuum tangent stiffness tensor $\mathbf{E}$ as

$$\dot{\boldsymbol{\sigma}} = \mathbf{E} : \dot{\boldsymbol{\epsilon}}\,, \quad \mathbf{E} = \begin{cases} \mathbf{E}^{ep} := \mathbf{E}^e - \dfrac{1}{h}\mathbf{E}^e : \mathbf{f}^* \otimes \mathbf{f} : \mathbf{E}^e & \text{if } \mathbf{f} : \mathbf{E}^e : \dot{\boldsymbol{\epsilon}} > 0 \ (P) \\ \mathbf{E}^e & \text{if } \mathbf{f} : \mathbf{E}^e : \dot{\boldsymbol{\epsilon}} \leq 0 \ (\mathrm{E})\,, \end{cases} \tag{15}$$

where $\mathbf{f} := \partial\phi/\partial\boldsymbol{\sigma}$ and $\mathbf{f}^* := \partial\phi^*/\partial\boldsymbol{\sigma}$, and (P) and (E) denote plastic and elastic loading, respectively. The plastic modulus h is defined as

$$h = \mathbf{f} : \mathbf{E}^e : \mathbf{f}^* + H\,, \quad H = -\frac{\partial \phi}{\partial K}\frac{\partial \bar{\Psi}}{\partial \kappa}\frac{\partial \phi}{\partial K}\,, \tag{16}$$

where H denotes the hardening modulus.

3 Regularized discontinuous fields

We shall investigate conditions for the existence of localized deformation for the hydro-mechanically coupled problem described above. To this end, we shall exploit the concept of regularized displacement discontinuity within the domain B, as shown in Figure 1. It is assumed that the discontinuities occurs across the internal surface Γ with unit normal $\mathbf{n}$. This surface subdivides B into the subdomains B_- and B_+ in such a way that $\mathbf{n}$ is pointing from B_- to B_+, as shown in Figure 1. The kinematics of a regularized strong discontinuity can be interpreted as a "weak discontinuity", i. e. the strain is bounded everywhere. We introduce a band width δ to construct regularized derivatives of the *Heaviside* function. These regularized functions are consistent in the sense that they describe qualitatively the true derivatives in a distributional sense as $\delta \to 0$. Consequently, δ should be small in relation to a characteristic dimension of the boundary value problem.

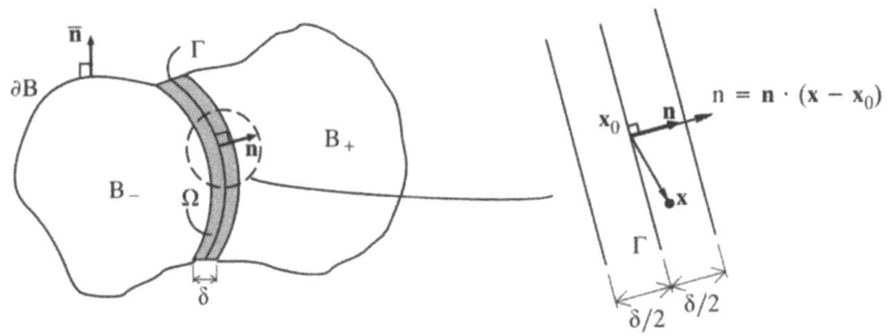

Fig. 1. Body B with boundary ∂B divided by the characteristic surface Γ. Regularization band Ω of width δ, centered at Γ.

We assume, in particular, that the velocity field of the skeleton has the structure

$$\dot{\mathbf{u}}(\mathbf{x},t) = \dot{\mathbf{u}}_c(\mathbf{x},t) + [\![\dot{\mathbf{u}}]\!](t)\, H_\Gamma(\mathbf{x})\,, \tag{17}$$

where $\mathbf{u}_c$ is the continuous portion of the solid displacement and $[\![\dot{\mathbf{u}}]\!]$ is the spatially constant jump of $\mathbf{u}$ across Γ. Moreover, H_Γ is the *Heaviside* function defined as

$$H_\Gamma = \begin{cases} 0 & \text{if} \quad \mathbf{x} \in B_- \\ 1 & \text{if} \quad \mathbf{x} \in B_+\,. \end{cases} \tag{18}$$

As to the gradient of H_Γ, we may formally define the *Dirac* delta function $\boldsymbol{\delta}_\Gamma = \nabla H_\Gamma$, only in a distributional sense. However, we define a regularized version of $\boldsymbol{\delta}_\Gamma$ upon introducing a band shaped zone Ω along Γ with the width δ, as shown in Figure 1. The strictly discontinuous *Heaviside* function is then replaced by a linear function across Ω, whereby the regularization $\boldsymbol{\delta}_{\Gamma,r}$ is expressed as

$$\boldsymbol{\delta}_{\Gamma,r} = \frac{f(n)}{\delta}\,\mathbf{n} \quad \text{with} \quad f\,(n(\mathbf{x})) = \begin{cases} 1 & \text{if} \quad \mathbf{x} = \mathbf{x}_0 \in \Gamma \\ 0 & \text{if} \quad \mathbf{x} = \mathbf{x}_\pm \in B\backslash\Gamma\,, \end{cases} \tag{19}$$

where we define $\boldsymbol{x}_0 \in \Gamma$ and $\boldsymbol{x} \in \Omega$ such that $\boldsymbol{x} = \boldsymbol{x}_0 + n\boldsymbol{n}(\boldsymbol{x})$, with $-\delta/2 \leq n \leq \delta/2$. Note also that the unit normal $\boldsymbol{n}$ is considered constant.

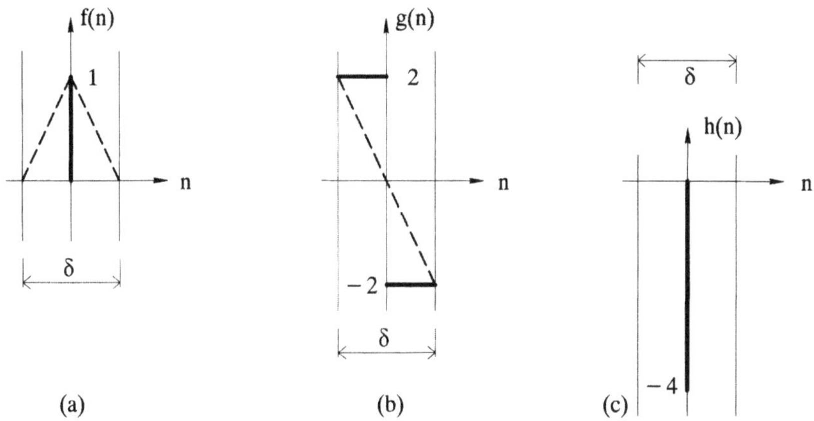

Fig. 2. Functions f, g, and h as the result of successive regularizations.

For later use, we also establish the regularization of the divergence $\nabla\cdot\boldsymbol{\delta}_{\Gamma,r}$ and the gradient $\nabla(\nabla\cdot\boldsymbol{\delta}_{\Gamma,r})$. In order to differentiate the function $\boldsymbol{\delta}_{\Gamma,r}$, we consider the function $f(n)$ regularized in terms of the dashed linear functions across Γ indicated in Figure 2 a. As a result, we obtain in view of (19) the relation

$$\nabla\cdot\boldsymbol{\delta}_{\Gamma,r} := \frac{1}{\delta^2}\,g(n) \quad \text{with} \quad g(n(\mathbf{x})) = \begin{cases} 2 & \text{if} \quad \mathbf{x} = \mathbf{x}_- \in B_-\backslash\Gamma \\ -2 & \text{if} \quad \mathbf{x} = \mathbf{x}_+ \in B_+\backslash\Gamma \\ 0 & \text{otherwise}\,. \end{cases} \tag{20}$$

As to the term $\nabla(\nabla\cdot\boldsymbol{\delta}_{\Gamma,r})$, we regularize the function $g(n)$ as in Figure 2 b, whereby we obtain

$$\nabla(\nabla\cdot\boldsymbol{\delta}_{\Gamma,r}) := \frac{1}{\delta^3}\,h(n)\,\mathbf{n} \quad \text{with} \quad h(n(\mathbf{x})) = \begin{cases} -4 & \text{if} \quad \mathbf{x} = \mathbf{x}_0 \in \Gamma \\ 0 & \text{if} \quad \mathbf{x} = \mathbf{x}_\pm \in B\backslash\Gamma\,. \end{cases} \tag{21}$$

We are now in the position to express the strain rate $\dot{\boldsymbol{\epsilon}}$ pertinent to the proposed regularization. Hence, in view of (17) and (19) we obtain

$$\dot{\boldsymbol{\epsilon}} = (\nabla \otimes \dot{\mathbf{u}})^{\mathrm{sym}} = \dot{\boldsymbol{\epsilon}}_c + \frac{f}{\delta}\left(\mathbf{n} \otimes [\![\dot{\mathbf{u}}]\!]\right)^{\mathrm{sym}} . \tag{22}$$

Moreover, upon invoking (22) into the constitutive relation (15), we find that the effective stress rate field $\dot{\boldsymbol{\sigma}}$ takes the structure

$$\dot{\boldsymbol{\sigma}} = \dot{\boldsymbol{\sigma}}_c + f[\![\dot{\boldsymbol{\sigma}}]\!] \quad \text{with} \quad [\![\dot{\boldsymbol{\sigma}}]\!] = \dot{\boldsymbol{\sigma}}(\mathbf{x}_0) - \dot{\boldsymbol{\sigma}}(\mathbf{x}_\pm) , \tag{23}$$

where

$$\dot{\boldsymbol{\sigma}}_c = \mathbf{E}(\mathbf{x}_\pm) : \dot{\boldsymbol{\epsilon}}_c , \quad [\![\dot{\boldsymbol{\sigma}}]\!] = [\![\mathbf{E}]\!] : \dot{\boldsymbol{\epsilon}}_c + \frac{1}{\delta}(\mathbf{E}(\mathbf{x}_0) \cdot \mathbf{n}) \cdot [\![\dot{\mathbf{u}}]\!] . \tag{24}$$

In (24), we introduced $[\![\mathbf{E}]\!] = \mathbf{E}(\mathbf{x}_0) - \boldsymbol{E}(\mathbf{x}_\pm)$ as the jump in the tangent stiffness. In the strain driven format, the condition for plastic loading is given by

$$\mathbf{f} : \mathbf{E}^e : \dot{\boldsymbol{\epsilon}}_c + \frac{f}{\delta}\mathbf{a} \cdot [\![\dot{\mathbf{u}}]\!] > 0 \tag{25}$$

with $\mathbf{a} := \mathbf{f} : \mathbf{E}^e \cdot \mathbf{n}$. As to the pore pressure, we are guided by the effective stress principle, i. e. $\bar{\boldsymbol{\sigma}} = \boldsymbol{\sigma} - p\,\mathbf{1}$, to employ the crucial assumption that p has the same regularity as the effective stress field. We thus assume

$$\dot{p} = \dot{p}_c + f\,[\![\dot{p}]\!] \quad \text{with} \quad [\![\dot{p}]\!] = \dot{p}(\boldsymbol{x}_0) - \dot{p}(\mathbf{x}_\pm) , \tag{26}$$

where p_c is the continuous portion of the pressure field. Hence, by combining (23) and (26), we can rewrite the total stress rate $\dot{\bar{\boldsymbol{\sigma}}}$ as

$$\dot{\bar{\boldsymbol{\sigma}}} = \dot{\bar{\boldsymbol{\sigma}}}_c + f\,[\![\dot{\bar{\boldsymbol{\sigma}}}]\!] \quad \text{with} \quad \dot{\bar{\boldsymbol{\sigma}}}_c = \dot{\boldsymbol{\sigma}} - \dot{p}_c\mathbf{1} \quad \text{and} \quad [\![\dot{\bar{\boldsymbol{\sigma}}}]\!] = [\![\dot{\boldsymbol{\sigma}}]\!] - [\![\dot{p}]\!]\,\mathbf{1} . \tag{27}$$

Moreover, from (9), (26), and (20) follows that the *Darcy*an velocity has the structure

$$\mathbf{v}_{,d} = \mathbf{v}_{d,c} + g\,[\![\mathbf{v}_d]\!] \quad \text{with} \quad \mathbf{v}_{d,c} = -k\,\nabla p_c \quad \text{and} \quad [\![\mathbf{v}_d]\!] = -\frac{k}{\delta}\,[\![p]\!]\,\mathbf{n} . \tag{28}$$

We note that the possible development of a jump in pressure in the post-localized regime may also be regarded as a consequence of mass balance: For the mass balance (5) to be satisfied, the two terms must have an identical structure. If the *Darcy* law is employed, the structure of the pore pressure in (26) becomes the obvious choice.

4 Localization condition for a two-phase porous medium

4.1 Condition for the existence of a regularized strong discontinuity

In order to establish the conditions that must be satisfied in order for a strong discontinuity to exist, we consider the momentum and mass conservation relations, as expressed in (4) and (5).

Provided that the body force is continuous, the condition of continuity of linear momentum conservation across Γ, as expressed in (4), becomes

$$\nabla \cdot \dot{\boldsymbol{\sigma}}|_{\mathbf{x}=\mathbf{x}_\pm} - \nabla \cdot \dot{\boldsymbol{\sigma}}|_{\mathbf{x}=\mathbf{x}_0} = \mathbf{0} \tag{29}$$

which leads to the (usual) traction continuity condition

$$\mathbf{n} \cdot [\![\dot{\boldsymbol{\sigma}}]\!] = \mathbf{0}\,. \tag{30}$$

Likewise, we formulate the condition for continuity of the mass conservation across Γ, in terms of (5), as

$$(\nabla \cdot \mathbf{u} + \nabla \cdot \mathbf{v}_d)\,|_{\mathbf{x}=\mathbf{x}_\pm} - (\nabla \cdot \dot{\mathbf{u}} + \nabla \cdot \mathbf{v}_d)|_{\mathbf{x}=\mathbf{x}_0} = 0 \tag{31}$$

which yields

$$-\frac{f}{\delta}\,\mathbf{n} \cdot [\![\dot{\mathbf{u}}]\!] - \frac{h}{\delta}\,\mathbf{n} \cdot [\![\mathbf{v}_d]\!] = 0\,. \tag{32}$$

Finally, upon introducing (24) into (30) and (28) into (32), we obtain a localization condition (= condition for the existence of regularized discontinuities) as follows:

$$\begin{aligned} &\frac{1}{\delta}\,\mathbf{Q}_\Gamma \cdot [\![\dot{\mathbf{u}}]\!] - [\![\dot{p}]\!]\,\mathbf{n} = -\mathbf{n} \cdot [\![\mathbf{E}]\!] : \dot{\boldsymbol{\epsilon}}_c\,, \\ &\mathbf{n} \cdot [\![\dot{\mathbf{u}}]\!] + \frac{4}{\delta}\,k\,[\![p]\!] = 0\,. \end{aligned} \tag{33}$$

We emphasize that this condition must be satisfied at the onset of localization as well as in the post-localized range. In (33), we introduced $\mathbf{Q}_\Gamma$ as the acoustic tensor associated with the effective material, i. e. $\mathbf{Q}_\Gamma = \mathbf{n} \cdot \mathbf{E}(\mathbf{x}_0) \cdot \mathbf{n}$.

Upon eliminating the displacement discontinuity in (33), we may express (33) in terms of a scalar evolution equation for the fluid pressure discontinuity as

$$[\![\dot{p}]\!] + c\,[\![p]\!] = b\,, \quad c = \frac{4\,k}{\delta^2}\frac{1}{S}\,, \quad b = \frac{B}{S}\,, \tag{34}$$

where

$$S := \mathbf{n} \cdot \mathbf{P}_\Gamma \cdot \mathbf{n}\,, \quad B := (\mathbf{n} \cdot \mathbf{P}_\Gamma) \cdot (\mathbf{n} \cdot [\![\mathbf{E}]\!] : \dot{\boldsymbol{\epsilon}}_c)\,, \quad \mathbf{P}_\Gamma := \mathbf{Q}_\Gamma^{-1}\,. \tag{35}$$

4.2 Conditions for the onset of localization for elasto-plasticity

Preliminaries

We consider the implications of the localization condition (33) in terms of the scalar equation (34) and the plasticity framework outlined in Subsection 2.2. As a result, the acoustic tensor $\mathbf{Q}_{\Gamma}$ takes on the values $\mathbf{Q}^e$ or $\mathbf{Q}^{ep}$ depending on whether elastic or plastic loading takes place:

$$\mathbf{Q}_{\Gamma} = \begin{cases} \mathbf{Q}^e = \mathbf{n} \cdot \mathbf{E}^e \cdot \mathbf{n} & \text{if } \mathbf{f} : \mathbf{E}^e : \dot{\epsilon}_c + \frac{f}{\delta} \mathbf{a} \cdot [\![\dot{\mathbf{u}}]\!] < 0 \\ \mathbf{Q}^{ep} = \mathbf{n} \cdot \mathbf{E}^{ep} \cdot \mathbf{n} = \mathbf{Q}^e - \frac{1}{h} \mathbf{a}^* \otimes \mathbf{a} & \text{if } \mathbf{f} : \mathbf{E}^e : \dot{\epsilon}_c + \frac{f}{\delta} \mathbf{a} \cdot [\![\dot{\mathbf{u}}]\!] > 0 \,, \end{cases} \tag{36}$$

where $\mathbf{a} := \mathbf{f} : \mathbf{E} \cdot \mathbf{n}$ and $\mathbf{a}^* := \mathbf{n} \cdot \mathbf{E} : \mathbf{f}^*$.

From (36), we also retrieve the localization conditions for the two (extreme) situations of fully drained, i. e. $[\![\dot{p}]\!] = 0$, and undrained behaviour, i. e. $k = 0$. It may be shown that singularity of $\mathbf{Q}_{\Gamma} = \mathbf{Q}^{ep}$ in these two different situations can be expressed as follows:

$$\begin{aligned} &\mathbf{Q}^{ep} \cdot [\![\dot{\mathbf{u}}]\!] = \mathbf{0} \qquad \Rightarrow \quad \mu = 1 - \frac{1}{h} \mathbf{a} \cdot \mathbf{P}^e \cdot \mathbf{a}^* = 0 \,, \\ &\mathbf{Q}^{ep} \cdot [\![\dot{\mathbf{u}}]\!] = \mathbf{0} \qquad \text{s.t. } \mathbf{n} \cdot [\![\dot{\mathbf{u}}]\!] = 0 \\ &\qquad \qquad \qquad \qquad \Rightarrow \quad \mu_u \overset{\text{def}}{=} \mu + \frac{1}{h} \frac{(\mathbf{n} \cdot \mathbf{P}^e \cdot \mathbf{a})(\mathbf{a}^* \cdot \mathbf{P}^e \cdot \mathbf{n})}{\mathbf{n} \cdot \mathbf{P}^e \cdot \mathbf{n}} = 0 \,, \end{aligned} \tag{37}$$

where μ is the smallest eigenvalue of $\mathbf{Q}^{ep}$.

Moreover, we note that in the neighbourhood of an arbitrary time $t = t_0$, where it is sufficient to consider the linearized response, the solution to (34) may be written as

$$\begin{aligned} [\![p]\!] &= [\![p_0]\!] \exp\left(-c\,(t - t_0)\right) - \frac{b}{c} \left[\exp\left(-c\,(t - t_0)\right) - 1\right] \,, \\ [\![\dot{p}]\!] &= \left(b - c\,[\![p_0]\!]\right) \exp\left(-c\,(t - t_0)\right) . \end{aligned} \tag{38}$$

It is recalled that all state variables are considered continuous up to the time $t = t_{loc} = t_0$ when onset of localization is possible. Hence, the initial condition is $[\![p_0]\!] = 0$, which gives the temporal evolution of the linearized response

$$[\![\dot{p}]\!] = b \exp\left(-c\,(t - t_0)\right) . \tag{39}$$

We conclude that any perturbation of $[\![\dot{p}]\!]$ will decay exponentially whenever $S > 0$, whereas whenever $S < 0$, $[\![\dot{p}]\!]$ will increase exponentially. Hence, it is of considerable importance to characterize the state when the value of S traverses from positive to negative.

To this end, we consider the localization condition (33) with due consideration to the various situations of loading/unloading at $\mathbf{x} = \mathbf{x}_0$ as compared to $\mathbf{x} = \mathbf{x}_\pm$. Only the situation at the onset of localization is considered, in which case we have $\mathbf{E}^{ep}(\mathbf{x}_\pm) = \mathbf{E}^{ep}(\mathbf{x}_0)$.

Case I: Elastic unloading inside and immediately outside the band

With $\mathbf{E}(\mathbf{x}_0) = \mathbf{E}(\mathbf{x}_\pm) = \mathbf{E}^e$, we obtain from (33)

$$\begin{aligned} &\frac{1}{\delta}\mathbf{Q}^e \cdot [\![\dot{\mathbf{u}}]\!] - [\![\dot{p}]\!]\,\mathbf{n} = \mathbf{0}\,, \\ &\mathbf{n}\cdot[\![\dot{\mathbf{u}}]\!] + \frac{4}{\delta}\,k\,[\![p]\!] = 0\,, \end{aligned} \tag{40}$$

from which $[\![\dot{\mathbf{u}}]\!]$ may be eliminated to give

$$[\![\dot{p}]\!] + c^e[\![p]\!] = 0\,, \quad c^e = \frac{4\,k}{\delta^2}\frac{1}{S^e}\,, \quad S^e = \mathbf{n}\cdot\mathbf{P}^e\cdot\mathbf{n}\,, \quad \mathbf{P}^e = (\mathbf{Q}^e)^{-1}\,. \tag{41}$$

Since the stiffness is continuous across Γ in this situation, we find that $b = 0$. Hence, $[\![\dot{p}]\!] = 0$ and $[\![\dot{\mathbf{u}}]\!] = 0$. Moreover, since $\mathbf{P}^e$ is positive definite, we conclude that $S^e > 0$. This means that any perturbation of $[\![\dot{p}]\!]$ will decay exponentially. In fact, this result is obtained whenever $\mathbf{Q}_\Gamma$ is positive definite (as will be discussed subsequently).

Case II: Plastic loading inside and immediately outside the band

We obtain

$$\begin{aligned} &\frac{1}{\delta}\mathbf{Q}^{ep} \cdot [\![\dot{\mathbf{u}}]\!] - [\![\dot{p}]\!]\,\mathbf{n} = \mathbf{0}\,, \\ &\mathbf{n}\cdot[\![\dot{\mathbf{u}}]\!] + \frac{4}{\delta}\,k\,[\![p]\!] = 0\,, \end{aligned} \tag{42}$$

from which $[\![\dot{\mathbf{u}}]\!]$ may be eliminated to give

$$[\![\dot{p}]\!] + c^{ep}\,[\![p]\!] = 0\,, \quad c^{ep} = \frac{4\,k}{\delta^2}\frac{1}{S^{ep}}\,, \quad S^{ep} = \mathbf{n}\cdot\mathbf{P}^{ep}\cdot\mathbf{n}\,. \tag{43}$$

In (43), the value of S^{ep} may be expressed more explicitly by noting that

$$\mathbf{P}^{ep} = \mathbf{P}^e + \frac{1}{h\,\mu}\mathbf{P}^e\cdot\mathbf{a}^*\otimes\mathbf{a}\cdot\mathbf{P}^e\,, \tag{44}$$

where we have introduced the "drained eigenvalue" μ, given in (37). By invoking $\mathbf{P}^{ep}$ from (44) into the expression for S^{ep}, we find that

$$S^{ep} = \frac{\mu_u}{\mu}\,\mathbf{n}\cdot\mathbf{P}^e\cdot\mathbf{n}\,. \tag{45}$$

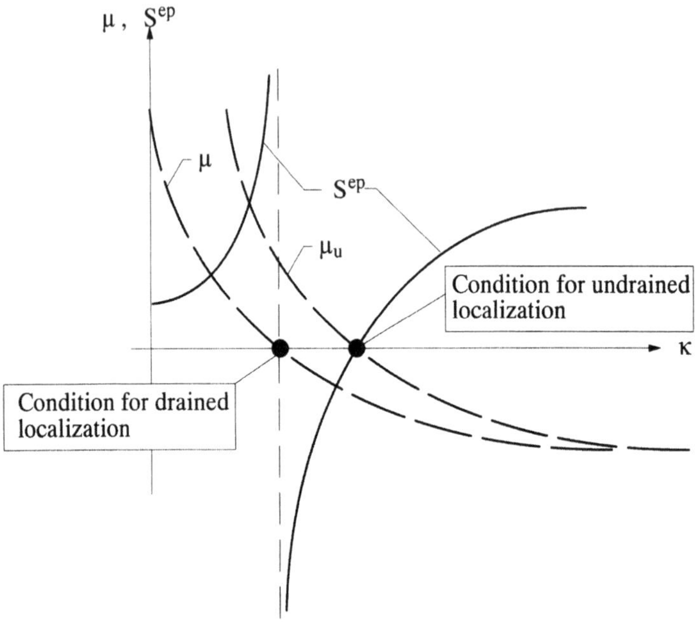

Fig. 3. Schematic sketch of the development of the values of μ, μ_u, and S during plastic deformation.

Pertinent to a ductile fracture process, as shown in Figure 3, the value of μ will decrease from the value $\mu = 1$ (at elastic response) in a smooth fashion during plastic loading as a function of the hardening/softening parameter κ. From (45), we note that a critical situation occurs when $\mathbf{Q}^{ep}$ becomes singular, i. e. μ traverses from positive to negative, whereby S^{ep} exhibits a singularity. In view of (39), localization in the sense that $[\![\dot{p}]\!]$ may grow exponentially in time is possible in the presence of perturbations from $b = 0$ when the drained localization criterion is satisfied, i. e. when $\mathbf{Q}^{ep}$ becomes singular. However, since $S^{ep} = 0$ is actually never attained with the present loading assumption, we obtain (formally) that $b = 0$ and the pressure and displacement discontinuities are zero in this case. On the other hand, if we approach a state where the undrained localization condition is satisfied, i. e. $\mu_u = 0$, then $S^{ep} = 0$ which means that we have a situation where the discontinuity may be unbounded.

5 Mixed variational format with embedded discontinuities

We consider the variational format for a domain B, as in Figure 1, with external boundary ∂B with outward unit normal $\bar{\mathbf{n}}$, which is divided into two parts in two different ways. On one hand, ∂B is divided into the two mutually exclusive parts $\partial_u B$ (prescribed skeleton displacement) and $\partial_t B$ (prescribed total traction) such that $\partial_u B \bigcup \partial_t B = \partial B$. On the other hand, ∂B is divided into the two mutually exclusive parts $\partial_p B$ (prescribed pore fluid pressure) and $\partial_q B$ (prescribed fluid volume flux) such that $\partial_p B \bigcup \partial_q B = \partial B$.

In order to establish the proper variational equations that incorporate embedded discontinuities of displacements and pressures, we generalize the developments for the undrained continuum in Larsson et al. [12] to the partly drained situation. To this end, we exploit the Enhanced Assumed Strain (EAS) approach, Simo and Rifai [21], in terms of the fields $\boldsymbol{\epsilon}' \in E$ for strains, $p' \in P$ for pressures and $\boldsymbol{\eta}' \in N$ for pressure gradients. These fields are further decomposed into the EAS-structure:

$$
\begin{aligned}
&\boldsymbol{\epsilon}' = \boldsymbol{\epsilon}'_c + \tilde{\boldsymbol{\epsilon}}' \in U \times \tilde{E} \quad \text{with} \quad \boldsymbol{\epsilon}'_c = (\nabla \otimes \mathbf{u}'_c)^{\text{sym}} , \\
&p' = p'_c + \tilde{p}' \in P_c \times \tilde{P} , \\
&\boldsymbol{\eta}' = \boldsymbol{\eta}'_c + \tilde{\boldsymbol{\eta}}' \in \nabla P_c \times \tilde{N} \quad \text{with} \quad \boldsymbol{\eta}'_c = \nabla p_c ,
\end{aligned}
\tag{46}
$$

where e. g. $\tilde{\boldsymbol{\epsilon}}' \in \tilde{E}$ represents the enhanced portion of the total strain. As to the enhanced quantities in (46), we are guided by the development in Section 3 to define $\tilde{\boldsymbol{\epsilon}}' \in \tilde{E}$, $\tilde{p}' \in \tilde{P}$, and $\tilde{\boldsymbol{\eta}}' \in \tilde{N}$ as

$$
\begin{aligned}
&\tilde{\boldsymbol{\epsilon}}' = \tilde{\boldsymbol{\epsilon}}'_c + \delta_\Gamma (\mathbf{n} \otimes \mathbf{v}')^{\text{sym}} , \\
&\tilde{p}' = \tilde{p}'_c + \delta_\Gamma q' , \\
&\tilde{\boldsymbol{\eta}}' = \tilde{\boldsymbol{\eta}}'_c + \nabla(\delta_\Gamma q') .
\end{aligned}
\tag{47}
$$

Here, $\mathbf{v}'$ is a displacement-like discontinuity and q' is a pressure-like discontinuity (defined per unit internal surface Γ). We note that these expressions have meaning only in a distributional sense. Moreover, in (46) we introduced the function spaces U and P_c of admissible displacement and pressure variations, defined in the standard way as

$$
\begin{aligned}
&U := \left\{ \mathbf{u}'_c \in H^1(B) : \mathbf{u}'_c = \mathbf{0} \text{ on } \partial_u B \right\} , \\
&P_c := \left\{ p'_c \in H^1(B) : p'_c = 0 \text{ on } \partial_p B \right\} .
\end{aligned}
\tag{48}
$$

We also introduce $\tilde{E}$, E_v, S, V, $\tilde{P}$, and $\tilde{N}$ as the spaces of admissible enhanced strain, volumetric strain, effective stress, *Darcy*an velocity, enhanced pressure and enhanced pressure gradient, respectively. We set tentatively

$$
\tilde{E} = E_v = S = V = \tilde{P} = \tilde{N} := L_2(B) . \tag{49}
$$

Following the developments in Larsson and Larsson [16], we formulate the variational form of momentum balance pertinent to the EAS-approach as

$$\begin{aligned}
&\int_B \boldsymbol{\epsilon}'_c : \bar{\boldsymbol{\sigma}}\,\mathrm{d}\Omega - W_u^{\mathrm{ext}}(\mathbf{u}'_c) = 0\,, \quad \forall\, \mathbf{u}'_c \in U \quad \text{with} \quad \bar{\boldsymbol{\sigma}} = \boldsymbol{\sigma}(\boldsymbol{\epsilon}) - p\,\mathbf{1}\,, \\
&\int_B \tilde{\boldsymbol{\epsilon}}'_c : \bar{\boldsymbol{\sigma}}\,\mathrm{d}\Omega + \int_\Gamma \mathbf{v}' \cdot \bar{\mathbf{t}}\,\mathrm{d}\Gamma = 0\,, \quad \forall\, \tilde{\boldsymbol{\epsilon}}' \in \tilde{E}\,.
\end{aligned} \tag{50}$$

Likewise, the variational form of mass balance may be written as

$$\begin{aligned}
&\int_B \boldsymbol{\eta}'_c \cdot \boldsymbol{v}_d\,\mathrm{d}\Omega - \int_B p'_c\,\dot{\epsilon}_v\,\mathrm{d}\Omega - W_p^{\mathrm{ext}}(p'_c) = 0\,, \quad \forall\, p'_c \in P_c\,, \\
&\int_B \tilde{\boldsymbol{\eta}}'_c \cdot \boldsymbol{v}_d\,\mathrm{d}\Omega - \int_\Gamma q'\,\nabla \cdot \mathbf{v}_d\,\mathrm{d}\Gamma - \\
&\qquad - \int_B \tilde{p}'_c\,\dot{\epsilon}_v\,\mathrm{d}\Omega - \int_\Gamma q'\,\dot{\epsilon}_v\,\mathrm{d}\Gamma = 0\,, \quad \forall\, \tilde{\boldsymbol{\eta}}' \in \tilde{N}\,, \quad \forall\, \tilde{p}' \in \tilde{P}\,,
\end{aligned} \tag{51}$$

which corresponds to the orthogonality conditions

$$\begin{aligned}
&\int_B \boldsymbol{\tau}' : \tilde{\boldsymbol{\epsilon}}'_c\,\mathrm{d}\Omega + \int_\Gamma \mathbf{n} \cdot \boldsymbol{\tau}' \cdot \mathbf{v}'\,\mathrm{d}\Omega \equiv 0\,, \quad \forall\, \boldsymbol{\tau}' \in S\,, \quad \forall\, \tilde{\boldsymbol{\epsilon}}' \in \tilde{E}\,, \\
&\int_B \gamma'\,\tilde{p}'_c\,\mathrm{d}\Omega + \int_\Gamma \gamma'\,q'\,\mathrm{d}\Gamma \equiv 0\,, \quad \forall\, \gamma' \in E_v\,, \quad \forall\, \tilde{p}' \in \tilde{P}\,, \\
&\int_B \mathbf{w}' \cdot \tilde{\boldsymbol{\eta}}'_c\,\mathrm{d}\Omega - \int_\Gamma \nabla \cdot \mathbf{w}'\,q'\,\mathrm{d}\Omega \equiv 0\,, \quad \forall\, \mathbf{w}' \in V\,, \quad \forall\, \tilde{\boldsymbol{\eta}}' \in \tilde{N}\,.
\end{aligned} \tag{52}$$

We emphasize that the balance of mass involves the volumetric strain field γ, the relative fluid flow $\mathbf{w}$, the total pressure gradient $\boldsymbol{\eta}$, and the volumetric strain $\epsilon_v = \mathrm{tr}\,(\boldsymbol{\epsilon})$. Moreover, the external virtual work quantities are given by

$$\begin{aligned}
&W_u^{\mathrm{ext}}(\mathbf{u}'_c) = \int_B \mathbf{u}'_c \cdot \hat{\rho}\,\mathbf{g}\,\mathrm{d}\Omega + \int_{\partial_t B} \mathbf{u}'_c \cdot \bar{\mathbf{t}}\,\mathrm{d}\Gamma\,, \\
&W_p^{\mathrm{ext}}(p'_c) = \int_{\partial_q B} p'_c\,\bar{q}\,\mathrm{d}\Gamma\,,
\end{aligned} \tag{53}$$

where $\bar{q} = \mathbf{v}_d \cdot \bar{\mathbf{n}}$ is the fluid volume flux (outflow) and $\bar{\mathbf{t}} = \bar{\boldsymbol{\sigma}} \cdot \bar{\mathbf{n}}$ is the total prescribed traction, where $\bar{\mathbf{n}}$ is the outward unit normal of ∂B.

6 Finite element procedure

6.1 Finite element formulation

The region B is considered discretized into finite elements Ω_e, $e = 1, \ldots$, NEL. For a specific element, the compatible displacement and fluid pressure fields along with the appropriate gradients are interpolated by using the standard compatible shape functions, i. e.

$$\begin{aligned}
&\mathbf{u}_{ce} = \sum_{I=1}^{\mathrm{NOEL}} N_u^I \, \hat{\mathbf{u}}_e^I \, , \quad \boldsymbol{\epsilon}_{ce} = (\mathbf{u}_{ce} \otimes \nabla)^{\mathrm{sym}} = \sum_{I=1}^{\mathrm{NOEL}} (\hat{\mathbf{u}}_e^I \otimes \mathbf{m}_u^I)^{\mathrm{sym}} \\
&\qquad \text{with} \quad \mathbf{m}_u^I = \nabla N_u^I \, , \\
&p_{ce} = \sum_{I=1}^{\mathrm{NOEL}} N_p^I \, \hat{p}_e^I \, , \quad \eta_{ce} = \nabla p_{ce} = \sum_{I=1}^{\mathrm{NOEL}} \hat{p}_e^I \, \mathbf{m}_p^I \\
&\qquad \text{with} \quad \mathbf{m}_p^I = \nabla N_p^I \, ,
\end{aligned} \tag{54}$$

where N_u^I and $\hat{\mathbf{u}}_e^I$ are the (displacement) shape function and nodal displacement of node I of the element, respectively. Likewise, N_p^I and $\hat{p}_e^I$ are the (pressure) shape function and nodal pressure of node I of the element, respectively. In the following, we shall restrict to a triangular element with quadratic shape functions for the displacement and shape functions linear for the pressure, i. e. NOEL = 6 and 3 for the displacement and pressure, respectively, cf. Figure 4.

Momentum balance

The FE-approximation of $\boldsymbol{\tau}$, $\tilde{\boldsymbol{\epsilon}}_c$, and $\boldsymbol{v}$ is proposed as

$$\boldsymbol{\tau} = \sum_{e=1}^{\mathrm{NEL}} \chi_e \, \boldsymbol{\tau}_e \, , \quad \tilde{\boldsymbol{\epsilon}}_c = \sum_{e=1}^{\mathrm{NEL}} \chi_e \, \tilde{\boldsymbol{\epsilon}}_{ce} \, , \quad \mathbf{v} = \sum_{e=1}^{\mathrm{NEL}} \chi_e \, \mathbf{v}_e \, , \tag{55}$$

where χ_e is defined as

$$\chi_e = \begin{cases} 1 & \text{if} \quad \mathbf{x} \in \Omega_e \\ 0 & \text{otherwise} \, . \end{cases} \tag{56}$$

In particular, we consider the situation when the quantities related to the enhanced strain are piecewise constant within each Ω_e. Upon inserting these FE-approximations into $(52)_1$ and $(50)_2$, we obtain the “orthogonality” con-

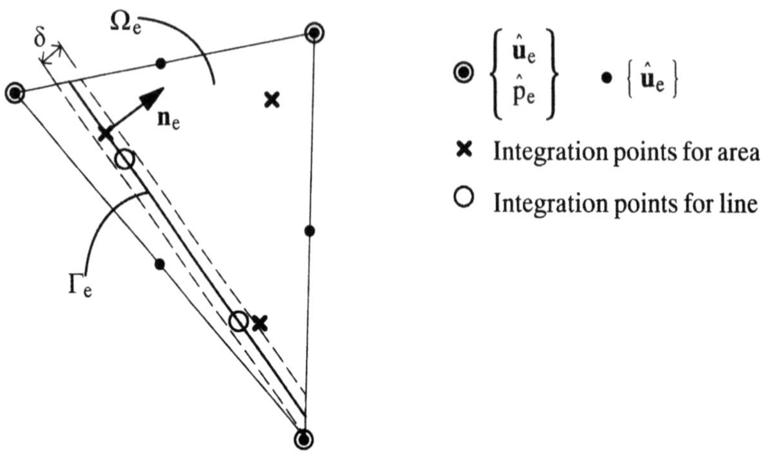

Fig. 4. Enhanced mixed triangular element. The base element has quadratic interpolation for displacement and linear interpolation for pressure. The enhanced fields are piece-wise constant.

dition that must be satisfied element-wise:

$$\begin{aligned}
&\int_{\Omega_e} \tilde{\boldsymbol{\epsilon}}'_{ce} : \boldsymbol{\tau}'_e \, \mathrm{d}\Omega + \int_{\Gamma_e} (\mathbf{n}_e \otimes \mathbf{v}'_e)^{\mathrm{sym}} : \boldsymbol{\tau}'_e \, \mathrm{d}\Gamma = 0 \,, \\
&\qquad \text{with} \quad e = 1, \ldots, \mathrm{NEL} \,, \\
&\int_{\Omega_e} \tilde{\boldsymbol{\epsilon}}'_{ce} : (\boldsymbol{\sigma} - p_e \mathbf{1}) \, \mathrm{d}\Omega + \int_{\Gamma_e} \mathbf{v}'_e \cdot (\boldsymbol{\sigma} - p_e \, \mathbf{1}) \, \mathbf{n}_e \, \mathrm{d}\Gamma = 0 \,, \\
&\qquad \text{with} \quad e = 1, \ldots, \mathrm{NEL} \,.
\end{aligned} \tag{57}$$

Since $\mathbf{n}_e$ is constant within each element, we obtain from $(57)_1$

$$\tilde{\boldsymbol{\epsilon}}'_{ce} = -\frac{1}{l_e} (\mathbf{v}'_e \otimes \mathbf{n}_e)^{\mathrm{sym}} \quad \text{with} \quad l_e = \frac{A_e}{L_e} \,, \; A_e = m(\Omega_e) \,, \; L_e = m(\Gamma_e) \,. \tag{58}$$

Hence, $(57)_2$ becomes

$$\mathbf{v}'_e \cdot \left(-\frac{1}{l_e} \int_{\Omega_e} (\boldsymbol{\sigma} - p_e \, \mathbf{1}) \cdot \mathbf{n}_e \, \mathrm{d}\Omega + \int_{\Gamma_e} (\boldsymbol{\sigma} - p_e \, \mathbf{1}) \cdot \mathbf{n}_e \, \mathrm{d}\Gamma \right) = 0 \tag{59}$$

$$\text{with} \quad e = 1, \ldots, \mathrm{NEL} \,.$$

Mass conservation

Quantities related to the enhanced pressure and pressure gradient are chosen piecewise constant within each Ω_e such that

$$\gamma = \sum_{e=1}^{\text{NEL}} \chi_e \, \gamma_e \,, \qquad \tilde{p}_c = \sum_{e=1}^{\text{NEL}} \chi_e \, \tilde{p}_{ce} \,, \quad q = \sum_{e=1}^{\text{NEL}} \chi_e \, q_e \,,$$
$$\tilde{\eta}_c = \sum_{e=1}^{\text{NEL}} \chi_e \, \tilde{\eta}_{ce} \,, \quad \mathbf{w} = \sum_{e=1}^{\text{NEL}} \chi_e \, \mathbf{w}_e \,. \tag{60}$$

Upon inserting these FE-approximations into $(52)_{2,3}$ and $(51)_2$, we obtain

$$\int_{\Omega_e} \gamma_e' \, \tilde{p}_e' \, \mathrm{d}\Omega = \int_{\Omega_e} \gamma_e' \, \tilde{p}_{ce}' \, \mathrm{d}\Omega + \int_{\Gamma_e} \gamma_e' \, q_e' \, \mathrm{d}\Gamma = 0$$
$$\text{with} \quad e = 1, \dots, \text{NEL}, \tag{61}$$

$$\int_{\Omega_e} \mathbf{w}_e' \cdot \tilde{\eta}_{ce}' \, \mathrm{d}\Omega - \int_{\Gamma_e} (\nabla \cdot \mathbf{w}_e') \, q_e' \, \mathrm{d}\Gamma = 0$$
$$\text{with} \quad e = 1, \dots, \text{NEL}, \tag{62}$$

$$\int_{\Omega_e} \tilde{\eta}_{ce}' \cdot \mathbf{v}_d \, \mathrm{d}\Omega - \int_{\Gamma_e} q_e' (\nabla \cdot \mathbf{v}_d) \, \mathrm{d}\Gamma \; -$$
$$- \int_{\Omega_e} \tilde{p}_{ce}' \, \dot{\epsilon}_{ve} \, d \, \mathrm{d}\Omega - \int_{\Gamma_e} q_e' \, \dot{\epsilon}_{ve} \, \mathrm{d}\Gamma = 0 \,. \tag{63}$$

Since $\tilde{p}_{ce}$, q_e, and $\mathbf{w}_e$ are element-wise constant fields, we obtain from (61) and (62) the relations

$$\tilde{p}_{ce}' = -\frac{1}{l_e} q_e' \,, \quad \tilde{\eta}_{ce}' = \mathbf{0} \,, \tag{64}$$

whereby (63) can be reduced to

$$q_e' \left(\frac{1}{l_e} \int_{\Omega_e} \dot{\epsilon}_{ve} \, \mathrm{d}\Omega - \int_{\Gamma_e} \dot{\epsilon}_{ve} \, \mathrm{d}\Gamma - \int_{\Gamma_e} \nabla \cdot \mathbf{v}_d \, \mathrm{d}\Gamma \right) = 0$$
$$\text{with} \quad e = 1, \dots, \text{NEL}. \tag{65}$$

Regularization of singular fields

In accordance with the development of the regularized strong discontinuities for the continuum in Section 3, the actual enhanced fields are chosen to possess the regularized form. In view of (47) and (64), we thus consider element approximation in terms of compatible and regularized enhanced fields such that

$$\begin{aligned}
&\boldsymbol{\epsilon}_e = \boldsymbol{\epsilon}_{ce} + \frac{1}{\delta}\left(f - \frac{\delta}{l_e}\right)(\mathbf{n}_e \otimes [\![\mathbf{u}]\!]_e)^e \\
&\qquad \Rightarrow \dot{\epsilon}_{ve} = \dot{\epsilon}_{vce} + \frac{1}{\delta}\left(f - \frac{\delta}{l_e}\right)\mathbf{n}_e \cdot [\![\dot{\mathbf{u}}]\!]_e, \\
&p_e = p_{ce} + \left(f - \frac{\delta}{l_e}\right)[\![p]\!]_e\,, \\
&\mathbf{v}_d = -k\left(\nabla p_{ce} + \frac{g}{\delta}\mathbf{n}_e [\![p]\!]_e\right) = \boldsymbol{v}_{d,c} + g\,[\![\mathbf{v}_d]\!] \\
&\qquad \text{with} \quad [\![\mathbf{v}_d]\!] = -\frac{k}{\delta}[\![p]\!]_e \mathbf{n}_e\,, \\
&\nabla \cdot \mathbf{v}_d = -k\left(\nabla \cdot \nabla p_{ce} + \frac{h}{\delta^2}[\![p]\!]_e\right) = \nabla \cdot \mathbf{v}_{d,c} + \frac{h}{\delta}\mathbf{n}_e \cdot [\![\mathbf{v}_d]\!]\,,
\end{aligned} \tag{66}$$

where $[\![\mathbf{u}]\!]_e$ and $[\![p]\!]_e$ are the displacement discontinuity and pressure discontinuity parameters of the element, respectively. Moreover, $f(n)$, $g(n)$, and $h(n)$ are the regularization functions given in Figure 2.

We emphasize that, up to this point, the formulation is general in terms of dimension of the problem and choice of displacement-pressure-based base element. In the subsequent developments we explicitly introduce properties of the chosen element in Figure 4. In particular, from $(66)_3$ and $(66)_4$, one obtains that $\boldsymbol{v}_{d,c}$ is constant whereby $\nabla \cdot \mathbf{v}_{d,c} = 0$. As a result, we may restate (65) as

$$q'_e\left(\frac{1}{l_e}\int_{\Omega_e} \dot{\epsilon}_{ve}\,\mathrm{d}\Omega - \int_{\Gamma_e} \dot{\epsilon}_{ve}\,\mathrm{d}\Gamma - \int_{\Gamma_e} \frac{h}{\delta}\mathbf{n}_e \cdot [\![\mathbf{v}_d]\!]\,\mathrm{d}\Gamma\right) = 0 \qquad \text{with} \quad e = 1, \ldots, \mathrm{NEL}\,. \tag{67}$$

6.2 Finite element equations

Upon inserting the preceding FE-discretizations into $(50)_1$, $(51)_1$, (59), and (65), we obtain the discretized formulation

$$\mathop{\mathbf{A}}_{e=1}^{\mathrm{NEL}}\left[\mathbf{b}_e - \mathbf{f}_{ue}^{\mathrm{ext}}\right] = \mathbf{0}\,, \tag{68$_1$}$$

$$\mathop{\mathbf{A}}_{e=1}^{\mathrm{NEL}} \left[\mathbf{c}_e - \mathbf{f}_{pe}^{\mathrm{ext}}\right] = \mathbf{0}\,, \tag{68$_2$}$$

$$\mathbf{r}_e = \mathbf{0}\,, \quad e = 1, \ldots, \mathrm{NEL}\,, \tag{68$_3$}$$

$$s_e = 0\,, \quad e = 1, \ldots, \mathrm{NEL}\,, \tag{68$_4$}$$

where $(68)_1$ and $(68)_2$ represent the global momentum and mass balance, whereas $(68)_3$ and $(68)_4$ represent momentum and mass balance across the element-embedded band. Subsequently, we introduce the backward *Euler* method for the temporal integration. The internal nodal forces and local traction/mass balance are given by

$$\begin{aligned}
\mathbf{b}_e &= \mathop{\boldsymbol{A}_u}_{I=1}^{6} \int_{\Omega_e} (\boldsymbol{\sigma} - p_e\,\mathbf{1}) \cdot \mathbf{m}_u^I \,\mathrm{d}\Omega\,, \\
\mathbf{c}_e &= \mathop{\boldsymbol{A}_p}_{I=1}^{3} \int_{\Omega_e} \left(\Delta t\,\mathbf{m}_e^I \cdot \mathbf{v}_d - N_p^I \Delta\epsilon_{ve}\right) \mathrm{d}\Omega\,, \\
\mathbf{r}_e &= \int_{\Gamma_e} (\boldsymbol{\sigma} - p_e\,\mathbf{1}) \cdot \mathbf{n}\,\mathrm{d}\Gamma - \frac{1}{l_e} \int_{\Omega_e} (\boldsymbol{\sigma} - p_e\,\mathbf{1})\,\mathrm{d}\Omega\,, \\
s_e &= \int_{\Gamma_e} \Delta\epsilon_{ve}\,\mathrm{d}\Gamma - \frac{1}{l_e} \int_{\Omega_e} \Delta\epsilon_{ve}\,\mathrm{d}\Omega - \Delta t \int_{\Gamma_e} \frac{4}{\delta} [\![\boldsymbol{v}_d]\!] \cdot \mathbf{n}\,\mathrm{d}\Gamma\,,
\end{aligned} \tag{69}$$

whereas the external nodal forces are defined as

$$\begin{aligned}
\mathbf{f}_{ue}^{\mathrm{ext}} &= \mathop{\boldsymbol{A}_u}_{I=1}^{6} \int_{\Omega_e} N_u^I \hat{\rho}\,\mathbf{b}\,\mathrm{d}\Omega + \mathop{\boldsymbol{A}_u}_{I=1}^{6} \int_{\Gamma_e} N_u^I \,\bar{\mathbf{t}}\,\mathrm{d}\Gamma\,, \\
\mathbf{f}_{pe}^{\mathrm{ext}} &= \mathop{\boldsymbol{A}_p}_{I=1}^{3} \int_{\Omega_e} \Delta t\, N_p^I \,\bar{q}\,\mathrm{d}\Omega\,.
\end{aligned} \tag{70}$$

In (68) and (69), we introduced the assembly operators $\mathbf{A}$, $\boldsymbol{A}_u$, and $\boldsymbol{A}_p$. The operator $\mathbf{A}$ determines the position in the global FE-vector based on the global nodal-element topology. The internal element operators, e. g. $\boldsymbol{A}_u$, determine the position of each local contribution within the element vector from the local node topology, i. e. $\boldsymbol{A}_u$ defines the identity

$$\sum_{I=1}^{\mathrm{NOEL}} (\mathbf{m}_u^I \otimes \mathbf{u}_e^I)^{\mathrm{sym}} : \boldsymbol{\sigma} \equiv \hat{\mathbf{u}}_e^t \left[\mathop{\boldsymbol{A}_u}_{I=1}^{\mathrm{NOEL}} \mathbf{m}_u^I \cdot \boldsymbol{\sigma}\right], \tag{71}$$

where the vector $\hat{\mathbf{u}}_e$ contains the element-nodal displacement variables. Likewise, the vector $\hat{\mathbf{p}}_e$ contains the element-nodal pressure variables.

6.3 Remarks on the solution procedure

The non-linear equations (68) are conveniently solved using *Newton-Raphson*'s method, whereby the proper linearization of (69) is required. As a result of the linearization process we obtain the element stiffness of the form

$$\begin{bmatrix} \mathrm{d}\mathbf{b}_e \\ \mathrm{d}\mathbf{c}_e \\ \mathrm{d}\mathbf{r}_e \\ \mathrm{d}s_e \end{bmatrix} = \begin{bmatrix} \mathbf{K}_e^u & \mathbf{C}_e & \mathbf{F}_e & \mathbf{P}_e \\ \mathbf{C}_e' & \mathbf{K}_e^p & \mathbf{G}_e & \mathbf{0} \\ \mathbf{F}_e' & \mathbf{G}_e' & \mathbf{H}_e & \mathbf{S}_e \\ \mathbf{P}_e' & \mathbf{0} & \mathbf{S}_e' & D_e \end{bmatrix} \begin{bmatrix} \mathrm{d}\hat{\mathbf{u}}_e' \\ \mathrm{d}\hat{\mathbf{p}}_e' \\ \mathrm{d}[\![\mathbf{u}]\!]_e \\ \mathrm{d}[\![p]\!]_e \end{bmatrix}, \tag{72}$$

where it is noted that the stiffness matrix is unsymmetric, due to the difference in the enhanced test and actual fields. In (72), the various matrix elements are defined as

$$\mathbf{K}_e^u = \mathop{\boldsymbol{A}_u}_{I=1}^{6} \mathop{\boldsymbol{A}_u}_{J=1}^{6} \int_{\Omega_e} \mathbf{m}_u^I \cdot \mathbf{E} \cdot \mathbf{m}_u^J \,\mathrm{d}\Omega\,, \tag{73$_1$}$$

$$\mathbf{K}_e^p = -\mathop{\boldsymbol{A}_p}_{I=1}^{3} \mathop{\boldsymbol{A}_p}_{J=1}^{3} \int_{\Omega_e} \Delta t\, k\, \mathbf{m}_p^I \cdot \mathbf{m}_p^J \,\mathrm{d}\Omega\,, \tag{73$_2$}$$

$$\mathbf{C}_e = -\mathop{\boldsymbol{A}_u}_{I=1}^{6} \mathop{\boldsymbol{A}_p}_{J=1}^{3} \int_{\Omega_e} \mathbf{m}_u^I N_p^J \,\mathrm{d}\Omega\,, \tag{73$_3$}$$

$$\mathbf{C}_e' = -\mathop{\boldsymbol{A}_p}_{I=1}^{3} \mathop{\boldsymbol{A}_u}_{J=1}^{6} \int_{\Omega_e} N_p^I \mathbf{m}_u^J \,\mathrm{d}\Omega = \mathbf{C}_e^T\,, \tag{73$_4$}$$

$$\mathbf{H}_e = \int_{\Gamma_e} \left(\frac{1}{\delta} - \frac{1}{l_e}\right) \mathbf{n}_e \cdot \mathbf{E} \cdot \mathbf{n}_e \,\mathrm{d}\Gamma + \int_{\Omega_e} \frac{1}{l_e^2} \mathbf{n}_e \cdot \mathbf{E} \cdot \mathbf{n}_e \,\mathrm{d}\Omega\,, \tag{73$_5$}$$

$$\mathbf{G}_e = \mathop{\boldsymbol{A}_p}_{I=1}^{3} \int_{\Omega_e} \frac{1}{l_e} N_p^I \mathbf{n}_e \,\mathrm{d}\Omega\,, \tag{73$_6$}$$

$$\mathbf{G}_e' = \mathop{\boldsymbol{A}_p}_{I=1}^{3} \left(-\int_{\Gamma_e} n_p^I \mathbf{n}_e \,\mathrm{d}\Gamma + \int_{\Omega_e} \frac{1}{l_e} n_p^I \mathbf{n}_e \,\mathrm{d}\Omega \right)\,, \tag{73$_7$}$$

$$\mathbf{F}_e = -\mathop{\boldsymbol{A}_u}_{I=1}^{6} \int_{\Omega_e} \frac{1}{l_e} \mathbf{m}_u^I \cdot \mathbf{E} \cdot \mathbf{n}_e \,\mathrm{d}\Omega\,, \tag{73$_8$}$$

$$\mathbf{F}'_e = \mathop{\boldsymbol{A}_u}_{I=1}^{6} \left(\int_{\Gamma_e} \mathbf{n}_e \cdot \mathbf{E} \cdot \mathbf{m}_u^I \, \mathrm{d}\Gamma - \int_{\Omega_e} \frac{1}{l_e} \mathbf{n}_e \cdot \mathbf{E} \cdot \mathbf{m}_u^I \, \mathrm{d}\Omega \right) , \tag{73$_9$}$$

$$D_e = \int_{\Gamma_e} \Delta t \frac{4\,k}{\delta^2} \, \mathrm{d}\Gamma , \tag{73$_{10}$}$$

$$\mathbf{S}_e = - \int_{\Gamma_e} \mathbf{n}_e \, \mathrm{d}\Gamma , \quad \mathbf{S}'_e = \int_{\Gamma_e} \frac{1}{\delta} \mathbf{n}_e \, \mathrm{d}\Gamma , \tag{73$_{11}$}$$

$$\mathbf{P}_e = \mathop{\boldsymbol{A}_u}_{I=1}^{6} \int_{\Omega_e} \frac{\delta}{l_e} \mathbf{m}_u^I \, \mathrm{d}\Omega , \tag{73$_{12}$}$$

$$\mathbf{P}'_e = \mathop{\boldsymbol{A}_u}_{I=1}^{6} \left(\int_{\Gamma_e} \mathbf{m}_u^I \, \mathrm{d}\Gamma - \int_{\Omega_e} \frac{1}{l_e} \mathbf{m}_u^I \, \mathrm{d}\Omega \right) . \tag{73$_{13}$}$$

In the actual FE-computations, displacement and pressure discontinuities are condensed at element level by partial inversion of (72) such that the resulting set of equations contains only displacement and pressure variables of the "base" element approximation. Moreover, as compared to a standard partial inversion of the local element problems, we adopt a staggered solution procedure where the local problems $(68)_3$ and $(68)_4$ are solved completely for every global iteration. The reason for this is to maintain control of the loading situation within an element. In the standard coupled procedure, the element condition will not be satisfied until global equilibrium is reached. This means that a certain loading situation must be assumed a priori (at the beginning of each time step) and the validity of the assumption can not be checked until the time step is solved. The algorithm to solve the element problems $(68)_3$ and $(68)_4$ is based on loading scenarios that can appear within the element, cf. Larsson and Larsson [14].

Remark: We may assess the condition for onset of localization for the discretized problem. To this end, consider an element with given equilibrated nodal variables $\hat{\boldsymbol{u}}_e$ and $\hat{\boldsymbol{p}}_e$ at the time $t = t^{n+1}$ from the incremental/iterative algorithm. In this situation the element response may be linearized in view of (72) as

$$\begin{bmatrix} \mathrm{d}\mathbf{r}_e \\ \mathrm{d}s_e \end{bmatrix} = \begin{bmatrix} \mathbf{H}_e & \mathbf{S}_e \\ \mathbf{S}'_e & D_e \end{bmatrix} \begin{bmatrix} \mathrm{d}[\![\mathbf{u}]\!]_e \\ \mathrm{d}[\![p]\!]_e \end{bmatrix} . \tag{74}$$

At the onset of localization the state is continuous within the element, and with the assumption of constant tangent stiffness, it is possible to evaluate the integrals in the expression for the tangent matrix in (74) such that

$$
\begin{aligned}
\mathrm{d}\mathbf{r}_e &= \frac{L_e}{\delta}\,\mathbf{Q}_\Gamma \cdot \mathrm{d}[\![\mathbf{u}]\!]_e - L_e\,\mathbf{n}\,\mathrm{d}[\![p]\!]_e = \mathbf{0}\,, \\
\mathrm{d}s_e &= \frac{L_e}{\delta}\,\mathbf{n}\cdot \mathrm{d}[\![\mathbf{u}]\!]_e - 4\,\Delta t\,\frac{L_e}{\delta}\,\mathbf{n}\cdot \mathrm{d}[\![\mathbf{v}_d]\!]_e = 0\,.
\end{aligned}
\tag{75}
$$

This is the discretized version of the condition for onset of localization for the continuum mixture. It is interesting to note, by comparison with (42), that the condition for onset of localization in the discretized problem coincides with the continuum situation if the ATS-tensor is replaced with the CTS-tensor.

7 Numerical example

7.1 Preliminaries

In the following example, a generalized *Mohr-Coulomb* model with the options of non-associated plastic flow and cohesive, isotropic hardening/softening is used to describe the solid phase response. A detailed description of the model and the consequent consistent linearization is given in Larsson and Runesson [13]. In the model, we specify the cohesion c, the angle of internal friction Φ and the dilatancy angle Φ^*. Hardening/softening is introduced via the cohesion. In the numerical simulations, we use (for simplicity) analytical expressions to determine bifurcation directions, whereby the vertices of the MC-yield surface are not taken into account.

7.2 Slope in plane strain – influence of the friction angle on limit load and localization mode

It is of interest to study the significance of the friction angle on the localization properties, such as orientation of the developing shear band and bearing capacity of the structure. To do this, we consider a steep slope with geometry, boundary conditions and material parameters as shown in Figure 5. Note that the softening behaviour is preceded by hardening, corresponding to a ductile material failure. The loading is defined by a prescribed vertical displacement of a point on the rigid footing. The position of this point is on the upper surface and 0.3 m to the right from the centre of the footing, cf. Figure 5. For simplicity, the contact between the footing and the soil is assumed to be perfectly rough.

Four simulations are performed with the friction angles $\Phi = 1.1°$, $11.5°$, $17.5°$, $23.6°$, respectively. The deformed meshes that were obtained are shown in Figure 7. As expected, with increased friction angle the "radius" of an inscribed circle segment in the shear band zone increases. For $\Phi = 23.6°$, it appears at first that a localization mode similar to that of a foundation

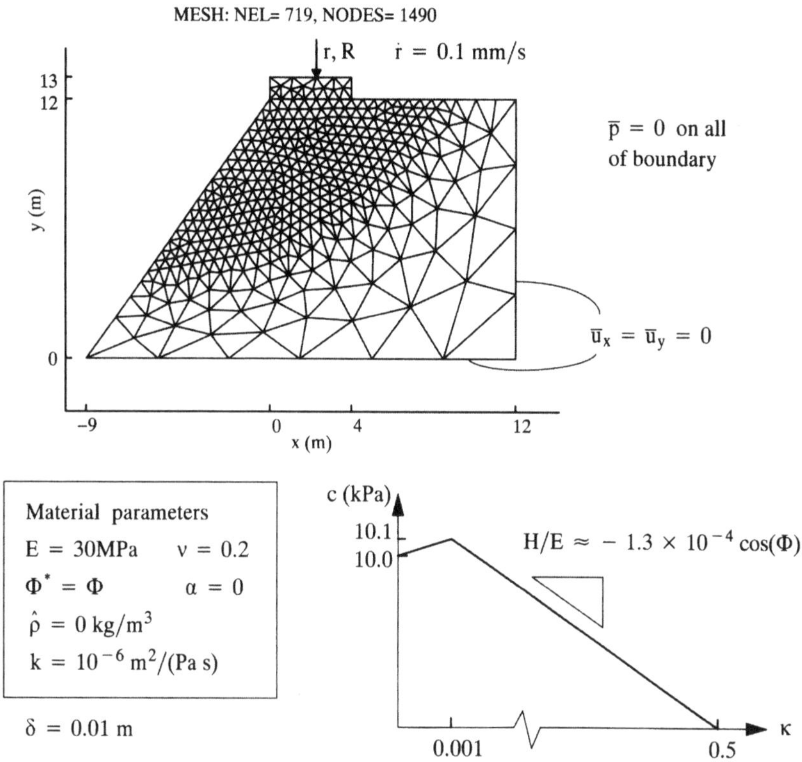

Fig. 5. Geometry, boundary conditions, and material parameters for analyzed steep slope in plane strain. The slope is analyzed for four different friction angles.

failure develops, but suddenly propagation of the final shear band takes over, cf. Figure 7 d. The corresponding load-time/displacement curves are given in Figure 6. We can see that the limit load increases as the friction angle becomes larger, and that the structural response is more brittle at failure for large friction angles.

8 Concluding remarks

In this contribution we analyzed the condition for deformation localization in the hydro-mechanically coupled problem. The concept of regularized discontinuity was extensively exploited, whereby a coupled localization condition was obtained. In the analytical study of the rate problem, we found that if the underlying drained material signals localization, i. e. the drained localization condition is satisfied, any existing discontinuity (or perturbation thereof) may grow exponentially. On the other hand, given a homogeneous

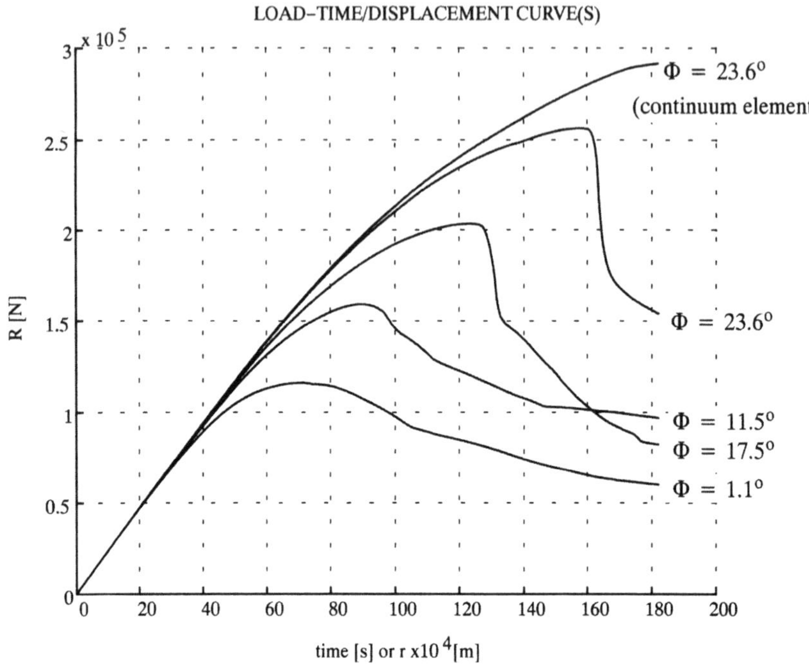

Fig. 6. Load-displacement curves obtained with different friction angles Φ for the steep slope in Figure 5.

plastic state, the condition for localization with unloading outside the band is that the undrained localization condition is satisfied.

We have also discussed a finite element method that can handle the condition for existence and development of regularized discontinuities in the displacement and the excess pore pressure fields. A key feature in the establishment of variational equations is the Enhanced Strain Approach, Simo and Rifai [21], whereby regularized discontinuities in displacement and pressure can be conveniently embedded in the finite elements. It was shown that the element behaviour basically recovers the continuum behaviour with respect to the discontinuity development, cf. Section 5. Based on staggering between the continuous structure problem and the discontinuous element problems, an algorithm was developed for the discontinuity evaluation. The algorithm can handle situations where onset of localization is preceded by diffuse plastic deformation, and the formulation is shown to work well in terms of the ability to capture localization of skeleton deformation and pore fluid pressure. The orientation of the embedded band is chosen so that its normal coincides with the bifurcation direction of the underlying "effective" material, described by the constitutive law for the effective stress. Furthermore, the embedded band is positioned within the element in such a way that it passes through the most stressed integration point at the onset of localization.

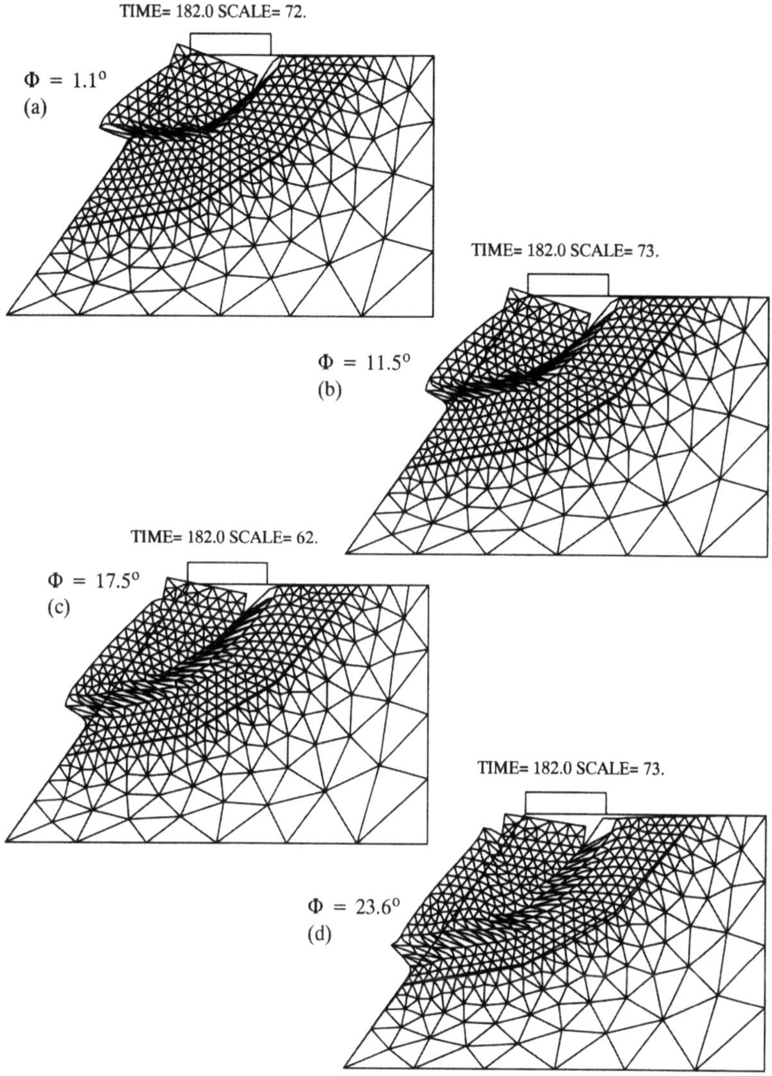

Fig. 7. Deformed meshes obtained with different internal friction angles Φ for the steep slope in Figure 5.

References

1. Armero, F., Callari, C.: An analysis of strong discontinuities in a saturated poro-plastic solid. *Int. J. Numer. Meth. Engng.* **46** (1999), 1673–1698.
2. Bluhm, J., de Boer, R.: Effective stresses – a clarification. *Arch. Appl. Mech.* **66** (1996), 479–492.
3. de Boer, R.: Highlights in the historical develoment of the porous media theory: Towards a consistent macroscopic theory. *Appl. Mech. Rev.* **49** (1996), 201–261.

4. de Boer, R., Ehlers, W.: The development of the concept of the effective stress. *Acta Mechanica* **83** (1989), 77–92.
5. de Borst, R., Sluys, L. J., Mühlhaus, H.-B., Pamin, J.: Fundamental issues in finite element analyses of localization of deformation. *Engineering Computations* **10** (1993), 99–121.
6. Bowen, R. M.: Compressible porous media theories by use of the theory of mixtures. *Int. J. Engng. Sci.* **20** (1982), 697–735.
7. Coussy, O.: *Mechanics of Porous Continua*. John Wiley & Sons, Chichester 1995.
8. Diebels, S.: Evolution of the volume fractions in compressible porous media. In Ehlers, W. (ed.): *IUTAM Symposium on Theoretical and Numerical Methods in Continuum Mechanics of Porous Materials*, Kluwer, Dordrecht 2000, pp. 21–26.
9. Ehlers, W., Volk, W.: On shear band localization phenomena of liquid-saturated granular elastoplastic porous solid materials accounting for fluid viscosity and micropolar solid rotations. *Mech. Cohes.-Frict. Mater.* **2** (1997), 301–320.
10. Han, C., Vardoulakis, I.: Plane strain compression experiments on water-saturated fine-grained sand. *Géotechnique* **41**(1) (1991), 49–78.
11. Larsson, R., Runesson, K., Ottosen, N. S.: Discontinuous displacement approximation for capturing plastic localization. *Int. J. Num. Meth. Engrg.* **36** (1993), 2087–2105.
12. Larsson, R., Runesson, K.: Element-embedded localization band based on regularized displacement discontinuity. *J. Engrg. Mech.* **122** (1996), 402–411.
13. Larsson, R., Runesson, K.: Implicit integration and consistent linearization for yield criteria of the Mohr-Coulomb type. *Mech. Cohes.-Frict. Mater.* **1** (1996), 1–17.
14. Larsson, J., Larsson, R.: Computational strategy for capturing localization in undrained soil. *Computational Mechanics* **24** (1999), 293–303.
15. Larsson, J., Larsson, R.: Localization analysis of a fluid saturated elastoplastic porous medium using regularized discontinuities. *Mech. Cohes.-Frict. Mater.* **5**(7) (2000), 565–582.
16. Larsson, J., Larsson, R.: Finite element analysis of localization of deformation and fluid pressure in an elastoplastic porous medium. *Int. J. Solids Structures* **37** (2000), 7231–7257.
17. Larsson, J., Larsson, R.: Non-linear analysis of nearly saturated porous media: Theoretical and numerical formulation. *Comp. Meth. Appl. Mech. Engng.* (2000), submitted.
18. Loret, B., Prevost, J. H.: Dynamic strain localization in fluid-saturated porous media. *J. Engrg. Mech.* **117** (1991), 907–922.
19. Needleman, A.: Material rate dependence and mesh sensitivity in localization problems. *Comput. Meth. Appl. Mech. Engrg.* **67** (1988), 69–85.
20. Simo, J. C., Oliver, J., Armero, F.: An analysis of strong discontinuities induced by strain-softening in rate-independent solids. *Computational Mechanics* **12** (1993), 277–296.
21. Simo, J. C., Rifai, M. S.: A class of mixed assumed strain methods and the method of incompatible modes. *Int. J. Num. Meth. Engrg.* **29** (1990), 1595–1638.
22. Vardoulakis, I.: Deformation of water-saturated sand: Part I–II. *Géotechnique* **46**(3) (1996), 441–472.

Geometrical and material non-linear analysis of fully and partially saturated porous media

Lorenzo Sanavia[1], Bernhard A. Schrefler[1], and Paul Steinmann[2]

[1] University of Padua, Department of Structural and Transportation Engineering, 35131 Padua, Italy
[2] University of Kaiserslautern, Chair of Applied Mechanics, 67653 Kaiserlautern, Germany

Abstract. A formulation for a partially saturated porous medium undergoing large elastic or elasto-plastic deformations is presented. The porous material is treated as a multiphase continuum with the pores of the solid skeleton filled by water and air, this last one at constant pressure. This pressure may either be the atmospheric pressure or the cavitation pressure. The governing equations at macroscopic level are derived in a spatial and a material setting. Solid grains and water are assumed to be incompressible at the microscopic level. The elasto-plastic behaviour of the solid skeleton is described by the multiplicative decomposition of the deformation gradient into an elastic and a plastic part. The effective stress state is limited by the *Drucker-Prager* yield surface. The water is assumed to obey *Darcy*'s law. Numerical examples of the *Liakopoulos*' test and of strain localization of dense or loose sand and of clay under undrained conditions conclude the paper.

1 Introduction

This paper presents a formulation for a saturated and partially saturated porous medium undergoing large elastic or elasto-plastic strains. Mechanics of porous materials has a wide spectrum of engineering applications and hence, in recent years, several porous media models and their numerical solutions have appeared in the literature (see [3, 20, 42] for a comprehensive state of the art). Most of these models are restricted to fluid saturated materials and have been developed using small strain assumptions. For soils, large strains result when ultimate or serviceability limit state is reached, as for example during slope instability or during the consolidation process in compressible clays. In laboratory, this can be the case of drained or undrained biaxial tests of sands, where axial logarithmic strains of the order of $0.12-0.15$ are reached [25, 38], or the case of triaxial tests of peats, where axial strains of the order of 0.15 are measured. Among others, the phenomenon of strain localization (i. e. strain accumulation in well defined narrow zones, also called shear bands) is most typical whereby large inelastic strains develop, at least inside the bands [25, 27, 38]. Moreover, it reveals also the strong coupling, which occurs between the solid skeleton and the fluids filling the voids of the porous material.

More recently, porous media models for fluid saturated materials have been extended to large strains, first in the framework of hypoelastoplasticity and, after the work of Simo for single-phase materials, also using hyperelastoplasticity [1, 6, 7]. Models based on an hyperelastic free energy function have been developed in [9, 12] for the dynamic and static case, respectively. To the authors' knowledge, a partially saturated model has been put forward only in the framework of hypoelastoplasticity, based on an updated *Lagrange*an approach, *Euler*ian strain rate tensor and *Jaumann* stress rate [23]. In the present contribution, a partially saturated porous media model is developed in the framework of hyperelastoplasticity [31], extending the previous work of Sanavia et al. [29, 30].

Conditions of partial saturation are of importance in engineering practice because many porous materials are in this natural state or can reach this state during deformations. Some simple examples can be found in soils or in concrete and in biological tissues, which can contain air or other gases in the pores together with liquids. For soils, this is the case of the zones above the free surface, or of deep reservoirs of hydrocarbon gas. The partially saturated state can also be reached during the deformation due, for instance, to earthquake in an earth dam or during the particular case of strain localization of dense sands under globally undrained conditions, where negative water pressures are measured and cavitation of the pore water was observed [25, 38].

In the model developed in this paper, the porous medium is treated as an isothermal multiphase continuum with the pores of the solid skeleton filled by water and air, this last one at constant pressure (passive air phase assumption). This pressure may either be the atmospheric pressure or the cavitation pressure (isothermal monospecies approach). Quasi-static loading conditions are considered. The governing equations at macroscopic level are derived in Section 3 in a spatial and a material setting and are based on averaging procedures (hybrid mixture theory). This model follows from the general thermo-hydro-mechanical model developed in [20], which is briefly recalled in Section 2 for the sake of completeness. Solid displacements and water pressures are the primary variables. The solid grains and water are assumed to be incompressible at microscopic level. The elasto-plastic behaviour of the solid skeleton is described by the multiplicative decomposition of the deformation gradient into an elastic and a plastic part, as described in Section 3.3. The modified effective stress in partially saturated conditions (*Bishop* like stress) in the form of *Kirchhoff* measure of the stress tensor and the logarithmic principal strains are used in conjunction with an hyperelastic free energy function. The effective stress state is limited by the *von Mises* or the *Drucker-Prager* yield surface. A particular "apex formulation" developed in [31] is used in the latter case. Water is assumed to obey *Darcy*'s law. In the partially saturated state, the water degree of saturation and the relative permeability are dependent on the capillary pressure by experimental functions. The spatial weak form of the governing equations, the temporal integration of the mixture mass balance equation, which is time dependent because of the seepage process of water, and the consistent linearization are described in Sections 4, 5, and 6, respectively. In particular, the generalized trapezoidal method is used for the time integration. Finally, the finite element discretization in space is obtained by applying a *Galerkin* procedure in the spatial setting, using different shape functions for solid and water (see Section 7).

Numerical examples of this research in progress on large elastic or inelastic strains in saturated and partially saturated porous media highlight the developments in Section 8. The first one is the simulation of the experimental *Liakopoulos*' test, for which the observed experimental desaturation of the column from the top to the bottom surface is described. In the second example, a stability problem on clay or sand is studied. In particular, in case of dense sand the equivalent plastic strain distribution shows the presence of a localized zone (*Drucker-Prager* model is used with non-associated flow rule) with negative water pressures induced by the dilatancy of the dense sand below the cavitation pressure at ambient temperature (of -96 kPa) inside the shear band.

For the aspects of the regularization properties of the multiphase model at localization, due to the presence of a *Laplace*an operator in the mass balance equation of the fluids when *Darcy*'s law is used, the interested reader is referred to [10, 32]. The internal length scale contained in the model is presented in [40].

As notation and symbols are concerned, bold-face letters denote tensors; capital or lower case letters are used for tensors in the reference or in actual configuration. The symbol '$\cdot$' denotes the scalar product between two vectors (e. g. $\mathbf{a} \cdot \mathbf{b} = a_i\, b_i$), while the symbol ':' denotes a double contraction of (adjacent) indices of two tensors of rank two or/and higher (e. g. $\mathbf{c} : \mathbf{d} = c_{ij}\, d_{ij}, \mathbf{e} : \mathbf{f} = e_{ijkl}\, f_{kl}$). Cartesian co-ordinates are used throughout.

2 General mathematical model of thermo-hydro-mechanical transient behaviour of geomaterials

The full mathematical model necessary to simulate thermo-hydro-mechanical transient behaviour of fully and partially saturated porous media is developed in [20] using averaging theories following Hassanizadeh and Gray [15–17]. The underlying physical model, thermodynamic relations and constitutive equations for the constituents, as well as governing equations are briefly summarized for sake of completeness in the present section. The governing equations of the simplified model used in the finite element discretization are described in Section 3.

The partially saturated porous medium is treated as multiphase system composed of $\pi = 1, \ldots, k$ constituents with the voids of the solid skeleton

(s) filled with water (w) and gas (g). The latter is assumed to behave as an ideal mixture of two species: dry air (non-condensable gas, ga) and water vapour (condensable one, gw). Using spatial averaging operators defined over a representative elementary volume R. E. V. (of volume $dv(\mathbf{x}, t)$ in the deformed configuration $B_t \subset \mathbb{R}^3$, see Figure 1, where $\mathbf{x}$ is the vector of the spatial co-ordinates and t is the current time), the microscopic equations are integrated over the R. E. V. giving the macroscopic balance equations. At the macroscopic level the porous media material is hence modelled by a substitute continuum of volume B_t with boundary ∂B_t that fills the entire domain simultaneously, instead of the real fluids and the solid which fill only a part of it. In these substitute continuum each constituent π has a reduced density which is obtained through the volume fraction $\eta^\pi(\mathbf{x}, t) = dv^\pi(\mathbf{x}, t)/dv(\mathbf{x}, t)$ (e. g. [2]) with the constraint

$$\sum_{\pi=1}^{k} \eta^\pi = 1\,, \tag{1}$$

where $dv^\pi(\mathbf{x}, t)$ is the π-phase volume inside the R. E. V. in the actual placement $\mathbf{x}$.

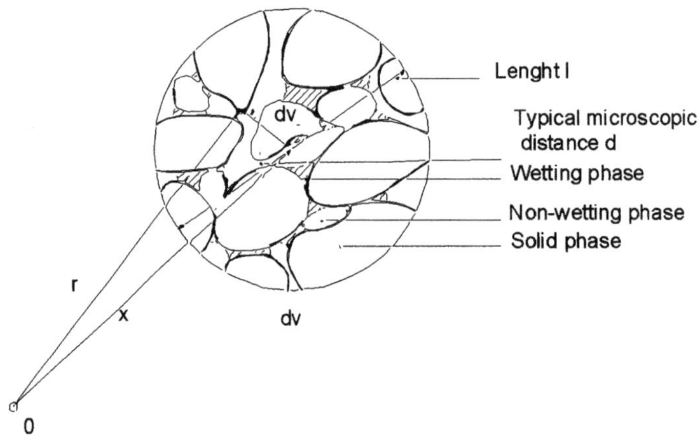

Fig. 1. Typical averaging volume $dv(\mathbf{x}, t)$ of a porous medium consisting of three constituents [20].

In this formulation heat conduction, vapour diffusion, heat convection, water flow due to pressure gradients or capillary effects and latent heat transfer due to water phase change (evaporation and condensation) inside the pores are taken into account. The solid is deformable and non-polar, and the fluid, the solid, and the thermal fields are coupled. All fluids are in contact with the solid phase. The constituents are assumed to be isotropic, homogeneous, immiscible except for dry air and vapour, and chemically non-reacting. Local thermal equilibrium between solid matrix, gas, and liquid phases is assumed so that the temperature is the same for all the constituents. The state of the medium is described by water pressure p^w, gas pressure p^g, temperature θ, and the displacement vector of the solid matrix $\mathbf{u}$.

Before summarizing the macroscopic balance equations, we specify the kinematics introducing the notion of initial and current configuration (Figure 2). In the following, the stress is defined as tension positive for the solid phase, while pore pressure is defined as compressive positive for the fluids.

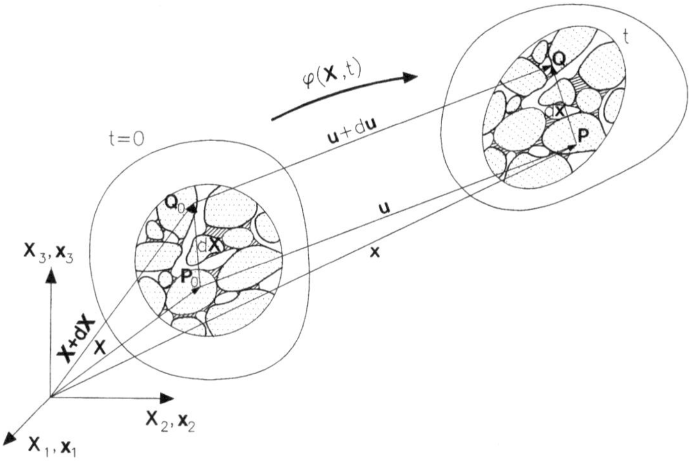

Fig. 2. Initial and current configuration of a multiphase medium.

2.1 Kinematic equations

At the macroscopic level the multiphase medium is described as the superposition of all π-phases, whose material points X^π with co-ordinates $\mathbf{X}^\pi$ in the reference configuration $B_0^\pi \subset \mathbb{R}^3$ at time $t = t_0$ can occupy simultaneously each spatial point $\mathbf{x}$ in the deformed configuration $B_t \subset \mathbb{R}^3$ at time t. In the *Lagrange*an description of the motion in terms of material co-ordinates the position of each material point in the actual configuration $\mathbf{x}$ is a function of its placement $\mathbf{X}^\pi$ in a chosen reference configuration B_0^π and of the current time t:

$$\mathbf{x} = \chi^\pi(\mathbf{X}^\pi, t) \tag{2}$$

with $\mathbf{x} = \mathbf{x}^\pi$, or it is given by the sum of the reference position $\mathbf{X}^\pi$ and the displacement $\mathbf{u}^\pi = (\mathbf{X}^\pi, t)$ at time t:

$$\mathbf{x} = \mathbf{X}^\pi + \mathbf{u}^\pi(\mathbf{X}^\pi, t)\,. \tag{3}$$

In (2), $\boldsymbol{\chi}^\pi(\mathbf{X}^\pi, t)$ is a continuous and bijective motion function (deformation map) of each phase because the *Jacobi*an J^π of each motion function

$$J^\pi = \det \frac{\partial \boldsymbol{\chi}^\pi(\mathbf{X}^\pi, t)}{\partial \mathbf{X}^\pi} > 0 \tag{4}$$

is restricted to be a positive value. The deformation gradient $\mathbf{F}^\pi(\mathbf{X}^\pi, t)$ is defined as

$$\mathbf{F}^\pi = \mathrm{Grad}^\pi \boldsymbol{\chi}^\pi(\mathbf{X}^\pi, t)\,, \tag{5}$$

where the differential operator 'Grad^π' denotes partial differentiation with respect to the reference position $\mathbf{X}^\pi$. Hence, from (4), $J^\pi = \det \mathbf{F}^\pi$.

The velocity and the acceleration of each constituent are given as

$$\mathbf{V}^\pi = \frac{\partial \boldsymbol{\chi}^\pi(\mathbf{X}^\pi, t)}{\partial t}\,, \qquad \mathbf{A}^\pi = \frac{\partial^2 \boldsymbol{\chi}^\pi(\mathbf{X}^\pi, t)}{\partial t^2}\,. \tag{6}$$

Due to the non-singularity of the *Lagrange*an relationship (2), the existence of its inverse function leads to the description of the motion in terms of spatial co-ordinates:

$$\mathbf{X}^\pi = (\boldsymbol{\chi}^\pi)^{-1}(\mathbf{x}, t)\,. \tag{7}$$

The inverse $(\mathbf{F}^\pi)^{-1}(\mathbf{x}, t)$ of the deformation gradient is given by

$$(\mathbf{F}^\pi)^{-1} = \mathrm{grad}\, \mathbf{X}^\pi(\mathbf{x}, t)\,, \tag{8}$$

where the differential operator 'grad' is now referred to spatial co-ordinates $\mathbf{x}$. The spatial parametrization of the velocity is given by

$$\mathbf{v}^\pi = \mathbf{v}^\pi(\mathbf{x}, t) = \mathbf{V}^\pi \circ (\boldsymbol{\chi}^\pi)^{-1}\,, \tag{9}$$

where '$\circ$' denotes the composition of functions. The parametrization of the spatial acceleration is related to the spatial velocity by the application of the chain rule to (9):

$$\mathbf{a}^\pi = \mathbf{a}^\pi(\mathbf{x}, t) = \frac{\partial \mathbf{v}^\pi}{\partial t} + (\mathrm{grad}\, \mathbf{v}^\pi)\, \mathbf{v}^\pi = \mathbf{A}^\pi \circ (\boldsymbol{\chi}^\pi)^{-1}\,. \tag{10}$$

Since the individual constituents follow in general different motions, different material time derivatives must be formulated. For an arbitrary scalar-valued function $f^\pi(\mathbf{x}, t)$, its material time derivative following the velocity of the constituents π is defined by [20]

$$\frac{\mathrm{D}^\pi f^\pi}{\mathrm{D}t} = \frac{\partial f^\pi}{\partial t} + \mathrm{grad}\, f^\pi \cdot \mathbf{v}^\pi\,, \tag{11}$$

where $f^\pi(\mathbf{x}, t)$ must be substituted by $\mathbf{f}^\pi(\mathbf{x}, t)$ in case of vector or tensor valued function $\mathbf{f}^\pi(\mathbf{x}, t)$. Thus, $\mathbf{a}^\pi = \mathrm{D}^\pi \mathbf{v}^\pi / \mathrm{D}t$.

In the theory of multiphase materials it is common to assume the motion of the solid as a reference and to describe the fluids in terms of motion relative to the solid. This means that a fluid relative velocity and the material time derivative with respect to the solid are introduced. The solid motion can be described both in terms of material or spatial co-ordinates. The second approach is now presented because the most natural numerical formulation of the elasto-plastic initial-boundary-value problem is based on the weak form of the balance equations in a spatial setting.

The fluid relative velocity $\mathbf{v}^{\pi s}(\mathbf{x},\, t)$ in spatial parametrization or diffusion velocity is given by

$$\mathbf{v}^{\pi s}(\mathbf{x},\, t) = \mathbf{v}^{\pi}(\mathbf{x},\, t) - \mathbf{v}^{s}(\mathbf{x},\, t)\,, \tag{12}$$

and the material time derivative of $f^{\pi}(\mathbf{x},\, t)$ with respect to the moving solid phase (s) is given by

$$\frac{\mathrm{D}^s f^{\pi}}{\mathrm{D}t} = \frac{\mathrm{D}^{\pi} f^{\pi}}{\mathrm{D}t} + \operatorname{grad} f^{\pi} \cdot \mathbf{v}^{s\pi} \quad \text{with} \quad \mathbf{v}^{s\pi} = -\mathbf{v}^{\pi s}\,. \tag{13}$$

For the section closure, the material and the spatial velocity gradient of the solid will be recalled. The first one is given as

$$\mathbf{L}^s = \operatorname{Grad}^s \mathbf{V}^s = \frac{\partial \mathbf{F}^s}{\partial t} = \mathbf{D}^s + \mathbf{W}^s\,, \tag{14}$$

where $\mathbf{D}^s(\mathbf{X}^s,\, t)$ and $\mathbf{W}^s(\mathbf{X}^s,\, t)$ are the symmetric and the skew-symmetric part of $\mathbf{L}^s(\mathbf{X}^s,\, t)$, while the spatial velocity gradient $\mathbf{l}^s(\mathbf{x},\, t)$ is defined as the gradient of the velocity (9) with respect to spatial co-ordinates, i. e.

$$\mathbf{l}^s = \operatorname{grad} \mathbf{v}^s = \frac{\partial \mathbf{F}^s}{\partial t}\,(\mathbf{F}^s)^{-1} = \mathbf{d}^s + \mathbf{w}^s\,, \tag{15}$$

where $\mathbf{d}^s(\mathbf{x},\, t)$ and $\mathbf{w}^s(\mathbf{x},\, t)$ are the symmetric and the skew-symmetric part of $\mathbf{l}^s(\mathbf{x},\, t)$, also called spatial rate of deformation tensor and spin tensor, respectively.

All strain measures and strain rates for each constituent follow similarly to classical non-linear continuum mechanics, but are not reported here because they are not useful for the approach developed in the sequel.

2.2 Mass balance equations

The averaged macroscopic balance equation for the solid phase is

$$\frac{\mathrm{D}^s \rho_s}{\mathrm{D}t} + \rho_s \operatorname{div} \mathbf{v}^s = \frac{\partial \rho_s}{\partial t} + \operatorname{div}(\rho_s\, \mathbf{v}^s) = 0\,, \tag{16}$$

where $\mathbf{v}^s(\mathbf{x},\, t)$ is the mass averaged solid velocity, $\rho_s(\mathbf{x}, t)$ is the averaged density of the solid related to the intrinsic averaged density $\rho^s(\mathbf{x}, t)$ by the volume fraction $\eta^s(\mathbf{x},\, t)$.

For the generic π-phase the relationship between the averaged density and the intrinsic averaged density is

$$\rho_\pi(\mathbf{x}, t) = \eta^\pi(\mathbf{x}, t)\, \rho^\pi(\mathbf{x}, t)\,, \tag{17}$$

where the intrinsic density $\rho^\pi(\mathbf{x}, t)$ is also named real or true density in the so-called Theory of Porous Media, e. g. [3, 4].

The mass balance equation for water is

$$\frac{\mathrm{D}^w \rho_w}{\mathrm{D}t} + \rho_w \operatorname{div} \mathbf{v}^w = \frac{\partial}{\partial t}(n\, S_w\, \rho^w) + \operatorname{div}(n\, S_w\, \rho^w\, \mathbf{v}^w) = \rho_w\, e^w\,, \tag{18}$$

where $\rho_w\, e^w(\mathbf{x}, t)$ is the quantity of water per unit time and volume lost through evaporation. The corresponding equations for dry air and vapour are, respectively,

$$\frac{\mathrm{D}^{ga} \rho_{ga}}{\mathrm{D}t} + \rho_{ga} \operatorname{div} \mathbf{v}^{ga} = \frac{\partial}{\partial t}(n\, S_g\, \rho^{ga}) + \operatorname{div}(n\, S_g\, \rho^{ga}\, \mathbf{v}^{ga}) = 0\,, \tag{19}$$

$$\frac{\mathrm{D}^{gw} \rho_{gw}}{\mathrm{D}t} + \rho_{gw} \operatorname{div} \mathbf{v}^{gw} = \frac{\partial}{\partial t}(n\, S_g\, \rho^{gw}) + \operatorname{div}(n\, S_g\, \rho^{gw}\, \mathbf{v}^{gw}) = \rho_{gw}\, e^{gw}\,, \tag{20}$$

where $n(\mathbf{x}, t)$ is the porosity of the medium, defined as

$$n = \frac{dv^w + dv^g}{dv} = \frac{dv^{\text{voids}}}{dv} = 1 - \eta^s\,, \tag{21}$$

and S_w and S_g are the water and gas degrees of saturation. The following relationships hold:

$$\begin{aligned} \eta^w &= n\, S_w \quad \text{with} \quad S_w = \frac{dv^w}{dv^w + dv^g}\,, \\ \eta^g &= n\, S_g \quad \text{with} \quad S_g = \frac{dv^g}{dv^w + dv^g}\,, \end{aligned} \tag{22}$$

with the saturation constraint $S_w + S_g = 1$. The right-hand side of (18) and (20) sum to zero.

2.3 Linear momentum balance equations

The linear momentum balance equation for the solid and the π-fluid are

$$\operatorname{div} \mathbf{t}^s + \rho_s\,(\mathbf{g} - \mathbf{a}^s) + \rho_s\, \hat{\mathbf{t}}^s = \mathbf{0} \tag{23}$$

and

$$\operatorname{div} \mathbf{t}^\pi + \rho_\pi\,(\mathbf{g} - \mathbf{a}^\pi) + \rho_\pi\,(\mathbf{e}^\pi + \hat{\mathbf{t}}^\pi) = \mathbf{0}\,, \tag{24}$$

respectively, where $\mathbf{t}^\pi(\mathbf{x}, t)$ is the partial *Cauchy* stress tensor defined via the constitutive equation presented in Section 2.6. $\hat{\mathbf{t}}^\pi(\mathbf{x}, t)$ accounts for the exchange of momentum due to mechanical interaction with other phases, $\rho_\pi\, \mathbf{a}^\pi$

for the volume density of the inertial force, $\rho_\pi \mathbf{g}$ for the volume density of gravitational force, and $\mathbf{e}^\pi(\mathbf{x}, t)$ takes into account the momentum exchange due to averaged mass supply or mass exchange between the fluid and the gas phases and the change of density. The linear momentum balance equation of the multiphase medium is subjected to the constraint [20]:

$$\sum_{\pi=1}^{k} \rho_\pi \left(\mathbf{e}^\pi + \hat{\mathbf{t}}^\pi \right) = \mathbf{0} \,. \tag{25}$$

2.4 Angular momentum balance equation

All the phases are considered microscopically non-polar and hence at macroscopic level the angular momentum balance equation states that the partial stress tensor is symmetric [20]:

$$\mathbf{t}^\pi = (\mathbf{t}^\pi)^T \,. \tag{26}$$

2.5 Energy balance equation and entropy inequality

These two relationships are simply quoted from [20], the second one is useful for the development of the constitutive equations.

The energy balance equation for the π-phase may be written as

$$\rho_\pi \frac{\mathrm{D}^\pi E^\pi}{\mathrm{D}t} = \mathbf{t}^\pi : \mathbf{d}^\pi + \rho_\pi h^\pi - \operatorname{div} \mathbf{q}^\pi + \rho_\pi R^\pi \,, \tag{27}$$

where $\rho_\pi R^\pi$ represents the exchange of energy between the π-phase and other phases of the medium due to phase change and mechanical interaction, $\mathbf{q}^\pi$ is the internal heat flux, h^π results from the heat sources, and $\mathbf{d}^\pi$ is the spatial rate of the deformation tensor. E^π accounts for the specific internal energy of the volume element.

The entropy inequality for the mixture is

$$\sum_\pi \left(\rho_\pi \frac{\mathrm{D}^\pi \lambda^\pi}{\mathrm{D}t} + \rho_\pi e^\pi \lambda^\pi + \operatorname{div} \frac{\mathbf{q}^\pi}{\theta^\pi} - \frac{\rho_\pi h^\pi}{\theta^\pi} \right) \geq 0 \,, \tag{28}$$

where θ^π is the absolute temperature, λ^π is the specific entropy of the constituent π, and $e^\pi \lambda^\pi$ the entropy supply due to mass exchange.

2.6 Constitutive equations

The momentum exchange term $\rho_\pi \hat{\mathbf{t}}^\pi$ of the linear momentum balance equation of the fluid can be expressed as [20]

$$\rho_\pi \hat{\mathbf{t}}^\pi = -\mu^\pi (\eta^\pi)^2 \mathbf{k}^{-1} \mathbf{v}^{\pi s} + p^\pi \operatorname{grad} \eta^\pi \qquad \text{with} \qquad \pi = g, w \,. \tag{29}$$

Here, $\mathbf{k} = k^{r\pi} \mathbf{k}^\pi$, where $\mathbf{k}^\pi(\mathbf{x}, t) = \mathbf{k}^\pi(\rho^\pi, \eta^\pi, T)$ is the intrinsic permeability tensor of dimension $[\mathrm{L}^2]$ depending in the isotropic case on the porosity of the medium, $k^{r\pi}(S_\pi)$ is the relative permeability parameter, and μ^π is the dynamic viscosity. The relative permeability is a function of the π-phase degree of saturation S_π and is determined in laboratory tests (see e. g. [20]).

The partial stress tensor in the fluid phase of the linear momentum balance equation (24) is related to the macroscopic pressure $p^\pi(\mathbf{x}, t)$ of the π-phase:

$$\mathbf{t}^\pi = -\eta^\pi \, p^\pi \, \mathbf{1}\,, \tag{30}$$

where $\mathbf{1}$ is the second order unit tensor.

From the entropy inequality it can also be shown that the spatial solid stress tensor $\mathbf{t}^s(\mathbf{x}, t)$ of the linear momentum balance equation (23) is decomposed as follows:

$$\mathbf{t}^s = \eta^s \, (\mathbf{t}^s_e - p^s \, \mathbf{1})\,, \tag{31}$$

and that the effective *Cauchy* stress tensor $\boldsymbol{\sigma}'(\mathbf{x}, t)$, which is responsible for all major deformation in the solid skeleton, is

$$\boldsymbol{\sigma}' = \eta^s \, \mathbf{t}^s_e\,. \tag{32}$$

In (31), $\mathbf{t}^s_e(\mathbf{x}, t)$ is the dissipative part [14] or effective stress tensor of the solid phase, while $p^s(\mathbf{x}, t)$ is the equilibrium part, also called solid pressure, with $p^s = S_w \, p^w + S_g \, p^g$.

From the previous equations, it follows that the total *Cauchy* stress tensor $\boldsymbol{\sigma} = \mathbf{t}^s + \mathbf{t}^w + \mathbf{t}^g$ can be written in the usual form used in soil mechanics:

$$\boldsymbol{\sigma} = \boldsymbol{\sigma}' - (S_w \, p^w + S_g \, p^g)\, \mathbf{1}\,. \tag{33}$$

The constitutive law for the solid skeleton used in this paper will be discussed in Section 3.

The pressure $p^g(\mathbf{x}, t)$ is given in the sequel. For a gaseous mixture of dry air and water vapour, the ideal gas law is introduced because the moist air is assumed to be a perfect mixture of two ideal gases. The equation of state of perfect gas (the *Clapeyron* equation) and *Dalton*'s law applied to dry air (ga), water vapour (gw) and moist air (g), yield:

$$p^{ga} = \rho^{ga} \, R\theta / M_a\,, \qquad p^{gw} = \rho^{gw} \, R\theta / M_w\,, \tag{34}$$

$$p^g = p^{ga} + p^{gw}\,, \qquad \rho^g = \rho^{ga} + \rho^{gw}\,. \tag{35}$$

In the partially saturated zones, water is separated from its vapour by a concave meniscus (capillary water). Due to the curvature of this meniscus, the sorption equilibrium equation gives the relationship between the capillary pressure $p^c(\mathbf{x}, t)$ and the gas $p^g(\mathbf{x}, t)$ and water pressure $p^w(\mathbf{x}, t)$ [14]:

$$p^c = p^g - p^w\,. \tag{36}$$

The equilibrium water vapour pressure $p^{gw}(\mathbf{x}, t)$ can be obtained from the *Kelvin-Laplace* equation:

$$p^{gw} = p^{gws}(\theta) \exp\left(\frac{p^c M_w}{\rho^w R \theta}\right), \tag{37}$$

where the water vapour saturation pressure p^{gws}, depending only upon the temperature $\theta(\mathbf{x}, t)$, can be calculated from the *Clausius-Clapeyron* equation or from an empirical correlation.

The saturation $S_\pi(\mathbf{x}, t)$ is an experimentally determined function of the capillary pressure p^c and the temperature θ:

$$S_\pi = S_\pi(p^c, \theta). \tag{38}$$

For the binary gas mixture of dry air and water vapour, *Fick*'s law gives the following relative velocities $\mathbf{v}_g^\pi = \mathbf{v}^\pi - \mathbf{v}^g$ $(\pi = ga, gw)$ of the diffusing species:

$$\mathbf{v}_g^{ga} = -\frac{M_a M_w}{M_g^2} \mathbf{D}_g \operatorname{grad}\left(\frac{p^{ga}}{p^g}\right) = -\mathbf{v}_g^{gw}, \tag{39}$$

where $\mathbf{D}_g$ is the effective diffusivity tensor and M_g is the molar mass of the gas mixture:

$$\frac{1}{M_g} = \frac{\rho^{gw}}{\rho^g} \frac{1}{M_w} + \frac{\rho^{ga}}{\rho^g} \frac{1}{M_a}. \tag{40}$$

2.7 Initial and boundary conditions

For model closure it is necessary to define the initial and boundary conditions. The initial conditions specify the full fields of gas pressure, water pressure, temperature, displacements, and velocity:

$$p^g = p_0^g, \quad p^w = p_0^w, \quad \theta = \theta_0, \quad \mathbf{u} = \mathbf{u}_0, \quad \dot{\mathbf{u}} = \dot{\mathbf{u}}_0 \quad \text{at} \quad t = t_0. \tag{41}$$

The boundary conditions can be imposed values on ∂B_π or fluxes on ∂B_π^q, where the boundary is $\partial B = \partial B_\pi \cup \partial B_\pi^q$. The imposed values on the boundary for gas pressure, water pressure, temperature, and displacements are as follows:

$$\begin{aligned} p^g &= \hat{p}^g \quad \text{on} \quad \partial B_g, \qquad & p^w &= \hat{p}^w \quad \text{on} \quad \partial B_w, \\ \theta &= \hat{\theta} \quad \text{on} \quad \partial B_\theta, \qquad & \mathbf{u} &= \hat{\mathbf{u}} \quad \text{on} \quad \partial B_u \quad \text{for} \quad t \geq t_0. \end{aligned} \tag{42}$$

The volume average flux boundary conditions for dry air and water species conservation equations and the energy equation to be imposed at the interface between the porous media and the surrounding fluid (the natural boundary conditions) are the following:

$$\begin{aligned} &\left(\rho^{ga} \mathbf{v}^g - \rho^g \mathbf{v}_g^{gw}\right) \cdot \mathbf{n} = q^{ga} && \text{on } \partial B_g^q, \\ &\left(\rho^{gw} \mathbf{v}^g + \rho^w \mathbf{v}^w + \rho^g \mathbf{v}_g^{gw}\right) \cdot \mathbf{n} = \beta_c \left(\rho^{gw} - \rho_\infty^{gw}\right) + q^{gw} + q^w && \text{on } \partial B_c^q, \\ &-\left(\rho^w \mathbf{v}^w \Delta h_{\text{vap}} - \lambda_{\text{eff}} \nabla \theta\right) \cdot \mathbf{n} = \alpha_c (\theta - \theta_\infty) + q^\theta && \text{on } \partial B_\theta^q \end{aligned} \tag{43}$$

for $t \geq t_0$, where $\mathbf{n}(\mathbf{x}, t)$ is the vector perpendicular to the surface of the porous medium, pointing towards the surrounding gas, $\rho_\infty^{gw}(\mathbf{x}, t)$ and $\theta_\infty(\mathbf{x}, t)$ are, respectively, the mass concentration of water vapour and temperature in the undisturbed gas phase distant from the interface, $\alpha_c(\mathbf{x}, t)$ and $\beta_c(\mathbf{x}, t)$ are convective heat and mass transfer coefficients, while $q^{ga}(\mathbf{x}, t)$, $q^{gw}(\mathbf{x}, t)$, $q^{w}(\mathbf{x}, t)$, and $q^{\theta}(\mathbf{x}, t)$ are the imposed dry air flux, imposed vapour flux, imposed liquid flux, and imposed heat flux, respectively.

The traction boundary conditions for the displacement field related to the total *Cauchy* stress tensor $\boldsymbol{\sigma}(\mathbf{x}, t)$ are

$$\boldsymbol{\sigma}\,\mathbf{n} = \bar{\mathbf{t}} \quad \text{on} \quad \partial B_u^q\,, \tag{44}$$

where $\bar{\mathbf{t}}(\mathbf{x}, t)$ is the imposed *Cauchy* traction vector.

3 Macroscopic balance equations for an isothermal saturated and partially saturated medium

In this section the macroscopic balance equations for mass and linear momentum of a simplified model that we shall use in the sequel are obtained. The constitutive equations for finite elasto-plasticity as well as their algorithmic counterpart will close the present section.

The following assumptions are now introduced in the general model previously presented:

- All the processes are isothermal. This means that the energy balance equation (27) is no more necessary.
- At the micro level, the porous medium is assumed to be constituted of incompressible solid and water constituents, while gas is considered compressible. The averaged intrinsic density $\rho^{\pi}(\mathbf{x}, t)$ ($\pi = s, w$) is hence constant, while the averaged density $\rho_{\pi}(\mathbf{x}, t)$ can vary due to the volume fraction $\eta^{\pi}(\mathbf{x}, t)$. Consequently, the density of the mixture $\rho(\mathbf{x}, t)$ (60) and the porosity $n(\mathbf{x}, t)$ can change during the deformation of the porous medium.
- The process is considered as quasi-static, so the solid and fluids accelerations are neglected.
- The passive air phase assumption will be introduced during the development of the mathematical model.

The formulation in terms of spatial co-ordinates is now presented.

3.1 Mass balance equation

Taking into account the incompressibility constraint of the solid and water constituents in (16) and (18), the mass balance equation for the solid and water phases becomes

$$\frac{\partial}{\partial t}(1-n) + \text{div}\,[(1-n)\,\mathbf{v}^s] = 0\,, \tag{45}$$

$$\frac{\partial}{\partial t}(n\,S_w) + \text{div}\,(n\,S_w\,\mathbf{v}^w) = 0\,, \tag{46}$$

where the definition of the phase average density (17) has been introduced, thus eliminating the intrinsic (constant) average density $\rho^\pi(\mathbf{x},\,t)$. Using the concept of the material time derivative (11), (45) is rewritten as

$$\frac{\mathrm{D}^s}{\mathrm{D}t}(1-n) + (1-n)\;\text{div}\,\mathbf{v}^s = 0\,, \tag{47}$$

where the classical relationship

$$\text{div}\,\mathbf{v}^s = \frac{\mathrm{D}^s J^s}{\mathrm{D}t}\,(J^s)^{-1} \tag{48}$$

can be introduced for the solid deformation [22]. The time integration of (47) gives the evolution law for the porosity $n(\mathbf{x},\,t)$ related to the determinant $J^s(\mathbf{X}^s,\,t)$ of the deformation gradient $\mathbf{F}^s(\mathbf{X}^s,\,t)$:

$$n = 1 - (1-n_0)\;(J^s)^{-1}\,, \tag{49}$$

where $n_0(\mathbf{X}^s)$ is the porosity in the reference configuration at $t = t_0$ (or initial porosity). Because of the relation $\eta^s(\mathbf{x},\,t) = 1 - n(\mathbf{x},\,t)$, (49) can be rewritten as

$$\eta^s = \eta_0^s\,(J^s)^{-1}\,, \tag{50}$$

where $\eta_0^s(\mathbf{X}^s)$ is the solid volume fraction in the reference configuration at $t = t_0$.

The gas mass balance equation

$$\frac{\partial}{\partial t}(n\,S_g\,\rho^g) + \text{div}\,(n\,S_g\,\rho^g\,\mathbf{v}^g) = 0 \tag{51}$$

is obtained by summation of the corresponding equations for the dry air and the water vapour, taking into account *Dalton*'s law (35) and introducing the mass averaged gas velocity $\mathbf{v}^g = (\,\rho^{ga}\,\mathbf{v}^{ga} + \rho^{gw}\,\mathbf{v}^{gw})\,/\,\rho^g$.

The sum of the mass balance equation of the three constituents (45), (46), and (51) produces the following mass balance equation for the mixture under consideration:

$$\text{div}\,[(1-n)\;\mathbf{v}^s + n\,S_w\,\mathbf{v}^w + n\,S_g\,\mathbf{v}^g] = 0\,, \tag{52}$$

in which the passive air phase assumption

$$p^g \cong \text{const.} \tag{53}$$

has been introduced, thus eliminating the terms of (51) depending on the spatial or the time variation of ρ^g. The gas pressure may either be the atmospheric pressure or the cavitation pressure at a certain temperature (e. g. the ambient temperature). The first case is a common assumption in soil mechanics because in many cases occurring in practice the air pressure is close to the atmospheric pressure as the pores are interconnected [42]. The second case can be derived from the experimental observations [25] and the obtained model is also called *Isothermal Monospecies Approach*, which can be used to simulate cavitation at localization in initially water saturated dense sands under globally undrained conditions, as first developed in [32] for the geometrically linear case. In fact, in this situation, neglecting air dissolved in water, only two fluid phases are present after cavitation: Liquid water and water vapour at cavitation pressure, which is then considered constant and is neglected because of its small value.

Introducing the water and the gas velocity relative to the solid, i. e. $\mathbf{v}^{ws} = \mathbf{v}^w - \mathbf{v}^s$ and $\mathbf{v}^{gs} = \mathbf{v}^g - \mathbf{v}^s$, the mixture mass balance equation (52) becomes

$$\operatorname{div}\left(\mathbf{v}^s + n\, S_w\, \mathbf{v}^{ws} + n\, S_g\, \mathbf{v}^{gs}\right) = 0\,. \tag{54}$$

The terms $n\, S_w\, \mathbf{v}^{ws}(\mathbf{x},\, t)$ and $n\, S_g\, \mathbf{v}^{gs}(\mathbf{x},\, t)$ represent the filtration water and gas velocity, respectively. The fluid velocity relative to the solid is related to the fluid pressure by the linear momentum balance equation for fluid phase after the introduction of the constitutive law (29), which gives *Darcy*'s law (63), as will be demonstrated in the sequel. Taking into account the expression of *Darcy*'s law for the gas, the passive air phase assumption implies that $\mathbf{v}^{gs}$ is negligible and hence the mass balance equation for the mixture (54) becomes

$$\operatorname{div}\left(\mathbf{v}^s + n\, S_w\, \mathbf{v}^{ws}\right) = 0\,. \tag{55}$$

In case of fully saturated conditions, $S_w = 1$ and hence the previous equation is reduced to the one of the saturated model.

3.2 Linear momentum balance equation

Neglecting the inertial term in (23) and (24), the linear momentum balance equation for the solid, water and gas phase are respectively

$$\operatorname{div}\mathbf{t}^s + (1-n)\,\rho^s\,\mathbf{g} + (1-n)\,\rho^s\,\hat{\mathbf{t}}^s = \mathbf{0}\,, \tag{56}$$

$$\operatorname{div}\mathbf{t}^w + n\, S_w\, \rho^w\, \mathbf{g} + n\, S_w\, \rho^w\, (\hat{\mathbf{t}}^w + \mathbf{e}^w) = \mathbf{0}\,, \tag{57}$$

$$\operatorname{div}\mathbf{t}^g + n\, S_g\, \rho^g\, \mathbf{g} + n\, S_g\, \rho^g\, (\hat{\mathbf{t}}^g + \mathbf{e}^g) = \mathbf{0}\,. \tag{58}$$

The linear momentum balance equation for the mixture

$$\operatorname{div}\left(\mathbf{t}^s + \mathbf{t}^w\right) + \rho\,\mathbf{g} = \mathbf{0} \tag{59}$$

is obtained by summation of (56), (57), and (58) taking into account the constraint (25) and the passive air phase assumption, thus neglecting the term $\operatorname{div} \mathbf{t}^g$ and the gas density ρ^g with respect to the water density.

In (59), $\rho(\mathbf{x}, t)$ is the density of the mixture:

$$\rho = (1 - n)\, \rho^s + n\, S_w\, \rho^w\,, \tag{60}$$

and $\mathbf{t}^s + \mathbf{t}^w = \boldsymbol{\sigma}$ is the total *Cauchy* stress, which can be decomposed into the effective and pressure (equilibrium) parts following the principle of effective stress:

$$\boldsymbol{\sigma} = \boldsymbol{\sigma}' - S_w\, p^w\, \mathbf{1}\,, \tag{61}$$

where $\boldsymbol{\sigma}'(\mathbf{x}, t)$ is the modified effective *Cauchy* stress tensor, also called *Bishop*'s stress tensor in soil mechanics. The linear momentum balance equation of the mixture in terms of total *Cauchy* stress assumes the form:

$$\operatorname{div} \boldsymbol{\sigma} + \rho\, \mathbf{g} = \mathbf{0}\,. \tag{62}$$

Using the constitutive equation (29) for $n\, S_\pi\, \rho^\pi\, \hat{\mathbf{t}}^\pi$ ($\pi = w,\, g$) and the definition (30) of $\mathbf{t}^\pi$, the linear momentum balance equation for water (57) and gas (58) give *Darcy*'s law:

$$n\, S_\pi\, \mathbf{v}^{\pi s} = -\frac{\mathbf{k}\, k^{r\pi}}{\mu^\pi}\, (\operatorname{grad} p^\pi - \rho^\pi\, \mathbf{g}) \tag{63}$$

for the π-fluid, where $k^{r\pi} = k^{r\pi}(\mathbf{x}, t)$ is the relative permeability which is an experimentally determined function of the capillary pressure. This law is valid for the transport of the π-fluid in slow phenomena when the thermal effects are negligible. In the following, due to the passive air phase assumption, only *Darcy*'s law for the water will be used. Moreover, the equilibrium equation for the fluid pressures (36) is simplified as follows:

$$p^c \cong -p^w\,, \tag{64}$$

which states that capillary pressures can be approximated as pore water tractions. Hence, the water pressure can change in sign, which means that a partially saturated zone is developing in the porous medium. The effect of the capillary pressure on the stiffness of the medium is taken into account by the constitutive laws for $S_w(p^c)$ and $k^{rw}(p^c)$.

As a consequence of the above assumptions, the independent fields of the model are the solid displacements $\mathbf{u}(\mathbf{x}, t)$ and the water pressure $p^w(\mathbf{x}, t)$.

In case of fully saturated conditions, $S_w = 1$ and $k^{rw} = 1$ and hence (60), (61), and (63) are reduced to those of the saturated model.

The *Lagrangean* counterpart of the mixture balance equations (52) and (59) in terms of material co-ordinates is now presented. The linear momentum balance equation is obtained using the properties that the total first *Piola-Kirchhoff* stress tensor $\mathbf{P}(\mathbf{X}^s, t)$ can be viewed as the *Piola* transform of the second leg of the total *Cauchy* stress tensor $\boldsymbol{\sigma}(\mathbf{x}, t)$ [22]:

$$P^{aB} = J^s \, [(\mathbf{F}^s)^{-1}]^B_b \, \sigma^{ab} \,. \tag{65}$$

Multiplying (59) by the *Jacobian* J^s and using the *Piola* identity $\mathrm{Div}^s \, \mathbf{P} = J^s \, \mathrm{div} \, \boldsymbol{\sigma}$ and the relation $J^s \rho = \rho_0$, the linear momentum balance equation of the mixture in material setting is

$$\mathrm{Div}^s \, \mathbf{P} + \rho_0 \, \mathbf{g} = \mathbf{0} \,, \tag{66}$$

where 'Div^s' is the divergence operator with respect to material co-ordinates of the solid and

$$\begin{aligned} \rho_0 \, (\mathbf{X}^s, t) = \rho_0^i \, (\mathbf{X}^s, t_0) + S_w \, (\mathbf{X}^s, t)[J^s \, (\mathbf{X}^s, t) - 1] \, \rho^w \, (\mathbf{X}^s, t_0) + \\ + \, n_0 \, (\mathbf{X}^s, t_0) \, [S_w \, (\mathbf{X}^s, t) - S_{w0} \, (\mathbf{X}^s, t_0)] \, \rho^w (\mathbf{X}^s, t_0) \end{aligned} \tag{67}$$

is the pull-back of the mass density of the mixture $\rho(\mathbf{x}, t)$, in which $n_0(\mathbf{X}^s, t_0)$ and $S_{w0}(\mathbf{X}^s, t_0)$ are the initial porosity and water saturation in the reference configuration.

The total first *Piola-Kirchhoff* stress tensor $\mathbf{P}(\mathbf{X}^s, t)$ results in an additive decomposition into effective $\mathbf{P}'(\mathbf{X}^s, t)$ and water pressure parts using *Terzaghi*'s principle in the form

$$\mathbf{P} = \mathbf{P}' - J^s \, S_w \, p^w \, (\mathbf{F}^s)^{-T} \qquad \text{with} \qquad \mathbf{P}' = J^s \, \boldsymbol{\sigma}' (\mathbf{F}^s)^{-T} \,. \tag{68}$$

The mass balance equation of the mixture in material co-ordinates

$$\mathrm{Div}^s \, (\bar{\mathbf{V}}^s + N \, S_W \, \bar{\mathbf{V}}^{ws}) = 0 \tag{69}$$

is obtained in a similar way multiplying the spatial equation (55) by the *Jacobian* J^s and making use of the *Piola* identity applied to the velocities $\mathbf{v}^s$ and $n \, S_w \mathbf{v}^{ws}$,

$$\mathrm{Div}^s \, \bar{\mathbf{V}}^s = J^s \, \mathrm{div} \, \mathbf{v}^s \,, \qquad \mathrm{Div}^s \, (N \, S_W \, \bar{\mathbf{V}}^{ws}) = J^s \, \mathrm{div} \, (n \, S_w \, \mathbf{v}^{ws}) \,, \tag{70}$$

where $\bar{\mathbf{V}}^{ws}$ is the *Piola* transform of $\mathbf{v}^{ws}$. $N(\mathbf{X}^s, t)$ and $S_W(\mathbf{X}^s, t)$ are the composition of $n(\mathbf{x}, t)$ and $S_w(\mathbf{x}, t)$ with $\boldsymbol{\chi}^s(\mathbf{X}^s, t)$, respectively, with $n(\mathbf{x}, t) = n(\mathbf{x}(\mathbf{X}^s), t) = N(\mathbf{X}^s, t)$ and $S_w(\mathbf{x}, t) = S_w(\mathbf{x}(\mathbf{X}^s), t) = S_W(\mathbf{X}^s, t)$.

3.3 Constitutive equation for the solid skeleton

The elasto-plastic behaviour of the solid skeleton at finite strain is based on the multiplicative decomposition of the deformation gradient $\mathbf{F}^s(\mathbf{X}^s, t)$ into an elastic and plastic part originally proposed by Lee [19] for crystals:

$$\mathbf{F}^s = \mathbf{F}^{se} \, \mathbf{F}^{sp} \,. \tag{71}$$

This decomposition states the existence of an intermediate stress free configuration (Figure 3) and its validity has been suggested for cohesive-frictional soils by Nemat-Nasser [26], where the plastic part of the deformation gradient is viewed as an internal variable related to the amount of slipping, crushing, yielding, and, for plate like particles, plastic bending of the granules comprizing the soil. A micro-macro interpretation of (71) can be found also in [18].

In this section, the superscript ' ^{s} ' will be neglected and the symbol ' $\dot{}$ ' will be used for the material time derivative with respect to the solid skeleton instead of D^s/Dt (as well as in the remaining part of the paper).

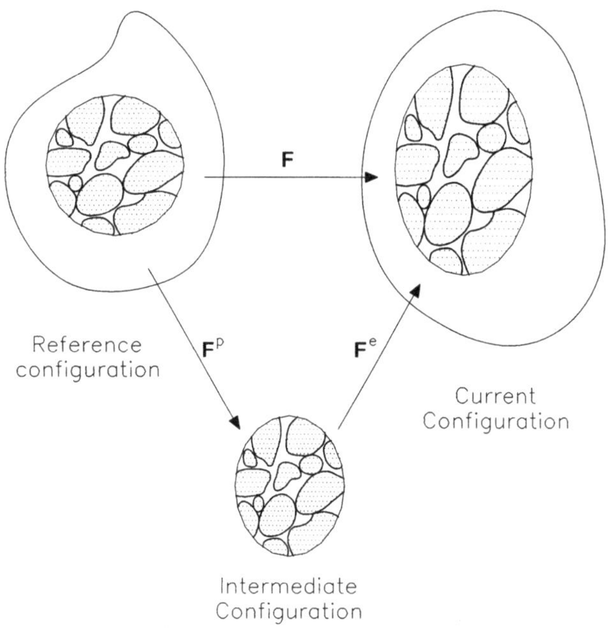

Fig. 3. Illustration of the local multiplicative decomposition of the solid deformation gradient **F**.

The spatial formulation is used in this section. The treatment of the isotropic elasto-plastic behaviour for the solid skeleton based on the product formula algorithm proposed for the single phase material by Simo [35] is now briefly summarized.

The effective *Kirchhoff* stress tensor $\boldsymbol{\tau}'(\mathbf{x}, t) = J\boldsymbol{\sigma}'(\mathbf{x}, t)$ and the logarithmic principal values of the elastic left *Cauchy-Green* strain tensor $\boldsymbol{\epsilon}_A(\mathbf{x}, t)$ are used. In the present sub-section also the prime ' $'$ ' for the effective stress tensor will be neglected. The yield function restricting the stress state is developed in the form of *von Mises* and *Drucker-Prager* [11], to take into account the behaviour of clays under undrained conditions and the dilatant/contractant behaviour of dense or loose sands, respectively. The return mapping and the consistent tangent operator is developed, solving the singular behaviour of the *Drucker-Prager* yield surface in the zone of the apex using the concept of multisurface plasticity.

The elastic behaviour of the solid skeleton is assumed to be governed by an hyperelastic free energy $\psi(\mathbf{x},\, t)$ function in the form

$$\psi = \psi(\mathbf{b}^e,\, \xi)\,, \tag{72}$$

dependent on the elastic left *Cauchy-Green* strain tensor $\mathbf{b}^e(\mathbf{x},\, t) = \mathbf{F}^e(\mathbf{F}^e)^{-1}$ and the internal strain like variable $\xi(\mathbf{x}, t)$, the equivalent plastic strain. The second law of thermodynamics yields, under the restriction of isotropy, the constitutive relations

$$\boldsymbol{\tau} = 2\,\frac{\partial \psi}{\partial \mathbf{b}^e}\,\mathbf{b}^e\,, \quad q = -\frac{\partial \psi}{\partial \xi}\,, \tag{73}$$

and the remaining dissipation inequality

$$-\frac{1}{2}\boldsymbol{\tau} : \left[(L_v\,\mathbf{b}^e)\,(\mathbf{b}^e)^{-1}\right] + q\,\dot{\xi} \geq 0\,, \tag{74}$$

where $L_v\,\mathbf{b}^e = \dot{\mathbf{b}}^e - \mathbf{l}\,\mathbf{b}^e - \mathbf{b}^e\,\mathbf{l}^T$ is the *Lie* derivative of the elastic left *Cauchy-Green* strain tensor $\mathbf{b}^e(\mathbf{x},\, t)$.

The evolution equations for the rate terms of the dissipation inequality (74) can be derived from the postulate of the maximum plastic dissipation in the case of associative flow rules [35]:

$$-\frac{1}{2}\,L_v\,\mathbf{b}^e = \dot{\gamma}\,\frac{\partial F}{\partial \boldsymbol{\tau}}\,\mathbf{b}^e\,, \tag{75}$$

$$\dot{\xi} = \dot{\gamma}\,\frac{\partial F}{\partial q}\,, \tag{76}$$

subjected to the classical loading-unloading conditions in *Kuhn-Tucker* form:

$$\dot{\gamma} \geq 0, \quad F = F(\boldsymbol{\tau},\, q) \leq 0, \quad \dot{\gamma}\,F = 0\,, \tag{77}$$

where $\dot{\gamma}$ is the plastic multiplier and $F = F(\boldsymbol{\tau},\, q)$ the isotropic yield function.

For the computation, classical elasto-plastic models have been selected. In particular, the *Drucker-Prager* [11] and the *von Mises* yield functions with linear isotropic hardening have been used in the form, respectively,

$$F(p,\, \mathbf{s},\, \xi) = 3\,\alpha_F\,p + ||\mathbf{s}|| - \beta_F\,\sqrt{\frac{2}{3}}\,(c_0 + h\,\xi) \tag{78}$$

and

$$F(\mathbf{s},\, \xi) = ||\mathbf{s}|| - \sqrt{\frac{2}{3}}\,(\sigma_0 + h\,\xi)\,, \tag{79}$$

in which $p = \frac{1}{3}(\boldsymbol{\tau} : \mathbf{1})$ is the mean effective *Kirchhoff* pressure, $||\mathbf{s}||$ is the L_2-norm of the deviator effective *Kirchhoff* stress tensor $\boldsymbol{\tau}$, c_0 is the initial apparent cohesion of the *Drucker-Prager* model, α_F and β_F are two parameters related to the friction angle ϕ of the soil,

$$\alpha_F = 2\,\frac{\sqrt{\frac{2}{3}}\,\sin\phi}{3-\sin\phi}\,, \quad \beta_F = \frac{6\,\cos\phi}{3-\sin\phi}\,, \tag{80}$$

h the hardening/softening modulus, and σ_0 is the yield stress in the *von Mises* law.

Remarks: In the present contribution, the effect of the capillary pressure p^c on the evolution of the yield surface is not taken into account. The interested reader can refer to [5] for a constitutive relationship function of the effective stress and the capillary pressure.

3.4 Algorithmic formulation for elasto-plasticity

The problem of the calculation of $\mathbf{b}^e$, ξ, and $\boldsymbol{\tau}$ is solved by an operator split into an elastic predictor and plastic corrector [36]. The calculation of the trial elastic state $(\cdot)^{\text{tr}}$ is based on freezing the plastic flow at time t_{n+1}. The $(\mathbf{b}^e_{n+1})^{\text{tr}}$ is hence the push forward of $\mathbf{b}^e_n$ by means of the relative deformation gradient $\mathbf{f}_{n+1} = \partial\boldsymbol{\chi}_{n+1}/\partial\mathbf{x}_n = \mathbf{1} + \operatorname{grad}\Delta\mathbf{u}_{n+1}$, i. e.

$$(\mathbf{b}^e_{n+1})^{\text{tr}} = \mathbf{f}_{n+1}\,\mathbf{b}^e_n\,\mathbf{f}^T_{n+1} \tag{81}$$

with $\xi^{\text{tr}}_{n+1} = \xi_n$, where $\Delta\mathbf{u}_{n+1}$ is the incremental displacement in the time interval $[t_n,\, t_{n+1}]$.

The same value can also be obtained from the reference configuration by the push-forward of $(\mathbf{C}^p_n)^{-1}$ by means of $\mathbf{F}_{n+1}$:

$$(\mathbf{b}^e_{n+1})^{\text{tr}} = \mathbf{F}_{n+1}\,(\mathbf{C}^p_n)^{-1}\,\mathbf{F}^T_{n+1}\,. \tag{82}$$

The corresponding trial elastic stress is obtained from the hyperelastic free energy function as

$$\boldsymbol{\tau}^{\text{tr}}_{n+1} = 2\left(\frac{\partial\psi}{\partial\mathbf{b}^e}\,\mathbf{b}^e\right)\Bigg|_{\mathbf{b}^e=(\mathbf{b}^e_{n+1})^{\text{tr}}} = 2\,\frac{\partial\psi}{\partial\mathbf{b}^e}\Bigg|_{\mathbf{b}^e=(\mathbf{b}^e_{n+1})^{\text{tr}}}(\mathbf{b}^e_{n+1})^{\text{tr}}\,. \tag{83}$$

If this trial state is admissible, it does not violate the inequality $F^{\text{tr}}_{n+1} = F(\boldsymbol{\tau}^{\text{tr}}_{n+1},\, q^{\text{tr}}_{n+1}) \leq 0$ and the stress state is hence already computed.

Otherwise, the return mapping or plastic corrector algorithm is applied to satisfy the condition $F_{n+1} = 0$. Since during this phase the spatial position $\boldsymbol{\chi} = \boldsymbol{\chi}^{\text{tr}}$ is held fixed and thus $\mathbf{l} \equiv \mathbf{0}$, the evolution equation for the elastic left *Cauchy-Green* strain tensor becomes

$$L_v\,\mathbf{b}^e = -2\,\dot{\gamma}\,\frac{\partial F}{\partial\boldsymbol{\tau}}\,\mathbf{b}^e \quad \text{with} \quad L_v\,\mathbf{b}^e\Big|_{\boldsymbol{\chi}=\boldsymbol{\chi}^{\text{tr}}} = \dot{\mathbf{b}}^e\,. \tag{84}$$

This first order differential equation is solved by the product formula algorithm (exponential approximation of the solution, having first order accuracy) during the time interval $[t_n, t_{n+1}]$ [35]:

$$\mathbf{b}^e_{n+1} \cong \exp\left(-2\,\Delta\gamma\,\frac{\partial F}{\partial \boldsymbol{\tau}}\right)\bigg|_{n+1} (\mathbf{b}^e_{n+1})^{\mathrm{tr}}\,. \tag{85}$$

It should now be noted that $\mathbf{b}^e_{n+1}$ commutes with $\boldsymbol{\tau}_{n+1}$ due to the assumption of isotropy and that $(\mathbf{b}^e_{n+1})^{\mathrm{tr}}$ and its principal axis are held fixed during the return mapping, the spectral decomposition of $(\mathbf{b}^e_{n+1})^{\mathrm{tr}}$, $\mathbf{b}^e_{n+1}$, and $\boldsymbol{\tau}_{n+1}$ can hence be written with the same eigenbases:

$$(\mathbf{b}^e)^{\mathrm{tr}} = \sum_{A=1}^{3} (\lambda^{\mathrm{tr}}_{Ae})^2\, \mathbf{n}^{\mathrm{tr}}_A \otimes \mathbf{n}^{\mathrm{tr}}_A\,, \qquad \mathbf{b}^e = \sum_{A=1}^{3} (\lambda_{Ae})^2\, \mathbf{n}^{\mathrm{tr}}_A \otimes \mathbf{n}^{\mathrm{tr}}_A\,,$$
$$\boldsymbol{\tau} = \sum_{A=1}^{3} \tau_A\, \mathbf{n}^{\mathrm{tr}}_A \otimes \mathbf{n}^{\mathrm{tr}}_A\,. \tag{86}$$

Using (86), the product formula (85) can be written in principal values in the form

$$(\lambda_{Ae})^2 = \exp\left(-2\,\Delta\gamma\,\frac{\partial F}{\partial \tau_A}\right)\bigg|_{n+1} (\lambda^{\mathrm{tr}}_{Ae})^2\,. \tag{87}$$

Taking the logarithm of (87), the following important *additive* decomposition of the logarithmic strain measure in elastic and plastic parts is obtained [35]:

$$\varepsilon^{\mathrm{tr}}_{Ae_{n+1}} = \varepsilon_{Ae_{n+1}} + \Delta\gamma\,\frac{\partial F}{\partial \tau_A}\bigg|_{n+1}\,, \tag{88}$$

in which ε_{Ae} is the principal logarithmic elastic strain $\varepsilon_{Ae} = \ln \lambda_A$. This is a very important consequence of the utilized model because it permits to use the return mapping of the elasto-plasticity developed for the linear case [36]. From the knowledge of $\Delta\gamma$ the equivalent plastic strain is computed by the backward *Euler* integration of (76):

$$\xi_{n+1} \cong \xi_n + \Delta\gamma\,\frac{\partial F}{\partial q}\bigg|_{n+1}\,. \tag{89}$$

The principal *Kirchhoff* stress components are then computed by the hyperelastic constitutive law

$$\tau_A = 2\,\lambda_{Ae}\,\frac{\partial \psi}{\partial \lambda_{Ae}} = \frac{\partial \psi}{\partial \varepsilon_{Ae}}\,, \tag{90}$$

where the free energy $\psi = \hat{\psi}(\varepsilon_{Ae}, \xi)$ is now written as a function of the principal elastic logarithmic strain components and the equivalent plastic strain (for isotropic linear hardening):

$$\hat{\psi} = \frac{L}{2}\,(\varepsilon_{1e} + \varepsilon_{2e} + \varepsilon_{3e})^2 + G\,(\varepsilon^2_{1e} + \varepsilon^2_{2e} + \varepsilon^2_{3e}) + \frac{1}{2}\,h\,\xi^2\,, \tag{91}$$

where L and G are the elastic *Lamé* constants, and h the linear hardening modulus.

The strain energy function $W = \rho_{s0}\,\hat{\psi}$ of the solid skeleton associated to $\hat{\psi}$ and used in this paper is the original one proposed by Simo [35]. It is valid only for moderate large elastic strain because it does not satisfy the convexity condition for very large elastic strains [8] but it has been used here because soils behave with moderate elastic strains. Some useful remarks related to the use of this strain energy function can be found e. g. in [28], in which it has been shown that the finite element computation loses stability at finite elastic strains.

The return mapping algorithm for the *Drucker-Prager* model with linear isotropic hardening and non-associated volumetric/deviatoric plastic flow is presented in [31], where a special treatment of the corner region using the concept of multi-surface plasticity is also formulated.

4 Weak form: Variational approach

The weak form of the spatial governing equations presented in the previous section is now derived obtaining the variational equations formally equivalent to the initial-boundary-value problem given by the governing equation and the boundary conditions. This means that the governing equations (52) and (59) are multiplied by independent weighting functions that vanish on the boundary in which *Dirichlet* boundary conditions are applied and are then integrated over the spatial domain B with boundary ∂B. The linear momentum balance equation of the binary porous media (59) is hence weighted on the domain by the test function $\delta \mathbf{u}_s$ corresponding to the solid displacement (or virtual displacement) in the form

$$\int_B (\operatorname{div}\boldsymbol{\sigma} + \rho\,\mathbf{g}) \cdot \delta\mathbf{u}_s \,\mathrm{d}v = 0 \quad \forall \;\; \delta\mathbf{u}_s \,. \tag{92}$$

Applying partial integration and *Green*'s theorem in the form (e. g. [8, 22])

$$\int_B \operatorname{div}\boldsymbol{\sigma} \cdot \delta\mathbf{u}_s \,\mathrm{d}v = -\int_B \boldsymbol{\sigma} : \operatorname{grad}\delta\mathbf{u}_s \,\mathrm{d}v + \int_{\partial B} \bar{\mathbf{t}} \cdot \delta\mathbf{u}_s \,\mathrm{d}s \tag{93}$$

to the divergence part of (92) and taking into account the boundary conditions, this equation is transformed into the weak form

$$\begin{aligned} -\int_B (\boldsymbol{\sigma}' - S_w\, p^w \mathbf{1}) : \operatorname{grad}\delta\mathbf{u}_s \,\mathrm{d}v + \int_B \rho\,\mathbf{g} \cdot \delta\mathbf{u}_s \,\mathrm{d}v \,+ \\ + \int_{\partial B} \bar{\mathbf{t}} \cdot \delta\mathbf{u}_s \,\mathrm{d}s = 0 \quad \forall \;\; \delta\mathbf{u}_s \,, \end{aligned} \tag{94}$$

where the effective stress principle (61) has been introduced. Using the relation $\operatorname{div} \delta \mathbf{u}_s = \operatorname{grad} \delta \mathbf{u}_s : \mathbf{1}$, the previous weak form is transformed into

$$\begin{aligned} - \int_B \boldsymbol{\sigma}' : \operatorname{grad} \delta \mathbf{u}_s \, dv + \int_B S_w \, p^w \operatorname{div} \delta \mathbf{u}_s \, dv + \int_B \rho \, \mathbf{g} \cdot \delta \mathbf{u}_s \, dv + \\ + \int_{\partial B} \bar{\mathbf{t}} \cdot \delta \mathbf{u}_s \, ds = 0 \quad \forall \ \delta \mathbf{u}_s \, . \end{aligned} \tag{95}$$

The weak form of the mixture mass balance equation (52) is obtained in a similar way, introducing *Darcy*'s law (63) and using the test function δp^w corresponding to p^w (or virtual water pressure):

$$\int_B \operatorname{div} \mathbf{v}^s \, \delta p^w \, dv + \underline{\int_B \operatorname{div} \left[\frac{\mathbf{k} \, k^{rw}}{\mu^w} \left(- \operatorname{grad} p^w + \rho^w \, \mathbf{g} \right) \right] \delta p^w \, dv} = 0 \ \forall \ \delta p^w . \tag{96}$$

Applying *Green*'s theorem to the underlined term of the previous equation, the following is obtained:

$$\begin{aligned} \int_B \operatorname{div} \mathbf{v}^s \, \delta p^w \, dv + \int_B \left[\frac{\mathbf{k} \, k^{rw}}{\mu^w} \left(\operatorname{grad} p^w - \rho^w \, \mathbf{g} \right) \right] \cdot \operatorname{grad} \delta p^w \, dv + \\ + \int_{\partial B} q^w \, \delta p^w \, ds = 0 \quad \forall \ \delta p^w \, , \end{aligned} \tag{97}$$

where $q^w(\mathbf{x}, t)$ is the water flow draining through the surface ∂B.

Remarks: It can be observed that the weak forms (95) and (97) are very similar to those of the geometrically linear theory, e. g. [20], by substituting the deformed integration domain B with the undeformed one B_0. Moreover, in the small strain theory $\operatorname{div} \mathbf{v}^s = \dot{\boldsymbol{\varepsilon}} : \mathbf{1}$, where $\boldsymbol{\varepsilon}$ is the small strain tensor of the solid skeleton, while in finite strain $\operatorname{div} \mathbf{v}^s = \dot{J}^s / J^s$. In the small strain theory the additive decomposition of the strain tensor $\boldsymbol{\varepsilon}$ in elastic and plastic parts is also possible, thus rendering the computation of the constitutive tangent operator in the linearization of the weak form particularly easy.

5 Time discretization

Time integration of the weak form of the mass balance equation (97) over a finite time step $\Delta t = t_{n+1} - t_n$ is necessary because of the time dependent term $\operatorname{div} \mathbf{v}^s$.

The generalized trapezoidal method is used here, as shown for instance in [20]. Because of the dependence of the integration domain on time, we rewrite the weak forms (95) and (97) with respect to the undeformed domain as follows:

$$
\begin{aligned}
&\int_{B_0} (\boldsymbol{\tau}' - J^s\, S_w\, p^w\, \mathbf{1}) : \operatorname{grad} \delta\mathbf{u}_s \,\mathrm{d}V - \int_{B_0} \rho_0\, \mathbf{g} \cdot \delta\mathbf{u}_s \,\mathrm{d}V - \\
&\quad - \int_{\partial B_0} \bar{\mathbf{T}} \cdot \delta\mathbf{u}_s \,\mathrm{d}A = 0 \quad \forall\ \delta\mathbf{u}_s\,,
\end{aligned}
\tag{98}
$$

$$
\begin{aligned}
&\int_{B_0} J^s\ \operatorname{div} \mathbf{v}^s\, \delta p^w \,\mathrm{d}V + \int_{B_0} \left[J^s\, \frac{\mathbf{k}\, k^{rw}}{\mu^w} \left(\operatorname{grad} p^w - \rho^w\, \mathbf{g}\right)\right] \cdot \operatorname{grad} \delta p^w \,\mathrm{d}V + \\
&\quad + \int_{\partial B_0} Q^w\, \delta p^w \,\mathrm{d}A = 0 \quad \forall\ \delta p^w\,,
\end{aligned}
\tag{99}
$$

where $\boldsymbol{\tau}'$ is the modified effective *Kirchhoff* stress tensor and $\bar{\mathbf{T}} = \mathbf{P}\,\mathbf{N}$ and $Q^w = N\, S_W\, \bar{\mathbf{V}}^{ws} \cdot \mathbf{N}$ are, respectively, the traction vector and the water flow computed with respect to the undeformed configuration. The form of (98) and (99) is also useful for the subsequent linearization because it will be easily performed with respect to the undeformed (fixed) domain.

Equation (99) is now rewritten at time t_{n+1} using the relationships

$$
\dot{J}^s_{n+\beta} = \frac{J^s_{n+1} - J^s_n}{\Delta t}\,,
\tag{100}
$$

$$
(\cdot)_{n+\beta} = (1-\beta)\,(\cdot)_n + \beta(\cdot)_{n+1} = (\cdot)_n + \beta\,[(\cdot)_{n+1} - (\,\cdot\,)_n]\,,
\tag{101}
$$

with $\beta \in [0,\, 1]$, obtaining

$$
\begin{aligned}
&\int_{B_0} \left(J^s_{n+1} - J^s_n\right) \delta p^w \,\mathrm{d}V - \Delta t \int_{B_0} \left(J^s\, \mathbf{v}^D \cdot \operatorname{grad} \delta p^w\right)_{n+\beta} \mathrm{d}V + \\
&\quad + \Delta t \int_{\partial B_0} Q^w_{n+\beta}\, \delta p^w \,\mathrm{d}A = 0 \quad \forall\ \delta p^w\,,
\end{aligned}
\tag{102}
$$

where $\mathbf{v}^D = -\mathbf{k}\, k^{rw}/\mu^w\, (\operatorname{grad} p^w - \rho^w\, \mathbf{g})$ is *Darcy*'s velocity of the water.

The weak form of the linear momentum balance equation (98) is directly written at time t_{n+1} because it is time independent:

$$
\begin{aligned}
&\int_{B_0} [(\boldsymbol{\tau}' - J^s\, S_w\, p^w\, \mathbf{1}) : \operatorname{grad} \delta\mathbf{u}_s]_{n+1} \,\mathrm{d}V - \int_{B_0} \rho_{0_{n+1}}\, \mathbf{g} \cdot \delta\mathbf{u}_s \,\mathrm{d}V - \\
&\quad - \int_{\partial B_0} \bar{\mathbf{T}}_{n+1} \cdot \delta\mathbf{u}_s \,\mathrm{d}A = 0 \quad \forall\ \delta\mathbf{u}_s\,.
\end{aligned}
\tag{103}
$$

Linearized analysis of accuracy and stability suggest the use of $\beta \geq \frac{1}{2}$. In the examples section, implicit one-step time integration has been performed ($\beta = 1$).

The weak forms (102) and (103) represent a non-linear coupled equations system where the non-linearities are introduced by the finite kinematics and the constitutive laws.

6 Consistent linearization

The non-linear equation system (102) and (103) can be written in the following compact form:

$$\mathbf{G}(\chi, \eta) = \mathbf{0}, \quad \text{where} \quad \chi = (\chi^s, p^w)^T \quad \text{and} \quad \eta = (\delta \mathbf{u}_s, \delta p^w)^T. \tag{104}$$

For its numerical solution, iterative methods have to be employed and the linearization at $\bar{\chi}$ is hence necessary:

$$\mathbf{G}(\bar{\chi}, \eta, \Delta\mathbf{u}) \cong \mathbf{G}(\bar{\chi}, \eta) + \mathrm{D}\mathbf{G}(\bar{\chi}, \eta) \cdot \Delta\mathbf{u} \cong \mathbf{0}, \tag{105}$$

where $\Delta\mathbf{u} = (\Delta\mathbf{u}_s, \Delta p^w)^T$ and $\mathrm{D}\mathbf{G} \cdot \Delta\mathbf{u} = \frac{\mathrm{d}}{\mathrm{d}\alpha}\mathbf{G}(\bar{\chi} + \alpha\Delta\mathbf{u})|_{\alpha=0}$ is the directional derivative or *Gateaux* derivative of $\mathbf{G}$ at $\bar{\chi}$ in the direction of $\Delta\mathbf{u}$ (e. g. [22, 39] for single-phase material). Since the equation system $\mathbf{G}$ is composed of the weak form of the linear momentum balance equation (G_{LBE}) and of the mass balance equation (G_{MBE}), then

$$\mathrm{D}\mathbf{G} \cdot \Delta\mathbf{u} = \begin{bmatrix} \mathrm{D}G_{\mathrm{LBE}} \cdot \Delta\mathbf{u}_s + \mathrm{D}G_{\mathrm{LBE}} \cdot \Delta p^w \\ \mathrm{D}G_{\mathrm{MBE}} \cdot \Delta\mathbf{u}_s + \mathrm{D}G_{\mathrm{MBE}} \cdot \Delta p^w \end{bmatrix}. \tag{106}$$

Using the symbol $(\cdot)^{k+1}_{n+1}$ to indicate the current iteration in the current time step, the linearization on the configuration $(\cdot)^k_{n+1}$ is written as

$$\mathrm{D}\mathbf{G}^k_{n+1} \cdot \Delta\mathbf{u}^{k+1}_{n+1} = -\mathbf{G}^k_{n+1}, \tag{107}$$

and the solution vector $\mathbf{u} = (\mathbf{u}_s, p^w)^T$ is then updated by the incremental relationship

$$\mathbf{u}^{k+1}_{n+1} = \mathbf{u}^k_{n+1} + \Delta\mathbf{u}^{k+1}_{n+1}. \tag{108}$$

For an efficient numerical performance of the scheme (107), the consistent linearization is applied [39] in which the linearization of the integrated constitutive equation (86) plays a central role (this concept was first pointed out in [34] for the geometrically linear case).

The linearization of (102) and (103), performed in the undeformed configuration B_0 and then pushed forward in the deformed configuration B, gives the following result:

- For the linear momentum balance equation:

$$\int_B \Big(\operatorname{grad}\delta\mathbf{u}_s : \mathbf{c}^{ep} : \operatorname{sym}(\operatorname{grad}\Delta\mathbf{u}_s) + \boldsymbol{\sigma}' : \operatorname{grad}^T\delta\mathbf{u}_s \operatorname{grad}\Delta\mathbf{u}_s\Big)\,\mathrm{d}v +$$
$$+ \int_B S_w\, p^w \operatorname{grad}\delta\mathbf{u}_s : \Big(\operatorname{grad}^T\Delta\mathbf{u}_s - \operatorname{div}\Delta\mathbf{u}_s\,\mathbf{1}\Big)\,\mathrm{d}v -$$
$$- \int_B \rho^w S_w\, \delta\mathbf{u}_s\cdot\mathbf{g}\operatorname{div}\Delta\mathbf{u}_s\,\mathrm{d}v - \int_B \left(p^w \frac{\partial S_w}{\partial p^w} + S_w\right)\operatorname{div}\delta\mathbf{u}_s\,\Delta p^w\,\mathrm{d}v\,. \tag{109}$$

- For the mass balance equation (in case of isotropic permeability):

$$\int_B \delta p^w \operatorname{div}\Delta\mathbf{u}_s\,\mathrm{d}v + \beta\,\Delta t\int_B \frac{k\,k^{rw}}{\mu^w}\operatorname{grad}\delta p^w\cdot\operatorname{grad}\Delta p^w\,\mathrm{d}v +$$
$$+\beta\,\Delta t\int_B \operatorname{grad}\delta p^w\cdot\left[\left(\frac{1-n}{k}\frac{\partial k}{\partial n}+1\right)\frac{k\,k^{rw}}{\mu^w}(\operatorname{grad}p^w -\right.$$
$$\left.-\rho^w\mathbf{g})\operatorname{div}\Delta\mathbf{u}_s\right]\mathrm{d}v -$$
$$-\beta\,\Delta t\int_B \operatorname{grad}\delta p^w\cdot\left[\frac{2\,k\,k^{rw}}{\mu^w}\operatorname{sym}(\operatorname{grad}\Delta\mathbf{u}_s)\operatorname{grad}p^w\right]\mathrm{d}v +$$
$$+\beta\,\Delta t\int_B \operatorname{grad}\delta p^w\cdot\left(\frac{k\,k^{rw}}{\mu^w}\rho^w\operatorname{grad}\Delta\mathbf{u}_s\,\mathbf{g}\right)\mathrm{d}v +$$
$$+\beta\,\Delta t\int_B \frac{k}{\mu^w}\frac{\partial k^{rw}}{\partial p^w}\operatorname{grad}p^w\cdot\operatorname{grad}\delta p^w\,\Delta p^w\,\mathrm{d}v\,. \tag{110}$$

In the directional derivative $\mathrm{D}G_{\mathrm{LBE}}\cdot\Delta\mathbf{u}_s$ the term

$$\int_B \Big(\operatorname{grad}\delta\mathbf{u}_s : \mathbf{c}^{ep} : \operatorname{sym}(\operatorname{grad}\Delta\mathbf{u}_s) + \boldsymbol{\sigma}' : \operatorname{grad}^T\delta\mathbf{u}_s \operatorname{grad}\Delta\mathbf{u}_s\Big)\,\mathrm{d}v \tag{111}$$

contains $\mathbf{c}^{ep}$, the spatial constitutive operator following the linearization of (86):

$$\begin{aligned}\mathbf{c}^{ep}_{n+1} = &\sum_{A=1}^{3}\sum_{B=1}^{3} a^{ep}_{AB_{n+1}}\,(\mathbf{n}^{\mathrm{tr}}_A\otimes\mathbf{n}^{\mathrm{tr}}_A)\otimes(\mathbf{n}^{\mathrm{tr}}_B\otimes\mathbf{n}^{\mathrm{tr}}_B) +\\ &+2\sum_{A=1}^{3}\tau_{A_{n+1}}\,\mathbf{c}^{\mathrm{tr}(A)}_{n+1}\,.\end{aligned} \tag{112}$$

It is useful to remark that in (112) only the second order tensor $\mathbf{a}^{ep} = \partial \tau_A / \partial \varepsilon_B^{\text{tr}}$ depends on the specific model of plasticity and the structure of the return mapping algorithm in principal stretches, while the tensors $\mathbf{c}_{n+1}^{\text{tr}(A)}$ and $\mathbf{n}_A^{\text{tr}} \otimes \mathbf{n}_A^{\text{tr}}$ are independent of the specific plastic model is use. Moreover, it is easy to proof that the moduli $\mathbf{a}^{ep}$ have a form identical to the algorithmic elasto-plastic tangent moduli of the infinitesimal theory [35]. The expression for $\mathbf{c}_{n+1}^{\text{tr}(A)}$ can be obtained by linearization of the eigenbases dyadic $\mathbf{n}_A^{\text{tr}} \otimes \mathbf{n}_A^{tr}$ in the spatial setting:

$$\mathbf{c}_{n+1}^{\text{tr}(A)} = \frac{\partial (\mathbf{n}_A^{\text{tr}} \otimes \mathbf{n}_A^{\text{tr}})}{\partial \mathbf{g}} , \tag{113}$$

where $\mathbf{g}$ is the spatial metric, or by pull-back [22] of $\mathbf{n}_A^{\text{tr}} \otimes \mathbf{n}_A^{\text{tr}}$, subsequent to linearization in the material setting using the relations e. g. of [24] and then by push-forward of the linearization in spatial setting.

The expressions for the algorithmic moduli $\mathbf{a}^{ep}$ of the *Drucker-Prager* model used in the computation are derived in [31], where a special treatment of the corner region using the concept of multi-surface plasticity has been developed.

7 Finite element discretization in space

The suitable spatial finite element formulation is derived by applying the well known *Galerkin* procedure, in which the weighting functions are approximated by the same shape functions used to approximate the driving variables (isoparametric finite elements). This means that the geometry $\mathbf{X}^s$, the current configuration $\mathbf{x}$, the displacement field $\mathbf{u}_s$, the water pressure p^w, the incremental generalized displacement $\Delta \mathbf{u} = (\Delta \mathbf{u}_s, \Delta p^w)^T$ and the variations $\eta = (\delta \mathbf{u}_s, \delta p^w)^T$, are interpolated within a finite element by the same type of functions. In the present setting, different shape functions are chosen for quantities associated respectively to the solid and the fluid, thus satisfying the LBB condition (*Ladyzhenskaya-Babuška-Brezzi* condition) for the locally undrained case. Standard procedures have been applied, following any text books on FEM. With respect to the small strain case, the discretization of the spatial form of the linearized system of equations is made taking into account that each quantity is referred to the spatial co-ordinates $\mathbf{x}$, instead of the co-ordinates of the undeformed configuration $\mathbf{X}^s$. (In the present formulation quadrilateral Q_2Q_1 or S_2Q_1 elements have been used for the *Liakopoulos*' test and the localization example, respectively). The solid displacement $\mathbf{u}_s(\mathbf{x}, t)$ and the water pressure $p^w(\mathbf{x}, t)$ are hence expressed in the whole domain by global shape function matrices $\mathbf{N}_u(\mathbf{x})$ and $\mathbf{N}_w(\mathbf{x})$ and the nodal value vectors $\bar{\mathbf{u}}(t)$ and $\bar{\mathbf{p}}(t)$:

$$\mathbf{u} = \mathbf{N}_u \bar{\mathbf{u}} , \qquad p^w = \mathbf{N}_w \bar{\mathbf{p}} . \tag{114}$$

The linearized system of equations (107) in matrix form can be expressed as

$$\begin{bmatrix} \mathbf{K}_T + \mathbf{K}_{sw}^{\text{geom}} & -c_{sw}\,\mathbf{Q}_{sw} \\ \mathbf{Q}_{ws} - \beta\,\Delta t\,\mathbf{Q}_{sw}^{\text{geom}} & \beta\,\Delta t\,\mathbf{H} \end{bmatrix} \begin{bmatrix} \Delta\bar{\mathbf{u}} \\ \Delta\bar{\mathbf{p}} \end{bmatrix} = - \begin{bmatrix} \mathbf{G}^u \\ \mathbf{G}^p \end{bmatrix}, \tag{115}$$

which is non-symmetric (details concerning the implementation as well as the matrices and the residuum vectors of (115) will be described in a future paper). Owing to the strong coupling between the mechanical and the pore fluid problem, a monolithic solution of (115) is preferred using a *Newton* scheme. As far as the numerical performance of the proposed implementation is concerned, it can be outlined that quadratic rate of convergence for the global *Newton* iteration in each step has been obtained in the computation. A typical rate of convergence is reported in Table 1.

Increment No. 50		
Iteration No. 0:	Residuum Norm:	$1.53800 \cdot 10^{-3}$
Iteration No. 1:	Residuum Norm:	$1.18553 \cdot 10^{1}$
Iteration No. 2:	Residuum Norm:	$1.40621 \cdot 10^{1}$
Iteration No. 3:	Residuum Norm:	$1.23343 \cdot 10^{0}$
Iteration No. 4:	Residuum Norm:	$6.19783 \cdot 10^{-2}$
Iteration No. 5:	Residuum Norm:	$2.09049 \cdot 10^{-5}$
Iteration No. 6:	Residuum Norm:	$6.28546 \cdot 10^{-9}$
Increment No. 75		
Iteration No. 0:	Residuum Norm:	$1.59534 \cdot 10^{-3}$
Iteration No. 1:	Residuum Norm:	$9.82091 \cdot 10^{0}$
Iteration No. 2:	Residuum Norm:	$1.42705 \cdot 10^{1}$
Iteration No. 3:	Residuum Norm:	$1.41873 \cdot 10^{0}$
Iteration No. 4:	Residuum Norm:	$6.81044 \cdot 10^{-2}$
Iteration No. 5:	Residuum Norm:	$2.85757 \cdot 10^{-5}$
Iteration No. 6:	Residuum Norm:	$8.62292 \cdot 10^{-9}$

Table 1. Rate of convergence for the implemented *Newton-Raphson* scheme (from the second example, using the *Drucker-Prager* law with $\varphi = 5°$).

8 Numerical examples

It is difficult to choose appropriate tests to validate the model developed in the previous sections and its implementation in the computer code. Indeed there are no analytical solutions for this type of coupled problems (to the author's knowledge), where deformations of the solid skeleton are studied with saturated-unsaturated flow of mass transfer. There are also very few documented laboratory experiments. One of these is the experiment conducted by Liakopoulos [21] on the isothermal drainage of water from a vertical column of water saturated sand (Figure 4). A column of perspex, 1 meter in height, was packed by *Del Monte* sand and instrumented to measure the moisture tension at several points along the column. Before starting the experiment ($t < 0$) water was continuously added from the top and was allowed to drain freely at the bottom through a filter. The flow was carefully regulated until the tensiometers read zero pore pressure. At $t = 0$ the water supply was ceased and the tensiometers reading were recorded. Only the porosity and the hydraulic properties of *Del Monte* sand were measured by [21] by an independent set of experiments, while the mechanical parameters are those of [33], where the *Liakopoulos*' test with a geometrically linear model is studied. The material parameters and the experimental constitutive laws for $S_w(p^c)$ and $k^{rw}(S_w)$ used in the computation are listed in Table 2.

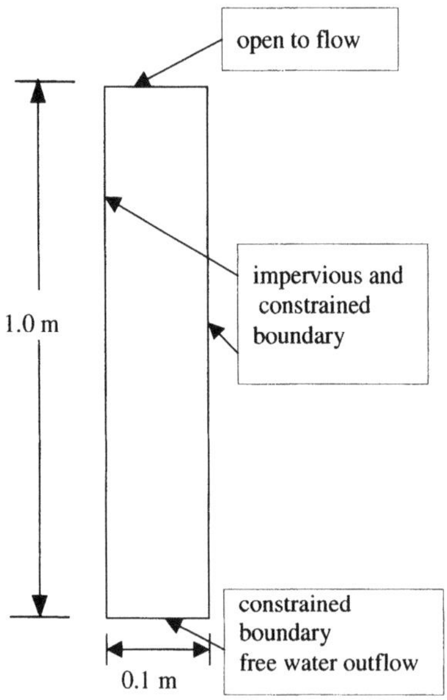

Fig. 4. Scheme of *Liakopoulos*' test.

The column is discretized with 10 Q_2Q_1 finite elements for the solid displacements and the water pressure, respectively (i. e. 9 nodes for the solid, 4 nodes for the fluid). The upper and lower boundaries are drained (water pressure is hence assumed to be at the atmospheric value), while the other boundaries are impervious. Horizontal displacements are constrained on the lateral surfaces while the bottom surface is also vertically constrained. Zero initial conditions are assumed and gravity acceleration is considered during the computation, since the gravity load is the driving force of the experiment. The hyperelastic constitutive law of (91) is utilized for the solid skeleton. The experiment reveals the desaturation of the column from the top to the bottom

Porosity	n	=	0.2975
Isotropic permeability	k	=	$4.3 \cdot 10^{-6}$ m/s
Solid grain density	ρ^s	=	2000 kg/m^3
Water density	ρ^w	=	1000 kg/m^3
Gravity acceleration	g	=	9.81 m/s^2
Water saturation	S_w	=	$1.0 - 1.9722 \cdot 10^{11} (p^c)^{2.4279}$
Relative permeability for water	k^{rw}	=	$1.0 - 2.207\,(1 - S_w)^{1.0121}$
Solid bulk modulus [33]	K	=	2166.77 kN/m^2
Solid shear modulus [33]	G	=	464.29 kN/m^2

Table 2. Material parameters used in the computation of *Liakopoulos*' test.

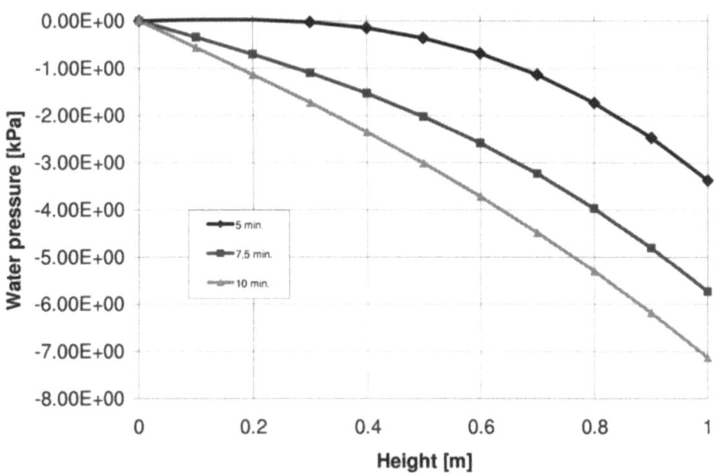

Fig. 5. Water pressure vs. column height.

surface. This behaviour is well described by the model, as it can be observed in Figure 5, where only capillary pressures (negative water pressures) appear.

The second example deals with the analysis of a square domain of water saturated porous material loaded by a rigid footing (Figure 6). This example was solved in [37] using the linear theory. The BVP is now solved numerically using the formulation developed in the previous sections. Plain strain conditions are assumed. The homogeneous soil domain has a side length of $l = 10$ m, with the rigid footing spanning over 5 m at the right part of the top surface. On the boundary, the horizontal and vertical displacements are constrained respectively on the left and the bottom surface. Drainage of the water is allowed only through the unloaded part of the top surface of the domain. The solid skeleton is assumed to obey the elasto-plastic *von Mises* or *Drucker-Prager* constitutive model, both with isotropic linear softening

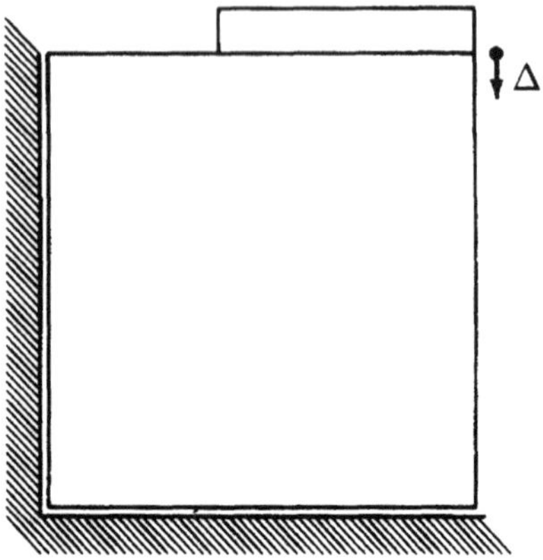

Fig. 6. Rigid footing on a square domain of porous elasto-plastic material.

Solid bulk modulus	K	=	$8333\ \mathrm{kN/m^2}$
Solid shear modulus	G	=	$3486\ \mathrm{kN/m^2}$
Initial yield stress	y_0	=	$100\ \mathrm{kN/m^2}$
Linear softening modulus	h	=	$-10\ \mathrm{kN/m^2}$
Initial apparent cohesion	c_0	=	$100\ \mathrm{kN/m^2}$
Angle of internal friction	ϕ	=	$20°$
Angle of dilatancy	φ	=	$-10°,\ 0°,\ +5°,\ +10°$
Isotropic permeability	k	=	$0.0001\ \mathrm{m/s}$
Solid grain density	ρ^s	=	$2000\ \mathrm{kg/m^3}$
Water density	ρ^w	=	$1000\ \mathrm{kg/m^3}$
Gravity acceleration	g	=	$9.81\ \mathrm{m/s^2}$
Water saturation	S_w	=	$0.084 + 0.916/(1.0 + 7.0\,p^c/\gamma^w)^2$
Relative permeability for water	k^{rw}	=	$1.0/[1.00 + (50.0\,p^c/\gamma^w)^4]^{0.9}$

Table 3. Material parameters used in the computation of the localization example.

behaviour as a phenomenological description of damage effects. The material parameters used in the computation are listed in Table 3.

The *von Mises* law has been selected as a reference material law used to test the implementation and to describe qualitatively the mechanical behaviour of clays under undrained conditions, while the *Drucker-Prager* law has been chosen to simulate the behaviour of dilatant/contractant geomaterials such as dense and loose sands, respectively. The loading is applied quasi-statically to the rigid footing by displacement control with a constant

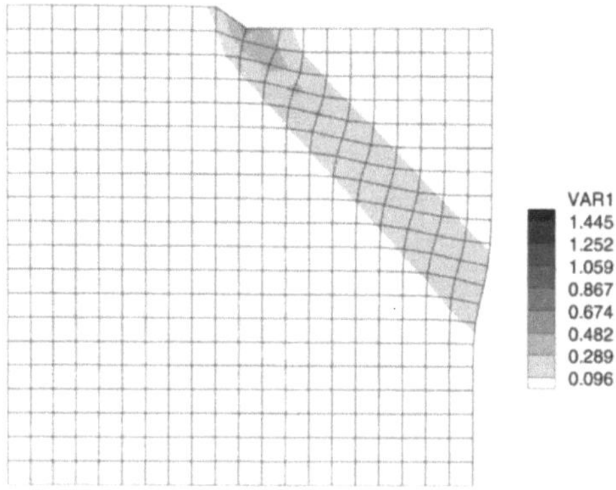

Fig. 7. Equivalent plastic strain contour using the *von Mises* law.

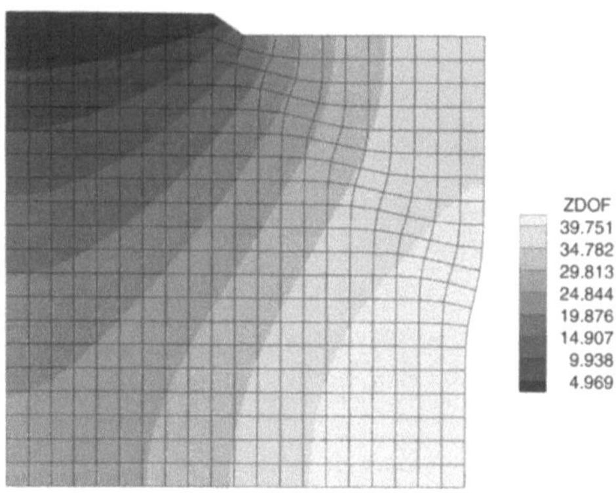

Fig. 8. Excess water pressure contour [kPa] using the *von Mises* law.

vertical velocity of $5 \cdot 10^{-3}$ m/s until the maximum displacement is obtained. The domain has been discretized using 20×20 S_2Q_1 elements for the solid and the fluid mesh (i. e. 8 nodes for the solid, 4 nodes for the fluid).

Figure 7 shows the distribution of the equivalent plastic strain at the end of the load history (0.5 m in this case) on the deformed configuration using the *von Mises* material model. No magnification of the displacements has been used in this and all the following figures. The plastic zone indicates the pronounced accumulation of inelastic strains in a narrow band, while the deformed configuration outlines the classical slip of a part of the domain on the other. Figure 8 shows the excess water pressures at the end of the load

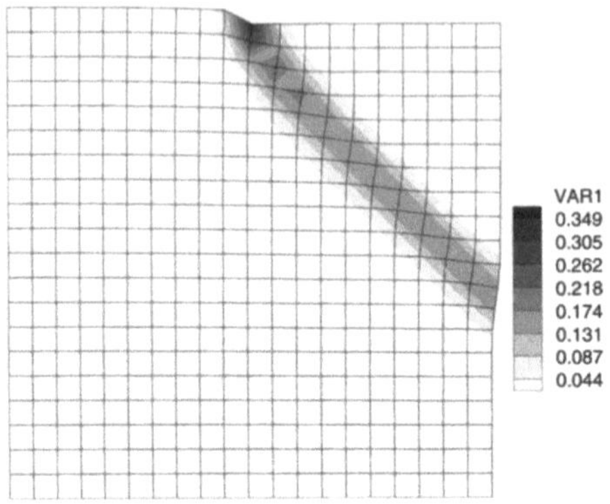

Fig. 9. Equivalent plastic strain contour using the *Drucker-Prager* law with $\varphi = 0°$.

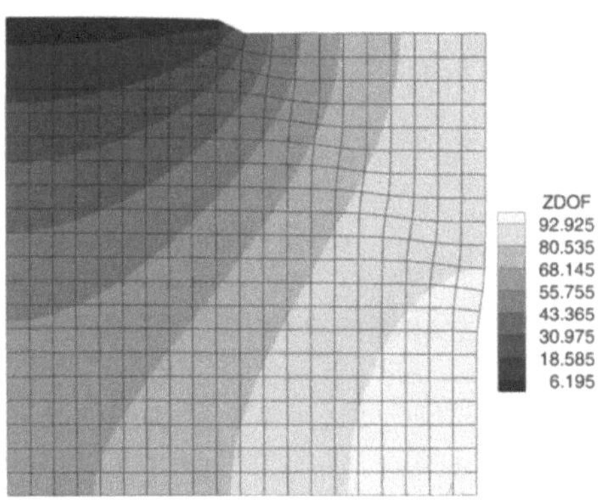

Fig. 10. Excess water pressure contour [kPa] using the *Drucker-Prager* law with $\varphi = 0^0$.

history (values expressed in kPa), where only positive pressure values can be observed due to the compressive load in a solid skeleton with isochoric plastic flow.

The effect of the plastic dilatancy/contractancy is shown by analyzing the square panel using the *Drucker-Prager* material law. The sample has been solved using different values of the angle of dilatancy: $0°, +5°, +10°$, and $-10°$. In particular, Figures 9 and 10 show the equivalent plastic strain and the excess water pressure distribution at the end of the load history in case of zero dilatancy (which means isochoric plastic flow). The resulting shear

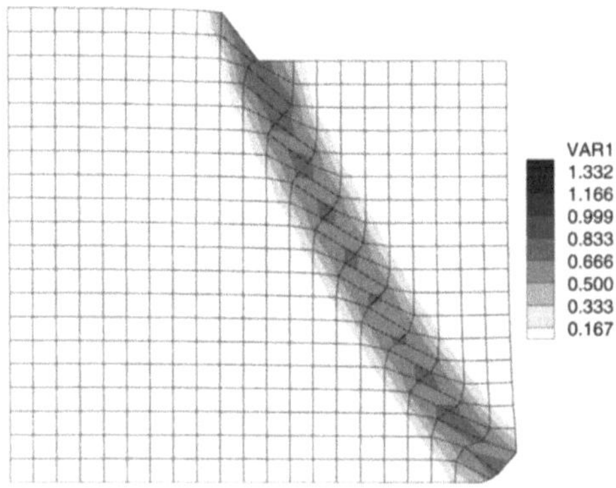

Fig. 11. Equivalent plastic strain contour using the *Drucker-Prager* law with $\varphi = 5°$.

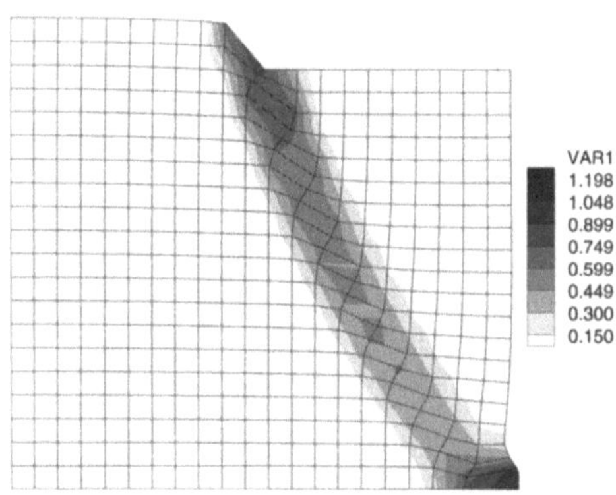

Fig. 12. Equivalent plastic strain contour using the *Drucker-Prager* law with $\varphi = 10°$.

band and the deformation pattern are hence very similar to those obtained using the *von Mises* material law, as it can be observed by comparison with Figures 7 and 8.

Increasing the value of the angle of dilatancy ($\varphi = 5°$ and $10°$, see Figures 11 and 12, respectively), an increase of the horizontal displacement of the right side of the panel, due to the increase of volumetric plastic strain with dilatancy, can be observed. The opposite behaviour appears in case of

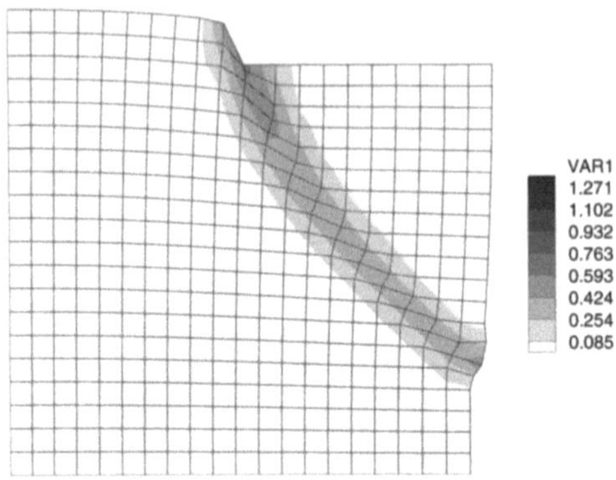

Fig. 13. Equivalent plastic strain contour using the *Drucker-Prager* law with $\varphi = -10°$.

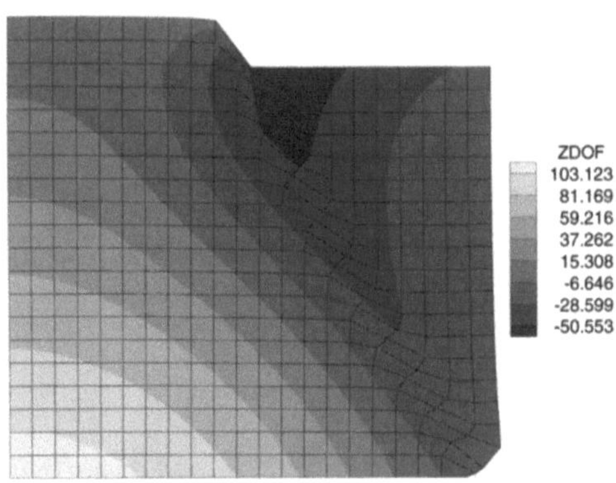

Fig. 14. Water pressure contour [kPa] using the *Drucker-Prager* law with $\varphi = 5°$.

negative value of the dilatancy angle, see Figure 13, where the equivalent plastic strain contour in case of dilatancy angle of $-10°$ is depicted.

The effect of the plastic dilatancy/contractany is evidenced also in the contour of the water pressures. In fact the variation of the porosity with the deformation of the medium, see (49) and the localization of the dilatant plastic strains imply the presence of negative water pressure, with the lowest values inside the plastic zones (Figures 14 and 15), as opposed to the case of contractant plastic flow (Figure 16). The presence of negative pressures is not surprizing. In fact, it was experimentally observed at localization by [25]

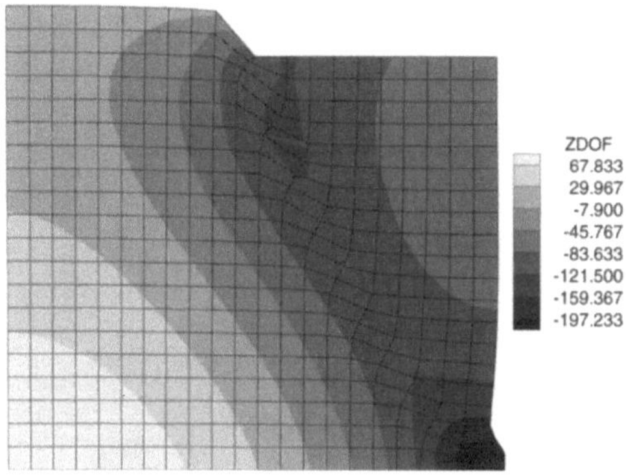

Fig. 15. Water pressure contour [kPa] using the *Drucker-Prager* law with $\varphi = 10°$.

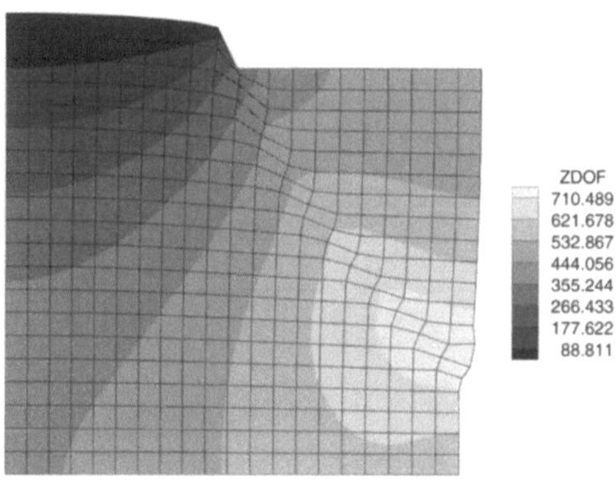

Fig. 16. Excess water pressure contour [kPa] using the *Drucker-Prager* law with $\varphi = -10°$.

and [38] during biaxial tests of globally undrained dense sands under imposed displacements. In particular, the value of -80 and -91 kPa was measured by the two authors, respectively. At those pressures, partially saturated condition due to cavitation of the pore water was observed, which means the presence of the vapour phase separated from the liquid phase by a meniscus. The values of negative water pressures below the cavitation pressure computed in the numerical example of Figure 15 suggest the presence of the cavitation phenomenon inside the shear band and close to it.

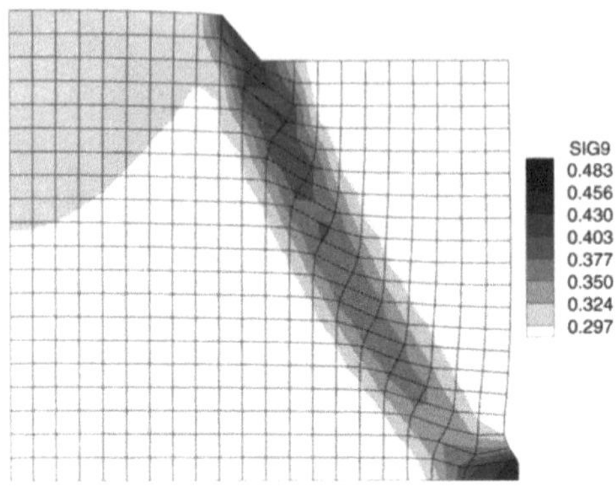

Fig. 17. Porosity contour using the *Drucker-Prager* law with $\varphi = 10°$.

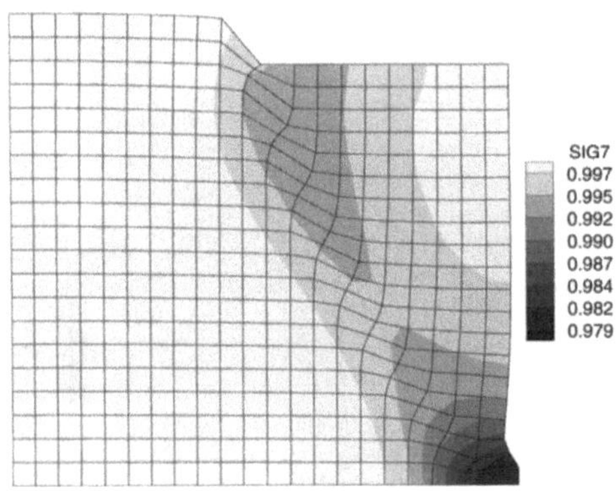

Fig. 18. Water saturation contour using the *Drucker-Prager* law with $\varphi = 10°$.

The dilatant behaviour of the used constitutive model for the solid skeleton is shown also in Figure 17 where the contour of the porosity for the case of *Drucker-Prager* with dilatant angle of 10° and friction angle of 20° is depicted. It can be observed that the porosity increases only inside the shear band, while decreases outside it (the initial value at $t = 0$ was 0.3). Figure 18 shows the contour of the water saturation for the same example, where the desaturation of the zones occupied by the shear band and of that close to it can be observed.

The effect of the plastic dilatancy/contractancy is evidenced also in Figures 19 and 20, where the water velocity in case of dilatant and contractant

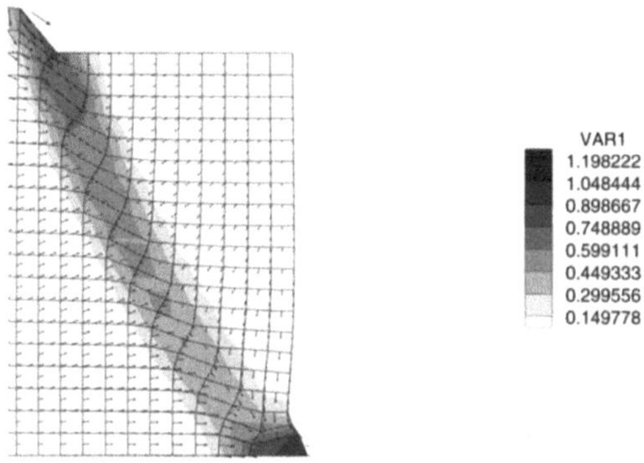

Fig. 19. Water velocity close the shear band using *Drucker-Prager* law with $\varphi = 10°$.

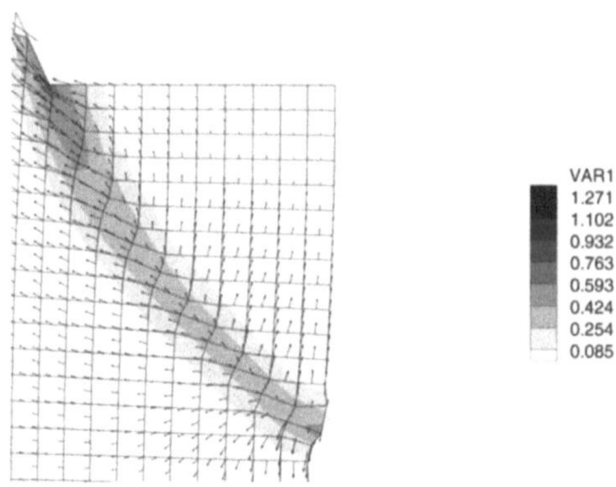

Fig. 20. Water velocity close the shear band using *Drucker-Prager* law with $\varphi = -10°$.

material is shown. Different directions of the fluid flow can be noted, flowing into the band in case of dilatant material, out of it in the other case.

Remarks: Negative water pressures start from the top surface, close to the left corner of the foundation, where the dilatant shear band first appears, and propagate inside the plastic zone following the evolution of the shear band. Then, they propagate also outside the band due to the filtration process generated by the high pressure gradient in conjunction with the quasi-static process, until they occupy almost all the domain, as it can be observed in Figure 15. This phenomenon is dependent on the value of the time step: Decreasing its value, a reduced zone with negative water pressure has been observed, but it is always present. The localization of the negative pressures could be probably more easily captured in the dynamic case, as shown in [13, 32, 41] using a geometrically linear model.

Conclusions

This paper shows a mathematical model and the related finite element discretization of quasi-static and isothermal inelastic saturated and partially saturated geomaterials, assuming incompressible constituents at microscopic level.

The governing equations are derived in material and spatial formulation. The elasto-plastic behaviour of the solid skeleton is based on the multiplicative decomposition of the deformation gradient in an elastic and plastic part and is developed in spatial setting. The solid effective stress is hyperelastic or limited by the *von Mises* or the *Drucker-Prager* yield criterion with isotropic linear softening. The water behaves following *Darcy*'s law. Consistent linearization of the non-linear equation system and a finite element formulation are derived.

Numerical results of this research in progress on large elastic or inelastic strains in saturated and partially saturated porous media are shown.

Acknowledgements

The authors would like to thank *Programma Vigoni 2001* from CRUI, Italy, for the financial support of the stay of the first author at the University of Kaiserslautern, Germany. This work has been carried out within the research project *Cofin 2000-2001 MM08323597* sponsored by the Italian Ministry of Scientific and Technological Research MURST.

References

1. Armero, F.: Formulation and Finite Element Implementation of a Multiplicative Model of Coupled Poro-Plasticity at Finite Strains under Fully Saturated Conditions. *Computer Methods Appl. Mech. Engrg.* **171** (1999), 205–241.
2. Bluhm, J., de Boer, R.: The Volume Fraction Concept in the Porous Media Theory. *ZAMM* **77** (1997), 8, 563–577.
3. de Boer, R.: *Theory of Porous Media: Highlight in Historical Development and Current State.* Springer-Verlag, Berlin 2000.
4. de Boer, R., Ehlers, W., Kowalski, S., Plischka, J.: *Porous media-a survey of different approaches.* Forschungsberichte aus dem Fachbereich Bauwesen der Universität Essen, 54, Essen 1991.

5. Bolzon, G., Schrefler, B. A., Zienkiewicz, O. C.: Elastoplastic soil constitutive laws generalized to partially saturated states. *Géotechnique* **46** (1996), 279–289.
6. Borja, R. I., Alarcon, E.: A Mathematical Framework for Finite Strain Elastoplastic Consolidation. Part 1: Balance Laws, Variational Formulation, and Linearization. *Computer Meth. Appl. Mech. Engrg.* **122** (1995), 145–171.
7. Borja, R. I., Tamagnini, C.: Numerical Implementation of a Mathematical Model for Finite Strain Elastoplastic Consolidation. In Owen, D. R. J., Onate, E., Hinton, E. (eds.): *Computational Plasticity – Fundamentals and Applications*, CIMNE, Barcelona 1997, pp. 1631–1640.
8. Ciarlet, P. G.: *Mathematical Elasticity. Volume I: Three-Dimensional Elasticity*. Elsevier Science, 1988.
9. Diebels, S., Ehlers, W.: Dynamic Analysis of a Fully Saturated Porous Medium Accounting for Geometrical and Material Non-linearities. *Int. J. Numer. Meth. Eng.* **39** (1996), 81–97.
10. Diebels, S., Ehlers, W., Ellsiepen, P., Volk, W.: On the Regularization of Shear Band Phenomena in Liquid-saturated and Empty Soils. In Brillard, A., Ganghoffer, J. F. (eds.): Proc. Euromech Colloquium 378 on Nonlocal Aspects in Solid Mechanics, University of Mulhouse 1998, pp. 58–63.
11. Drucker, D. C., Prager, W.: Soil Mechanics and Plastic Analysis or Limit Design. *Quart. Appl. Math.* **10** (1952), 2, 157–165.
12. Ehlers, W., Eipper, G.: Finite Elastic Deformation in Liquid-Saturated and Empty Porous Solids. *Transport in Porous Media* **34** (1999), 179–191.
13. Gawin, D., Sanavia, L., Scherfler, B. A.: Cavitation Modelling in Saturated Geomaterials with Application to Dynamic Strain Localization. *Int. J. Num. Methods in Fluids* **27** (1998), 109–125.
14. Gray, W. G., Hassanizadeh, M.: Unsaturated Flow Theory including Interfacial Phenomena. *Water Resources Res.* **27** (1991), (8), 1855–1863.
15. Hassanizadeh, M., Gray, W. G.: General Conservation Equations for Multiphase System: 1. Averaging technique. *Adv. Water Res.* **2** (1979), 131–144.
16. Hassanizadeh, M., Gray, W. G.: General Conservation Equations for Multi-Phase System: 2. Mass, Momenta, Energy and Entropy Equations. *Adv. Water Res.* **2** (1979), 191–201.
17. Hassanizadeh, M., Gray, W. G.: General Conservation Equations for Multi-Phase System: 3. Constitutive Theory for Porous Media Flow. *Adv. Water Res.* **3** (1980), 25–40.
18. Jeremić, B., Runesson, K., Sture, S.: Finite Deformation Analysis of Geomaterials. *Int. J. Numer. Anal. Meth. Geomech.* **25** (2001), 741–756.
19. Lee, E. H.: Elastic-Plastic Deformation at Finite Strains. *J. Appl. Mech.* **1**(6) (1969).
20. Lewis, R. W., Schrefler, B. A.:*The Finite Element Method in the Static and Dynamic Deformation and Consolidation of Porous Media.* John Wiley & Sons, Chichester 1998.
21. Liakopoulos, A. C.:*Transient flow through unsaturated porous media.* Ph. D. Thesis, University of California, Berkeley 1965.
22. Marsden, J. E., Hughes, T. J. R.: *Mathematical Foundations of Elasticity.* Prentice-Hall, Englewood Cliffs 1983.
23. Meroi, E., Schrefler, B. A., Zienkiewicz, O. C.: Large Strain Static and Dynamic Semi-Saturated Soil Behaviour. *Int. J. Num. Analytical Methods Geomech.* **19**(2) (1995), 81–106.

24. Miehe, C.: Computation of Isotropic Tensor Functions. *Comm. Num. Meth. Eng.* **9** (1993), 889–896.
25. Mokni, M., Desrues, J.: Strain Localization Measurements in Undrained Plane-strain Biaxial Tests on Hostun RF Sand. *Mech. Cohes-Frict. Mater.* **4** (1998), 419–441.
26. Nemat-Nasser, S.: On Finite Plastic Flow of Crystalline Solids and Geomaterials. *Transactions of ASME.* **50** (1983), 1114–1126.
27. Peters, J. F., Lade, P. V., Bro, A.: Shear Band Formation in Triaxial and Plane Strain tests. In Donaghe, R. T., Chaney, R. C., Silver, M. L. (eds.): *Advanced Triaxial Testing of Soil and Rock*, ASTM STP **977**, Am. Soc. Testing and Materials, Philadelphia 1988, p. 604.
28. Reese, S.: Elastoplastic Material Behaviour with Large Elastic and Large Plastic Deformation. *ZAMM Z. Angew. Math. Mech.* **77** (1997), 277–278.
29. Sanavia, L., Schrefler, B. A., Stein, E., Steinmann, P.: Modelling of localization at finite inelastic strain in fluid saturated porous media. In Ehlers, W. (ed.): *IUTAM Symposium on Theoretical and Numerical Methods in Continuum Mechanics of Porous Materials.* Kluwer, Dordrecht 2001, pp. 239–244.
30. Sanavia, L., Schrefler, B. A., Steinmann, P.: A mathematical and numerical model for finite elastoplastic deformations in fluid saturated porous media. In Capriz, G., Ghionna, V. N. Giovine, P. (eds.): *Modeling and Mechanics of Granular and Porous Materials*, Series of Modeling and Simulation in Science, Engineering and Technology. Birkhäuser, Boston (in print), pp. 297–346. Also UKL-LTM Report J2001-10, University of Kaiserslautern, Germany, Chair of Applied Mechanics.
31. Sanavia, L., Schrefler, B. A., Steinmann, P.: A formulation for an unsaturated porous medium undergoing large inelastic strains. *Computational Mechanics* (in print). Also UKL-LTM Report J2001-12 University of Kaiserslautern, Germany, Chair of Applied Mechanics.
32. Schrefler, B. A., Sanavia, L., Majorana, C. E.: A Multiphase Medium Model for Localization and Postlocalization Simulation in Geomaterials. *Mech. Cohes.-Frict. Mater.* **1** (1996), 95–114.
33. Schrefler, B. A., Simoni, L.: A unified approach to the analysis of saturated-unsaturated elastoplastic porous media. In *Proc. Numerical Methods in Geomechanics*, Balkema, 1988.
34. Simo, J. C., Taylor, R.: Consistent Tangent Operators for Rate-Independent Elastoplasticity. *Comp. Meth. Applied Mech. Eng.* **48** (1985), 101–118.
35. Simo, J. C.: Numerical Analysis and Simulation of Plasticity. In Ciarlet, P. G., Lions, J. L. (eds.): *Numerical Methods for Solids*, Part 3, Vol. 6 of Handbook of Numerical Analysis, North-Holland, 1998.
36. Simo, J. C., Hughes, T. J. R.: *Computational Inelasticity.* Springer-Verlag, 1998.
37. Steinmann, P.: A finite element formulation for Strong Discontinuities in Fluid-Saturated Porous Media. *Mech. Cohes.-Frict. Mater.* **4** (1999), 133–152.
38. Vardoulakis, J., Sulem J.: *Bifurcation Analysis in Geomechanics.* Blakie Academic and Professional, London 1995.
39. Wriggers, P.: Continuum Mechanics, Non-linear Finite Element Techniques and Computational Stability. In Stein, E. (ed.): *Progress in Computational Analysis of Inelastic Structures*, CISM Courses and Lectures No. 321, Springer-Verlag, Wien 1993.

40. Zhang, H. W., Sanavia, L., Schrefler, B. A.: An Internal Length Scale in Strain Localization of Multiphase Porous Media. *Mech. Cohes.-Frict. Mater.* **4** (1999), 433–460.
41. Zhang, H. W., Sanavia, L., Schrefler, B. A.: Numerical analysis of dynamic strain localization in initially water saturated dense sand with a modified generalized plasticity model. *Comp. Struct.* **79** (2000), 441–459.
42. Zienkiewicz, O. C., Chan, A., Pastor, M., Schrefler, B. A., Shiomi T.: *Computational Geomechanics with special Reference to Earthquake Engineering.* John Wiley & Sons, Chichester 1999.

Waves in poroelastic half space: Boundary element analyses

Martin Schanz and Heinz Antes

Technical University Braunschweig, Institute of Applied Mechanics,
38023 Braunschweig, Germany

Abstract. For simulating wave propagation in fluid-saturated semi-infinite porous continua, a boundary element formulation in time domain is presented. Usually, for such an integral formulation, the respective time domain fundamental solution is needed. This solution does only exist for *Biot*'s theory of poroelastic continua and, moreover, it is not given in closed form contrary to its *Laplace* domain representation. The recently developed "Convolution Quadrature Method", proposed by *Lubich*, utilizes this *Laplace* transformed fundamental solution. Hence, applying this quadrature formula to the time dependent boundary integral equation, a time stepping procedure is obtained based only on the *Laplace* domain fundamental solution and a linear multistep method.

Finally, as an application, wave propagation in poroelastic half space is considered. Especially, the *Rayleigh* wave and the *Love* wave is studied. Also, the so-called slow compressional wave, a wave only existing in two-phase materials as poroelastic is, is confirmed by numerical studies.

1 Introduction

For a wide range of fluid infiltrated materials, such as water saturated soils, oil impregnated rocks, or air filled foams, the elastic theory is a crude approximation for investigating wave propagation in such media. Due to their porosity, a different theory is necessary.

A historical view on this subject identifies two theories which have been developed and are used nowadays. For more details, the reader is directed to the work of de Boer, see [9, 10] or his recently published book [8]. The first works on porous media are attributed to Fillunger in 1913 [30]. In this paper and in subsequent ones, Fillunger was concerned with the question of buoyancy of barrages. Another, more intuitive theory has been developed by Terzaghi in 1923 [51]. These two works form the basis of two theories used up to day.

Based on Terzaghi, a theory of porous materials containing a viscous fluid was presented by Biot [3]. This has generally been attributed as the starting point of the theory of poroelasticity. In the following years, Biot extended his theory to anisotropic cases [4] and also to poroviscoelasticity [5]. The dynamic extension was published in two papers, one for low frequency range [6] and the other for high frequency range [7]. One of its significant findings was the identification of three waves for a 3-d continuum, two compressional waves and one shear wave. This extra compressional wave, known as the slow wave, has been experimentally confirmed [40]. In *Biot*'s theory a fully saturated material is assumed. The extension to a nearly saturated poroelastic solid was presented by Vardoulakis and Beskos [53].

Based on the work of Fillunger, a different approach, the Theory of Porous Media, has been developed. This theory is based on the axioms of continuum theories of mixtures [14, 52] extended by the concept of volume fractions by Ehlers [26–28], thus proceeding from the assumption of immiscible and superimposed continua with internal interactions. It has been demonstrated that under small deformations, and some other restrictions, this and *Biot*'s theory lead to the same governing equations [29]. Although *Biot*'s theory is more based on physical intuition, it has the widest acceptance in geophysics and geomechanics.

Beside some one-dimensional analytical solutions, e. g. [11] or [49], in general, a numerical method has to be applied to tackle the behaviour of a poroelastic continuum. Because wave propagation phenomena are often observed in semi-infinite media, e. g. earthquake motion or propagation of machine foundation excitations in the soil and their effect on neighbouring buildings, here, the Boundary Element Method (BEM) is used. A two-dimensional (2-d) quasi-static BE formulation has been developed by Cheng and Ligget for consolidation problems [21] and for fracture [20]. Later, a three-dimensional (3-d) quasi-static formulation was published by Badmus et al. [2]. A complete overview on the different available quasi-static formulations can be found in [19]. Those and also the following BE formulations are based on *Biot*'s theory. For the Theory of Porous Media, no boundary element formulation exists since no fundamental solutions have been found yet.

In case of dynamic BE formulations, no closed form time-dependent fundamental solution is available. Therefore, first poroelastodynamic BE formulations based on *Biot*'s theory have been published in *Laplace* domain by Manolis and Beskos [37] expressed in terms of solid and fluid displacements. However, it can be shown that only the solid displacements and one additional variable, the fluid pressure, are independent [12]. Based on these four (three) unknowns in 3-d (2-d), formulations in frequency domain have been published by Cheng et al. [18] and Domínguez [24]. In these formulations, the transient response of a poroelastic continuum has be determined with an inverse transformation. However, it is more natural to work in the real time domain and observe the phenomenon as it evolves. Such a time domain formulation was developed by Wiebe and Antes [54], but with the restriction of vanishing damping between the solid skeleton and the fluid. Another time dependent formulation was proposed by Chen and Dargush based on analytical inverse transformation of the *Laplace* domain fundamental solutions [17], but, as the author admits, this formulation is highly CPU-time demanding. Recently, the author developed a poroelastic time stepping BE procedure similar to the viscoelastic case based on the convolution quadrature method [43, 45, 46] which is detailed discussed in [44].

In the following, the governing equations based on *Biot*'s constitutive equations for poroelastic media are first recalled. After deriving fundamental solutions, subsequently, a boundary integral formulation is deduced. Introducing spatial discretization and applying the convolution quadrature for time discretization gives a time stepping procedure for poroelastic continua. The two body waves, compression wave and shear wave, as well as the surface and interface waves, *Rayleigh* wave and *Love* wave, will be presented. Finally, the appearance of the slow compressional wave is studied.

2 *Biot*'s theory – governing equations

Following *Biot*'s approach to model the behaviour of porous media, an elastic skeleton with a statistical distribution of interconnected pores is considered [4]. This porosity is denoted by

$$\phi = \frac{V^f}{V}, \tag{1}$$

where V^f is the volume of the interconnected pores contained in a sample of bulk volume V. Contrary to these pores, the sealed pores will be considered as part of the solid. As mentioned above, full saturation is assumed leading to $V = V^f + V^s$ with the volume of the solid V^s, i. e. a two-phase material is given.

One possible representation of poroelastic constitutive equation is obtained using the total stress $\sigma_{ij} = \sigma_{ij}^s + \sigma^f \delta_{ij}$ and the pore pressure p as independent variables [3]. Introducing *Biot*'s effective stress coefficient α and the solid strain ε_{ij}, the constitutive equation reads

$$\sigma_{ij} = 2G\,\varepsilon_{ij} + \left(K - \frac{2}{3}G\right)\varepsilon_{kk}\,\delta_{ij} - \alpha\,\delta_{ij}\,p \tag{2}$$

with the shear modulus and the compression modulus of the solid frame G and K, respectively. In this equation, and in the following, Latin indices take the values 1, 2, 3 in 3-d and 1, 2 in 2-d, where summation convention is implied over repeated indices. Additional to the total stress σ_{ij}, as a second constitutive equation, the variation of fluid volume per unit reference volume ζ is introduced:

$$\zeta = \alpha\,\varepsilon_{kk} + \frac{\phi^2}{R}p\,. \tag{3}$$

This variation ζ of the fluid is defined by the mass balance over a reference volume, i. e. by the continuity equation

$$\frac{\partial \zeta}{\partial t} + q_{i,i} = a \tag{4}$$

with the specific flux $q_i = \phi\, \partial v_i / \partial t$, the relative fluid to solid displacement v_i, and a source term $a\,(t)$. (4) identifies ζ as a kind of strain describing the motion of the fluid relative to the solid which takes a source in the fluid into account. In equation (4) and in the following, the spatial derivative with respect to the coordinate x_i is denoted by $(\,\cdot\,)_{,i}$.

Additional to the fluid balance (4), the balance of momentum for the bulk material must be fulfilled. This dynamic equilibrium is given by

$$\sigma_{ij,j} + F_i = \varrho\, \frac{\partial^2 u_i}{\partial t^2} + \phi\, \varrho_f\, \frac{\partial^2 v_i}{\partial t^2} \tag{5}$$

with the bulk body force per unit volume F_i, the solid displacements u_i, and the bulk density $\varrho = \varrho_s\,(1-\phi) + \phi\,\varrho_f$. The density of the solid and the fluid is denoted by ϱ_s and ϱ_f, respectively. The relation of the solid strain to the solid displacement is chosen linear

$$\varepsilon_{ij} = \frac{1}{2}\,(u_{i,j} + u_{j,i}) \tag{6}$$

assuming small deformation gradients.

Next, the fluid transport in the interstitial space expressed by the specific flux q_i is modelled with a generalized *Darcy*'s law

$$q_i = -\kappa \left(p_{,i} + \varrho_f\, \frac{\partial^2 u_i}{\partial t^2} + \frac{\varrho_a + \phi\,\varrho_f}{\phi}\, \frac{\partial^2 v_i}{\partial t^2} \right), \tag{7}$$

where κ denotes the permeability. In (7), an additional density, the apparent mass density ϱ_a is introduced by Biot [6] to describe the interaction between fluid and skeleton. It can be written as $\varrho_a = C\,\phi\,\varrho_f$, where C is a factor depending on the geometry of the pores and the frequency of excitation. At low frequency, Bonnet and Auriault [13] measured $C = 0.66$ for a sphere assembly of glass bead. In higher frequency ranges, a certain functional dependence of C on frequency has been proposed based on conceptual porosity structures, e. g. in [7] and [13]. In the following, $C = 0.66$ is assumed.

Aiming at the equation of motion, the above balance laws and constitutive equations have to be combined. To do this, first, the degrees of freedom must be determined. Here, there are several possibilities: i) to use the solid displacements u_i and the relative fluid to solid displacement v_i (six unknowns in 3-d) or ii) a combination of the pore pressure p and the solid displacements u_i (four unknowns in 3-d). As shown in [12], it is sufficient to use the latter choice, i. e. the solid displacements u_i and the pore pressure p will become basic variables to describe a poroelastic continuum. Therefore, the above equations are reduced to these four unknowns. First, *Darcy*'s law (7) is rearranged to obtain v_i. Since v_i is given as second time derivative in (7), this is only possible in *Laplace* domain. After transformation to *Laplace* domain, denoted by $\mathscr{L}\{f\,(t)\} = \hat{f}\,(s)$ with the complex *Laplace* variable s, the relative fluid to solid displacement is

$$\hat{v}_i = -\underbrace{\frac{\kappa\,\varrho_f\,\phi^2\,s^2}{\phi^2\,s + s^2\,\kappa\,(\varrho_a + \phi\,\varrho_f)}}_{\beta}\;\frac{1}{s^2\,\phi\,\varrho_f}\left(\hat{p}_{,i} + s^2\,\varrho_f\,\hat{u}_i\right). \tag{8}$$

In (8), the abbreviation β is defined for further usage. Moreover, vanishing initial conditions for u_i and v_i are assumed here and in the following. Now, the final set of differential equations for the displacement $\hat{u}_i$ and the pore pressure $\hat{p}$ is obtained by inserting the constitutive equations (2) and (3) in the *Laplace* transformed dynamic equilibrium (5) and continuity equation (4) with $\hat{v}_i$ from (8). This leads to the final set of differential equations for the displacement $\hat{u}_i$ and the pore pressure $\hat{p}$:

$$G\,\hat{u}_{i,jj} + \left(K + \frac{1}{3}G\right)\hat{u}_{j,ij} - (\alpha-\beta)\,\hat{p}_{,i} - s^2\,(\varrho - \beta\,\varrho_f)\,\hat{u}_i = -\hat{F}_i \tag{9}$$

$$\frac{\beta}{s\,\varrho_f}\hat{p}_{,ii} - \frac{\phi^2\,s}{R}\hat{p} - (\alpha-\beta)\,s\,\hat{u}_{i,i} = -\hat{a}\,, \tag{10}$$

or in operator notation

$$\mathbf{B}\begin{bmatrix}\hat{u}_i\\ \hat{p}\end{bmatrix} = -\begin{bmatrix}\hat{F}_i\\ \hat{a}\end{bmatrix} \quad \text{with the not self-adjoint operator}$$

$$\mathbf{B} = \begin{bmatrix} G\,\nabla^2 + \left(K + \frac{1}{3}G\right)\partial_i\partial_j - s^2\,(\varrho - \beta\,\varrho_f) & -(\alpha-\beta)\,\partial_i \\ -s\,(\alpha-\beta)\,\partial_j & \frac{\beta}{s\,\varrho_f}\nabla^2 - \frac{\phi^2\,s}{R} \end{bmatrix}. \tag{11}$$

This set of equations describes the behaviour of a poroelastic continuum completely. However, an analytical representation in time domain is only possible for $\kappa \to \infty$. This case would represent a negligible friction between solid and interstitial fluid [54].

3 Poroelastic boundary integral formulation

For a poroelastodynamic boundary element formulation, fundamental solutions are necessary, especially, to treat wave propagation problems, time-dependent solutions. In general, these are not available in closed form, except for the special case of an inviscid fluid, $\kappa \to \infty$ [54]. Fortunately, a time-dependent boundary element formulation based on the convolution quadrature method needs only *Laplace* transformed fundamental solutions. These have been presented in the literature for 2-d [15] and 3-d [16] using solid displacements and pore pressure as unknowns. For the other case, solid and fluid displacements as unknowns, the fundamental solutions can be found in [37].

However, there is some possible misunderstanding about the relation which fundamental solution corresponds to which integral formulation, due to the fact that the poroelastodynamic operator is not self-adjoint. It is necessary to recall the derivation of fundamental solutions to gain confidence and resolve inconsistencies in some papers as reported in [23]. In principle, two possibilities exist: i) using the analogy between thermo- and poroelasticity in *Laplace* or *Fourier* domain to convert the thermoelastic solutions to poroelastic ones [24], or ii) the method of Hörmander [33]. The latter will be used here.

3.1 Fundamental solutions

From the mathematical theory of *Green*'s formula it is known that the fundamental solutions should satisfy the adjoint operator equation [50]. Opposite to elasticity, the governing operator in poroelasticity is not self-adjoint. Therefore, here, the solution for the adjoint operator $\mathbf{B}^*$

$$\mathbf{B}^*\,\mathbf{G} + \mathbf{I}\,\delta\,(\mathbf{x}-\mathbf{y}) = \mathbf{0} \tag{12}$$

is required with the *Dirac* distribution $\delta\,(\mathbf{x}-\mathbf{y})$ and the operator matrix

$$\mathbf{B}^* = \begin{bmatrix} A + B\,\partial_{11} & B\,\partial_{12} & B\,\partial_{13} & s\,C\,\partial_1 \\ B\,\partial_{12} & A + B\,\partial_{22} & B\,\partial_{23} & s\,C\,\partial_2 \\ B\,\partial_{13} & B\,\partial_{23} & A + B\,\partial_{33} & s\,C\,\partial_3 \\ C\,\partial_1 & C\,\partial_2 & C\,\partial_3 & D \end{bmatrix},$$

$$A = G\,\nabla^2 - s^2\left(\varrho - \beta\,\varrho_f\right), \qquad B = K + \frac{1}{3}\,G\,,$$

$$C = \alpha - \beta\,, \qquad D = \frac{\beta}{s\,\varrho_f}\nabla^2 - \frac{\phi^2\,s}{R}\,.$$

Above, ∇^2 denotes the *Laplace* operator and ∂_i the spatial derivative with respect to x_i.

Having in mind the definition $\mathbf{B}^{*-1} = \mathbf{B}^{*co}/\det\left(\mathbf{B}^*\right)$ of an inverse matrix to $\mathbf{B}^*$ with the matrix of co-factors $\mathbf{B}^{*co}$, the ansatz

$$\mathbf{G} = \mathbf{B}^{*co}\,\varphi \tag{13}$$

for the matrix of fundamental solutions with a unknown scalar function φ yields a more convenient representation of (12):

$$\begin{aligned} &\mathbf{B}^*\,\mathbf{B}^{*co}\,\varphi + \mathbf{I}\,\delta\,(\mathbf{x}-\mathbf{y}) = \det\left(\mathbf{B}^*\right)\mathbf{I}\,\varphi + \mathbf{I}\,\delta\,(\mathbf{x}-\mathbf{y}) = \mathbf{0} \\ &\quad\rightsquigarrow\ \det\left(\mathbf{B}^*\right)\varphi + \delta\,(\mathbf{x}-\mathbf{y}) = 0\,. \end{aligned} \tag{14}$$

So, first, the unknown function φ has to be determined. For this, the determinant of the operator matrix $\mathbf{B}^*$ is calculated:

$$\det(\mathbf{B}^*) = A^2 \left[(B\,D - C^2\,s)\,\nabla^2 + A\,D \right] = \frac{\beta}{s\,\varrho_f} \left(K + \frac{4}{3} G \right) \cdot \\ \cdot\, G^2 \left(\nabla^2 - \frac{s^2\,(\varrho - \beta\,\varrho_f)}{G} \right)^2 \left[\nabla^4 - (a+b+c)\,\nabla^2 + a\,b \right] \tag{15}$$

with the abbreviations $a = \phi^2\,s^2\,\varrho_f / (\beta\,R)$, $b = s^2\,(\varrho - \beta\,\varrho_f) / (K + 4/3\,G)$, and $c = s^2\,\varrho_f\,(\alpha - \beta)^2 / [\beta\,(K + 4/3\,G)]$. This determinant has obviously three roots: The two resulting from the brackets above

$$\lambda_{1,2}^2 = \frac{1}{2} \left[a + b + c \pm \sqrt{(a+b+c)^2 - 4\,a\,b} \right] \tag{16$_1$}$$

and the double root

$$\lambda_3^2 = \frac{s^2\,(\varrho - \beta\,\varrho_f)}{G}\,. \tag{16$_2$}$$

These three roots correspond to the three expected waves – the fast and slow compressional wave to $\lambda_{1,2}$ and the shear wave to λ_3. Using them yields a representation of the determinant

$$\det(\mathbf{B}^*) = \frac{G^2\,\beta}{s\,\varrho_f} \left(K + \frac{4}{3} G \right) \left(\nabla^2 - \lambda_3 \right) \left(\nabla^2 - \lambda_3 \right) \left(\nabla^2 - \lambda_1 \right) \left(\nabla^2 - \lambda_2 \right). \tag{17}$$

This expression is inserted in (14) to determine φ. With the abbreviation $\psi = G^2\,\beta / (s\,\varrho_f)\,(K + 4/3\,G)\,(\nabla^2 - \lambda_3)\,\varphi$ it becomes obvious that ψ and subsequently φ is found by solving a higher order *Helmholtz* equation:

$$\left(\nabla^2 - \lambda_3 \right) \left(\nabla^2 - \lambda_1 \right) \left(\nabla^2 - \lambda_2 \right) \psi + \delta(\mathbf{x} - \mathbf{y}) = 0\,. \tag{18}$$

The solution for either 2-d and 3-d is given in [22]. Here, only the 3-d solution

$$\psi = \frac{1}{4\pi r} \left[\frac{\mathrm{e}^{-\lambda_1 r}}{(\lambda_1^2 - \lambda_2^2)(\lambda_1^2 - \lambda_3^2)} + \frac{\mathrm{e}^{-\lambda_2 r}}{(\lambda_2^2 - \lambda_1^2)(\lambda_2^2 - \lambda_3^2)} + \right. \\ \left. + \frac{\mathrm{e}^{-\lambda_3 r}}{(\lambda_3^2 - \lambda_1^2)(\lambda_3^2 - \lambda_2^2)} \right] \tag{19}$$

is presented.

Now, following (13), the definition of ψ yields φ and applying the matrix of differential operators $\mathbf{B}^{*co}$ on φ as given in (13), the fundamental solutions of poroelastodynamics are obtained:

$$\mathbf{G} = \begin{bmatrix} \hat{U}^s_{ij} & \hat{U}^f_i \\ \hat{P}^s_j & \hat{P}^f \end{bmatrix} = \frac{s\,\varrho_f}{G\beta\left(K+\frac{4}{3}G\right)} \begin{bmatrix} \left(F\,\nabla^2 + A\,D\right)\delta_{ij} - F\,\partial_{ij} & -A\,C\,s\,\partial_i \\ -A\,C\,\partial_i & A\,\left(B\,\nabla^2 + A\right) \end{bmatrix} \psi\,. \tag{20}$$

The explicit expressions of the elements of $\mathbf{G}$ are listed in Appendix A.

3.2 Boundary integral equation

The boundary integral equation for dynamic poroelasticity in *Laplace* domain can be obtained using either the corresponding reciprocal work theorem [18] or the weighted residuals formulation [24]. A comparison of both technologies in the case of poroelasticity is presented in [44], whereas, here, the latter is used.

The poroelastodynamic integral equation can be derived directly by equating the inner product of (9) and (10), written in matrix form with matrix $\mathbf{B}$ defined in (11), and the matrix of the fundamental solutions $\mathbf{G}$ to a null vector, i. e.

$$\int_\Omega \mathbf{G}^T\,\mathbf{B}\begin{bmatrix} \hat{u}_i \\ \hat{p} \end{bmatrix} \mathrm{d}\Omega = \mathbf{0} \qquad \text{with} \qquad \mathbf{G} = \begin{bmatrix} \hat{U}^s_{ij} & \hat{U}^f_i \\ \hat{P}^s_j & \hat{P}^f \end{bmatrix}, \tag{21}$$

where the integration is performed over a domain Ω with boundary Γ and vanishing body forces F_i and sources a are assumed. By this inner product, essentially, the error in satisfying the governing differential equations (9) and (10) is forced to be orthogonal to $\mathbf{G}$. According to the theory of *Green*'s formula and using partial integration the operator $\mathbf{B}$ is transformed from acting on the vector of unknowns $[\hat{u}_i\,\hat{p}]^T$ to the matrix of fundamental solutions $\mathbf{G}$. These steps are easier understood looking at (21) written in index notation. This results in three (two) integral equations for the solid ($j = 1, 2, 3$ in 3-d and $j = 1, 2$ in 2-d):

$$\begin{aligned} \int_\Omega \Bigg[& G\,\hat{u}_{i,kk}\,\hat{U}^s_{ij} + \left(K + \frac{1}{3}G\right)\hat{u}_{k,ik}\,\hat{U}^s_{ij} - (\alpha-\beta)\,\hat{p}_{,i}\,\hat{U}^s_{ij} - \\ & - s^2\left(\varrho - \beta\,\varrho_f\right)\hat{u}_i\,\hat{U}^s_{ij} + \frac{\beta}{s\,\varrho_f}\,\hat{p}_{,kk}\,\hat{P}^s_j - \frac{\phi^2 s}{R}\,\hat{p}\,\hat{P}^s_j - \\ & - (\alpha-\beta)\,s\,\hat{u}_{k,k}\,\hat{P}^s_j \Bigg] \mathrm{d}\Omega = 0 \end{aligned} \tag{22}$$

and one integral equation for the fluid

$$\int_{\Omega} \Big[G\,\hat{u}_{i,kk}\,\hat{U}_i^f + \Big(K + \frac{1}{3}G\Big)\,\hat{u}_{k,ik}\,\hat{U}_i^f - (\alpha - \beta)\,\hat{p}_{,i}\,\hat{U}_i^f - \\ - s^2\,(\varrho - \beta\,\varrho_f)\,\hat{u}_i\,\hat{U}_i^f + \frac{\beta}{s\,\varrho_f}\,\hat{p}_{,kk}\,\hat{P}^f - \frac{\phi^2\,s}{R}\,\hat{p}\,\hat{P}^f - \\ - (\alpha - \beta)\,s\,\hat{u}_{k,k}\,\hat{P}^f \Big]\,\mathrm{d}\Omega = 0\,. \tag{23}$$

In the above integral equations, either one or two differentiations have to be transformed by either one or two partial integrations. Two exemplary parts of integral equations (22) and (23) are presented in detail to show the principal procedure. All other partial integrations for the other parts in integral equations (22) and (23) can be performed analogously.

First, an integral with one differentiation in the kernel leads to (n_k is the outward normal vector)

$$\int_{\Omega} (\alpha - \beta)\,s\,\hat{u}_{k,k}\,\hat{P}^f\,\mathrm{d}\Omega = \\ = \int_{\Gamma} (\alpha - \beta)\,s\,\hat{u}_k\,n_k\,\hat{P}^f\,\mathrm{d}\Gamma - \int_{\Omega} (\alpha - \beta)\,s\,\hat{u}_k\,\hat{P}^f_{,k}\,\mathrm{d}\Omega \tag{24}$$

while an integral with two differentiation is transformed to

$$\int_{\Omega} G\,\hat{u}_{i,kk}\,\hat{U}_{ij}^s\,\mathrm{d}\Omega = \int_{\Gamma} G\,\hat{u}_{i,k}\,n_k\,\hat{U}_{ij}^s\,\mathrm{d}\Gamma - \int_{\Omega} G\,\hat{u}_{i,k}\,\hat{U}_{ij,k}^s\,\mathrm{d}\Omega = \\ = \int_{\Gamma} G\,\hat{u}_{i,k}\,n_k\,\hat{U}_{ij}^s\,\mathrm{d}\Gamma - \int_{\Gamma} G\,\hat{u}_i\hat{U}_{ij,k}^s\,n_k\,\mathrm{d}\Gamma + \int_{\Omega} G\,\hat{u}_i\hat{U}_{ij,kk}^s\,\mathrm{d}\Omega\,. \tag{25}$$

In both integrations by parts the divergence theorem is used. Obviously, one integration by parts changes the sign of the resulting domain integral while it remains unchanged in the case of two integration by parts, i. e. the operator **B** is transformed into its adjoint operator **B***. This yields the following system of integral equations given in matrix notation as

$$\int_{\Gamma} \begin{bmatrix} \hat{U}_{ij}^s & -\hat{P}_j^s \\ \hat{U}_i^f & -\hat{P}^f \end{bmatrix} \begin{bmatrix} \hat{t}_i \\ \hat{q} \end{bmatrix} \mathrm{d}\Gamma - \int_{\Gamma} \begin{bmatrix} \hat{T}_{ij}^s & \hat{Q}_j^s \\ \hat{T}_i^f & \hat{Q}^f \end{bmatrix} \begin{bmatrix} \hat{u}_i \\ \hat{p} \end{bmatrix} \mathrm{d}\Gamma = \begin{bmatrix} \hat{u}_j \\ \hat{p} \end{bmatrix}, \quad \mathbf{y} \in \Omega\,. \tag{26}$$

To obtain (26) as well as the partial integration the definition of fundamental solutions (12) and the property of the *Dirac* distribution is used. Additionally, the flux $\hat{q} = -\beta/(s\,\varrho_f)\,\big(\hat{p}_{,i} + \varrho_f\,s^2\,\hat{u}_i\big)\,n_i$ and the traction vector $\hat{t}_i = \hat{\sigma}_{ij}\,n_j$

is introduced, and the abbreviations

$$\begin{aligned}
\hat{T}^s_{ij} &= \left\{\left[\left(K - \frac{2}{3}G\right)\hat{U}^s_{kj,k} + \alpha\, s\, \hat{P}^s_j\right]\delta_{i\ell} + G\left(\hat{U}^s_{ij,\ell} + \hat{U}^s_{\ell j,i}\right)\right\} n_\ell\,, \\
\hat{Q}^s_j &= \frac{\beta}{s\,\varrho_f}\left(\hat{P}^s_{j,i} - \varrho_f\, s\, \hat{U}^s_{ji}\right) n_i\,, \\
\hat{T}^f_i &= \left\{\left[\left(K - \frac{2}{3}G\right)\hat{U}^f_{k,k} + \alpha\, s\, \hat{P}^f\right]\delta_{i\ell} + G\left(\hat{U}^f_{i,\ell} + \hat{U}^f_{\ell,i}\right)\right\} n_\ell\,, \\
\hat{Q}^f &= \frac{\beta}{s\,\varrho_f}\left(\hat{P}^f_{,j} - \varrho_f\, s\, \hat{U}^f_j\right) n_j
\end{aligned} \tag{27}$$

are used, where $(27)_1$ and $(27)_2$ can be interpreted as being the adjoint term to the traction vector $\hat{t}_i$ and the flux $\hat{q}$, respectively. With the fundamental solutions calculated in Section 3.1 or the explicit form given in Appendix A, the integral representation is completely given.

When moving $\mathbf{y}$ to the boundary Γ to determine the unknown boundary data, it is necessary to know the behaviour of the fundamental solutions when $r = |\mathbf{y} - \mathbf{x}|$ tends to zero, i. e. when an integration point $\mathbf{x}$ approaches a collocation point $\mathbf{y}$. Six of the eight fundamental solutions, four in $\mathbf{G}$ and four calculated by (27), are singular. The order of their singularity can be determined by series representations [44]. Beside the regular fundamental solutions $\hat{P}^s_i$ and $\hat{U}^f_i$, the fundamental solutions $\hat{U}^s_{ij}$, $\hat{P}^f$, $\hat{Q}^s_j$, and $\hat{T}^f_i$ are weakly singular, and the fundamental solutions $\hat{T}^s_{ij}$ and $\hat{Q}^f$ are strongly singular. The strongly singular parts in the kernel functions of $\hat{T}^s_{ij}$ and $\hat{Q}^f$ are known from elastostatics and acoustics, respectively. Therefore, shifting in (26) load point $\mathbf{y}$ to the boundary Γ results in the boundary integral equation

$$\int\limits_\Gamma \begin{bmatrix} \hat{U}^s_{ij} & -\hat{P}^s_j \\ \hat{U}^f_i & -\hat{P}^f \end{bmatrix} \begin{bmatrix} \hat{t}_i \\ \hat{q} \end{bmatrix} \mathrm{d}\Gamma = \oint\limits_\Gamma \begin{bmatrix} \hat{T}^s_{ij} & \hat{Q}^s_j \\ \hat{T}^f_i & \hat{Q}^f \end{bmatrix} \begin{bmatrix} \hat{u}_i \\ \hat{p} \end{bmatrix} \mathrm{d}\Gamma + \begin{bmatrix} c_{ij} & 0 \\ 0 & c \end{bmatrix} \begin{bmatrix} \hat{u}_i \\ \hat{p} \end{bmatrix} \tag{28}$$

with the integral free terms c_{ij} and c known from elastostatics and acoustics, respectively. A transformation to time domain gives, finally, the time-dependent integral equation for poroelasticity

$$\int\limits_0^t \int\limits_\Gamma \begin{bmatrix} U^s_{ij}(t-\tau,\mathbf{y},\mathbf{x}) & -P^s_j(t-\tau,\mathbf{y},\mathbf{x}) \\ U^f_i(t-\tau,\mathbf{y},\mathbf{x}) & -P^f(t-\tau,\mathbf{y},\mathbf{x}) \end{bmatrix} \begin{bmatrix} t_i(\tau,\mathbf{x}) \\ q(\tau,\mathbf{x}) \end{bmatrix} \mathrm{d}\Gamma\, \mathrm{d}\tau = \tag{29}$$

$$= \int_0^t \oint_\Gamma \begin{bmatrix} T_{ij}^s(t-\tau,\mathbf{y},\mathbf{x}) & Q_j^s(t-\tau,\mathbf{y},\mathbf{x}) \\ T_i^f(t-\tau,\mathbf{y},\mathbf{x}) & Q^f(t-\tau,\mathbf{y},\mathbf{x}) \end{bmatrix} \begin{bmatrix} u_i(\tau,\mathbf{x}) \\ p(\tau,\mathbf{x}) \end{bmatrix} \mathrm{d}\Gamma\,\mathrm{d}\tau +$$

$$+ \begin{bmatrix} c_{ij}(\mathbf{y}) & 0 \\ 0 & c(\mathbf{y}) \end{bmatrix} \begin{bmatrix} u_i(t,\mathbf{y}) \\ p(t,\mathbf{y}) \end{bmatrix}.$$

3.3 Boundary element formulation

A boundary element formulation is achieved following the usual procedure. First, the boundary surface Γ is discretized by E iso-parametric elements Γ_e, where F polynomial shape functions $N_e^f(\mathbf{x})$ are defined. Hence, the following ansatz functions with the time-dependent nodal values $u_i^{ef}(t)$, $t_i^{ef}(t)$, $p^{ef}(t)$, and $q^{ef}(t)$ are used to approximate the boundary states

$$\begin{aligned} u_i(\mathbf{x},t) &= \sum_{e=1}^{E}\sum_{f=1}^{F} N_e^f(\mathbf{x})\, u_i^{ef}(t), \quad & t_i(\mathbf{x},t) &= \sum_{e=1}^{E}\sum_{f=1}^{F} N_e^f(\mathbf{x})\, t_i^{ef}(t), \\ p(\mathbf{x},t) &= \sum_{e=1}^{E}\sum_{f=1}^{F} N_e^f(\mathbf{x})\, p^{ef}(t), \quad & q(\mathbf{x},t) &= \sum_{e=1}^{E}\sum_{f=1}^{F} N_e^f(\mathbf{x})\, q^{ef}(t). \end{aligned} \tag{30}$$

In (30), the shape functions of all four variables are denoted by the same function $N_e^f(\mathbf{x})$ indicating the same approximation level for all variables. This is not mandatory but usual. Inserting these ansatz functions (30) in the time dependent integral equation (29) yields

$$\begin{bmatrix} c_{ij}(\mathbf{y}) & 0 \\ 0 & c(\mathbf{y}) \end{bmatrix} \begin{bmatrix} u_i(\mathbf{y},t) \\ p(\mathbf{y},t) \end{bmatrix} =$$

$$= \sum_{e=1}^{E}\sum_{f=1}^{F} \left\{ \int_\Gamma \begin{bmatrix} U_{ij}^s(r,t) & -P_j^s(r,t) \\ U_i^f(r,t) & -P^f(r,t) \end{bmatrix} N_e^f(\mathbf{x})\,\mathrm{d}\Gamma * \begin{bmatrix} t_i^{ef}(t) \\ q^{ef}(t) \end{bmatrix} - \right. \tag{31}$$

$$\left. - \oint_\Gamma \begin{bmatrix} T_{ij}^s(r,t) & Q_j^s(r,t) \\ T_i^f(r,t) & Q^f(r,t) \end{bmatrix} N_e^f(\mathbf{x})\,\mathrm{d}\Gamma * \begin{bmatrix} u_i^{ef}(t) \\ p^{ef}(t) \end{bmatrix} \right\}.$$

Next, a time discretization has to be introduced. Since no time-dependent fundamental solutions are known, the convolution quadrature method (briefly summarized in Appendix B) is the most effective method compared to the possibility inverting the *Laplace* domain fundamental solutions at every collocation point in every time step using a series expansion [17].

Hence, after dividing time period t in N time steps of equal duration Δt, so that $t = N\,\Delta t$, the convolution integrals between the fundamental solutions and the nodal values in (31) are approximated by the convolution quadrature method, i. e. the quadrature formula (45) is applied to the integral equation (31). This results in the following boundary element time stepping formulation (n = 0, 1, ..., N):

$$\begin{bmatrix} c_{ij}(\mathbf{y}) & 0 \\ 0 & c(\mathbf{y}) \end{bmatrix} \begin{bmatrix} u_i(\mathbf{y}, n\,\Delta t) \\ p(\mathbf{y}, n\,\Delta t) \end{bmatrix} =$$

$$= \sum_{e=1}^{E} \sum_{f=1}^{F} \sum_{k=0}^{n} \left\{ \begin{bmatrix} \omega_{n-k}^{ef}(\hat{U}_{ij}^{s}, \mathbf{y}, \Delta t) & -\omega_{n-k}^{ef}(\hat{P}_{j}^{s}, \mathbf{y}, \Delta t) \\ \omega_{n-k}^{ef}(\hat{U}_{i}^{f}, \mathbf{y}, \Delta t) & -\omega_{n-k}^{ef}(\hat{P}^{f}, \mathbf{y}, \Delta t) \end{bmatrix} \begin{bmatrix} t_i^{ef}(k\,\Delta t) \\ q^{ef}(k\,\Delta t) \end{bmatrix} - \right.$$

$$\left. - \begin{bmatrix} \omega_{n-k}^{ef}(\hat{T}_{ij}^{s}, \mathbf{y}, \Delta t) & \omega_{n-k}^{ef}(\hat{Q}_{j}^{s}, \mathbf{y}, \Delta t) \\ \omega_{n-k}^{ef}(\hat{T}_{i}^{f}, \mathbf{y}, \Delta t) & \omega_{n-k}^{ef}(\hat{Q}^{f}, \mathbf{y}, \Delta t) \end{bmatrix} \begin{bmatrix} u_i^{ef}(k\,\Delta t) \\ p^{ef}(k\,\Delta t) \end{bmatrix} \right\} \tag{32}$$

with the integration weights corresponding to (47), e. g.

$$\omega_n^{ef}\left(\hat{U}_{ij}^{s}, \mathbf{y}, \Delta t\right) = \frac{\mathscr{R}^{-n}}{L} \sum_{\ell=0}^{L-1} \int_{\Gamma} \hat{U}_{ij}^{s}\left(\mathbf{x}, \mathbf{y}, \frac{\gamma\left(\mathscr{R}e^{i\ell\frac{2\pi}{L}}\right)}{\Delta t}\right) N_e^f(\mathbf{x})\, \mathrm{d}\Gamma\, \mathrm{e}^{-in\ell\frac{2\pi}{L}}. \tag{33}$$

Note, the calculation of the integration weights is only based on the *Laplace* transformed fundamental solutions. Therefore, with this time stepping procedure (32), a boundary element formulation for poroelastodynamics is given without time-dependent fundamental solutions.

To calculate the integration weights ω_{n-k}^{ef} in (32), spatial integration over the boundary Γ has to be performed. Because the essential constituents of the *Laplace* transformed fundamental solutions are exponential functions, i. e. the integrand is smooth, the regular integrals are evaluated by standard *Gaußian* quadrature rule. The weakly singular parts of the integrals in (32) are regularized by polar coordinate transformation. The strongly singular integrals in (32) are equal to those of elastostatics or acoustics, respectively, and, hence, the regularization methods known from these theories can be applied, e. g. the method suggested by Guiggiani and Gigante [32]. Moreover, to obtain for (32) a system of algebraic equations, collocation is used at every node of the shape functions $N_e^f(\mathbf{x})$.

According to $t - \tau = (n-k)\,\Delta t$, the integration weights ω_{n-k}^{ef} are only dependent on the difference $n-k$. This property is analogous to elastodynamic time domain BE formulations (see, e. g. [25]) and can be used to establish a recursion formula for $n = 1, 2, \ldots, \mathrm{N}$ $(m = n - k)$:

$$\omega_0 (\mathbf{C}) \, \mathbf{d}^n = \omega_0 (\mathbf{D}) \, \bar{\mathbf{d}}^n + \sum_{m=1}^{n} \left(\omega_m (\mathbf{U}) \, \mathbf{t}^{n-m} - \omega_m (\mathbf{T}) \, \mathbf{u}^{n-m} \right) \tag{34}$$

with the time dependent integration weights ω_m containing the *Laplace* transformed fundamental solutions **U** and **T**, respectively (see (33)). Similarly, $\omega_0 (\mathbf{C})$ and $\omega_0 (\mathbf{D})$ are the corresponding integration weights of the first time step related to the unknown boundary data $\mathbf{d}^n$ and the known boundary data $\bar{\mathbf{d}}^n$ in time step n, respectively. Finally, a direct equation solver is applied.

4 Wave propagation in poroelastic solids

As in an elastic continuum, in a poroelastic continuum two body waves and in presence of a free surface, i. e. a semi-infinite half space, also the *Rayleigh* wave are observed. Further, in a layered half space the *Love* wave is present [31]. Additionally, to these waves in a poroelastic continuum a second so-called slow compressional wave exists. The behaviour of these waves will be analyzed next. Studies concerning spatial and temporal discretization regarding the presented method can be found in [44]. First, some remarks on the adequate numerical handling have to be added.

4.1 Numerical preliminaries

In all calculations, the underlying multistep method $\gamma(z)$ is a BDF 2 and $L = N$ is chosen due to the experiences in the comparable visco- and elastodynamic formulations [44]. Further, numerical stable solutions are only achieved if dimensionless variables are introduced. As suggested in [17], the dimensionless spatial and temporal variables are

$$x_i \to \frac{x_i}{\varrho \kappa V}, \quad t \to \frac{t}{\varrho \kappa} \qquad \text{with} \qquad V = \sqrt{\frac{K + \frac{4}{3} G + \alpha^2 \frac{R}{\phi^2}}{\varrho}}, \tag{35}$$

where the velocity V is the compression wave velocity of a poroelastic solid with an inviscid interstitial fluid. These non-dimensional variables are connected with dimensionless material parameters

$$\begin{aligned} K &\to \frac{K}{K + \frac{4}{3} G + \alpha^2 \frac{R}{\phi^2}}, & G &\to \frac{G}{K + \frac{4}{3} G + \alpha^2 \frac{R}{\phi^2}}, & R &\to \frac{R}{K + \frac{4}{3} G + \alpha^2 \frac{R}{\phi^2}}, \\ \kappa &\to \kappa = 1, & \varrho &\to \frac{\varrho}{\varrho} = 1, & \varrho_f &\to \frac{\varrho_f}{\varrho}. \end{aligned} \tag{36}$$

The condition number of the equation system confirms the necessity of dimensionless variables. It has an order of 10^9 using the dimensionless variables instead of 10^{22} else.

4.2 *Rayleigh* surface wave

Dealing with wave propagation in a half space, surface waves are one of the most interesting effects. Especially, the *Rayleigh* wave is of interest due to its disastrous consequence in earthquakes. This surface wave caused by wave reflections at the free surface was first investigated by Rayleigh [41], who has shown that its effect decreases rapidly with depth, and its velocity of propagation is smaller than that of a body wave, but contains most of the energy [31]. According to an analytical solution of the vertical displacement at the half space surface presented by Pekeris [39], a pole arise at the arrival time of the *Rayleigh* wave. To capture this pole for a poroelastic half space, the proposed boundary element formulation will be used. Since in the boundary element formulation applied here, a full space fundamental solution is used, the free surface has to be discretized. In the considered example (see Figure 1), a long strip (6 × 30 m) is discretized with 396 triangular linear elements on 242 nodes. The used time step size is $\Delta t = 0.0006$ s. The modelled half space is loaded on area A (1 m^2) by a vertical total stress vector $t_z = -1000$ N/m^2 $H(t)$ (shaded area in Figure 1) and the remaining surface is traction free. The pore pressure is assumed to be zero all over the surface, i. e. the surface is permeable. The material properties (see Table 1) are those of a soil (coarse sand) taken from measurements presented in [34].

	K $\left[\frac{\mathrm{N}}{\mathrm{m}^2}\right]$	G $\left[\frac{\mathrm{N}}{\mathrm{m}^2}\right]$	ϱ $\left[\frac{\mathrm{kg}}{\mathrm{m}^3}\right]$	ϕ	K_s $\left[\frac{\mathrm{N}}{\mathrm{m}^2}\right]$	ϱ_f $\left[\frac{\mathrm{kg}}{\mathrm{m}^3}\right]$	K_f $\left[\frac{\mathrm{N}}{\mathrm{m}^2}\right]$	κ $\left[\frac{\mathrm{m}^4}{\mathrm{N\,s}}\right]$
rock	$8 \cdot 10^9$	$6 \cdot 10^9$	2458	0.19	$3.6 \cdot 10^{10}$	1000	$3.3 \cdot 10^9$	$1.9 \cdot 10^{-10}$
soil	$2.1 \cdot 10^8$	$9.8 \cdot 10^7$	1884	0.48	$1.1 \cdot 10^{10}$	1000	$3.3 \cdot 10^9$	$3.55 \cdot 10^{-9}$

Table 1. Material data of *Berea* sandstone and a soil.

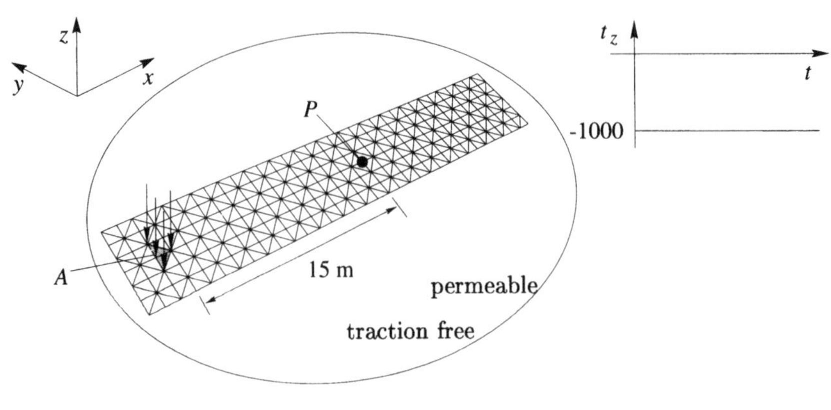

Fig. 1. Half space under vertical load: Discretization and load history.

In the following, the wave propagation is studied not only for a poroelastic modelled soil but also for an elastic modelling. The two considered elastic material models, drained and undrained, are both listed in Table 2 using the same shear modulus as the poroelastic material but either drained *Poisson*'s ratio or undrained *Poisson*'s ratio. These two elastic media represent a possibility to model a water saturated soil as a one-phase material, where the undrained case assumes a strong influence of the interstitial fluid and the drained case less influence. Therefore, results for the two elastic media should give an upper and a lower bound for the poroelastic soil.

	E $\left[\frac{\text{N}}{\text{m}^2}\right]$	ν	$\varrho\left[\frac{\text{kg}}{\text{m}^3}\right]$
rock drained	$1.4 \cdot 10^{10}$	0.2	2458
rock undrained	$1.6 \cdot 10^{10}$	0.33	2458
soil drained	$2.5 \cdot 10^{8}$	0.298	1884
soil undrained	$2.9 \cdot 10^{8}$	0.49	1884

Table 2. Material data of *Berea* sandstone (rock) and a soil (coarse sand) modelled elastic.

The presented results can better be understood when the wave velocities of the above materials are known to identify the arrival time of the different waves at point P (see the mesh in Figure 1). In Table 3, the compression wave velocity c_1, the shear wave velocity c_2, and the *Rayleigh* wave velocity c_R are given together with their arrival time, where the *Rayleigh* wave velocity is approximately determined by $c_R = (0.87 + 1.12\,\nu) \,/\, (1+\nu)\; c_2$ [31]. For the poroelastic modelled soil, no wave velocities can be given due to their strong time dependence, since no simple approximation is available [1].

	c_1 $\left[\frac{\text{m}}{\text{s}}\right]$	c_2 $\left[\frac{\text{m}}{\text{s}}\right]$	c_R $\left[\frac{\text{m}}{\text{s}}\right]$	t_1 [s]	t_2 [s]	t_R [s]
drained	425	228	211	0.035	0.066	0.071
undrained	1629	228	217	0.009	0.066	0.069

Table 3. Wave velocities and corresponding arrival times for the soil.

In Figure 2, the time history of the vertical and horizontal displacement component at point P is depicted versus time for the poroelastic soil and both elastic modelled soils. Since waves in poroelastic media are dispersive, i. e.

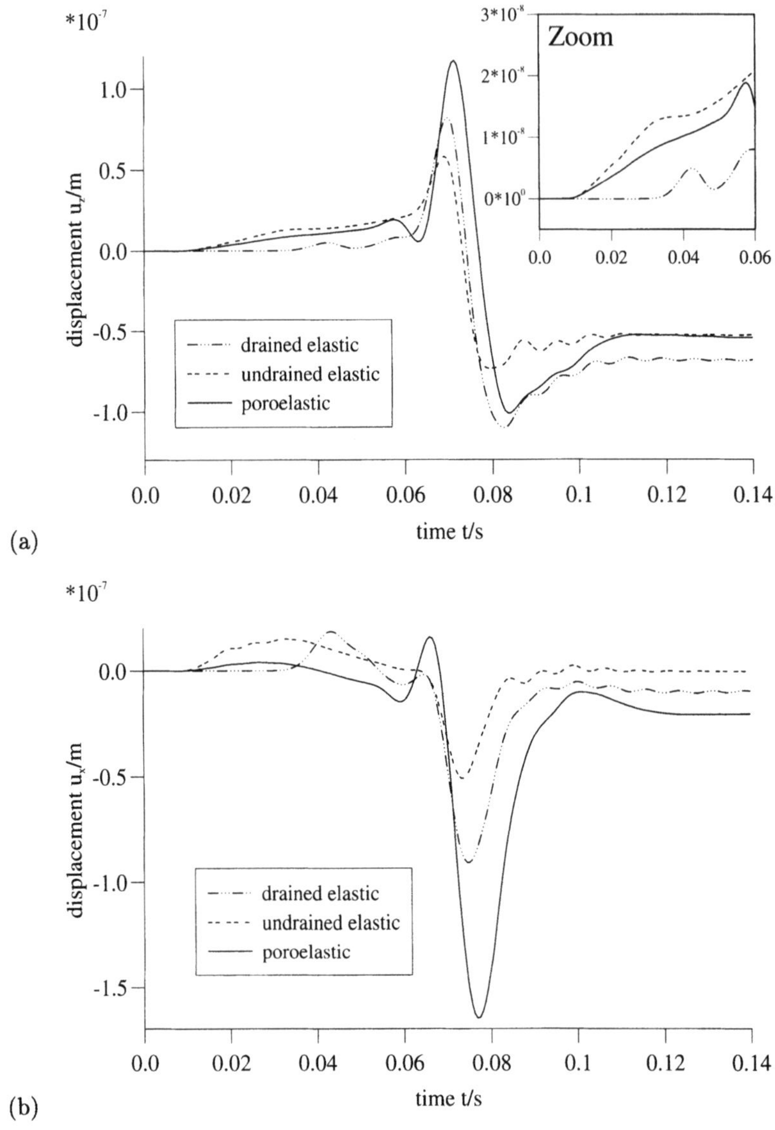

Fig. 2. Vertical and horizontal displacement at point P versus time. Comparison poroelastic and elastodynamic modelling of the soil: (a) vertical, (b) horizontal.

the wave velocities are time-dependent, the waves are already damped when arriving at point P. This may be the reason for the modest increase of the displacement components u_z and u_x at the compression wave arrival ($t \approx 0.01$ s). The arrival time is that of the compression wave of the undrained elastic medium which coincides with the value given in Table 3. The compression wave of the drained elastic medium arrives later ($t \approx 0.035$ s).

The shear wave of both elastic media arrives at the same time since both materials have the same shear modulus. The dissipation of the poroelastic material is expressed not only in the modest increase at the wave front but also in the small absolute displacement values before the shear wave arrives. Even before this arrival ($0.05 < t < 0.06$ s), negative values are visible contrary to the elastic medium, where the negative displacement values are found at $t > 0.06$ s. Summarizing, the compression wave in the poroelastic medium has less influence than in an elastic medium.

Contrary, the *Rayleigh* pole in the vertical component and in the horizontal component of the poroelastic displacement solution is more pronounced than in the elastic cases. The displacement values are nearly twice that of the elastic values. After the arrival of the *Rayleigh* wave ($t > 0.09$ s), the vertical poroelastic displacement component lies in between the elastic solutions, i. e. they are an upper and a lower bound, whereas the horizontal poroelastic component is larger than both corresponding elastic values, i. e. the two elastic cases are no upper and lower bound for the horizontal component.

As mentioned at the beginning of this section, the effect of the *Rayleigh* wave decreases with depth. To visualize this physical property, the displacements at several points under the surface are considered. The observation points are put on a circle with radius of 15 m measured from the excitation point in area A. This ensures that the body waves, i. e. the compression wave and the shear wave, arrive at all points at the same time. The geometry is shown in the upper left of Figure 3. In the same figure, the time history of the vertical displacement is mapped while in Figure 4 the pressure is presented.

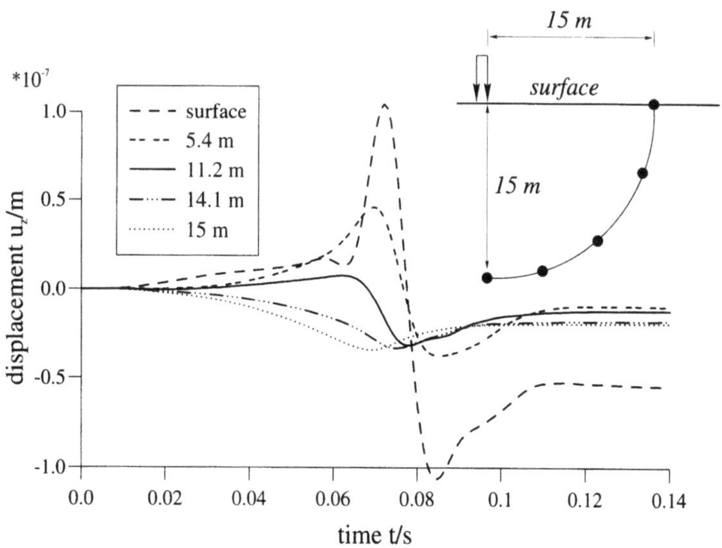

Fig. 3. Vertical displacement at points below the surface: Decrease of the effects of the *Rayleigh* wave with depth.

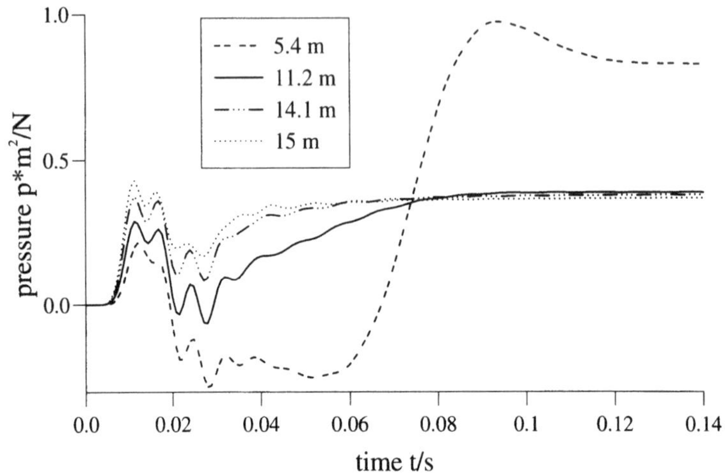

Fig. 4. Pore pressure at points below the surface: Decrease of the effects of the *Rayleigh* wave with depth.

First, the displacement solution (Figure 3) is discussed. There, the decrease of the *Rayleigh* wave influence with depth is observed: at 14 m, all effects are disappeared, whereas at depths until 5.4 m the *Rayleigh* pole is found. For larger times $t > 0.1$ s, the displacement amplitudes vary with depth only slightly but the surface solution is much larger than the other values. This is partly caused by the *Rayleigh* wave but also and probably much stronger influenced by the geometrical damping.

The pressure solution (Figure 4) has a special interesting effect. In the solutions for the depths of 11.2, 14.1, and 15 m, at $t \approx 0.01$ s the compression wave produces a pressure jump with a subsequent creeping to the final value. Contrary, for the depth 5.4 m, the pressure solution has negative values after the jump. At this time, a lifting of the surface (i. e. positive displacement values, see Figure 3) due to the compression wave causes negative pressure in the surrounding because water is sucked in. But, this is only a local effect and for larger depth not visible.

As the fluid itself is modelled without shear components there exists no shear wave, and, consequently, no *Rayleigh* wave. However, the pressure solution has a jump at $t \approx 0.07$ s in the depth of 5.4 m. This must be an effect of the *Rayleigh* wave because this effect diminishes with depth and is exactly at the arrival time of the *Rayleigh* wave. This shows that the *Rayleigh* wave in the solid skeleton induces a compression wave in the fluid.

4.3 Interface waves: The *Love* wave

Next, a layered half space with an upper layer of soil resting on a rock (*Berea* sandstone) foundation is considered. Material data of both are given in Table 1. The surface and the interface is discretized on a strip (6 × 15 m) with 180 linear triangular elements on 112 nodes, respectively. The geometry, loading history, and discretization is sketched in Figure 5. The layer depth is first taken to be 10 m. Further, the modelled layered half space is loaded by a vertical total stress vector $t_z = -1000\ \mathrm{N/m^2}\ H(t)$. The remaining surface is assumed to be traction free and permeable, i. e. the pore pressure is assumed to vanish. The two domains, the layer and the semi-infinite foundation, are coupled using the equilibrium of traction and flux as well as the compatibility of displacements and pressure.

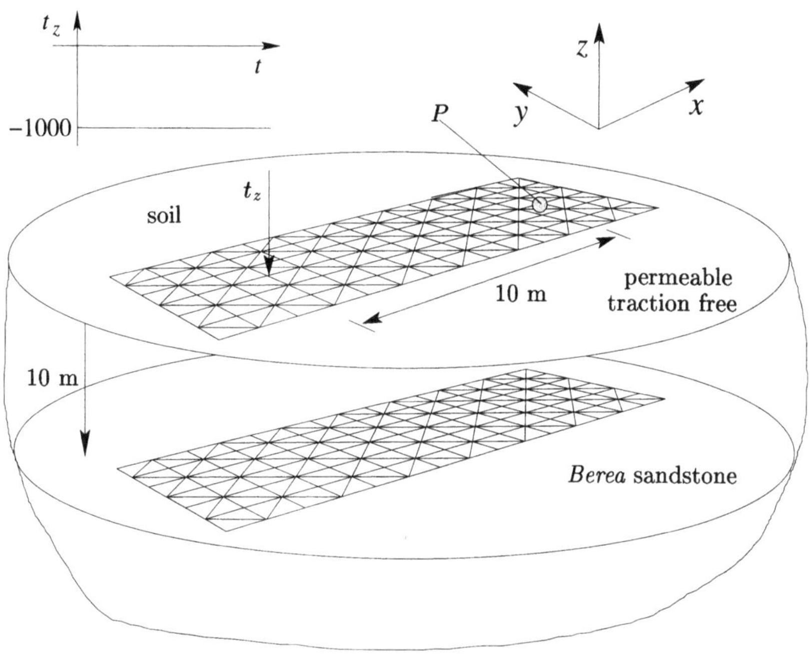

Fig. 5. Layered half space under vertical load: Discretization and load history.

As for the above *Rayleigh* wave analysis, three different modellings of the layered half space are considered: poroelastic, undrained elastic, and drained elastic. To gain confidence, the displacement at point P for all three cases is compared as depicted versus time in Figure 6. Clearly, in the short time range $t < 0.06$ s the same behaviour as in the *Rayleigh* wave study is observed. The zoom shows in detail the arrival of the compression wave ($t \approx 0.006$ s) followed by the shear/*Rayleigh* wave ($t \approx 0.04$ s). The arrival time of the compression wave for the drained elastic case is later ($t \approx 0.02$ s). This must be equal to the case discussed above with the semi-infinite half space because no reflection from the layer is possible in this time range.

Then, at $t > 0.06$ s, a disturbance becomes visible which coincides with the time which the compression wave takes to propagate to the layer and, after

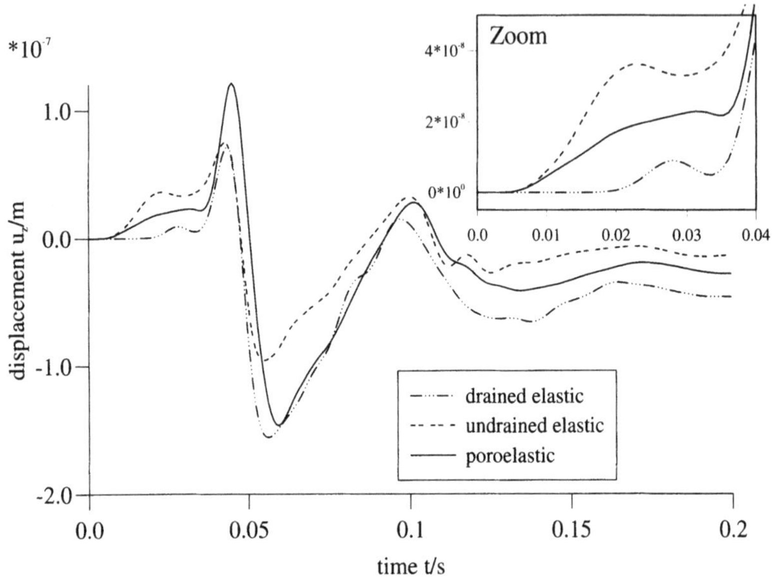

Fig. 6. Displacement at point P versus time: Modelling both layers poroelastic, undrained elastic, and drained elastic.

reflecting, back to the surface point P. Comparing the poroelastic case with both elastic cases, the dispersive behaviour is obvious. As discussed above, the first arrival of the directly propagating compression wave is similar to that of the undrained elastic case. Now, considering the arrival of the reflected compression wave, the poroelastic result coincides with the behaviour of the slower drained elastic model. Finally, both elastic models are either an upper or lower bound for the poroelastic result.

Changing the material data vice-versa, i. e. taking the top layer as rock and the lower semi-infinite layer as soil, the situation and the result is quite different. In Figure 7, the displacement at point P is depicted versus time for this changed situation. In the zoom, the early time behaviour $t < 0.006$ s is similar to the previous case. The compression wave behaves in poroelastic modelling similar to the undrained elastic model. However, after passing of the *Rayleigh* wave ($t > 0.01$ s) the displacement amplitudes of the poroelastic layered half space are only a third of the values of the elastic half space. Then, in all three cases, the compression wave being reflected at the interface arrives at point P ($t \approx 0.018$ s). For $t > 0.02$ s, the displacement in both elastic modelled layered half spaces stays on different but constant values. Whereas, the result for the poroelastic case shows a strong consolidation effect. This is surprising because in the switched configuration, i. e. soil layer on rock foundation, no consolidation was observed. One reason may be the smaller stiffness of the soil foundation compared to the rock foundation.

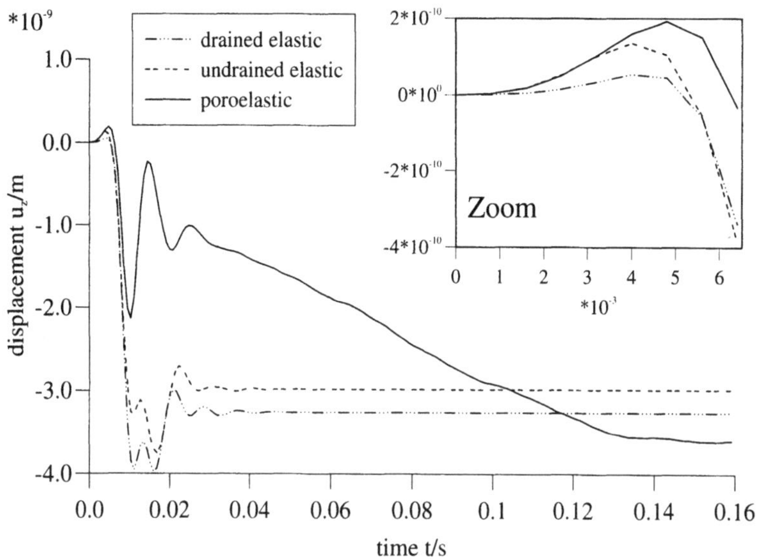

Fig. 7. Displacement at point P versus time: Upper layer *Berea* sandstone and lower semi-infinite layer soil.

Finally, the depth of the layer is varied. In Figure 8, displacement results at point P for four different layer depths (5, 10, 20, and 30 m) are presented. For this test, the configuration is changed back to the first more realistic case of a soil layer on a rock foundation. For the larger depths 20 and 30 m, the reflected compression wave arrives after the *Rayleigh* wave and shows only a small influence. Contrary, for a layer depth of 5 m even the solution between the arrival time of the directly propagated compression wave ($t \approx 0.006$ s) and the arrival time of the shear wave ($t \approx 0.04$ s) is changed. The effect of the *Rayleigh* wave is reduced and strong oscillations after this time ($t > 0.06$ s) are observed. In this case, a wave guide is modelled by the layer producing those strong effects as it is also observed in earthquake areas with a layered structure [31].

4.4 Slow compressional wave in poroelastic half space

One of the main differences between wave propagation in an elastic and a poroelastic solid is the second compressional wave, the so-called slow compressional wave. In theory, Biot has found this wave in 1956 [6, 7] while, in experiment, this wave has been verified by Plona in 1980 [40].

Here, now, this slow wave will be captured in a poroelastic half space taking material data of soil (see Table 1). The discretization is truncated behind 4 m around the center where 684 linear triangles on 397 nodes are used (see Figure 9). The half space is loaded by total stress $t_z = -1\,\mathrm{N/m^2}\,H(t)$ and

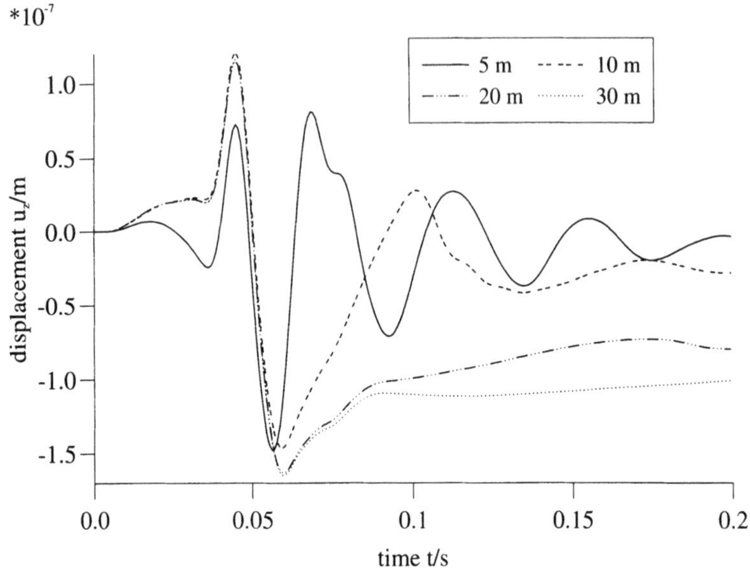

Fig. 8. Displacement at point P versus time: Different depth of the layer.

is taken to be permeable (i. e. $p = 0$ N/m^2) in area A (shaded in Figure 9, 2 m radius) whereas the remaining surface (not shaded) is assumed to be impermeable and traction free, i. e. zero flux and zero traction is given.

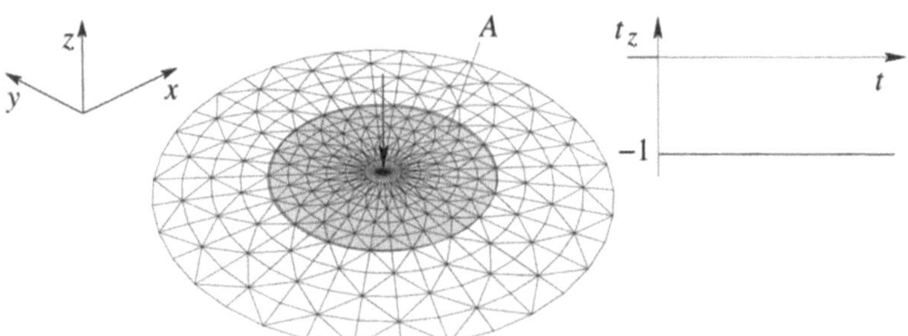

Fig. 9. Half space under vertical load: Discretization, loading area A, and load history.

At first, before discussing the results obtained with the proposed boundary element formulation, the analytical 1-d solution of an infinite long column [49] is discussed. The infinite extension of the column can be approximated by a model of 1000 m length and an observation time short enough that no waves are reflected at the not excited end. But, this will not model geometrical damping as in a half space. In Figure 10, the analytical pressure solution in a distance of 3 m behind the stress excitation point is depicted versus time for *Berea* sandstone and soil (material data see Table 1). Additionally to the results calculated with realistic permeability (solid lines: *Berea* sandstone $\kappa = 1.9 \cdot 10^{-10}$ $m^4/(N\,s)$, soil $\kappa = 3.55 \cdot 10^{-9}$ $m^4/(N\,s)$), the pressure solution for an increased permeability (dashed lines: *Berea* sandstone $\kappa = 1.9 \cdot 10^{-6}$ $m^4/(N\,s)$, soil $\kappa = 3.55 \cdot 10^{-6}$ $m^4/(N\,s)$ is presented. Keeping the intrinsic permeability constant, this is equivalent to reducing viscosity, i. e. in the limit an inviscid fluid is assumed. In principle, both materials have the same behaviour. A first jump in the pressure solution indicates the arrival of the fast compressional wave. Then, in case of realistic permeability, a constant pressure value is observed. In case of the increased permeabilities, a second jump with negative sign indicates the slow compressional wave. This is in accordance with Biot [6] who has shown that the slow compressional wave has a phase shift of 90° to the fast compressional wave.

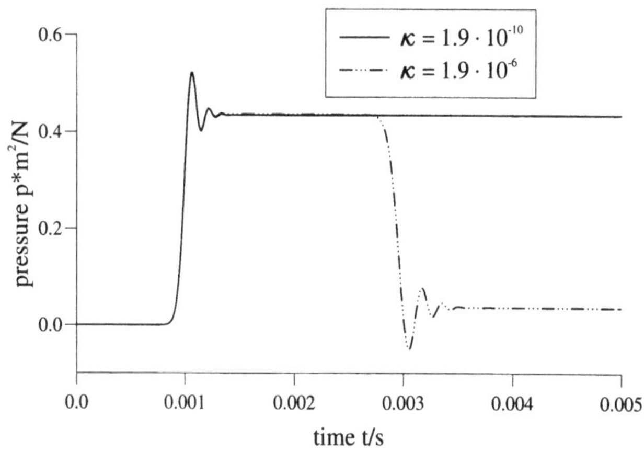

(a)

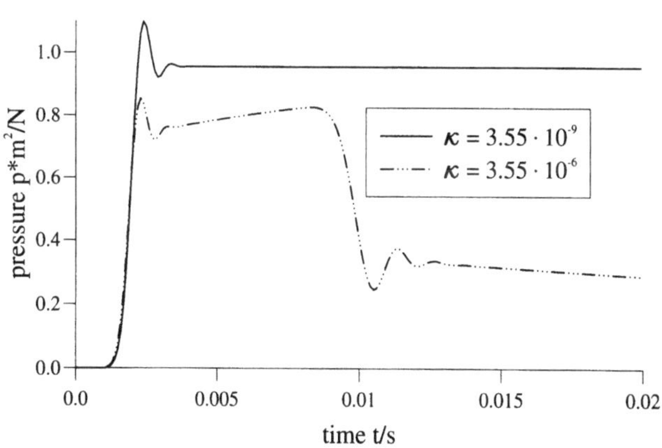

(b)

Fig. 10. Analytical solution: Pore pressure 3 m behind excitation in infinite long column: (a) *Berea* sandstone, (b) soil.

Now, the same physical effects will be captured with the 3-d poroelastic boundary element formulation. In Figure 11, the pressure solution of a half space at 3 m below the surface versus time is given for *Berea* sandstone and for soil. As before, two solutions, one for a realistic value of κ (solid line) and the other for an increased value of κ (dashed line), are presented, respectively.

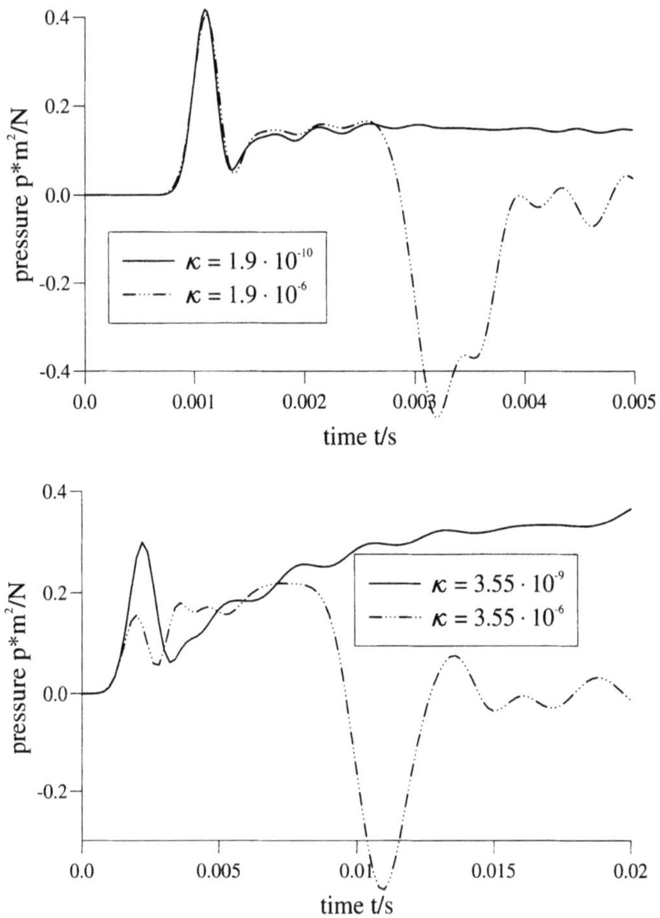

(a)

(b)

Fig. 11. Numerical BE solution: Pore pressure 3 m below surface versus time: (a) *Berea* sandstone, (b) soil.

Identical to the analytical solution, two wave fronts are observed arriving at the same time. In case of *Berea* sandstone, the fast wave arrives at $t \approx 0.001$ s and the slow wave at $t \approx 0.003$ s, while in case of soil the arrival times are $t \approx 0.002$ s and $t \approx 0.009$ s, respectively. Again, the slow compressional wave is only visible for increased values of κ. In this case, the pressure amplitudes in front of and behind of the slow wave of the 3-d boundary element solution

(Figure 11) differ from the amplitudes in the analytical solution (Figure 10). These differences are caused by the geometrical damping. In the 3-d model, the geometrical damping is correctly modelled and, therefore, the amplitudes decrease with increasing distance of the surface. As mentioned above, the 1-d column can not model this behaviour leading to same amplitudes at different locations. However, regardless of the model, the ratio between the pressure values in *Berea* sandstone and in soil is the same. Both calculations give a doubled pressure for soil compared to *Berea* sandstone.

The most significant difference between 1-d analytical solution and 3-d boundary element solution appears in the case of soil. In the 3-d boundary element solution, a kind of creep behaviour is observed for the realistic κ value (see solid line in Figure 11 b). The pressure solution increases asymptotically to a constant value approximately twice the value as for *Berea* sandstone. This creep behaviour is caused by the dispersion of the compression wave. Whereas for soil this effect is obvious indicating strong damping effects, for *Berea* sandstone (Figure 11 a) nearly no creep is observed indicating very small damping. This is in accordance with larger porosity ϕ of soil compared to *Berea* sandstone, i. e. more interstitial fluid leads to larger contact areas between solid and fluid and, finally, to an increased dissipated energy. This effect is found in the analytical solution for a slightly increased permeability [49].

When considering the wave front, in both, the analytical and the boundary element solution, "overshooting" is visible. However, this overshooting in Figure 10 is not as strong as in Figure 11. From the analytical solution it is known that this is caused by the time stepping algorithm. Similarly, in the boundary element formulation this large overshooting is caused by numerics and can be influenced concerning duration and amplitude by the applied multistep method and by the chosen discretization in space and time.

A Fundamental solutions

The explicit expressions of the poroelastodynamic fundamental solutions are given in the following. The four elements of the matrix $\mathbf{G}$ are the displacements caused by a *Dirac* force in the solid:

$$\hat{U}^s_{ij} = \frac{1}{4\pi r\left(\varrho - \beta\,\varrho_f\right)s^2}\left[R_1\,\frac{\lambda_4^2-\lambda_2^2}{\lambda_1^2-\lambda_2^2}\,\mathrm{e}^{-\lambda_1 r} - R_2\,\frac{\lambda_4^2-\lambda_1^2}{\lambda_1^2-\lambda_2^2}\,\mathrm{e}^{-\lambda_2 r} + \right. \\ \left. + \left(\delta_{ij}\,\lambda_3^2 - R_3\right)\mathrm{e}^{-\lambda_3 r}\right] \tag{37}$$

with $R_k = \left(3\,r_{,i}\,r_{,j} - \delta_{ij}\right)/r^2 + \lambda_k\left(3\,r_{,i}\,r_{,j} - \delta_{ij}\right)/r + \lambda_k^2 r_{,i} r_{,j}$ and $\lambda_4^2 = s^2\left(\varrho - \beta\,\varrho_f\right)/\left(K + 4/3\,G\right)$. The pressure caused by the same load is

$$\hat{P}^s_j = \frac{\left(\alpha-\beta\right)s\,\varrho_f\,r_{,j}}{4\pi\beta\left(K+\frac{4}{3}G\right)r\left(\lambda_1^2-\lambda_2^2\right)}\left[\left(\lambda_1+\frac{1}{r}\right)\mathrm{e}^{-\lambda_1 r} - \left(\lambda_2+\frac{1}{r}\right)\mathrm{e}^{-\lambda_2 r}\right]. \tag{38}$$

For a *Dirac* source in the fluid, the respective displacement solution is

$$\hat{U}^f_i = s\,\hat{P}^s_i \tag{39}$$

and the pressure

$$\hat{P}^f = \frac{s\,\varrho_f}{4\pi r\beta\left(\lambda_1^2-\lambda_2^2\right)}\left[\left(\lambda_1^2-\lambda_4^2\right)\mathrm{e}^{-\lambda_1 r} - \left(\lambda_2^2-\lambda_4^2\right)\mathrm{e}^{-\lambda_2 r}\right]. \tag{40}$$

The roots λ_i, $i = 1,2,3$ of the characteristic equation (15) are given in (16). In the derivation of the poroelastodynamic boundary integral equation (26) several abbreviations (27) corresponding to an "adjoint" traction or flux are introduced. First, the "adjoint" traction solution is presented. However, due to the extensive expression only parts are given:

$$\hat{T}^s_{ij} = \left\{\left[\left(K - \frac{2}{3}G\right)\hat{U}^s_{kj,k} + \alpha\,s\,\hat{P}^s_j\right]\delta_{i\ell} + G\left(\hat{U}^s_{ij,\ell} + \hat{U}^s_{\ell j,i}\right)\right\}n_\ell\,, \tag{41}$$

$$\hat{U}^s_{kj,k}\,\delta_{i\ell}\,n_\ell = \frac{r_{,j}\,n_i}{4\pi r\,s^2\left(\varrho-\beta\,\varrho_f\right)\left(\lambda_1^2-\lambda_2^2\right)}\left[\mathrm{e}^{-\lambda_1 r}\left(\frac{1}{r}+\lambda_1\right)\cdot\right. \\ \left.\cdot\,\lambda_1^2\left(\lambda_2^2-\lambda_4^2\right) - \mathrm{e}^{-\lambda_2 r}\left(\frac{1}{r}+\lambda_2\right)\lambda_2^2\left(\lambda_1^2-\lambda_4^2\right)\right],$$

$$
\begin{aligned}
&\left(\hat{U}^s_{ij,\ell} + \hat{U}^s_{\ell j,i}\right) n_\ell = \\
&= \frac{1}{4\,\pi\, r\, s^2\,(\varrho - \beta\,\varrho_f)} \left[\frac{R_5\, 6}{r^3} \left(\frac{\lambda_4^2 - \lambda_2^2}{\lambda_1^2 - \lambda_2^2}\, \mathrm{e}^{-\lambda_1 r} - \frac{\lambda_4^2 - \lambda_1^2}{\lambda_1^2 - \lambda_2^2}\, \mathrm{e}^{-\lambda_2 r} - \mathrm{e}^{-\lambda_3 r} \right) + \right. \\
&\quad + \frac{R_5\, 6}{r^2} \left(\frac{\lambda_4^2 - \lambda_2^2}{\lambda_1^2 - \lambda_2^2}\, \lambda_1\, \mathrm{e}^{-\lambda_1 r} - \frac{\lambda_4^2 - \lambda_1^2}{\lambda_1^2 - \lambda_2^2}\, \lambda_2\, \mathrm{e}^{-\lambda_2 r} - \lambda_3\, \mathrm{e}^{-\lambda_3 r} \right) + \\
&\quad + \frac{R_6\, 2}{r} \left(\frac{\lambda_4^2 - \lambda_2^2}{\lambda_1^2 - \lambda_2^2}\, \lambda_1^2\, \mathrm{e}^{-\lambda_1 r} - \frac{\lambda_4^2 - \lambda_1^2}{\lambda_1^2 - \lambda_2^2}\, \lambda_2^2\, \mathrm{e}^{-\lambda_2 r} - \lambda_3^2\, \mathrm{e}^{-\lambda_3 r} \right) - \\
&\quad - 2\, r_{,n}\, r_{,i}\, r_{,j} \left(\frac{\lambda_4^2 - \lambda_2^2}{\lambda_1^2 - \lambda_2^2} \lambda_1^3\, \mathrm{e}^{-\lambda_1 r} - \frac{\lambda_4^2 - \lambda_1^2}{\lambda_1^2 - \lambda_2^2} \lambda_2^3\, \mathrm{e}^{-\lambda_2 r} - \lambda_3^3\, \mathrm{e}^{-\lambda_3 r} \right) - \\
&\quad \left. - \lambda_3^2 \left(\delta_{ij} r_{,n} + r_{,i} n_j\right) \left(\lambda_3 + \frac{1}{r} \right) \mathrm{e}^{-\lambda_3 r} \right]
\end{aligned}
$$

with $R_5 = r_{,j}\, n_i + r_{,i}\, n_j + r_{,n}\,(\delta_{ij} - 5 r_{,i}\, r_{,j})$ and $R_6 = r_{,j}\, n_i + r_{,i}\, n_j + r_{,n}\,(\delta_{ij} - 6\, r_{,i}\, r_{,j})$. The other explicit expressions are

$$
\begin{aligned}
\hat{Q}^s_j = \frac{n_i}{4\,\pi\, r\,(\varrho - \beta\,\varrho_f)\, s^2} & \left[\frac{\mathrm{e}^{-\lambda_1 r}}{\lambda_1^2 - \lambda_2^2}\, R_1 \left(\beta\,\lambda_2^2 - \alpha\,\lambda_4^2\right) - \right. \\
& \left. - \frac{\mathrm{e}^{-\lambda_2 r}}{\lambda_1^2 - \lambda_2^2}\, R_2 \left(\beta\,\lambda_1^2 - \alpha\,\lambda_4^2\right) + \beta\, \mathrm{e}^{-\lambda_3 r} \left(R_3 - \delta_{ij}\,\lambda_3^2\right) \right],
\end{aligned}
\tag{42}
$$

$$
\begin{aligned}
\hat{T}^f_i = \frac{s^2\, \varrho_f}{4\,\pi\, r\, \beta\,(\lambda_1^2 - \lambda_2^2)} & \left\{ \frac{n_j\,(\alpha - \beta)\, 2\, G}{K + \frac{4}{3} G} \left(R_2\, \mathrm{e}^{-\lambda_2 r} - R_1\, \mathrm{e}^{-\lambda_1 r}\right) + \right. \\
& + n_i\, \mathrm{e}^{-\lambda_2 r} \left[\frac{(\alpha - \beta)\,(K - \frac{2}{3} G)}{K + \frac{4}{3} G} \left(\frac{2}{r^2} + \frac{2\,\lambda_2}{r} + \lambda_2^2 \right) - \alpha \left(\lambda_2^2 - \lambda_4^2\right) \right] - \\
& \left. - n_i\, \mathrm{e}^{-\lambda_1 r} \left[\frac{(\alpha - \beta)\,(K - \frac{2}{3} G)}{K + \frac{4}{3} G} \left(\frac{2}{r^2} + \frac{2\,\lambda_1}{r} + \lambda_1^2 \right) - \alpha \left(\lambda_1^2 - \lambda_4^2\right) \right] \right\},
\end{aligned}
\tag{43}
$$

$$
\begin{aligned}
\hat{Q}^f = \frac{r_{,n}}{4\,\pi\, r\, (\lambda_1^2 - \lambda_2^2)} & \left[\left(\lambda_2 + \frac{1}{r}\right) \left(\lambda_2^2 - \lambda_4^2 \frac{\varrho - \alpha\,\varrho_f}{\varrho - \beta\,\varrho_f}\right) \mathrm{e}^{-\lambda_2 r} - \right. \\
& \left. - \left(\lambda_1 + \frac{1}{r}\right) \left(\lambda_1^2 - \lambda_4^2 \frac{\varrho - \alpha\,\varrho_f}{\varrho - \beta\,\varrho_f}\right) \mathrm{e}^{-\lambda_1 r} \right].
\end{aligned}
\tag{44}
$$

B Convolution Quadrature Method

The "Convolution Quadrature Method" developed by *Lubich* numerically approximates a convolution integral for $n = 0, 1, \ldots, N$,

$$y(t) = \int_0^t f(t-\tau)\, g(\tau)\, \mathrm{d}\tau \quad \rightarrow \quad y(n\,\Delta t) = \sum_{k=0}^{n} \omega_{n-k}(\Delta t)\, g(k\,\Delta t)\ , \tag{45}$$

by a quadrature rule whose weights are determined by the *Laplace* transformed function $\hat{f}$ and a linear multistep method. This method was originally published in [35] and [36]. Application to the boundary element method may be found in [48]. Here, a brief overview of the method is given.

In (45), the time t is divided in N equal steps Δt. The weights $\omega_n(\Delta t)$ are the coefficients of the power series

$$\hat{f}\left(\frac{\gamma(z)}{\Delta t}\right) = \sum_{n=0}^{\infty} \omega_n(\Delta t)\, z^n \tag{46}$$

with the complex variable z. The coefficients of a power series are usually calculated with *Cauchy*'s integral formula. After a polar coordinate transformation, this integral is approximated by a trapezoidal rule with L equal steps $2\pi/L$. This leads to

$$\begin{aligned} \omega_n(\Delta t) &= \frac{1}{2\pi i} \int_{|z|=\mathscr{R}} \hat{f}\left(\frac{\gamma(z)}{\Delta t}\right) z^{-n-1}\, \mathrm{d}z \\ &\approx \frac{\mathscr{R}^{-n}}{L} \sum_{\ell=0}^{L-1} \hat{f}\left(\frac{\gamma\left(\mathscr{R}\, \mathrm{e}^{i\ell 2\pi/L}\right)}{\Delta t}\right) \mathrm{e}^{-in\ell 2\pi/L} \end{aligned} \tag{47}$$

where $\mathscr{R}$ is the radius of a circle in the domain of analyticity of $\hat{f}(z)$.

The function $\gamma(z)$ is the quotient of the characteristic polynomials of the underlying multistep method, e. g. for a BDF 2, $\gamma(z) = 3/2 - 2z + 1/2z^2$. The used linear multistep method must be $A(\alpha)$-stable and stable at infinity [36]. Experience shows that the BDF 2 is the best choice [42]. Therefore, it is used in all calculations in this paper.

If one assumes that the values of $\hat{f}(z)$ in (47) are computed with an error bounded by ϵ, then the choice $L = N$ and $\mathscr{R}^N = \sqrt{\epsilon}$ yields an error in ω_n of size $\mathscr{O}(\sqrt{\epsilon})$ [35]. Several tests conducted by the author lead to the conclusion that the parameter $\epsilon = 10^{-10}$ is the best choice for the kind of functions dealt with in this paper [47]. The assumption $L = N$ leads to an order of complexity $\mathscr{O}(N^2)$ for calculating the N coefficients $\omega_n(\Delta t)$. Due to the exponential function at the end of (47) this can be reduced to $\mathscr{O}(N \log N)$ using the technique of the Fast *Fourier* Transformation (FFT).

References

1. Badiey, M., Cheng, A. H.-D., Mu, Y.: From Geology to Geoacoustics – Evaluation of Biot-Stoll Sound Speed and Attenuation for Shallow Water Acoustics. *Journal of the Acoustical Society of America* **103**(1) (1998), 309–320.
2. Badmus, T., Cheng, A. H.-D., Grilli, S.: A Laplace-Transform Based Three-Dimensional BEM for Poroelasticity. *International Journal for Numerical Methods in Engineering* **36**(1) (1993), 67–85.
3. Biot, M. A.: General Theory of Three-Dimensional Consolidation. *Journal of Applied Physics* **12** (1941), 155–164.
4. Biot, M. A.: Theory of Elasticity and Consolidation for a Porous Anisotropic Solid. *Journal of Applied Physics* **26** (1955), 182–185.
5. Biot, M. A.: Theory of Deformation of a Porous Viscoelastic Anisotropic Solid. *Journal of Applied Physics* **27**(5) (1956), 459–467.
6. Biot, M. A.: Theory of Propagation of Elastic Waves in a Fluid-Saturated Porous Solid. I. Low-Frequency Range. *Journal of the Acoustical Society of America* **28**(2) (1956), 168–178.
7. Biot, M. A.: Theory of Propagation of Elastic Waves in a Fluid-Saturated Porous Solid. II. Higher Frequency Range. *Journal of the Acoustical Society of America* **28**(2) (1956), 179–191.
8. de Boer, R.: *Theory of Porous Media: Highlights in the Historical Development and Current State*. Springer-Verlag, Berlin 2000.
9. de Boer, R., Ehlers, W.: A Historical Review of the Formulation of Porous Media Theories. *Acta Mechanica* **74** (1988), 1–8.
10. de Boer, R., Ehlers, W.: The Development of the Concept of Effective Stresses. *Acta Mechanica* **83** (1990), 77–92.
11. de Boer, R., Ehlers, W., Liu, Z.: One-Dimensional Transient Wave Propagation in Fluid-Saturated Incompressible Porous Media. *Archive of Applied Mechanics* **63** (1993), 59–72.
12. Bonnet, G.: Basic Singular Solutions for a Poroelastic Medium in the Dynamic Range. *Journal of the Acoustical Society of America* **82**(5) (1987), 1758–1762.
13. Bonnet, G., Auriault, J. L.: Dynamics of Saturated and Deformable Porous Media: Homogenization Theory and Determination of the Solid-Liquid Coupling Coefficients. In Boccara, N., Daoud, M. (eds.): *Physics of Finely Divided Matter*, Springer-Verlag, Berlin 1985, pp. 306–316.
14. Bowen, R. M.: Theory of Mixtures. In Eringen, A. C. (ed.): *Continuum Physics*, Vol. III, Academic Press, New York 1976, pp. 1–127.
15. Chen, J.: Time Domain Fundamental Solution to Biot's Complete Equations of Dynamic Poroelasticity. Part I: Two-Dimensional Solution. *International Journal of Solids and Structures* **31**(10) (1994), 1447–1490.
16. Chen, J.: Time Domain Fundamental Solution to Biot's Complete Equations of Dynamic Poroelasticity. Part II: Three-Dimensional Solution. *International Journal of Solids and Structures* **31**(2) (1994), 169–202.
17. Chen, J., Dargush, G. F.: Boundary Element Method for Dynamic Poroelastic and Thermoelastic Analysis. *International Journal of Solids and Structures* **32**(15) (1995), 2257–2278.
18. Cheng, A. H.-D., Badmus, T., Beskos, D. E.: Integral Equations for Dynamic Poroelasticity in Frequency Domain with BEM Solution. *Journal of Engineering Mechanics* **117**(5) (1991), 1136–1157.

19. Cheng, A. H.-D., Detournay, E.: On Singular Integral Equations and Fundamental Solutions of Poroelasticity. *International Journal of Solids and Structures* **35**(34-35) (1998), 4521–4555.
20. Cheng, A. H.-D., Ligget, J. A.: Boundary Integral Equation Method for Linear Porous-elasticity with Applications to Fracture Propagation. *International Journal for Numerical Methods in Engineering* **20**(2) (1984), 279–296.
21. Cheng, A. H.-D., Ligget, J. A.: Boundary Integral Equation Method for Linear Porous-elasticity with Applications to Soil Consilidation. *International Journal for Numerical Methods in Engineering* **20**(2) (1984), 255–278.
22. Cheng, A. H.-D., Antes, H.: On Free Space Green's Function for High Order Helmholtz Equations. In Kobayashi, S., Nishimura, N. (eds.): *Boundary Element Methods: Fundamentals and Applications*, Springer-Verlag, Berlin 1991, pp. 67–71.
23. Domínguez, J.: An Integral Formulation for Dynamic Poroelasticity. *Journal of Applied Mechanics* **58** (1991), 588–591.
24. Domínguez, J.: Boundary Element Approach for Dynamic Poroelastic Problems. *International Journal for Numerical Methods in Engineering* **35**(2) (1992), 307–324.
25. Domínguez, J.: *Boundary Elements in Dynamics.* Computational Mechanics Publication, Southampton 1993.
26. Ehlers, W.: Poröse Medien – ein kontinuumsmechanisches Modell auf der Basis der Mischungstheorie. *Forschungsbericht aus dem Fachbereich Bauwesen* **47**, Universität - GH Essen, 1989.
27. Ehlers, W.: Compressible, Incompressible and Hybrid Two-phase Models in Porous Media Theories. *ASME: AMD-Vol.* **158** (1993), 25–38.
28. Ehlers, W.: Constitutive Equations for Granular Materials in Geomechanical Context. In Hutter, K. (ed.): *Continuum Mechanics in Environmental Sciences and Geophysics*, CISM Courses and Lecture Notes **337**, Springer-Verlag, Wien 1993, pp. 313–402.
29. Ehlers, W., Kubik, J.: On Finite Dynamic Equations for Fluid-Saturated Porous Media. *Acta Mechanica* **105** (1994), 101–117.
30. Fillunger, P.: Der Auftrieb von Talsperren, Teil I-III. *Österr. Wochenschrift für den öffentlichen Baudienst* (1913), 532–570.
31. Graff, K. F.: *Wave Motion in Elastic Solids.* Oxford University Press, 1975.
32. Guiggiani, M., Gigante, A.: A General Algorithm for Multidimensional Cauchy Principal Value Integrals in the Boundary Element Method. *Journal of Applied Mechanics* **57** (1990), 906–915.
33. Hörmander, L.: *Linear Partial Differential Operators.* Springer-Verlag, 1963.
34. Kim, Y. K., Kingsbury, H. B.: Dynamic Characterization of Poroelastic Materials. *Experimental Mechanics* **19** (1979), 252–258.
35. Lubich, C.: Convolution Quadrature and Discretized Operational Calculus, I. *Numerische Mathematik* **52** (1988), 129–145.
36. Lubich, C.: Convolution Quadrature and Discretized Operational Calculus, II. *Numerische Mathematik* **52** (1988), 413–425.
37. Manolis, G. D., Beskos, D. E.: Integral Formulation and Fundamental Solutions of Dynamic Poroelasticity and Thermoelasticity. *Acta Mechanica* **76** (1989), 89–104. Errata [38].
38. Manolis, G. D., Beskos, D. E.: Corrections and Additions to the Paper "Integral Formulation and Fundamental Solutions of Dynamic Poroelasticity and Thermoelasticity". *Acta Mechanica* **83** (1990), 223–226.

39. Pekeris, C. L.: The Seismic Surface Pulse. *Proc. of the National American Society* **41** (1955), 469–480.
40. Plona, T. J.: Observation of a Second Bulk Compressional Wave in Porous Medium at Ultrasonic Frequencies. *Applied Physics Letters* **36**(4) (1980), 259–261.
41. Rayleigh, J. W. S.: On Waves Propagated Along the Plane Surface of an Elastic Solid. *Proceedings of the London Mathematical Society* **17** (1887), 4–11.
42. Schanz, M.: A Boundary Element Formulation in Time Domain for Viscoelastic Solids. *Communications in Numerical Methods in Engineering* **15** (1999), 799–809.
43. Schanz, M.: Boundary Element Calculation of Wave Propagation in 3-d Poroelastic Solids. In Atluri, S. N., Brust, F. W. (eds.): *Advances in Computational Engineering & Sciences*, Vol. I, Tech. Science Press, Palmdale 2000, pp. 118–123.
44. Schanz, M.: Wave Propagation in Viscoelastic and Poroelastic Continua: A Boundary Element Approach. In: *Lecture Notes in Applied Mechanics*, Springer-Verlag, Berlin 2001.
45. Schanz, M.: Application of 3-d Boundary Element Formulation to Wave Propagation in Poroelastic Solids. In: *Engineering Analysis with Boundary Elements*, in press.
46. Schanz, M.: Dynamic Poroelasticity Treated by a Time Domain Boundary Element Method. In: *IUTAM Symposium on Advanced Mathematical and Computational Mechanic Aspects of the Boundary Element Method*, Kluwer Academic Publishers, in press.
47. Schanz, M., Antes, H.: Application of 'Operational Quadrature Methods' in Time Domain Boundary Element Methods. *Meccanica* **32**(3) (1997), 179–186.
48. Schanz, M., Antes, H.: A New Visco- and Elastodynamic Time Domain Boundary Element Formulation. *Computational Mechanics* **20**(5) (1997), 452–459.
49. Schanz, M., Cheng, A. H.-D.: Transient Wave Propagation in a One-Dimensional Poroelastic Column. *Acta Mechanica* **145** (2000), 1–18.
50. Stakgold, I.: *Green's Functions and Boundary Value Problems. Pure and Applied Mathematics.* 2nd ed., John Wiley & Sons, 1998.
51. Terzaghi, K.: Die Berechnung der Durchlässigkeit des Tones aus dem Verlauf der hydromechanischen Spannungserscheinungen. *Sitzungsbericht der Akademie der Wissenschaften (Wien): Mathematisch-Naturwissenschaftlichen Klasse* **132** (1923), 125–138.
52. Truesdell, C., Toupin, R. A.: The Classical Field Theories. In Flügge, S. (ed.): *Handbuch der Physik*, Vol. III/1, Springer-Verlag, Berlin 1960, pp. 226–793.
53. Vardoulakis, I., Beskos, D. E.: Dynamic Behavior of Nearly Saturated Porous Media. *Mechanics of Composite Materials* **5** (1986), 87–108.
54. Wiebe, Th., Antes, H.: A Time Domain Integral Formulation of Dynamic Poroelasticity. *Acta Mechanica* **90** (1991), 125–137.

Multicomponent reactive transport modelling: Applications to ore body genesis and environmental hazards

Reem Freij-Ayoub, Hans-Bernd Mühlhaus, and Laurent Probst

CSIRO, Division of Exploration and Mining,
Nedlands 6009, Western Australia

Abstract. The interplay of chemical and mechanical effects is of crucial importance for understanding and solving key problems of geology, engineering and environmental science. We begin with an outline of the governing equations for a coupled multicomponent reactive transport model. Thermal coupling and heat flow is also considered. The reaction part of the model is kept very general and applies to a wide variety of problems including simulation of in situ leaching, mineralization (ore body genesis), environmental hazards and calcite leaching to name just a few. We briefly introduce the CSIRO symbolic finite element solver FASTFLO (http://www.cmis.csiro.au/fastflo) and illustrate the theory by a number of finite element solutions. While focussing initially on reactive transport dominated problems (deformation and damage negligible) we close with a limit load problem involving chemically induced reduction of the yield strength. We consider a stiff, smooth strip foundation on an infinite half plane (*Prandtl* case). The substrate deforms and partially plastifies (*von Mises* yield, *Prandtl-Reuss* flow rule). We then prescribe a constant concentration of a leachant on the surface of the half plane. The concentration of the leachant diffuses through the half plane, lowering the yield strength and ultimately causing unacceptably large deformation of the structure.

1 Introduction

Computer simulation of coupled mechanical-hydrothermal-reactive transport processes enhances our understanding of geochemical problems. The feedback between the various coupled processes and phases is often so strong decoupling is not possible. As an example for positive (unstable) feedback we consider coupling between chemical reactions and flow. The flow of an under saturated fluid causes the dissolution of a mineral in a porous rock and hence increases the permeability of the rock. This enhanced permeability causes an increased flux of the unsaturated fluid (reagent) leading to further dissolution and further advancement of the reaction front. Positive feedback between transport and geochemical processes can lead to magnification of fluctuations and hence to symmetry breaking patterns that, once they reach the macroscopic level, are considered as instabilities or dissipative structures like reaction front fingering (e. g. Ortoleva [14], Mühlhaus et al. 2001).

The boundary value problems of rock alteration and ore body genesis contain a variety of fascinating bifurcation problems. As a small sampling of such problems we mention the instability of initially straight reaction fronts associated with mineral dissolution (Chadam et al. [3]), the propagation of solitary porosity waves (Connolly [4]), the breakup of a simply connected precipitation zone into banded zones of mineralization like cement bands and metamorphic layering (Ortoleva et al. [15]). An interesting account of geological reaction-transport phenomena, including instability problems, can be found in Ortoleva (1989).

Multicomponent reactive transport modelling is a key tool for simulating hydrothermal rock alteration and ore body deposition processes. Computer simulation enables the testing of various flow scenarios, as certain flow regimes promote the occurrence of ore body forming reactions more than the others. Commonly deposition of base and precious metals in rocks is promoted by chemical gradients induced by fluid mixing. In geological scenarios the mixing process may be driven by natural convection and non-uniform flows associated with preexisting geological structures (e. g. faults, seals). Mixing of fluids of different chemical composition during transport through rocks of various physical and chemical properties creates chemical gradients (e. g. in oxidation and acidity). Such gradients lead to the precipitation of new minerals at places where these fluids focus (e. g. Freij-Ayoub et al. [5]).

In engineering applications material strength is usually taken as independent, i. e. uninfluenced by chemical processes. Environmental hazards (e. g. a corrosive reagent leaking into the ground as the byproducts of pesticides manufacturing processes) can change the way rock will respond to loading in an often unexpected manner. The prediction of the flow path and reaction products and their effects on rock strength is crucial for the assessment of the stability of the rock supporting an engineering structure. The same holds true for the stability of a borehole used to inject such chemicals. Numerical simulations of the coupled mechanical and reactive transport processes can predict the altered rock load bearing capacity and its resulting deformation.

In the following we give an outline of the governing equations for a multicomponent reactive transport model for convective 2-phase flow (Section 2). Such a model is suited for the description of the effect of nuclear waste stored in a partially saturated porous media on the saturation state of the rock due to the heat generation accompanying the decay of the nuclear waste (e. g. Lichtner and Walton [9]). It can be used for the modelling of in situ leaching (a mining technique) (e. g. Mühlhaus et al. [11]).

Reactive flow simulations are also a potentially powerful predictive tool for rock alteration and ore body genesis (e. g. Freij-Ayoub et al. [5]). In Section 3 we discuss numerical issues and the onset of convection is investigated for a one-phase flowing in a 3-d block. In Section 4 we discuss the interplay of reaction kinetics and transport parameters producing spatial patterning of the dissolution front. We describe a geochemical application, an example of a multicomponent reactive 2-phase convective flow problem. Finally we consider the effect of mechano-chemical coupling, the effect of reagent infiltration on the strength degradation of the rock supporting a strip foundation.

2 Governing equations

2.1 2-phase flow equations

We describe the interaction between the two phases, a gas and a liquid by the empirical van Genuchten equations [6]. Saturation is modelled as follows:

$$S_g + S_l = 1\,, \quad S_\pi = \varphi_\pi / \varphi_T\,, \quad \varphi_l = S_l\, \varphi_T\,, \quad \varphi_g = (1 - S_l)\, \varphi_T\,, \tag{1}$$

where S_π is the phase (g: gas, l: liquid) saturation, φ_π is the phase pore volume, and φ_T is the total porosity. The effective saturation is related to the residual saturation S_l^r and to the capillary pressure h as follows:

$$\begin{aligned} &S_l^{\text{eff}} = (S_l - S_l^r)/(1 - S_l^r)\,, \quad S_l^{\text{eff}} = [1 + (\alpha\, h)^m]^{-\lambda}\,, \\ &m > 2\,, \quad \lambda = 1 - 1/m\,, \quad h = P_g - P_l\,, \end{aligned} \tag{2}$$

α is an empirical constant, P_g and P_l are the pressures. The relative permeabilities k_π^r are expressed as follows:

$$k_l^r = \sqrt{S_l^{\text{eff}}}\, \left\{1 - \left[1 - (S_l^{\text{eff}})^{1/\lambda}\right]^\lambda\right\}^2\,, \quad k_g^r = 1 - k_l^r\,. \tag{3}$$

The *Darcy* law is used to model the flow of each phase π,

$$\mathbf{q}_\pi = -k\, k_\pi^r / \mu_\pi \nabla (P_\pi - \rho_\pi\, \mathbf{g}\, z)\,, \tag{4}$$

where k is the intrinsic permeability, μ_π is the viscosity, ρ_π is the density, $\mathbf{g}$ is the direction of gravity, where $\|\, \mathbf{g}\, \| = g$, g is the gravitational constant, and z is the coordinate along $\mathbf{g}$. The density is temperature dependent for both phases and pressure dependent for the gaseous phase only

$$\rho_l = \rho_{l0}\, (1 - \alpha T)\,, \quad \rho_g = \rho_{g0}\, (1 + P/P_0 - T/T_0)\,, \tag{5}$$

α is the volumetric thermal expansion coefficient for the liquid, the subscript "0" marks the reference values of each quantity.

2.2 Heat equation

Thermal equilibrium between the three phases π (gas, liquid, and solid) is assumed and the heat conservation equation takes the following form:

$$\left(\sum_{\pi=1}^{3} \varphi_\pi\, \rho_\pi\, C_{p\pi}\, T\right)_{,t} + (q_{\pi i}\, C_{p\pi}\, T)_{,i} - \bar{\gamma}\, T_{,ii} = 0\,, \tag{6}$$

where φ_π, ρ_π, $C_{p\pi}$, $q_{\pi i}$ are the phase volume fraction, density, specific heat and discharge vector, respectively. $\bar{\gamma}$ is the heat conductivity of the rock and T is the temperature.

2.3 Constitutive equations

Boyle's law is used to model the gas behaviour which is assumed ideal in this case. Equilibrium chemical reactions are governed by non-linear mass action relations as follows:

$$K = \sum_{c=1}^{\hat{c}} \gamma_c^{\nu_{cs}} M_c^{\nu_{cs}} . \tag{7}$$

K is the reaction equilibrium constant. It is defined by the multiplication of the activities (which are the products of the activity coefficients and concentrations) of the components that appear in the reaction. M_c is the molal concentration of species c, γ_c is the activity coefficient and ν_{cs} is the number of moles of species c in the formula. The coefficients ν_{cs} are negative for species on the left-hand side of the reaction. All activity coefficients are considered as equal to one implying a diluted (ideal) solution. Solid activities are taken as unity and hence do not appear in such an expression. The product on the right-hand side of (7) is called the solubility product. The equality (7) is valid for equilibrium reactions. In this model the equilibrium constants are coupled to T and P as follows:

$$\log K = A + BT + C/T + D \log T + E\,P/T , \tag{8}$$

where A, B, C, and D and E are constants which are reaction specific and are obtained from interpolating a large set of data for K obtained using a thermodynamic package (e. g. Shvarov [16]). When minerals are far from equilibrium with the solution (as is assumed for K-feldspar and the silicate mineral considered in example 1 in this paper) their dissolution and precipitation are kinetic reactions controlled by a first order rate expression of the type

$$\begin{aligned} R &= A/V\,k(a_H)^n\,[1-(Q/K)^m] \quad &\text{if} \quad \varphi_{\textbf{feldspar}} > 0 , \\ R &= 0 &\text{if} \quad \varphi_{\textbf{feldspar}} = 0 , \end{aligned} \tag{9}$$

where A is the mineral surface area, V is the fluid volume in the pores in contact with the surface area A, k is the rate constant, a_H is the hydrogen ion activity, n is a constant between 0 and 1 (for first order kinetics n is equal to 1), Q and K are the solubility product and equilibrium constant for the hydrolysis reaction of the mineral (discussed below), m is a constant between 0.5 and 2.5 (Lasaga [7]; Steefel and Lasaga [17]), and $\varphi_{\textbf{feldspar}}$ is the volume fraction occupied by the mineral, K-feldspar in our example. The rate of mineral precipitation/dissolution is sensitive to both temperature and the pH (pH is defined as the negative of the log of the hydrogen ion concentration and is a measure of acidity). The temperature dependence of the rate constant

k can be expressed in terms of an *Arrhenius* equation (Steefel and Lasaga [17]):

$$k(T) = k_{25} \exp[-E/R(1/T - 1/298.15)], \tag{10}$$

where k_{25} is the rate constant at the reference temperature 25 °C (298.15 K), T is the absolute temperature, E is the activation energy, R is the gas constant.

2.4 Coupled convective fluid-heat flow equations with chemical reactions

The chemistry of ore body genesis is characterized by massive pore-water flows and mixing of fluids through and within minerals bearing rock. In geological systems one of the key drivers of such flows is natural convection. Here, we lay out the equations for the convective flow in porous media for a 2-phase fluid.

Using (4) and (5), the non-dimensional steady state forms of the fluid and gas continuity equations are obtained as

$$\begin{aligned} &\left[k_l^r\, \varphi^3\, (P_{l,i}/\gamma_l - \mathrm{Ra}_l\, T\, \delta_{3i})\right]_{,i} = 0\,, \\ &\left\{(1-k_l^r)\, \varphi^3\, \left[P_{g,i} + \delta_{3i}\, \mathrm{Ra}_g (P_g/(\alpha_g\, \alpha_l\, \beta_g\, \Delta T) - T)\right]\right\}_{,i} = 0\,. \end{aligned} \tag{11}$$

Inserting the *Darcy* velocities into the heat conservation equation (6) yields in a non-dimensional form

$$\begin{aligned} &\beta_1\, T_{,t} - \{\beta_2\, (T - \delta_{3i})\, [(1-k_l^r)\, \varphi^3\, (P_{g,i} + \\ &\qquad + \delta_{3i}\, \mathrm{Ra}_g (P_g/(\alpha_g\, \alpha_l\, \beta_g\, \Delta T) - T))]\,\}_{,i} - \\ &\qquad - \{\beta_3\, (T - \delta_{3i})\, [k_l^r\, \varphi^3\, (P_{l,i}/\gamma_l - \delta_{3i}\, \mathrm{Ra}_l\, T)]\,\}_{,i} - T_{,ii} = 0\,, \\ &\mathrm{Ra}_\pi \;= k_0\, \rho_\pi^0\, \Delta T\, g\, H\, \overline{\rho\, C_p}\, \beta_\pi / \mu_\pi\, \bar{\gamma}\,, \\ &P_n \;= \mu_g\, \bar{\gamma} / k_0\, \overline{\rho\, C_p}\,, \quad \overline{\rho\, C_p} = \sum \varphi_\pi^0\, \rho_\pi^0\, C_{P\pi}^0\,, \\ &\alpha_g \;= P_{0g}/P_n\,, \quad \beta_g = 1/T_0\,, \quad \alpha_l = 1/P_0\,, \quad \gamma_l = \mu_L/\mu_g\,, \\ &\beta_1 \;= (\varphi_g\, \rho_g\, C_{pg} + \varphi_l\, \rho_l\, C_{pl} + \varphi_s\, \rho_s\, C_{ps})/(\overline{\rho\, C_p})\,, \\ &\beta_2 \;= \rho_g\, C_{pg}/(\overline{\rho\, C_p})\,, \quad \beta_3 = \rho_l\, C_{pl}/(\overline{\rho\, C_p})\,. \end{aligned} \tag{12}$$

ΔT represents the temperature variation across the vertical block dimension (assuming a rectangular block domain). The equations are expressed in terms of the convective temperature and pressures, i. e. we have decomposed the pressure and temperature into no-flow and convective parts. The no-flow part corresponds to the conduction state where heat is transferred by conduction only. The convective part corresponds to the state where the system parameters (*Rayleigh* numbers Ra_π) trigger natural flow. The temperature that shows in these equations is the convective temperature and is normalized by ΔT. The convective pressure is normalized by P_n defined in (12). δ_{ij} is the *Kronecker* delta defined by $\delta_{ij} = 0$ if $i \neq j$ and $\delta_{ij} = 1$ if $i = j$. The model is intended for natural convection but is also suitable for other flow situations provided the pressure temperature decomposition is considered in the formulation of the boundary conditions. We observe here the presence of two *Rayleigh* numbers, one for the gas and one for the liquid. The influence of other factors that can trigger convection, like salinity for example, is not considered in this model.

In this coupled model if we consider a solute (aqueous species) of concentration M_C then the mass conservation equation for the aqueous species is as follows:

$$
\begin{aligned}
\mathrm{L}(M_C) &= R/\phi_0\,, \\
\mathrm{L}(M_C) &= (\phi\, M_C)_{,t} + 1/\varphi_0 \left[k_l^r\, \varphi^3 (P_{l,i}/\gamma_l - \mathrm{Ra}_l\, T\, \delta_{3\,i}) M_C\right]_{,i} - \\
&\quad - 1/\mathrm{Le}\left(\phi^{m+1}\, M_{C,i}\right)_{,i}\,, \\
\mathrm{Le} &= \bar{\gamma}/(\overline{\rho\, C_p})/D_0\,.
\end{aligned}
\tag{13}
$$

D_0 is the chemical diffusion of the species in the fluid. Le is the *Lewis* number and it expresses the ratio between thermal and chemical diffusion. The term R on the right-hand side of (13) is the rate of change of the concentration of the aqueous species due to mineral dissolution or precipitation. The volume change produced by the dissolution or precipitation of minerals is coupled to the total porosity as follows:

$$
\varphi = \varphi_f + \varphi_g = 1 - \sum_{m=1}^{\hat{m}} \bar{V}(M_m)\,. \tag{14}
$$

φ_f and φ_g are liquid and gaseous porosity. $\bar{V}$ is the molar volume and M_m is the mineral molar concentration. The intrinsic permeability is coupled to the porosity as follows:

$$
k = k_0\, (\phi/\phi_0)^3\,. \tag{15}
$$

2.5 Mass conservation equations for primary components

A system that has many interacting chemical species can be described in various ways. We identify an independent set of primary components that we use to uniquely describe the chemical reactions taking place in the system (e. g. Lichtner [10]). A mass conservation equation is written for every primary component. (16) is the sum of the mass fractions (in moles per liter) of the primary components available in a free form (c), as aqueous or gases secondary complexes (s) or (g), or as minerals (m):

$$M_c^T = M_c + \sum_{j=1}^{\hat{s}} \nu_{cj}^s\, M_j^s + \sum_{j=1}^{\hat{g}} \nu_{cj}^g\, M_j^g + \sum_{j=1}^{\hat{m}} \nu_{cj}^m\, M_j^m\,, \tag{16}$$

M_c, M^s, and M^g are molar concentrations of the primary, secondary, and gaseous species, respectively. M^m is the number of moles of the mineral in a unit volume of rock. ν_{cj}^s, ν_{cj}^g, and ν_{cj}^m stand for the total number of secondary aqueous, or gaseous, and mineral species, respectively. The presence of secondary aqueous species requires adding a term $\mathrm{L}\left(\sum_{j=1}^{\hat{s}} \nu_{cj}^s\, M_j^s\right)$ on the right-hand side of (13), L is the operator defined in (13). The presence of minerals leads to the source term R of the form $\left(\sum_{j=1}^{\hat{m}} \nu_{cj}^m\, M_j^m\right)_{,t}$. An equation (13) is written for every primary component. If the mineral (assuming there is only one mineral in the equation) is not saturated then this equation is used to calculate the primary component concentration. Upon saturation the mineral/aqueous species concentrations are determined from the mass conservation equation/mass action relations, respectively. This will be further explained in Section 4.

3 Numerical aspects

We have implemented the model equations using the finite element based PDE solver *Fastflo* (see http://www.cmis.csiro.au/fastflo/presentations/ for details). This has the advantage that the rather complex coupled equations can be written down in a high level language, which simplifies experimentation and error tracking. The governing equations are transformed in the usual way into a suitable weak form.

The three equations (11 and 12) are solved simultaneously implicitly for the convective parts of the temperature, liquid and gas pressures applying a fully explicit time integration scheme. The resulting pressures are used to update the saturation, relative permeabilities and the relative pore space of the two fluid phases. These porosities, pressures and temperatures are used then in (13) to calculate the primary species or mineral concentrations. A mass conservation equation for one primary species (as an example consider the mass conservation of Ca^{2+} in (31)) is used to calculate its concentration unless, according to the state of the solution, the mineral calcite $CaCO_3$ saturates and precipitates, an equation will then be needed to determine its quantity in the rock. In such a case, (31) is used to calculate calcite concentration and Ca^{2+} is calculated using the mass action relation of calcite of the type (7). A mass conservation equation of the type (13) that contains a primary and a secondary species of vastly different orders of magnitude is used to calculate the species with the largest concentration to minimize the error.

The vast differences in the order of magnitudes of the concentrations encountered in multicomponent reactive transport problems is a well known issue. One approach to manage this problem is to solve equations for the total concentrations defined in (16) and then speciate in a different algorithm based on the *Newton-Raphson* iteration method that calculates the change in the log of the primary species and solid concentrations rather than of the concentrations themselves, see Walsh [18].

3.1 The onset of convection

The general case, where there are two *Rayleigh* numbers, one for the gas and one for the liquid, is not benchmarked. The relation between the phases' *Rayleigh* numbers need to be examined. Here, we consider the case of one phase only: A water saturated block of porous rock (Figure 1). In this case, $k_l^r = 1$ and $\gamma_l = 1$ in (11) and (12) and the liquid pressure are normalized with respect to $P_n = \mu_l \bar{\gamma}/k_0 \overline{\rho C_p}$. The side $x_3 = 0$ is heated so that the temperature difference over the height of the block is equal to ΔT. Convection sets in if ΔT is larger than a certain critical value. The critical condition for convection is obtained in the usual way (e. g. Nield [13]) from a linear instability analysis.

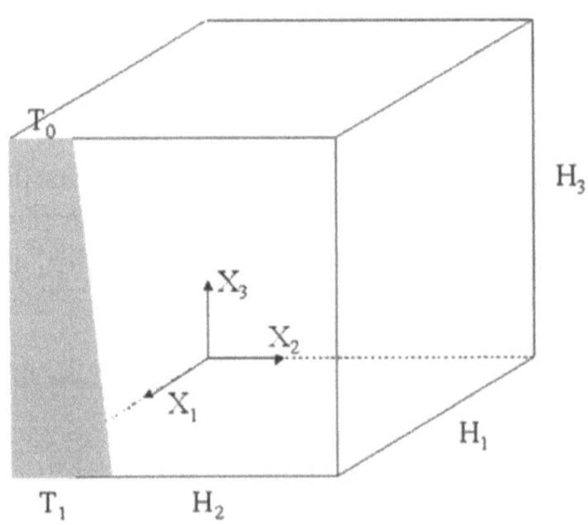

Fig. 1. A 3-d block with temperature and pressure gradient.

In the linear stability analysis we investigate the behaviour of the system with respect to small perturbations $P_l + \delta P_l$, $T + \delta T$ of the no-flow state. Linearization of (11) and (12) with respect to δP and δT and insertion of the modes

$$\begin{bmatrix} \delta P \\ \delta T \end{bmatrix} = \begin{bmatrix} \hat{P} \\ \hat{T} \end{bmatrix} \exp(\mathrm{i}\, n_i\, \pi\, x_i + \omega\, t)\,, \tag{17}$$

where P and T are the pressure and temperature amplitudes, yields the dispersion relationship

$$\omega = \frac{\mathrm{Ra}\,\beta_3 (n_2\,\pi)^2}{(n_2\,\pi)^2 + (n_3\,\pi)^2} - [(n_2\,\pi)^2 + (n_3\,\pi)^2] \tag{18}$$

for convection rolls with axes parallel to the x_1-axis; the integers n_2 and n_3 are the wave numbers in the x_2- and x_3-directions, respectively.

A mode with the wave numbers n_2 and n_3 is unstable if

$$\mathrm{Ra} > \frac{1}{\beta_3 (n_2\,\pi)^2} [(n_2\,\pi)^2 + (n_3\,\pi)^2]^2\,. \tag{19}$$

The critical *Rayleigh* number is the one for which the right-hand side of (19) assumes a minimum over (n_2, n_3). Differentiation with respect to n_2 yields $n_2 = n_3$ and hence

$$\mathrm{Ra}_{\mathrm{min}} = \frac{4}{\beta_3} (n_3\,\pi)^2\,. \tag{20}$$

For the case of zero velocity and temperature bc's at $x_3 = 0$ and H, respectively, and periodic bc's at the sides, we find $\mathrm{Ra}_{\mathrm{min}} = 4\,\pi^2/\beta_3$. If the boundary $x_3 = 0$ is water permeable and insulating while $x_3 = H$ is impermeable with $T = 0$ then $n_2 = 1$, $n_3 = 1/2$. In this case, $\mathrm{Ra} = (5/4\,\pi)^2/\beta_3$. These lower limits for the *Rayleigh* numbers have been tested and found valid for these two cases. We confirm that no convection takes place for subcritical *Rayleigh* numbers and that steady state modes are insensitive with respect to the initial perturbation.

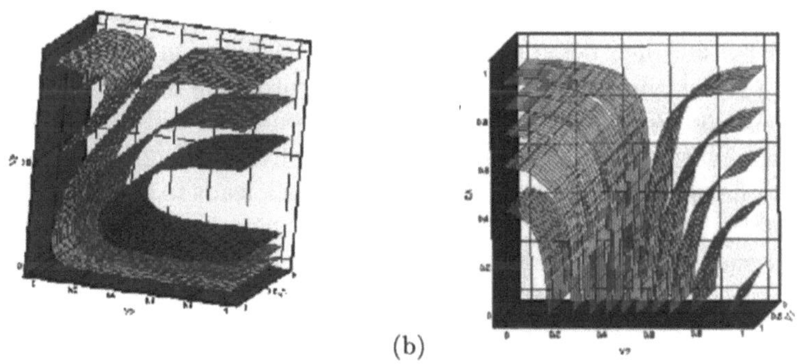

(a) (b)

Fig. 2. Isotherms in natural convection, 20×20×20 8-noded quadrilateral elements, $\Delta t = 0.001$, $p_3 = 1$: (a) Ra = 100, zero velocity and temperature on top and bottom and natural bc's on the sides; (b) Ra = 40, zero velocity and temperature on top and zero pore pressure and zero temperature gradient on bottom.

Figure 2 presents two cases: Case (a), where Ra = 100 and zero convective temperature and velocity conditions are imposed on upper and lower boundaries. Case (b), where Ra = 40 and zero convective temperature and velocity conditions are imposed on upper boundary while zero pore pressure and temperature gradient conditions are imposed on the lower boundary. In both cases the vertical boundaries of the block has the periodic boundary conditions. In the first case steady state is reached quickly unlike case (b), where at non-dimensional time step 50 steady state was not yet reached.

4 Examples

We present three examples, one that depicts a two-dimensional infiltration instability problem where a solute bearing fluid dissolves the parent rock. The system parameters allows perturbations in the advance of the dissolution front in the rock causing eventual porosity fingering.

The second example is devoted to predict the chemical changes resulting from mixing of fluids of different chemical compositions (due to equilibrium conditions under thermal and pressure gradients) by the onset of convection. These reactions are encouraged also by the injection of highly acidic solution at the base of a 3-d block. This example is of significance to geochemists and to the community interested in rock alteration processes.

The third example demonstrates how the seepage of a corrosive reagent under a foundation may alter the mechanical response of the stable foundation to loading and render it unstable by undermining the rock yield strength. This example demonstrates a situation of interest to geotechnical engineers.

4.1 Infiltration instability model

Pore water flow assisted chemical dissolution of minerals is usually accompanied by so-called infiltration instabilities. Dissolution starts at the front of the propagating solvent X. If for some reason the alteration front is locally more advanced, then solvent bearing pore-water will be focused at this location, causing further advancement of the front around the location. For advection controlled flows, this positive feedback leads to a porosity fingering instability.

We model this problem mathematically by means of the continuity equation

$$[\phi^n(P_{,i})]_{,i} = \phi^0\, \phi_{,t}\,. \tag{21}$$

P is the pressure, ϕ^0 is the initial porosity and n is the exponent arising from the permeability porosity relation as in (15).

The mass conservation equation for the solute X that is infiltrating in a porous rock containing a dissolving mineral A of volume fraction ϕ_A is as follows:

$$(\phi\, X)_{,t} - \frac{1}{\phi^0}\,[\phi^n(P_{,i})X]_{,i} = \frac{1}{\mathrm{Sc}}\,(\varphi^{m+1}\, X_{,i})_{,i} - \frac{\rho_A}{\phi^0\, \bar{X}}\, \phi_{A,t}\,. \tag{22}$$

The *Schmidt* number $\mathrm{Sc} = \bar{v}\, H / D_0^X = H^2 / (t_{\text{fluid}}\, D_0^X)$ expresses the relative importance of advective versus diffusive transport. $\bar{X}$ is a reference concentration. The last term in the above equation is the rate of dissolution of mineral A (Mühlhaus et al. [12]). This rate is proportional to the concentration of the infiltrating solute X and the surface area of the mineral available for the reaction and is related to the volume fraction $\phi_A = \frac{1}{6}\pi\, n_A\, d_A^3$ (where n_A is the number of grains and d_A is the grain diameter that decreases as the mineral dissolves) as follows:

$$\frac{\rho_A}{\phi^0\, \bar{X}}\, \phi_{A,t} = \frac{t_{\text{fluid}}}{t_A}\, \phi_A^{2/3}\, X\,, \tag{23}$$

where t_{fluid} is used to non-dimensionalize the time and t_A is the time necessary to dissolve all mineral A (Mühlhaus et al. 2001) as follows:

$$t = t_{\text{fluid}}\, \tilde{t}\,, \qquad t_{\text{fluid}} = \frac{\phi^0\, H^2}{K_0\, \bar{P}} \tag{24}$$

$\bar{P}$ is a reference pore pressure and H is a suitable reference length. The time necessary for the complete dissolution of the mineral is inversely proportional to the initial mineral surface area and the rate constant of the reaction k_A.

$$t_A = \phi^0\, [k_A\, (36\, \pi\, n_A)^{1/3} (\phi_A^0)^{2/3}]^{-1}\,. \tag{25}$$

It can be seen that there are vast differences in orders of magnitude between the coefficients of the transport terms and the reaction terms in the conservation equations (22). For instance the ratio $\varepsilon = \bar{X}\, \phi^0 / (\rho_A\, \phi_A^0)$ is usually of the order of 10^{-3} to 10^{-10} or even less. This means that the complete dissolution of mineral A from $\phi_A^0 \to 0$ occurs quickly and over a short distance from the front of the propagating reagent X. The dissolution is instantaneous in the limit and/or $k_A \to \infty$. In this limit the front $S(x_i) = 0$ with $X_{S=0} = 0$ becomes an internal boundary. The mass conservation of mineral A is then expressed by a boundary condition for the gradient of X on $S = 0$. In the present case this so-called *Stefan* condition reads

$$\frac{1}{\mathrm{Sc}}\, \phi^{m+1}\, X_{,i}\, n_i = \frac{\rho_A\, \phi_A^0}{\bar{X}\, \phi^0}\, \bar{v}_i\, n_i\,, \qquad n_i = \frac{S_{,i}}{\| S_{,j} \|}\,, \tag{26}$$

where $\bar{v}_i$ is the propagation speed of the front $S = 0$. In many cases, the only computationally tractable way to formulate problems involving propagating dissolution fronts is in the form of a *Stefan* problem.

In this example, we consider fluid infiltrating a rectangular block (H_1, H_2). At the boundary $x_2 = 0$, we prescribe the pore-water velocity component $\phi^f\, v_2 = \phi^f\, v_{\text{inlet}}$ and we define the reference pore pressure as $\bar{P} = \phi^0\, v_{\text{inlet}}\, \frac{H}{K_0}$; ϕ^f and ϕ^0 are the values of the porosity after and before the dissolution of mineral A at the front of X.

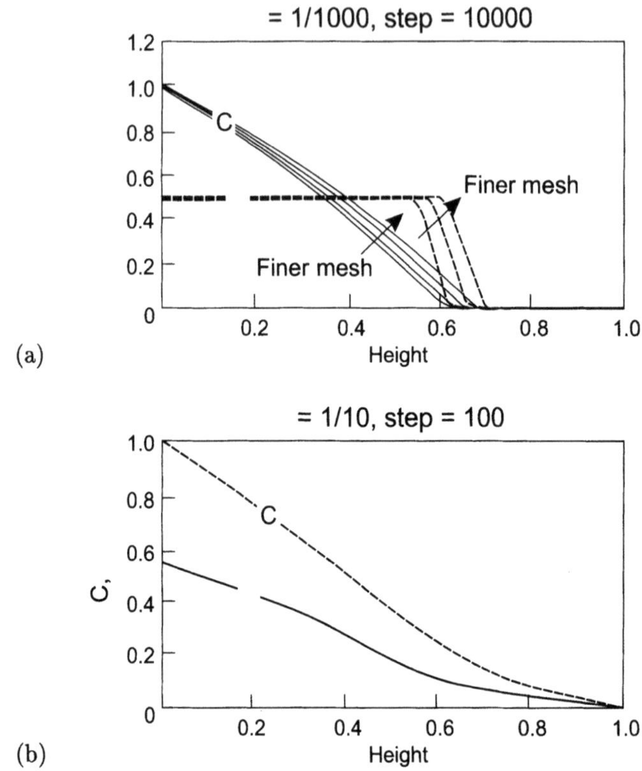

Fig. 3. Porosity alteration ($\phi^f/\phi^0 = 0.5$) and reagent concentration fronts ($X_{\text{inlet}} = 1$): (a) $\varepsilon = 1/1000$; (b) $\varepsilon = 1/10$.

In the computations, we use $\bar{v} = v_{\text{inlet}}$ as the reference fluid velocity. Also prescribed at $x_2 = 0$ is the concentration $X = X_{\text{inlet}}$ and we use X_{inlet} as the reference concentration. At $x_2 = H_2$ we assume $P = 0$ and $X_{,2} = 0$. On $x_1 = 0$ and H_1 we assume periodic boundary conditions, viz. $P_{,1} = 0$ and $X_{,1} = 0$. In the examples in Figures 3 a and 3 b, we assume that the front of the reagent X is initially straight; that $\varepsilon = \bar{X}\,\phi^0/(\rho_A\,\phi_A^0) = 1/1000$, and $\varepsilon = 1/10$, respectively, and $\Delta t = \varepsilon/100$, Sc $= 10$ and $H_1 = H_2 = 1$ in both cases. Results are shown for 500, 800, 1000 and 2000 linear, triangular elements. The dilemma illustrated by these examples is that there exist two length scales; there is the relatively large scale associated with the front of the reagent and the short length scale (determining the accuracy of the solution) associated with the porosity change. As we mentioned in some cases the problem with the two length scales can be resolved by reformulation as a *Stefan* problem. However this is not always possible. The techniques developed for advection dominated flows may be helpful in such cases (e. g. Brooks and Hughes [2]).

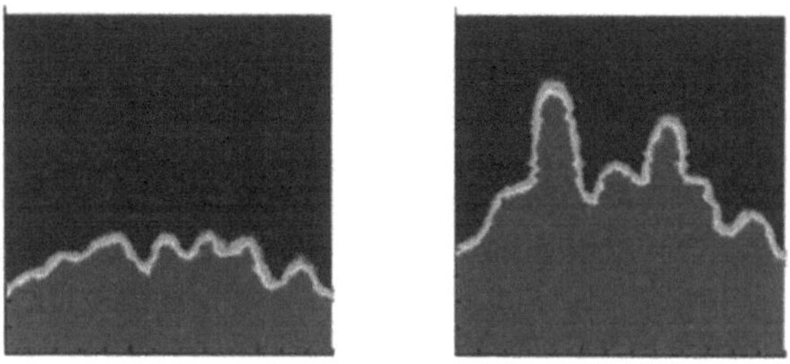

Fig. 4. The evolution of porosity fingering; 2000 linear triangular elements.

In the examples shown in Figure 4, we assume a perturbation in the form of a small harmonic deviation from $\varphi = \varphi^f$ on $x_2 = 0$. In this case we assume $H_1 = H_2 = 13$. This choice was based on the linear instability analysis of Orteleva et al. [15]. The *Schmidt* number is again set equal to 10 and $\varepsilon = 1/100$. Snapshots of the porosity distribution are shown in Figure 4. The fingering instability evolves as expected for this choice of the parameters.

4.2 Multicomponent reactive transport convective flow example

We consider thermal convection in a block of rock composed mainly of calcite $CaCO_3$ and a minor amount of K-feldspar $KAlSi_3O_8$. The background temperature T_0 is 275 °C with a vertical temperature gradient of $\Delta T = 100$ °C. The block contains a fault in the upper layer. The in-plane permeability of the fault is five times the permeability of the surrounding region. The initial value of the ratio of the liquid to gas pressures at the upper surface of the block is 0.66. We assume zero boundary values for the gas/liquid normal velocities and the convective temperature on top and the base and periodic boundary conditions on the sides of the block. In addition, a low pH liquid is injected at the base of the block. The gaseous and liquid fluid phases are CO_2 and H_2O, respectively. The chemical system can be described in terms of the following primary components: Al^{3+}, H^+, K^+, H_2O, H_4SiO_4, CO_2, Ca^{2+}, Mg^{2+}. The hydrolysis reactions for calcite $CaCO_3$, quartz SiO_2, dolomite $CaMg(CO_3)_2$ and K-feldspar $KAlSi_3O_8$ are expressed in terms of these components as follows:

$$\mathrm{CaCO_3} + 2\,\mathrm{H^+} \rightleftharpoons \mathrm{Ca^{2+}} + \mathrm{CO_2} + \mathrm{H_2O}\,, \tag{27}$$

$$\mathrm{CaMg(CO_3)_2} + 4\,\mathrm{H^+} \rightleftharpoons \mathrm{Ca^{2+}} + \mathrm{Mg^{2+}} + 2\,\mathrm{CO_2} + 2\,\mathrm{H_2O}\,, \tag{28}$$

$$\mathrm{KAlSi_3O_8} + 4\,\mathrm{H^+} + 4\,\mathrm{H_2O} \rightleftharpoons \mathrm{K^+} + \mathrm{Al^{3+}} + 3\,\mathrm{H_4SiO_4}\,, \tag{29}$$

$$\mathrm{SiO_2} + 2\,\mathrm{H_2O} \rightleftharpoons \mathrm{H_4SiO_4}\,. \tag{30}$$

Calcite and dolomite reactions are considered as equilibrium reactions governed by the mass action relation (7); K-feldspar and quartz reactions are modelled as kinetic reactions as described by expression (9). The concentrations of K^+ and Al^{3+} are assumed as constant; the Mg^{2+} concentration is calculated from the dolomite saturation. As an example for a conservation equation of type (13) we consider Ca^{2+} conservation; we obtain

$$\begin{aligned}&(\phi\, C_{\mathrm{Ca}})_{,t} + 1/\phi_0 \left[k_l^r\, \phi^3\, (P_{l,i}/\gamma_l - \mathrm{Ra}_l\, T\, \delta_{3\,i})\, C_{\mathrm{Ca}}\right]_{,i} - \\ &- 1/\,\mathrm{Le}\, (\phi^{m+1}\, C_{\mathrm{Ca},\,i})_{,i} = 1/\phi_0\, (M_{\mathrm{MgCa(CO_3)_2}} + M_{\mathrm{CaCO_3}})_{,t}\,,\end{aligned} \tag{31}$$

where C_{Ca} is the concentration of Ca^{2+}.

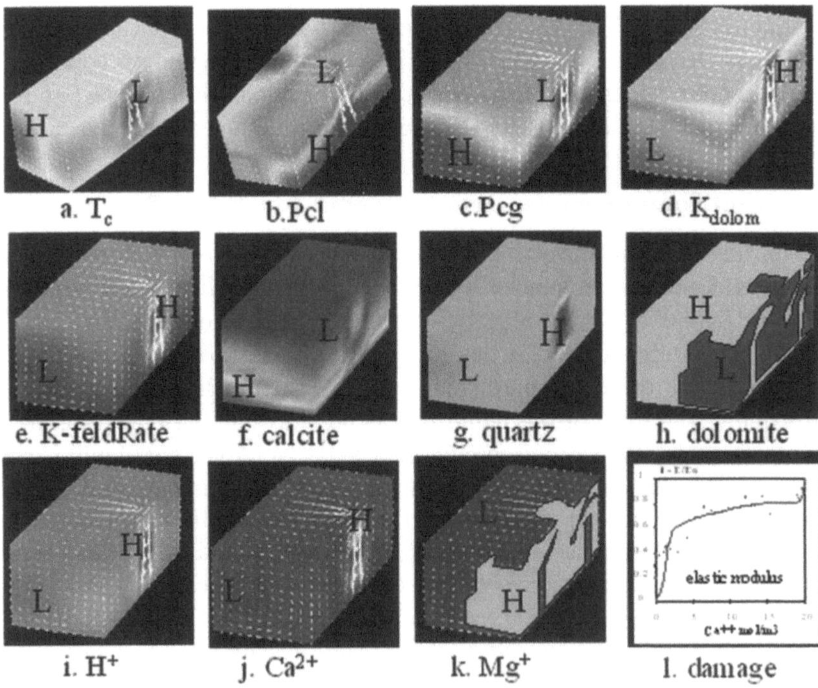

Fig. 5. (a – c) Convective temperature, liquid and gas pressures; (d) dolomite equilibrium constant; (e) rate of feldspar dissolution; (f – k) concentration of minerals and aqueous species; (l) damage vs. Ca^{2+} concentration. H indicates high value and L indicates low value.

We present results of a transient analysis after 100 non-dimensional time steps (time scale: $\bar{t} = \overline{\rho\, C\, p}\, H^2/\lambda$). An initial harmonic temperature perturbation was assumed to trigger clockwise convective flow. Figures 5 a – c show the convective temperature, liquid and gas pressure distributions, respectively.

The liquid velocity arrows indicate the focussing effect of the fault zone. Figure 5 d shows the value of the pressure/temperature dependent equilibrium constant for dolomite. Figures 5 e and 5 f show the regions of low calcite concentration (corresponding to dissolution) and high K-feldspar dissolution are the regions where hydrogen ion is focused (Figure 5 i). Calcite dissolution leads to dolomite precipitation as presented in Figure 5 h. The region of high quartz concentration (Figure 5 g) corresponds to the region where hydrogen ion (acidity) is high and K-feldspar dissolution rate is highest. Silicic acid, a product of feldspar hydrolysis, precipitates as quartz. The dissolution of calcite and the resulting precipitation of dolomite lead to a net solid volume decrease (Shvarov [16]) and an increased porosity in these regions. In Figure 5 j, the Ca^{2+} concentration is presented. The dissolution of calcite and the absence of dolomite in the fault due to its high K (equilibrium constant) allowed focussing the Ca^{2+} and Mg^{2+} in the fault. In a completely different context, detecting the concentrations of Ca^{2+} was important for mechanical problems. Increasing concentrations of Ca^{2+} are reported by Le Belligo et al. [8] to cause strength degradation of concrete (see Figure 5 l).

4.3 Coupled reactive transport and mechanical deformation example

We consider a stiff, smooth foundation on an infinite half plane (*Prandtl* case) as shown in Figure 6. The substrate deforms and partially plastifies (*von Mises* yield, *Prandtl-Reuss* flow rule). We then prescribe a constant concentration of a leachant on the surface of the half plane, outside the foundation. The concentration of the leachant diffuses through the half plane, lowering the yield strength and ultimately causing unacceptably large deformation of the structure.

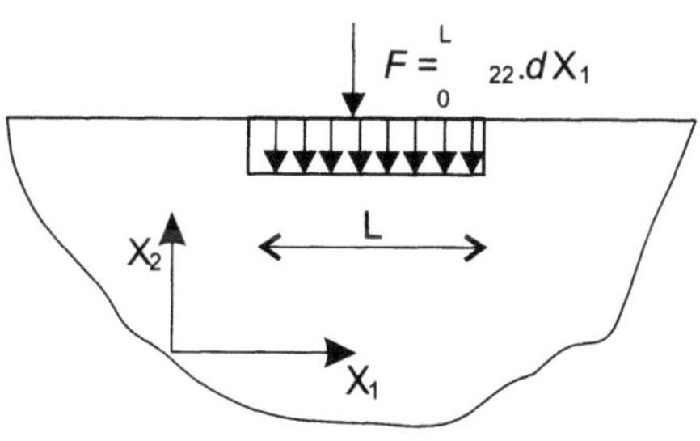

Fig. 6. A rigid, smooth foundation on an infinite half plane.

The load bearing capacity of the strip foundation of length L laid on a rock bed of yield strength τ_y is expressed as follows:

$$F_C = (2 + \pi)\, \tau_y\, L\,. \tag{32}$$

The tensor C_{ijkl} of the incremental moduli relates the strain to the stress increment as follows:

$$\mathrm{d}\sigma_{ij} = C_{ijkl}\,\mathrm{d}\varepsilon_{kl}\,, \quad \mathrm{d}\varepsilon_{ij} = \frac{1}{2}\left(\mathrm{d}u_{i,j} + \mathrm{d}u_{j,i}\right). \tag{33}$$

In the absence of volume forces and assuming hydrostatic equilibrium, we have the equilibrium equation as follows:

$$\mathrm{d}\sigma_{ij,j} = 0\,. \tag{34}$$

The components of C_{ijkl} are defined by the well known relationships

$$C^{e}_{ijkl} = \mu\left(\delta_{ik}\,\delta_{jl} + \delta_{il}\,\delta_{jk}\right), \tag{35}$$

$$C_{ijkl} = C^{e}_{ijkl} - \frac{\alpha}{\mu}\left(\frac{\mu}{\tau}\right)^{2}\sigma'_{ij}\,\sigma'_{kl}\,, \tag{36}$$

where σ'_{ij} is defined as

$$\sigma'_{ij} = \sigma_{ij} - \frac{1}{3}\sigma_{kk}\,\delta_{ij} \tag{37}$$

and

$$\tau = \sqrt{\frac{1}{2}\sigma'_{ij}\,\sigma'_{ij}}\,, \quad F = \tau - \tau_y\,, \quad \alpha = \begin{cases} 0 & \text{if } F < 0 \\ 1 & \text{else} \end{cases}\,. \tag{38}$$

The injection of a reagent changes the yield strength $\tau_y = \tau_0$ of the rock. This reduction in the yield strength can be made dependent on the leaching concentration or on the porosity change accompanying a chemical reaction as in the previous example. One possible form of dependence of the yield strength τ_y on the concentration C is shown below:

$$\tau_y = \tau_{\text{residual}} - (\tau_0 - \tau_{\text{residual}})\exp(-\alpha\,C)\,, \tag{39}$$

where α is an arbitrary coefficient larger than 1, and t_{residual} is the residual yield strength. In general terms, the injected reagent is assumed to dissolve part of the constituents of the rock and deposit a new mineral that at least has different molar volume (and usually different cohesive characteristics). This process changes the solid volume and here we assume it increases the pore space, a similar situation happens in in situ leaching processes. The rate of porosity change, which here reflects the extent of the reaction and is equal to the rate of mineral dissolution, can be expressed as follows (Orteleva [14]):

$$\phi_{,t} = K\,(\phi_f - \phi)^{2/3}\,(C - 1)\,, \tag{40}$$

where C is the concentration, φ is the current porosity and φ_f is the residual or final porosity when the reaction ceases. Note that the power 2/3 relates to the surface area of the dissolving mineral. K is a rate constant. A diffusive mass balance equation is used to calculate the leachant concentration:

$$C_{,t} = (D\,C_{,i})_{,i} - K\,(\phi_f - \phi)^{2/3}\,(C-1)\,, \tag{41}$$

where D is the diffusion coefficient which is taken as zero for the foundation material.

The numerical algorithm for solving the mechanical deformation problem is outlined in (42):

$$\begin{aligned}
&1. \quad \sigma_{ij} = \sigma'_{ij} + 2\,\mu\,\mathrm{d}\varepsilon_{ij} - \delta_{ij}\,\mathrm{d}P\\
&2. \quad \sigma'_{ij} = \sigma_{ij} - \frac{1}{3}\,\sigma_{kk}\,\delta_{ij}\\
&3. \quad \tau = \sqrt{\frac{1}{2}\,\sigma'_{ij}\,\sigma'_{ij}}\\
&4. \quad F = \tau - \tau_y \begin{cases} F \geq 0 & \text{goto } 5\,\text{b}\\ F < 0 & \text{goto } 5\,\text{a}\end{cases}\\
&5\,\text{a}. \quad C_{ijkl} = C^e_{ijkl} \quad \text{goto momentum (34)}\\
&5\,\text{b}. \quad \sigma'_{ij} \leftarrow \sigma'_{ij} - \dot{\lambda}\,\frac{\sigma'_{ij}}{\tau/\mu}\,, \quad \dot{\lambda} = \frac{1}{\mu}\,(\tau - \tau_y)\,,\\
&\qquad \text{establish } C_{ijkl} \text{ as in (36), goto momentum (34).}
\end{aligned} \tag{42}$$

In this example we use a mesh of 3 000 nodes of two regions 1 and 2 to simulate the concrete footing and the supporting rock, Figure 7 a. The concrete foundation is considered as rigid. The strip footing is loaded incrementally with a displacement $X_0 = 0.0005$ until the yield strength is approached. The base of the block has zero displacement. The vertical sides of the block have the natural boundary conditions. The resulting load displacement curve is shown in Figure 7 b. The normalized (by the initial yield stress of the region) shear stress distribution is shown in Figure 7 c. It can be seen that yielding occurs in the rock almost at the edges where the footing meets the supporting rock. The vertical displacement is shown in Figure 7 d. The velocity vectors are depicted in Figure 7 e.

In order to model the effect of the corrosive reagent on the stability of the strip footing, at the stage of loading when the vertical displacement in the Figure 7 b reaches 0.055 no further displacement is applied to the foundation top surface and the stress is kept constant corresponding to this total displacement of 0.055. At that stage, a reagent infiltrates in the region around the foundation. This is achieved by specifying a non-dimensional concentration of unity of the reagent at the top surface around the footing. Figures 8 (a – d) show the resulting shear stress, vertical displacement, velocity arrows and reagent concentration one time step after injecting the reagent. It can be seen that a large domain under the foundation that was far from yielding in Figure 7 c has yielded now. The foundation has immediately settled more, the maximum downward vertical displacement has increased from 0.055 in Figure 7 d to 0.06 in the case with the infiltrating reagent Figure 8 b, the displacement contours are identical to those in Figure 7 d with a larger value. We observe a change in the velocity direction. Soon after injecting the reagent, we observe a quick loss of stability with time as Figure 8 e suggests.

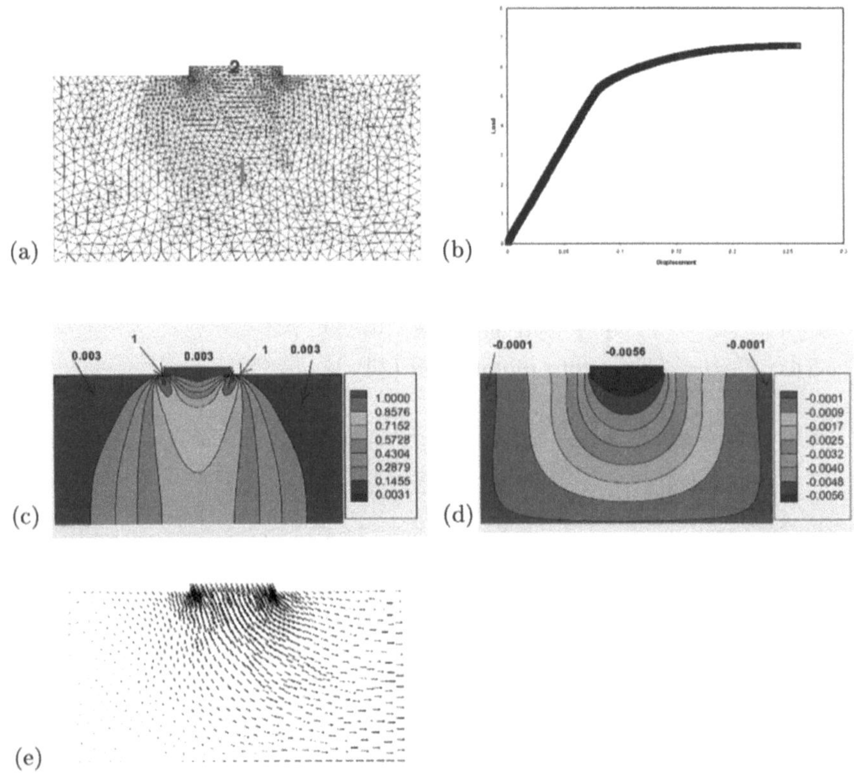

Fig. 7. Loading of the strip footing: (a) mesh; (b) load-displacement curve; (c) shear stress τ; (d) vertical displacement; (e) velocity arrows.

The situation observed in Figure 8 is due to the chemical damage, reduction of the yield strength of the system due the leaching reagent. The stage in loading at which this reduction of strength occurs ie the safety factor or distance from yield is critical to the final stability or instability of the foundation.

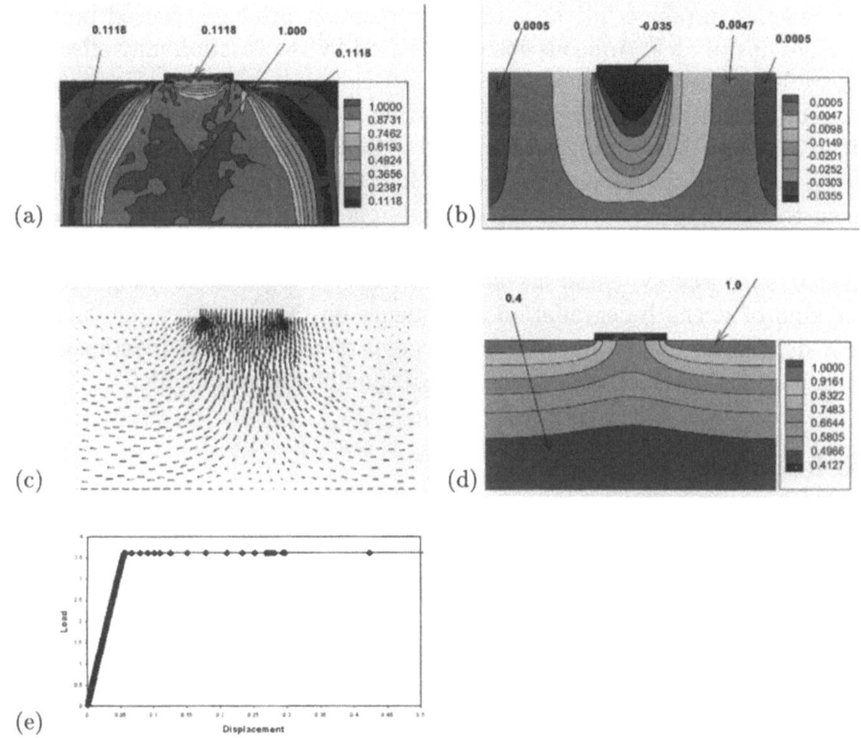

Fig. 8. (a) Shear stress due to adding chemistry after loading the footing; (b) vertical displacement; (c) velocity arrows; (d) leachant concentration; (e) load-displacement curve.

5 Conclusions

In this paper we presented a simulator for coupled reactive transport processes in porous rock. The infiltration instability model highlighted the different scales involved in reactive transport modelling. In addition to the relative importance of diffusive and advective transport phenomena there is also the time scale of the kinetic reaction rate which can be orders of magnitude different from that of the transport processes. Such a situation imposes a numerical difficulty and requires special treatment. The model has also been used to produce porosity fingering as an infiltration instability once the correct ratios of parameters based on linear instability analysis are used. The potential of the simulator was illustrated by means of the multicomponent reactive transport convective flow example of carbonate rock containing silicate reacting with an acidic fluid. As a mixing mechanism we assumed natural, thermal convection. Fluid flow was focused by a preexisting fault localizing the reaction products in the fault. The reaction products caused porosity changes of the rock. Aqueous species released by the reaction may affect the strength properties of the rock. In the last example the effect of a corrosive reagent infiltrating into the rock supporting a stable strip footing was investigated. A *von Mises* material was used with a *Prandtl-Reuss* flow rule. The deterioration of the supporting rock yield strength resulted in higher settlement to the rigid foundation. The example demonstrates the importance of taking the effect of such reagents when considering the safety of structures built in sites with potential chemical activities. It is worth pointing out that the kind of strength degradation dependence on the chemistry can vary but it is dependent on the nature and progress of the chemical reactions. The examples demonstrate the capabilities of the finite element code.

References

1. Bethke, C. M.: *Geochemical Reaction Modeling*. Oxford University Press, New York 1996.
2. Brooks, A. N., Hughes, T. J. R.: Streamline upwind/Petrov-Galerkin formulations for convection dominated flows with particular emphasis on the incompressible Navier-Stokes equations. *Computer Methods in Applied Mechanics and Engineering* **32** (1982), 199–259.
3. Chadam, J., Hoff, D., Merino, E., Ortoleva, P., Sen, A.: Reactive infiltration instabilities. *Journal of Applied Mathematics* **36** (1986), 207–221.
4. Connolly, J. A. D.: Devolatilization-generated fluid pressure and deformation-propagated fluid flow during prograde regional metamorphism. *Journal of Geophysical Research* **102**(B8) (1997), 18149–18173.
5. Freij-Ayoub, R., Walshe, J. L., Mühlhaus, H.-B.: Coupled multispecies reactive transport modeling and rock alteration. *Australian Journal of Earth Sciences* **47** (2000), 885–894.
6. van Genuchten, M.: A closed form equation for predicting the hydraulic conductivity of unsaturated soils. *Soil Science society of American Journal* **44** (1980), 892–898.
7. Lasaga, A.: Chemical kinetics of water-rock interactions. *Journal of Geophysical Research* **89**(B6) (1984), 4009–4025.
8. Le Belligo, C., Pijaudier-Cabot, G., Bruno, G.: Coupled Mechanical and Chemical Damage in Concrete Structures. In Benallal, A. (ed.): *Continuous Damage and Fracture*, Elsevier, 2000, pp. 373–407.
9. Lichtner, P. C., Walton, J.: *Near-Field Liquid-Vapor Transport in Partially Saturated High Level Nuclear Waste Repository*. Centre for Nuclear Waste Regulatory Analyses, CNWRA94-022, Texas 1994.
10. Lichtner, P. C.: Continuum Formulation of Reactive Transport. *Reactive Transport in Porous Media, Reviews in Mineralogy, Mineralogical Society of America* **34** (1996), 5–81.
11. Mühlhaus, H.-B., Liu, J., Hobbs, B.: A Porosity Evolution Model for In situ Leaching Processes. In: *Proceedings of Second Australian Congress on Applied Mechanics*, Canberra 1999.

12. Mühlhaus, H.-B., Hobbs, B., Freij-Ayoub, R., Walshe, J. L., Ord, A., Stokes, N.: Multiple Scales in Rock Alterations. In: *International Workshop on Localization and Bifurcation in Geomechanics, Proc. IWBL99 Conference*, 2000.
13. Nield, D. A., Bejan, A.: *Convection in Porous Media*. Springer-Verlag, New York 1992.
14. Ortoleva, P. J.: *Geochemical Self-Organization*. Oxford University Press, 1994.
15. Ortoleva, P. J., Merino, E., Moore, C., Chadam, J.: Geochemical self-organization I: Reaction- transport feedbacks and modelling approach. *American Journal of Science* **287** (1987), 979–1007.
16. Shvarov, Y. V.: Algorithmization of Numerical Equilibrium Modeling of Dynamical Geochemical Processes. *Geochemistry International* **37**(6) (1999), 571–576.
17. Steefel, C., Lasaga, A.: A coupled model for transport of multiple species and kinetic precipitation/dissolution reactions with application to reactive flow in single phase hydrothermal systems. *American Journal of Science* **294** (1994), 529–592.
18. Walsh, M.: *Geochemical Flow Modeling*. Ph. D. thesis, The University of Texas at Austin 1983.

A numerical model and its finite element solution for multiphase flow: Application to pulp and paper processing

D. R. J. Owen, S. Y. Zhao, and E. A. de Souza Neto

Department of Civil Engineering, University of Wales Swansea,
Swansea SA2 8PP, United Kingdom

Abstract. This paper presents a general framework for the large scale finite element simulation of rolling and compression of multiphase deformable porous media. The main objective is the simulation of pulp and paper processing operations. On the theoretical side, the generalized *Biot* theory is extended and modified to derive the governing equations of general multiphase flow through porous media. These equations are specialised to the case of two-fluid flow having the velocity of the solid skeleton, the pressure, and saturation of the wetting fluid as the primary unknowns. On the numerical side, the finite element method is used for spatial discretization of the relevant equations. This is combined with a *Newmark* scheme discretisation in time. A direct solution scheme is adopted in the treatment of the coupling between the solid, liquid, and gas phases. To demonstrate the effectiveness of the developed framework, the simulation of a compression and a rolling process of relevance to the paper industry are presented.

1 Introduction

Extrusion and rolling of multiphase materials are important industrial operations that have been used in various manufacturing processes for centuries. Typical applications include the dewatering of pulp in paper production, the extraction of juice from prepared sugar cane in sugar production, and a wide range of food processing operations. In spite of its relevance, the technology of processing multiphase materials has relied almost exclusively on empirical knowledge acquired through experience. Only a limited number of studies have been carried out to provide a more rigorous scientific approach to the analysis and design of the relevant industrial operations [8, 9, 15]. This fact is probably justified by the complexity of the physical phenomena involved in such processes. Extremely large strains, strong and highly nonlinear coupling between solid and fluid phases, frictional contact between tools and multiphase bodies as well as thermal coupling with possible phase changes are highly complex phenomena typically interacting in operations of this kind.

The objective of the present paper is to present a general framework for the finite element simulation of compression and rolling of multiphase deformable porous media. Our main motivation is the current industrial need for optimisation of pulp and paper processing operations. On the theoretical side, a macroscopic approach is adopted to model the porous medium, with the overall stress linked to the solid skeleton stress and fluid pressures through the principle of effective stress. The solid phase is modelled as a compressible highly nonlinear rate-independent elasto-plastic material. It is assumed that the two immiscible fluid phases fill up the entire pore space and are separated from one another by interfaces. The interface between the two fluid phases form a capillary system and their effects can be embodied by means of the capillary pressure between the fluids. The resulting formulation has the velocity of the solid skeleton, the liquid pressure, and saturation as the primary variables.

On the computational side, the finite element method is used in the spatial discretisation of the relevant equations. This is combined with a *Newmark* scheme for time domain discretisation. In addition, under quasi-static conditions, a reduced rank formulation can be used which allows the use of equal finite element interpolation order for all unknowns.

The framework presented is applied to the simulation of a rolling and a compression operation relevant to the paper processing industry.

2 The multiphase model equations

In this section, the basic equations that model the behaviour of a generic porous medium with an arbitrary number m of fluid phases are reviewed and specialised for the case of interest where only two fluids are present. We recall that the main motivation behind the present work is precisely the simulation of processes involving two fluid phases (water and steam). As opposed to the case of an arbitrary number of fluid phases, considerable simplifications are possible in the presence of only two fluid components. As we shall see, the simplified two-fluid-only equations can be treated by an efficient computational procedure which allows the effective simulation of large scale problems of interest within reasonable time scales. The equations presented in this section are the basis of the finite element solution procedure discussed in Section 3.

2.1 Generic multiphase equilibrium equations

We start by stating the differential or local form of the overall equilibrium equation for a multiphase porous medium with m fluid phases. These can be written as

$$\mathbf{L}^T[\boldsymbol{\sigma}] + \rho\,\mathbf{g} = \rho\,\ddot{\mathbf{u}} + \sum_{k=1}^{m} \rho_k\, n\, S_k\, \frac{\mathrm{D}}{\mathrm{D}t}\left(\frac{\dot{\boldsymbol{\omega}}_k}{n\, S_k}\right), \tag{1}$$

where $\mathbf{L}$ is the differential operator with the following matrix representation:

$$\mathbf{L} = \begin{bmatrix} \partial/\partial x_1 & 0 & 0 \\ 0 & \partial/\partial x_2 & 0 \\ 0 & 0 & \partial/\partial x_3 \\ \partial/\partial x_2 & \partial/\partial x_1 & 0 \\ 0 & \partial/\partial x_3 & \partial/\partial x_2 \\ \partial/\partial x_3 & 0 & \partial/\partial x_1 \end{bmatrix}, \tag{2}$$

and $\boldsymbol{\sigma}$ is the array of total stress components:

$$\boldsymbol{\sigma} = [\sigma_{11} \ \sigma_{22} \ \sigma_{33} \ \sigma_{12} \ \sigma_{23} \ \sigma_{13}]^T. \tag{3}$$

The term $\mathbf{L}^T[\boldsymbol{\sigma}]$ in (1) is the divergence of the stress tensor. The vector $\mathbf{g}$ is the gravitational acceleration and $\mathbf{u}$ and $\ddot{\mathbf{u}}$ are, respectively, the displacement and acceleration vectors of the solid phase. The scalar n is the porosity of the porous medium, S_k, ρ_k, and $\dot{\boldsymbol{\omega}}_k$ are, respectively, the saturation, density, and *Darcy* velocity of fluid phase k. The symbol

$$\frac{\mathrm{D}}{\mathrm{D}t}(\cdot)$$

stands for the material time derivative of $(\cdot)$ and ρ is the density of the solid-fluid mixture which can be expressed as

$$\rho = \rho_s(1-n) + \sum_{k=1}^{m} \rho_k \, n \, S_k\,, \tag{4}$$

with ρ_s denoting the density of the solid.

Note that the saturations of the m fluids must satisfy the trivial constraint

$$\sum_{k=1}^{m} S_k = 1\,. \tag{5}$$

2.2 Effective stress principle

The *principle of effective stress* (Terzaghi [13, 14]) is the starting point of continuum modelling of multiphase media. This principle relates the total stress $\boldsymbol{\sigma}$, acting on the multiphase medium, to the effective stress $\boldsymbol{\sigma}_e$, which acts on the solid skeleton, and the pore pressure P by means of the following expression:

$$\boldsymbol{\sigma} = \boldsymbol{\sigma}_e - \mathbf{m}\,P\,, \tag{6}$$

where

$$\mathbf{m} = [1 \quad 1 \quad 1 \quad 0 \quad 0 \quad 0]^T \tag{7}$$

is the array representation of the identity tensor. Note that, here, the stress is defined as tension positive, whereas the pore pressure is assumed compression positive.

The above relationship can be extended to account for the compressibility of the solid skeleton, as shown in [1, 16]:

$$\boldsymbol{\sigma} = \boldsymbol{\sigma}'' - \mathbf{m}\,\alpha\,P\,. \tag{8}$$

The *real effective stress* $\boldsymbol{\sigma}''$ depends exclusively on the constitutive equations of the solid phase and, in the present context, it is the outcome of the constitutive integration algorithm referred to in Section 5. If multiple fluids are present, the pore pressure P in the mixture of fluids can be postulated as a weighted average pressure of the individual fluid phases [5]:

$$P = \sum_{k=1}^{m} S_k\, P_k\,. \tag{9}$$

For the particular case of only two porous fluids (liquid and gas in the present) we have

$$P \;=\; S_w\, P_w \;+\; S_g\, P_g\,, \tag{10}$$

where, in view of (5),

$$S_w + S_g = 1\,. \tag{11}$$

In the above, the subscripts w and g denote quantities associated, respectively, with the liquid (water) and gas (steam) phase. The parameter α in (8) is derived in [3] for isotropic media as a constant:

$$\alpha = 1 - \frac{K_t}{K_s}\,, \tag{12}$$

where K_s and K_t are, respectively, the bulk moduli of the solid material and the multiphase medium. The above constant value of α can be used under small strain conditions. However, processing operations such as rolling and compression are characterised by very large strains. Under such conditions α can no longer be regarded as constant. The porosity and saturation of the porous medium change considerably affecting the bulk modulus and, consequently, the value of α. In such cases (as adopted here), α is an experimentally determined function of porosity and liquid saturation. Typically, in paper processing applications α assumes values between 0.6 and 1.0.

2.3 Overall equilibrium with two fluid phases only

We now introduce two simplifications into the original overall equilibrium equation (1):

1. Firstly, in view of the specific industrial problems we have in mind, we shall limit our model to the particular case, where only two porous fluids (liquid and gas in our case) are present.

2. In addition, it is assumed that the relative acceleration of the fluids with respect to the solid phase are not significant as compared to the acceleration of the solid. In this case, the material time derivative of the *Darcy* velocity in equation (1) can be neglected. This assumption is compatible with the actual conditions observed in the industrial problems of interest.

Under such circumstances, the equilibrium equation (1) for the porous medium can be reduced to [17, 18]

$$\mathbf{L}^T[\boldsymbol{\sigma}''] + \rho\,\mathbf{g} = \rho\,\ddot{\mathbf{u}} + \alpha\,[\nabla P_w + \nabla(P_c\,S_g)]\,, \tag{13}$$

where P_c is the capillary pressure between the two fluid phases:

$$P_c = P_g - P_w\,. \tag{14}$$

The capillary pressure is a function of the liquid saturation S_w, i. e.

$$P_c = P_c(S_w)\,. \tag{15}$$

This function is determined experimentally and depends on the particular solid and fluid phases involved. Examples of such functions have been determined by Bloch [2] in the context of paper processing.

2.4 Equilibrium equations for the fluid phases

The equilibrium equations for each fluid phase k in the porous medium are given by the generalized *Darcy* law [4]:

$$\mathbf{k}_k^{-1}\,\dot{\boldsymbol{\omega}}_k \;=\; -\nabla P_k + \rho_k\left[\mathbf{g} - \ddot{\mathbf{u}} - \frac{\mathrm{D}}{\mathrm{D}t}\left(\frac{\dot{\boldsymbol{\omega}}_k}{n\,S_k}\right)\right], \tag{16}$$

where $\mathbf{k}_k$ is the *Darcy* coefficient matrix of permeability of the k-th fluid.

2.5 Mass conservation

The mass conservation principle is expressed by the following mass conservation equations for the solid and the k-th fluid:

$$\frac{\mathrm{D}}{\mathrm{D}t}M_s = \frac{\mathrm{D}}{\mathrm{D}t}\int_v (1-n)\,\rho_s\,\mathrm{d}v = 0\,, \tag{17}$$

$$\frac{\mathrm{D}}{\mathrm{D}t} M_k = \frac{\mathrm{D}}{\mathrm{D}t} \int_v n\, S_k\, \rho_k\, \mathrm{d}v = 0\,, \tag{18}$$

where M_s and M_k are the masses of the solid and the k-th fluid, respectively. The above integral equations lead, after some algebra [6], to the following point-wise expressions:

$$\frac{1-n}{\rho_s}\frac{\mathrm{D}\rho_s}{\mathrm{D}t} + \frac{n}{\rho_k}\frac{\mathrm{D}\rho_k}{\mathrm{D}t} + \frac{n}{S_k}\frac{\mathrm{D}S_k}{\mathrm{D}t} + \dot{u}_{i,i} + \frac{1}{S_k}\dot{\omega}_{ki,i} + \frac{\dot{\omega}_{ki}}{S_k\,\rho_k}\frac{\partial \rho_k}{\partial x_i} = 0\,, \tag{19}$$

where the following notation is employed:

$$(\,\cdot\,)_{,i} = \frac{\partial(\,\cdot\,)}{\partial x_i}\,, \tag{20}$$

and summation is implied over the repeated index i. The mass conservation equations for the solid and the k-th fluid can also be written in differential form as

$$\frac{\mathrm{D}(\rho_s\, V_s)}{\mathrm{D}t} = 0\,, \tag{21}$$

$$\frac{\mathrm{D}(\rho_k\, V_k)}{\mathrm{D}t} = 0\,, \qquad k = 1\,, \dots\,, m\,. \tag{22}$$

The above equations give [19]

$$\frac{1}{\rho_s}\frac{\mathrm{D}\rho_s}{\mathrm{D}t} = \frac{1}{1-n}\left[(\alpha - n)\frac{1}{K_s}\frac{\mathrm{D}P}{\mathrm{D}t} - (1-\alpha)\,\dot{u}_{i,i} - (\alpha - n)\,\beta_s\,\dot{T}\right]\,, \tag{23}$$

$$\frac{1}{\rho_k}\frac{\mathrm{D}\rho_k}{\mathrm{D}t} = \frac{1}{K_k}\frac{\mathrm{D}P_k}{\mathrm{D}t} - \beta_k\,\dot{T}\,, \qquad k = 1\,, \dots\,, m\,, \tag{24}$$

where V_s and V_k are the volumes of the solid and the k-th fluid respectively, β_s and β_k are the thermal expansion coefficients, T is the temperature, and K_k is the bulk modulus of the k-th fluid.

From equations (19), (23), and (24), and considering (5) and (9), the mass conservation equation of the k-th fluid phase in the general multiphase flow can be given as

$$\begin{aligned} &C_{0k}\,\dot{P}_1 + C_{1k}\,\dot{S}_1 + \dots + C_{n-1,k}\,\dot{S}_{n-1} + \\ &\quad + \alpha\,S_k\,\dot{u}_{i,i} - \beta_{sk}\,\dot{T} + n\,\dot{S}_k + \dot{\omega}_{ki,i} + \frac{\dot{\omega}_{ki}}{\rho_k}\frac{\partial \rho_k}{\partial x_i} = 0\,. \end{aligned} \tag{25}$$

The two-fluid case

The mass conservation equations for a two-fluid situation are obtained as a particular case of the above equations as

$$\begin{aligned} &(C_{ww} + C_{wg})\,\dot{P}_w + C_{wg}\frac{\mathrm{d}P_c}{\mathrm{d}S_w}\,\dot{S}_w + \alpha\,S_w\,\dot{u}_{i,i} - \beta_{sw}\,\dot{T} + \\ &\quad + \dot{\omega}_{wi,i} + n\,\dot{S}_w - C_{wg}\frac{P_c}{S_g}\,\dot{S}_w + \frac{\dot{\omega}_{wi}}{\rho_w}\frac{\partial \rho_w}{\partial x_i} = 0\,, \end{aligned} \tag{26}$$

$$(C_{gw} + C_{gg})\,\dot{P}_w + C_{gg}\,\frac{\mathrm{d}P_c}{\mathrm{d}S_w}\,\dot{S}_w + \alpha\,S_g\,\dot{u}_{i,i} - \beta_{sg}\,\dot{T} + \\ + \dot{\omega}_{gi,i} + n\,\dot{S}_g - C_{wg}\,\frac{P_c}{S_w}\,\dot{S}_w + \frac{\dot{\omega}_{gi}}{\rho_g}\,\frac{\partial \rho_g}{\partial x_i} = 0\,, \tag{27}$$

where

$$C_{ww} = \frac{S_w^2\,(\alpha - n)}{K_s} + \frac{S_w\,n}{K_w}\,, \tag{28}$$

$$C_{gg} = \frac{S_g^2\,(\alpha - n)}{K_s} + \frac{S_g\,n}{K_g}\,, \tag{29}$$

$$C_{wg} = C_{gw} = \frac{S_w\,S_g\,(\alpha - n)}{K_s}\,, \tag{30}$$

with ρ_g and K_g denoting the density and the bulk modulus of the gas.

3 Finite element solution

The derivation of a finite element scheme for simulation of multiphase flow through deformable porous media requires two basic steps:

1. Finite element discretisation of the equations of equilibrium and mass conservation.

2. Time discretisation of the relevant rate quantities.

These are briefly described in the following.

3.1 Finite element discretization

Following standard finite element procedures [20], the differential form (13) together with (26) and (27) combined with (16) are re-written in integral form and the relevant functional sets are replaced with finite dimensional counterparts generated through standard finite element interpolation functions. The resulting spatially discrete equations are

$$\mathbf{P}(\bar{\mathbf{u}}) - \mathbf{Q}_p^T\,\bar{\boldsymbol{P}}_w + (\mathbf{Q}_s^*)^T\bar{\boldsymbol{S}}_w + \mathbf{M}_u\,\ddot{\bar{\mathbf{u}}} = \mathbf{F}^u\,, \tag{31}$$

$$\mathbf{M}_p\,\ddot{\bar{\mathbf{u}}} + \mathbf{Q}_p\,\dot{\bar{\mathbf{u}}} + \mathbf{C}_{pp}\,\dot{\bar{\boldsymbol{P}}}_w + \mathbf{C}_{ps}\,\dot{\bar{\boldsymbol{S}}}_w + \mathbf{H}_{pp}\,\bar{\boldsymbol{P}}_w + \mathbf{H}_{ps}\,\bar{\boldsymbol{S}}_w = \mathbf{F}^p\,, \tag{32}$$

$$\mathbf{M}_s\,\ddot{\bar{\mathbf{u}}} + \mathbf{Q}_s\,\dot{\bar{\mathbf{u}}} + \mathbf{C}_{sp}\,\dot{\bar{\boldsymbol{P}}}_w + \mathbf{C}_{ss}\,\dot{\bar{\boldsymbol{S}}}_w + \mathbf{H}_{sp}\,\bar{\boldsymbol{P}}_w + \mathbf{H}_{ss}\,\bar{\boldsymbol{S}}_w = \mathbf{F}^s\,. \tag{33}$$

In the above, $\mathbf{P}$ is the internal force vector:

$$\mathbf{P}(\bar{\mathbf{u}}) = \int_{\Omega} \mathbf{B}^T \boldsymbol{\sigma}''(\bar{\mathbf{u}})\,\mathrm{d}\Omega\,, \tag{34}$$

where **B** is the standard discrete symmetric gradient operator (or strain-displacement matrix) of finite element analysis:

$$\mathbf{B} = \mathbf{L}\,\mathbf{N}_{\dot{u}}\,, \tag{35}$$

with $\mathbf{N}_{\dot{u}}$ denoting the global shape functions for interpolation of the solid skeleton velocity. As already mentioned above, $\boldsymbol{\sigma}''$ is obtained as the outcome of an algorithm for numerical integration of the constitutive equations of the solid material (referred to in Section 5). This implies that $\boldsymbol{\sigma}''$ is an implicit function of the strain or, in the finite element context, an implicit function of the vector of nodal displacements of the solid skeleton $\bar{\mathbf{u}}$:

$$\boldsymbol{\sigma}'' = \boldsymbol{\sigma}''(\bar{\mathbf{u}})\,. \tag{36}$$

The other terms taking part in (31) – (33) are given by

$$\mathbf{Q}_p^T = \int_\Omega \mathbf{B}^T \mathbf{m}\,\alpha\,\mathbf{N}_P\,\mathrm{d}\Omega\,, \tag{37}$$

$$(\mathbf{Q}_s^*)^T = \int_\Omega \mathbf{B}^T \mathbf{m}\,\alpha \left(S_n \frac{\mathrm{d}P_c}{\mathrm{d}S_w} - P_c \right) \mathbf{N}_s\,\mathrm{d}\Omega\,, \tag{38}$$

$$\mathbf{C}_{pp} = \int_\Omega \mathbf{N}_P^T \left(\frac{\alpha - n}{K_s} + \frac{n\,S_w}{K_w} + \frac{n\,S_g}{K_g} \right) \mathbf{N}_P\,\mathrm{d}\Omega\,, \tag{39}$$

$$\mathbf{H}_{pp} = \int_\Omega (\nabla \mathbf{N}_P)^T (\mathbf{k}_w + \mathbf{k}_g)\,\nabla \mathbf{N}_P\,\mathrm{d}\Omega\,, \tag{40}$$

$$\mathbf{C}_{ps} = \int_\Omega \mathbf{N}_P^T \left[\left(\frac{\alpha - n}{K_s} + \frac{n}{K_g} \right) S_g \frac{\mathrm{d}P_c}{\mathrm{d}S_w} - \frac{(\alpha - n)\,P_c}{K_s} \right] \mathbf{N}_s\,\mathrm{d}\Omega\,, \tag{41}$$

$$\mathbf{H}_{sp}^T = \mathbf{H}_{ps} = \int_\Omega (\nabla \mathbf{N}_P)^T\,\mathbf{k}_g \frac{\mathrm{d}P_c}{\mathrm{d}S_w}\,\nabla \mathbf{N}_s\,\mathrm{d}\Omega\,, \tag{42}$$

$$\mathbf{Q}_s^T = \int_\Omega \mathbf{B}^T \mathbf{m}\,\alpha\,S_g \frac{\mathrm{d}P_c}{\mathrm{d}S_w} \mathbf{N}_s\,\mathrm{d}\Omega\,, \tag{43}$$

$$\mathbf{C}_{sp} = \int_\Omega \mathbf{N}_s^T \left(\frac{\alpha - n}{K_s} + \frac{n}{K_g} \right) S_g \frac{\mathrm{d}P_c}{\mathrm{d}S_w} \mathbf{N}_P\,\mathrm{d}\Omega\,, \tag{44}$$

$$\begin{aligned} \mathbf{C}_{ss} = \int_\Omega \mathbf{N}_s^T \frac{\mathrm{d}P_c}{\mathrm{d}S_w} \Bigg[& \frac{n\,S_g}{K_g} \frac{\mathrm{d}P_c}{\mathrm{d}S_w} - n + \\ & + \frac{(\alpha - n)\,S_g}{K_s} \left(S_g \frac{\mathrm{d}P_c}{\mathrm{d}S_w} - P_c \right) \Bigg] \mathbf{N}_s\,\mathrm{d}\Omega\,, \end{aligned} \tag{45}$$

$$\mathbf{H}_{ss} = \int_{\Omega} (\nabla \mathbf{N}_s)^T \, \mathbf{k}_g \, \frac{\mathrm{d}P_c}{\mathrm{d}S_w} \frac{\mathrm{d}P_c}{\mathrm{d}S_w} \, \nabla \mathbf{N}_s \, \mathrm{d}\Omega \,, \tag{46}$$

$$\mathbf{F}^u = \int_{\Omega} \mathbf{N}_{\dot{u}}^T \, \rho \, \mathbf{g} \, \mathrm{d}\Omega + \int_{\partial\Omega_t} \mathbf{N}_{\dot{u}}^T \, \bar{\mathbf{t}} \, \mathrm{d}S \,, \tag{47}$$

$$\mathbf{F}^P = \mathbf{f}^P - \mathbf{J}_P \, \dot{\mathbf{T}} \,, \tag{48}$$

$$\mathbf{J}_P = - \int_{\Omega} \mathbf{N}_P^T \, (\beta_{sw} + \beta_{sg}) \, \mathbf{N}_t \, \mathrm{d}\Omega \,, \tag{49}$$

$$\begin{aligned} \mathbf{f}^P = & \int_{\Omega} (\nabla \mathbf{N}_P)^T \, (\mathbf{k}_w \, \rho_w + \mathbf{k}_g \, \rho_g) \, \mathbf{g} \, \mathrm{d}\Omega - \int_{\partial\Omega_w} \mathbf{N}_P^T \, \dot{\boldsymbol{\omega}}_w^T \, n \, \mathrm{d}S \, - \\ & - \int_{\partial\Omega_g} \mathbf{N}_P^T \, \dot{\boldsymbol{\omega}}_g^T \, n \, \mathrm{d}S - \int_{\Omega} \mathbf{N}_P^T \Big[\mathbf{k}_w \Big(- \nabla \mathbf{P}_w + \rho_w \, \mathbf{g} \, - \\ & - \rho_w \, \ddot{\mathbf{u}} \Big) \frac{\nabla \rho_w}{\rho_w} + \mathbf{k}_g \, \left(- \nabla \mathbf{P}_w - \nabla \mathbf{P}_c + \rho_g \, \mathbf{g} - \rho_g \, \ddot{\mathbf{u}} \right) \frac{\nabla \rho_g}{\rho_g} \Big] \, \mathrm{d}\Omega \,, \end{aligned} \tag{50}$$

$$\mathbf{F}^s = \mathbf{f}^s - \mathbf{J}_s \, \dot{\mathbf{T}} \,, \tag{51}$$

$$\mathbf{J}_s = - \int_{\Omega} \mathbf{N}_s^T \, \frac{\mathrm{d}P_c}{\mathrm{d}S_w} \, \beta_{sg} \, \mathbf{N}_t \, \mathrm{d}\Omega \,, \tag{52}$$

$$\begin{aligned} \mathbf{f}^s = & \int_{\Omega} (\nabla \mathbf{N}_s)^T \, \frac{\mathrm{d}P_c}{\mathrm{d}S_w} \, \mathbf{k}_g \, \rho_g \, \mathbf{g} \, \mathrm{d}\Omega - \int_{\partial\Omega_g} \mathbf{N}_s^T \, \frac{\mathrm{d}P_c}{\mathrm{d}S_w} \, \dot{\boldsymbol{\omega}}_g^T \, n \, \mathrm{d}S \, - \\ & - \int_{\Omega} \mathbf{N}_s^T \, \frac{\mathrm{d}P_c}{\mathrm{d}S_w} \, \mathbf{k}_g \left(- \nabla \mathbf{P}_w - \frac{\mathrm{d}P_c}{\mathrm{d}S_w} \, \nabla \mathbf{S}_w + \rho_g \, \mathbf{g} - \rho_g \, \ddot{\mathbf{u}} \right) \frac{\nabla \rho_g}{\rho_g} \, \mathrm{d}\Omega \,, \end{aligned} \tag{53}$$

where $\mathbf{N}_{\dot{u}}$, $\mathbf{N}_P$, $\mathbf{N}_s$, and $\mathbf{N}_t$ are the global shape functions for the velocity of the solid skeleton, liquid pressure, liquid saturation, and temperature, respectively. The terms $\bar{\dot{\boldsymbol{u}}}$, $\bar{\boldsymbol{P}}_w$, and $\bar{\boldsymbol{S}}_w$ are the vectors of nodal velocities of the solid skeleton, liquid pressure, and liquid saturation. Different interpolation and integration orders are necessary to ensure the stability of the finite element solution for the coupled problem.

In the above, K_s, K_w, and K_s are the bulk moduli of the solid, liquid, and gas phases. For the case of only two fluids we have

$$\rho = \rho_s \, (1 - n) + n \, (\rho_w \, S_w + \rho_g \, S_g) \,, \tag{54}$$

$$\beta_{sw} = [(\alpha - n) \, \beta_s + n \, \beta_w] \, S_w \,, \tag{55}$$

$$\beta_{sg} = [(\alpha - n) \, \beta_s + n \, \beta_g] \, S_g \,. \tag{56}$$

3.2 Time discretization. The *Newmark* scheme

The *Newmark* scheme is adopted here in the discretisation of the rate quantities taking part in the basic finite element equations (31) – (33). The solid skeleton velocity $\dot{\bar{\boldsymbol{u}}}$ and displacement $\bar{\boldsymbol{u}}$ at two adjacent time stations n (at t) and $n+1$ (at $t+\Delta t$) are then linked by the following equations:

$$\dot{\bar{\boldsymbol{u}}}_{n+1} = \dot{\bar{\boldsymbol{u}}}_n + \ddot{\bar{\boldsymbol{u}}}_n \, \Delta t + \Delta \ddot{\bar{\boldsymbol{u}}}_n \, \beta_2 \, \Delta t \,, \tag{57}$$

$$\bar{\boldsymbol{u}}_{n+1} = \bar{\boldsymbol{u}}_n + \dot{\bar{\boldsymbol{u}}}_n \, \Delta t + \frac{1}{2} \ddot{\bar{\boldsymbol{u}}}_n \, \Delta t^2 + \frac{1}{2} \Delta \ddot{\bar{\boldsymbol{u}}}_n \, \beta_1 \, \Delta t^2 \,, \tag{58}$$

where

$$\Delta \ddot{\bar{\boldsymbol{u}}}_n = \ddot{\bar{\boldsymbol{u}}}_{n+1} - \ddot{\bar{\boldsymbol{u}}}_n \,, \tag{59}$$

and β_1 and β_2 are the *Newmark* parameters. Here, these parameters are typically chosen within the range 0.6 – 1.0. Similarly, for the vectors $\bar{\boldsymbol{P}}_w$ and $\bar{\boldsymbol{S}}_w$ of nodal liquid pressures and liquid saturations we have

$$\bar{\boldsymbol{P}}_{w,n+1} = \bar{\boldsymbol{P}}_{w,n} + \dot{\bar{\boldsymbol{P}}}_{w,n} \, \Delta t + \frac{1}{2} \ddot{\bar{\boldsymbol{P}}}_{w,n} \, \Delta t^2 + \frac{1}{2} \Delta \ddot{\bar{\boldsymbol{P}}}_{w,n} \, \beta_1 \, \Delta t^2 \,, \tag{60}$$

$$\bar{\boldsymbol{S}}_{w,n+1} = \bar{\boldsymbol{S}}_{w,n} + \dot{\bar{\boldsymbol{S}}}_{w,n} \, \Delta t + \frac{1}{2} \ddot{\bar{\boldsymbol{S}}}_{w,n} \, \Delta t^2 + \frac{1}{2} \Delta \ddot{\bar{\boldsymbol{S}}}_{w,n} \, \beta_1 \, \Delta t^2 \,, \tag{61}$$

where

$$\Delta \ddot{\bar{\boldsymbol{P}}}_{w,n} = \ddot{\bar{\boldsymbol{P}}}_{w,n+1} - \ddot{\bar{\boldsymbol{P}}}_{w,n} \,, \tag{62}$$

$$\Delta \ddot{\bar{\boldsymbol{S}}}_{w,n} = \ddot{\bar{\boldsymbol{S}}}_{w,n+1} - \ddot{\bar{\boldsymbol{S}}}_{w,n} \,. \tag{63}$$

The *Newton* iteration

Introduction of the above approximations into (31) – (33) results in a system of non-linear algebraic equations whose solution via the *Newton* scheme requires at a typical iteration k the solution of the following linear system:

$$\begin{bmatrix} \mathbf{M}_u + \mathbf{K}_T \beta_1 \Delta t^2 & -\mathbf{Q}_p^T \beta_1 \Delta t^2 & (\mathbf{Q}_s^*)^T \beta_1 \Delta t^2 \\ \mathbf{M}_p + \mathbf{Q}_p \beta_2 \Delta t & \mathbf{C}_{pp} \beta_2 \Delta t + \mathbf{H}_{pp} \beta_1 \Delta t^2 & \mathbf{C}_{ps} \beta_2 \Delta t + \mathbf{H}_{ps} \beta_1 \Delta t^2 \\ \mathbf{M}_s + \mathbf{Q}_s \beta_2 \Delta t & \mathbf{C}_{sp} \beta_2 \Delta t + \mathbf{H}_{ps}^T \beta_1 \Delta t^2 & \mathbf{C}_{ss} \beta_2 \Delta t + \mathbf{H}_{ss} \beta_1 \Delta t^2 \end{bmatrix} \begin{bmatrix} \delta \ddot{\bar{\boldsymbol{u}}}_n \\ \delta \ddot{\bar{\boldsymbol{P}}}_{w,n} \\ \delta \ddot{\bar{\boldsymbol{S}}}_{w,n} \end{bmatrix} = \begin{bmatrix} \mathbf{F}_{n+1}^{u^*} \\ \mathbf{F}_{n+1}^{P^*} \\ \mathbf{F}_{n+1}^{s^*} \end{bmatrix} , \tag{64}$$

where $\delta(\cdot)$ denotes the iterative increment of $(\cdot)$,

$$\Delta(\cdot)^{k+1} = \Delta(\cdot)^k + \delta(\cdot)\,, \tag{65}$$

and $\mathbf{K}_T$ the tangential stiffness matrix, defined as

$$\mathbf{K}_T = \frac{\partial \mathbf{P}(\bar{\boldsymbol{u}})}{\partial \bar{\boldsymbol{u}}} = \frac{\partial}{\partial \bar{\boldsymbol{u}}} \int_\Omega \mathbf{B}^T \boldsymbol{\sigma}''(\bar{\boldsymbol{u}})\, \mathrm{d}\Omega\,. \tag{66}$$

Its computation here will be based on the exact linearization of the implicit function alluded to in (36) which results from the constitutive integration algorithm referred to in Section 5.

Quasi-static case. Reduced rank formulation

Under quasi-static conditions, the following approximations can be made:

$$\ddot{u} \approx 0\,, \qquad \frac{D}{Dt}\left(\frac{\dot{\boldsymbol{\omega}}}{n\, S_l}\right) \approx 0\,. \tag{67}$$

Significant simplifications result from the introduction of this assumption into the above equations. In particular, it can be seen that in the quasi-static case the order of time differentiation is the same for the three variables. Therefore, the same lowest permissible order of time integration schemes for the three variables can be employed. *Taylor*'s expansion can be used for **u**, **P**, and **S** and the expanded series truncated in the simplest way as

$$\bar{\boldsymbol{u}}_{n+1} = \bar{\boldsymbol{u}}_n + \dot{\bar{\boldsymbol{u}}}_n\, \Delta t + \Delta\dot{\bar{\boldsymbol{u}}}_n\, \beta\, \Delta t\,, \tag{68}$$

$$\bar{\boldsymbol{P}}_{w,n+1} = \bar{\boldsymbol{P}}_{w,n} + \dot{\bar{\boldsymbol{P}}}_{w,n}\, \Delta t + \Delta\dot{\bar{\boldsymbol{P}}}_{w,n}\, \gamma\, \Delta t\,, \tag{69}$$

$$\bar{\boldsymbol{S}}_{w,n+1} = \bar{\boldsymbol{S}}_{w,n} + \dot{\bar{\boldsymbol{S}}}_{w,n}\, \Delta t + \Delta\dot{\bar{\boldsymbol{S}}}_{w,n}\, \theta\, \Delta t\,, \tag{70}$$

where

$$\Delta\dot{\bar{\boldsymbol{u}}}_n = \dot{\bar{\boldsymbol{u}}}_{n+1} - \dot{\bar{\boldsymbol{u}}}_n\,, \tag{71}$$

$$\Delta\dot{\bar{\boldsymbol{P}}}_{w,n} = \dot{\bar{\boldsymbol{P}}}_{w,n+1} - \dot{\bar{\boldsymbol{P}}}_{w,n}\,, \tag{72}$$

$$\Delta\dot{\bar{\boldsymbol{S}}}_{w,n} = \dot{\bar{\boldsymbol{S}}}_{w,n+1} - \dot{\bar{\boldsymbol{S}}}_{w,n}\,. \tag{73}$$

Substitution of the above into equations (31 – 33) yields, after some algebra, a reduced rank system of algebraic equations with the corresponding *Newton* iteration scheme requiring the solution of the system:

$$\begin{bmatrix} \mathbf{K}_T\, \beta\, \Delta t & -\mathbf{Q}_p^T\, \gamma\, \Delta t & (\mathbf{Q}_s^*)^T\, \theta\, \Delta t \\ \mathbf{Q}_p & \mathbf{C}_{pp} + \mathbf{H}_{pp}\, \gamma\, \Delta t & \mathbf{C}_{ps} + \mathbf{H}_{ps}\, \theta\, \Delta t \\ \mathbf{Q}_s & \mathbf{C}_{sp} + \mathbf{H}_{ps}^T\, \gamma\, \Delta t & \mathbf{C}_{ss} + \mathbf{H}_{ss}\, \theta\, \Delta t \end{bmatrix} \begin{bmatrix} \delta\dot{\bar{\boldsymbol{u}}}_n \\ \delta\dot{\bar{\boldsymbol{P}}}_{w,n} \\ \delta\dot{\bar{\boldsymbol{S}}}_{w,n} \end{bmatrix} = \begin{bmatrix} \mathbf{F}_{n+1}^{u^*} \\ \mathbf{F}_{n+1}^{P^*} \\ \mathbf{F}_{n+1}^{s^*} \end{bmatrix}, \tag{74}$$

where

$$\begin{aligned}\mathbf{F}^{u*}_{n+1} &= \mathbf{F}^{u}_{n+1} - \mathbf{K}_T(\bar{\boldsymbol{u}}_n + \dot{\bar{\boldsymbol{u}}}_n\,\Delta t) + \\ &\quad + \mathbf{Q}_p^T(\bar{\boldsymbol{P}}_{w,n} + \dot{\bar{\boldsymbol{P}}}_{w,n}\,\Delta t) - (\mathbf{Q}_s^*)^T(\bar{\boldsymbol{S}}_{w,n} + \dot{\bar{\boldsymbol{S}}}_{w,n}\,\Delta t)\,,\end{aligned} \tag{75}$$

$$\begin{aligned}\mathbf{F}^{P*}_{n+1} &= \mathbf{F}^{P}_{n+1} - \mathbf{Q}_p\,\dot{\bar{\boldsymbol{u}}}_n - \mathbf{C}_{pp}\,\dot{\bar{\boldsymbol{P}}}_{w,n} - \\ &\quad - \mathbf{C}_{ps}\,\dot{\bar{\boldsymbol{S}}}_{w,n} - \mathbf{H}_{pp}(\bar{\boldsymbol{P}}_{w,n} + \dot{\bar{\boldsymbol{P}}}_{w,n}\,\Delta t) - \mathbf{H}_{ps}(\bar{\boldsymbol{S}}_{w,n} + \dot{\bar{\boldsymbol{S}}}_{w,n}\,\Delta t)\,,\end{aligned} \tag{76}$$

$$\begin{aligned}\mathbf{F}^{S*}_{n+1} &= \mathbf{F}^{S}_{n+1} - \mathbf{Q}_s\,\dot{\bar{\boldsymbol{u}}}_n - \mathbf{C}_{sp}\,\dot{\bar{\boldsymbol{P}}}_{w,n} - \\ &\quad - \mathbf{C}_{ss}\,\dot{\bar{\boldsymbol{S}}}_{w,n} - \mathbf{H}_{sp}(\bar{\boldsymbol{P}}_{w,n} + \dot{\bar{\boldsymbol{P}}}_{w,n}\,\Delta t) - \mathbf{H}_{ss}(\bar{\boldsymbol{S}}_{w,n} + \dot{\bar{\boldsymbol{S}}}_{w,n}\,\Delta t)\,.\end{aligned} \tag{77}$$

The above reduced rank formulae will be exclusively adopted in the simulations presented in this paper. We remark that the equation solution techniques discussed in [15] can be applied to solve the above equations.

4 Boundary and initial conditions

In the definition of boundary conditions, we will limit ourselves here to the quasi-static case.

Natural boundary conditions

Under the quasi-static assumption, the natural boundary conditions imposed on a deforming porous multiphase body with two fluid phases are:

1. The surface traction $\mathbf{t}$ applied to the portion $\partial\Omega_t$ of the boundary. The strong form equation associated to this boundary condition is

$$\boldsymbol{\sigma}\,\mathbf{n} = \mathbf{t} \qquad \text{on} \quad \partial\Omega_t\,. \tag{78}$$

2. The *Darcy* velocity $\dot{\bar{\boldsymbol{\omega}}}_w$ of the liquid phase imposed on the portion $\partial\Omega_w$ of $\partial\Omega$. The corresponding equation is

$$\mathbf{k}_w\,(-\nabla\mathbf{P}_w + \rho_w\,\mathbf{g}) = \dot{\bar{\boldsymbol{\omega}}}_w \qquad \text{on} \quad \partial\Omega_w\,. \tag{79}$$

3. The *Darcy* velocity $\dot{\bar{\boldsymbol{\omega}}}_g$ of the gas phase imposed on $\partial\Omega_g$. The associated strong form is

$$\mathbf{k}_g\,(-\nabla\mathbf{P}_g + \rho_g\,\mathbf{g}) = \dot{\bar{\boldsymbol{\omega}}}_g \qquad \text{on} \quad \partial\Omega_g\,. \tag{80}$$

Forced boundary conditions

The forced boundary conditions in the present case are the following:

4. The velocity and displacement of the solid skeleton imposed on the portion $\partial\Omega_u$ of the boundary:

$$\dot{\mathbf{u}} = \bar{\dot{\mathbf{u}}}\,, \qquad \mathbf{u} = \bar{\mathbf{u}} \qquad \text{on} \quad \partial\Omega_u\,. \tag{81}$$

5. The liquid pressure:

$$\mathbf{P}_w = \bar{\mathbf{P}}_w \qquad \text{on} \quad \partial\Omega_P\,. \tag{82}$$

6. The liquid saturation:

$$\mathbf{S}_w = \bar{\mathbf{S}}_w \qquad \text{on} \quad \partial\Omega_S\,. \tag{83}$$

Boundary conditions and contact

Note for instance that, as a particular case of no. 2 and 3 above, an impermeable boundary corresponds to the conditions

$$\dot{\bar{\boldsymbol{\omega}}}_w = \mathbf{o}\,, \qquad \dot{\bar{\boldsymbol{\omega}}}_g = \mathbf{o}\,, \tag{84}$$

imposed on the relevant portions of the boundary. A fully permeable boundary, on the other hand, is characterised by

$$\mathbf{P}_w = \mathbf{0}\,. \tag{85}$$

For the industrial problems we envisage, these two distinct conditions may apply on the same portion of the boundary at different stages of processing. A typical example is an initially fully permeable surface which becomes impermeable after contact with an impermeable tool. In order to simulate this, the finite element code has to be adequately prepared to handle such boundary condition changes.

Initial conditions

In addition to the above boundary conditions, initial saturations, pore pressures, effective stress, and initial values of other relevant state variables must be known before a finite element analysis can be made. In paper processing problems, the appropriate initialisation of the liquid saturations is essential. In this case, saturation distributions have to be experimentally obtained. Initialisation of non-zero effective stress, for instance, can be important in situations where high solid skeleton stresses are present prior to the analysis. This is typically the case in problems of multiphase flow through deep cracked rock masses where, in its natural state, the rocks are subjected to high compressive stresses.

5 Constitutive modelling of the solid skeleton

Another important requirement in the modelling of multiphase media is the accurate description of the behaviour of the solid skeleton. This issue becomes even more crucial for the processes we intend to simulate. Since the multiphase medium is subject to extremely high strains, operations such as compression and rolling, an appropriate constitutive model, capable of handling the strain magnitudes involved, has to be selected in order to preserve the accuracy of the overall computational scheme. In the present case, we adopt a rate-independent elasto-plastic law to model the solid skeleton. The key features of the model include: a nonlinear porous elasticity pressure law and a non-associative compressible plasticity response. A *Cam-Clay* type yield surface is adopted in conjunction with a flow potential that produces compressive (expansive) plastic flow under compressive (tensile) pressure. The model and the relevant constitutive integration algorithm are briefly described in the following. Further details can be found in [7].

Nonlinear elasticity law

Due to the porous nature of the skeleton material, high elastic compressibility may be observed at low levels of compressive hydrostatic stress. With increasing compressive pressures (as the pores decrease in size) the material becomes progressively incompressible. This behaviour suggests the adoption of a non-linear elasticity law for the hydrostatic stress. Here, the following constitutive function for the hydrostatic pressure is employed to describe this phenomenon:

$$p = \hat{p}\left(\epsilon_v^e\right) = p_t + (p_t - p_0)\exp\left[\frac{1+e_0}{\kappa}\left(1 - \exp\epsilon_v^e\right)\right], \tag{86}$$

where p is the effective pressure stress,

$$p = \frac{1}{3}\operatorname{tr}\boldsymbol{\sigma}'', \tag{87}$$

κ is the logarithmic bulk modulus, p_t is the strength of the material to tensile pressure, p_0 and e_0 are the pressure and voids ratio at the reference configuration respectively; ϵ_v^e is the elastic volumetric stain. The above law is complemented with the introduction of a linear relationship between the deviatoric effective stress and strain:

$$\boldsymbol{\sigma}_d'' = 2\,G\,\boldsymbol{\epsilon}_d^e, \tag{88}$$

where $\boldsymbol{\sigma}_d''$ is the effective stress deviator and $\boldsymbol{\epsilon}_d^e$ is the elastic strain deviator and the constant G is the shear modulus.

Yield surface

The yield surface for the present model is defined by means of the following *Cam-Clay* type yield function:

$$\phi = \frac{1}{b^2}\left(p - p_c + a\right)^2 + \left(\frac{q}{M}\right)^2 - a^2 \,, \tag{89}$$

where q is the *von Mises* equivalent effective stress; M and b are material constants; a is the hardening force defining the size of the elastic region in stress space. Isotropic hardening/softening of the skeleton is accounted for by defining a as a given (experimentally determined) function of the volumetric plastic strain ϵ_v^p, i. e.

$$a = a\left(\epsilon_v^p\right) . \tag{90}$$

Plastic flow

The plastic flow rule is defined by adopting the normality condition, which gives

$$\dot{\boldsymbol{\epsilon}}^p = \dot{\gamma}\,\frac{\partial \Psi}{\partial \boldsymbol{\sigma}''} \,, \tag{91}$$

where Ψ is the flow potential here defined as

$$\Psi = p^2 + q^2 \,, \tag{92}$$

and $\dot{\gamma}$ is the plastic multiplier which satisfies the usual loading/unloading conditions of rate-independent plasticity:

$$\phi \leq 0 \,, \qquad \dot{\gamma} \geq 0 \,, \qquad \dot{\gamma}\,\phi = 0 \,. \tag{93}$$

We remark that the above constitutive model can be modified by rephrasing its equations in terms of *Jaumann* rate of *Kirchhoff* stress. In this case, an elastic law with variable shear modulus can be adopted allowing greater flexibility in matching experimental results.

Large strain implementation and algorithmic treatment

The large strain implementation of the above constitutive model is carried out following the logarithmic strain-based procedure described in [11]. This procedure has been shown to coincide with conventional *Jaumann* rate type formulations under small elastic strains. A fully hyperelastic-based implementation of the model is also possible. This has been described in [12].

An operator-split based algorithm is adopted in the numerical integration of the constitutive relations. The algorithm has the standard format where in each step, an elastic predictor stage is performed, where the material is assumed purely elastic, followed by a plastic corrector where the plasticity equations are solved implicitly. Within the typical time step $[t_n, t_{n+1}]$, given the incremental deformation gradient

$$\mathbf{F}_\Delta = \mathbf{F}_{n+1}\,\mathbf{F}_n^{-1}\,, \tag{94}$$

the application of polar decomposition gives

$$\mathbf{F}_\Delta = \mathbf{V}_\Delta\,\mathbf{R}_\Delta\,, \tag{95}$$

where $\mathbf{V}_\Delta$ and $\mathbf{R}_\Delta$ are, respectively, the incremental left stretch tensor and incremental rigid rotation. The incremental (logarithmic) strain is then computed as

$$\Delta\boldsymbol{\epsilon} = \ln \mathbf{V}_\Delta\,, \tag{96}$$

where $\ln(\cdot)$ denotes the tensor logarithm of $(\cdot)$. With the incremental strain at hand, the elastic trial effective stress deviator is computed as

$$\boldsymbol{\sigma}_d^{\prime\prime\,\mathrm{trial}} = \mathbf{R}_\Delta\,(\boldsymbol{\sigma}_d^{\prime\prime})_n\,\mathbf{R}_\Delta^T + 2\,G\,\Delta\boldsymbol{\epsilon}\,. \tag{97}$$

The trial hydrostatic effective stress is computed from the volumetric component of the elastic trial volumetric strain as

$$p_{n+1}^{\mathrm{trial}} = \hat{p}\,(\epsilon_v^{e\,\mathrm{trial}})\,, \qquad \epsilon_v^{e\,\mathrm{trial}} = (\epsilon_v^e)_n + \Delta\epsilon_v\,. \tag{98}$$

Having computed the elastic trial stresses, the return mapping procedure to determine the updated effective stress $\boldsymbol{\sigma}_{n+1}^{\prime\prime}$ follows the same steps as the classical procedures for infinitesimal elasto-plasticity [7, 12].

6 Numerical applications

Numerical examples are presented here to illustrate the effectiveness of the framework described in the previous sections. Before proceeding to the examples, it is worth remarking that, in addition to the developments presented above, the simulation of such processes requires state-of-the-art knowledge in many areas of finite element modelling. In particular, a robust and efficient frictional contact algorithm must be used to simulate the dynamic compression and rolling of the very thin wet paper medium. Here, the penalty approach described in [10] is adopted and extended to control maximum penetration. The penetration control procedure relies on a penalty updating scheme. It is essential in order to maintain the non-physical overlap between processed media and tools within reasonable tolerances.

6.1 Permeability and capillary pressure

For the above model and solution procedure to be applied to the simulation of compression and rolling of wet paper sheet, the permeabilities of water and gas and a capillary pressure function must be given.

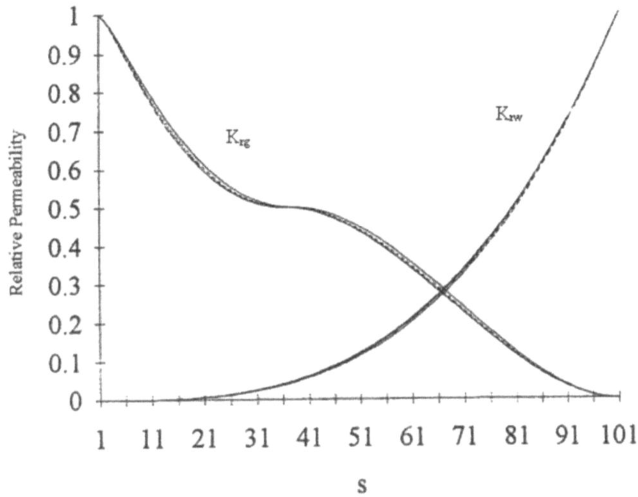

Fig. 1. Relative permeability vs. saturation.

In the paper industry, the following expressions are often used for the permeability matrices of water and gas:

$$\mathbf{k}_w = K_{rw}\,\mathbf{K}_{aw}\,, \tag{99}$$

$$\mathbf{k}_g = K_{rg}\,\mathbf{K}_{ag}\,, \tag{100}$$

where K_{rw} and K_{rg} are the relative permeabilities of water and gas, both defined by experimentally determined curves (see Figure 1), $\mathbf{K}_{aw}$ and $\mathbf{K}_{ag}$ are the absolute permeabilities of water and gas given as

$$\mathbf{K}_{aw} = F_{cw}\,\mathbf{K_a}\,, \tag{101}$$

$$\mathbf{K}_{ag} = F_{cg}\,\mathbf{K_a}\,. \tag{102}$$

Here, we have

$$\mathbf{K_a} = (\phi_e)^{F_3}\begin{bmatrix} 1/(F_1 - F_2\,\phi_a) & 0 \\ 0 & 1 \end{bmatrix}, \tag{103}$$

where F_{cw}, F_{cg}, F_1, F_2, F_3 are material constants, ϕ_e is the extrafibre porosity and ϕ_a is the absolute porosity.

The capillary pressure between water and gas is regarded as a general function of the liquid saturation defined in (15). In the following numerical examples the curves shown in Figure 2 are adopted. These have been obtained by Bloch [2].

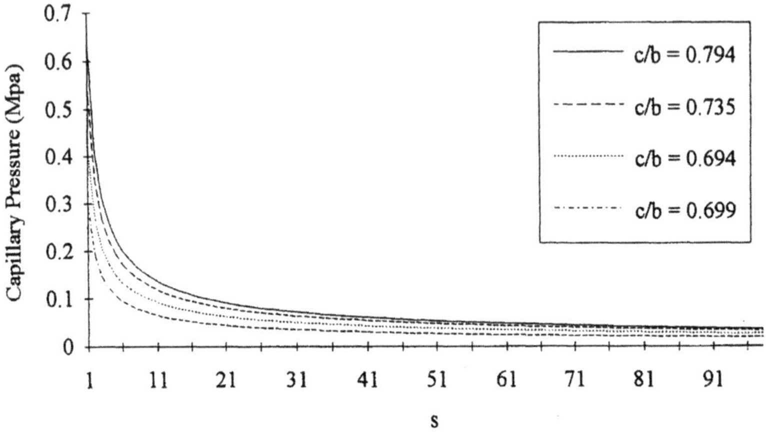

Fig. 2. Capillary pressure vs. saturation.

6.2 Simulation of wet paper compression

The numerical example presented in this section simulates the compression of a wet pulp sheet. A wet pulp sheet with 0.2 mm initial thickness is compressed to the maximum compression of 75 % of the initial thickness within 0.2 seconds. The pulp has uniform initial water saturation of 0.8. The material parameters used in the simulation are listed in Table 1. A portion of the discretized sheet in its initial configuration and after full compression is shown in Figure 3. The high compression rate is clearly illustrated. The pore pressure and water saturation distribution across the sheet thickness predicted at the end of the simulation are plotted in the graphs of Figure 4.

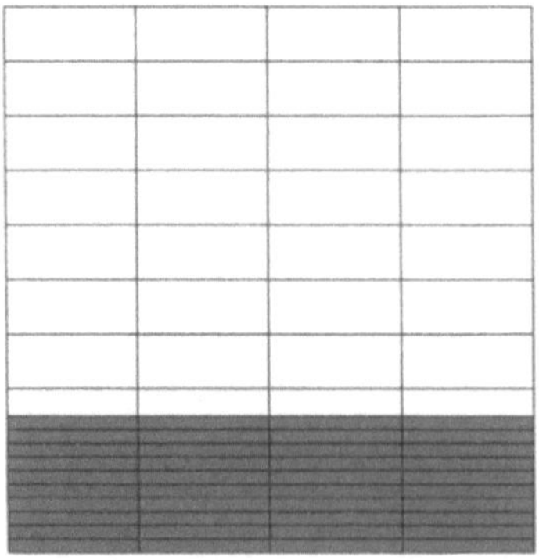

Fig. 3. Pulp compression. Portion of the finite element mesh in its initial and final compressed states.

Parameter	Value
G	1.3 MPa
κ	0.01
M	3.8
b	1.0
e_0	7.197
P_t	0.1 MPa
F_{cw}	$3.069 \cdot 10^{-06}$
F_{cg}	$1.1 \cdot 10^{-07}$
F_1	1.3
F_2	0.3
F_3	2.22
ϕ_a	0.877
ϕ_e	0.7899

Table 1. Data used in the analysis.

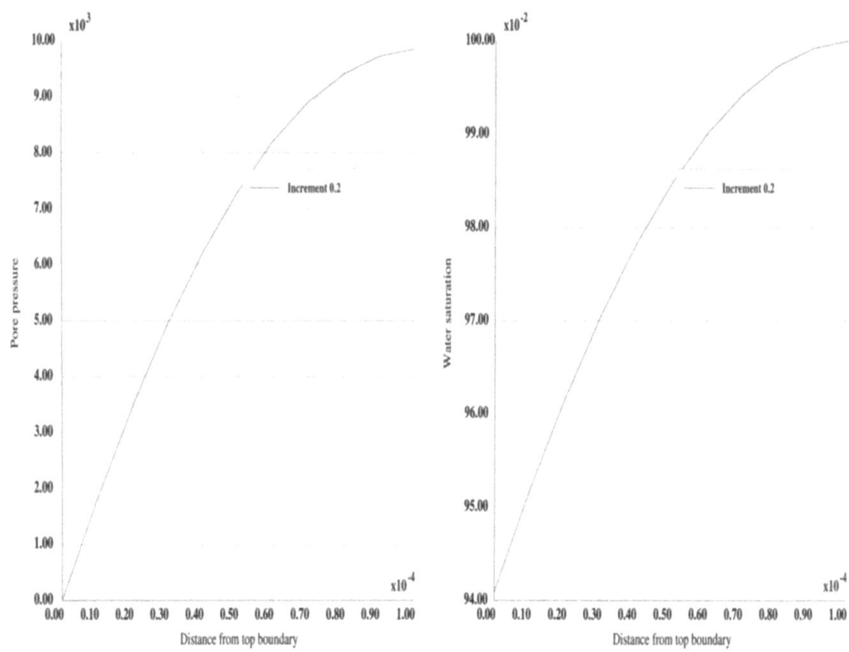

Fig. 4. Pulp compression. Pore pressure [Pa] and water saturation distributions across the sheet thickness.

6.3 Simulation of wet paper rolling

The finite element simulation of wet paper rolling in a paper production process is presented in this section. The process consists of the extraction

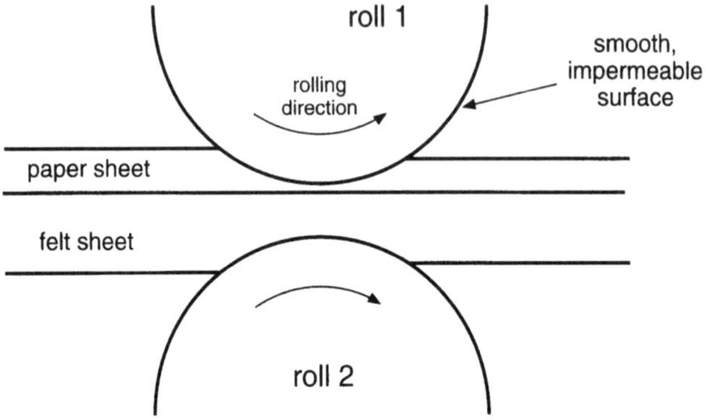

Fig. 5. Paper rolling process. Schematic illustration.

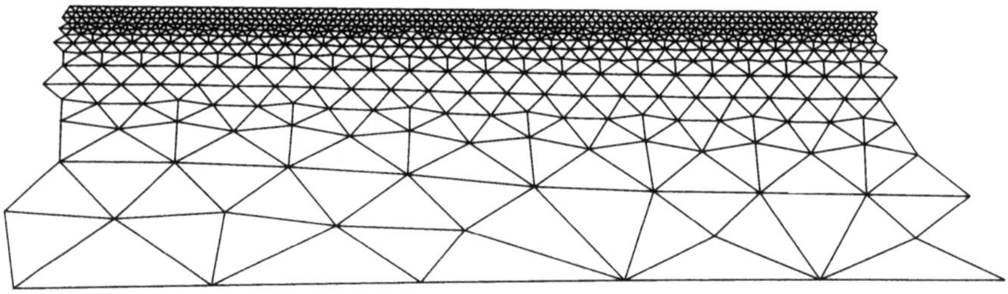

Fig. 6. Paper rolling. FE mesh for felt and paper sheets.

of water from a wet paper sheet laid on felt in a two roll mill. A schematic illustration of the process in shown in Figure 5. The two rolls are identical with 0.6 m diameter and the gap between the roll surfaces at the narrowest section is set as 1.07 mm. The thickness of the paper and felt sheets are 0.1 and 20 mm, respectively. The rotation of the rolls are prescribed to produce a high surface speed of 10 m/s. Frictional contact conditions are assumed between the rolls and the rolled materials. Again, the material parameters of Table 1 are adopted for the paper sheet. A portion of the mesh discretizing the felt and paper sheet is plotted in Figure 6. The paper sheet corresponds to a thin layer of small elements on top of the felt mesh. This layer is amplified in Figure 7 which illustrates the distribution of water pressure in the paper sheet after passing through the rolls. The final distribution of water saturation in the wet paper is shown in Figure 8.

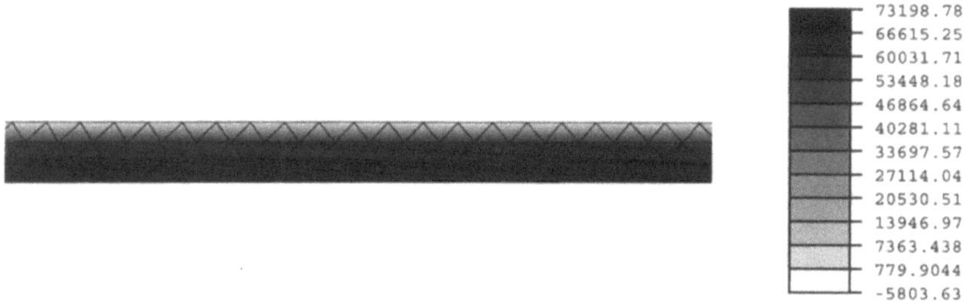

Fig. 7. Paper rolling. Pore pressure [Pa] in the paper sheet.

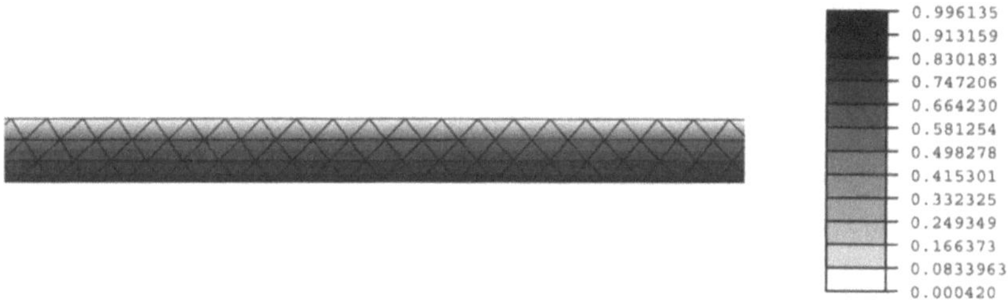

Fig. 8. Paper rolling. Water saturation in the paper sheet.

7 Concluding remarks

In this paper, a general framework for the finite element simulation of the processing of multiphase media has been presented. Since our main interest is the simulation of pulp and paper processing, the developments presented have been focused on the particular case of only two immiscible fluids. We remark, however, that the present approach can be easily extended to accommodate more fluid phases and allow the simulation of a wider range of industrial operations. Numerical examples of relevance to the paper processing industry have been presented demonstrating the effectiveness of the proposed scheme in the simulation of realistic large scale problems. The incorporation of phase change phenomena into the present scheme, with thermo-mechanical coupling treated by a staggered scheme, is currently under investigation. This will allow the simulation of paper processing operations with high temperature rolls and will be addressed in a forthcoming publication.

References

1. Biot, M. A., Willis, P. G.: The elastic coefficients of the theory of consolidation. *J. Appl. Mechanics* **24** (1957), 594–601.
2. Bloch, J.-F.: *Transferts de masse et Chaleur dans les Milieux Poreux Deformables non Satures: Application au Pressage du Papier*. Ph. D. Thesis, Institut National Polytechnique de Grenoble, Grenoble, France, October 1995.

3. Chan, A. H. C.: *A Unified Finite Element Solution to Static and Dynamic Problems in Geomechanics*. Ph. D. Thesis, C/Ph/106/88, University College of Swansea, 1988.
4. Darcy, H.: *Les fontaines publiques de la ville de Dijon*. Dalmont, Paris 1856.
5. Lewis, R. W., Schrefler, B. A.: *The Finite Element Method in the Deformation and Consolidation of Porous Media*. Wiley, London 1987.
6. Lui, X., Zienkiewicz, O. C.: Multiphase flow in deforming porous media and finite element solutions. *Comp. Struct.* **45**(2) (1992), 211–227.
7. Owen, D. R. J., de Souza Neto, E. A., Zhao, S. Y., Peric, D., Loughran, J.: Finite element simulation of the rolling and extrusion of multi-phase materials – application to the rolling of prepared sugar cane. *Computer Methods in Applied Mechanics and Engineering* **151** (1998), 479–495.
8. Owen, D. R. J., Zhao, S. Y., Loughran, J. G.: Application of porous media mechanics to the numerical simulation of the rolling of sugar cane. *Engineering Computations* **12** (1995), 281–303.
9. Owen, D. R. J., Zhao, S. Y., Loughran, J. G: Determination of a Constitutive Relation for Prepared Cane Using Computer Simulation. In: *Proc. ASSCT Conference*, Townsville, Queensland 1994.
10. Peric, D., Owen, D. R. J.: Computational model for 3-d contact problems with friction based on the penalty method. *Int. J. Numer. Methods Engrg.* **35** (1992), 1289–1309.
11. Peric, D., Owen, D. R. J., Honnor, M. E.: A model for finite strain elasto-plasticity based on logarithmic strains: Computational issues. *Comp. Meth. Appl. Mech. Engng.* **94** (1992), 35–61.
12. de Souza Neto, E. A., Peric, D., Dutko, M., Owen, D. R. J.: Finite strain implementation of an elastoplastic model for crushable foam. In: *Proc. of the European Workshop on Computational Mechanics*, Goteborg, Sweden, November 1995.
13. Terzaghi, K.: Die Berechnung der Durchlässigkeitsziffer des Tones aus dem Verlauf der hydrodynamischen Spannungserscheinungen. *Sitz. Akad. Wiss. Wien, Austria* **132** (1923).
14. Terzaghi, K.: *Theoretical Soil Mechanics*. Wiley, New York, 1943.
15. Zhao, S. Z.: *Finite Element Solution of Saturated-Unsaturated Porous Materials with Application to the Rolling of Prepared Cane*. Ph. D. Thesis, C/Ph/173/93, University College of Swansea, 1993.
16. Zienkiewicz, O. C.: Basic formulation of static and dynamic behaviour of soil and other porous media. In Martins, J. B. (ed.): *Numerical Methods in Geomechanics*, D. Reidel Publishing Co., 1982, pp. 39–57.
17. Zienkiewicz, O. C., Bettess, P.: Soil and other porous media under transient, dynamic conditions; general formulation and the validity of various simplifying assumptions. In Pande, G. N., Zienkiewicz, O. C. (eds.): *Soil Mechanics–Transient Loads*, Wiley, New York 1982, pp. 1–16.
18. Zienkiewicz, O. C., Chang, C. T., Bettess, P.: Drained, undrained, consolidating and dynamic behaviour assumption in soils; limits of validity. *Géotechnique* **30** (1980), 385–395.
19. Zienkiewicz, O. C., Shiomi, T.: Dynamic behaviour of saturated porous media: the generalized Biot formulation and its numerical solution. *Int. J. Numer. Methods Geomech.* **8** (1985), 71–96.
20. Zienkiewicz, O. C., Taylor, R. L.: *The Finite Element Method*. Butterworth Heinemann, 2000.

Author Index

MIX
Papier aus verantwortungsvollen Quellen
Paper from responsible sources
FSC® C105338

If you have any concerns about our products,
you can contact us on
ProductSafety@springernature.com

In case Publisher is established outside the EU,
the EU authorized representative is:
Springer Nature Customer Service Center GmbH
Europaplatz 3, 69115 Heidelberg, Germany

Printed by Libri Plureos GmbH
in Hamburg, Germany